An Introduction to

Abstract
Algebra

Sets, Groups, Rings, and Fields

An Introduction to

Abstract Algebra

Sets, Groups, Rings, and Fields

Steven H Weintraub
Lehigh University, USA

NEW JERSEY · LONDON · SINGAPORE · BEIJING · SHANGHAI · HONG KONG · TAIPEI · CHENNAI · TOKYO

Published by

World Scientific Publishing Co. Pte. Ltd.

5 Toh Tuck Link, Singapore 596224

USA office: 27 Warren Street, Suite 401-402, Hackensack, NJ 07601

UK office: 57 Shelton Street, Covent Garden, London WC2H 9HE

Library of Congress Control Number: 2022020362

British Library Cataloguing-in-Publication Data
A catalogue record for this book is available from the British Library.

ISBN 978-981-124-666-1 (hardcover)
ISBN 978-981-124-755-2 (paperback)
ISBN 978-981-124-667-8 (ebook for institutions)
ISBN 978-981-124-668-5 (ebook for individuals)

For any available supplementary material, please visit
https://www.worldscientific.com/worldscibooks/10.1142/12539#t=suppl

Desk Editors: Soundararajan Raghuraman/Yumeng Liu

Typeset by Stallion Press
Email: enquiries@stallionpress.com

Printed in Singapore

To Judy, 30 years on

Preface

This book is a textbook for a semester-long or a year-long introductory course in abstract algebra.

There is a lot of information in that sentence, so let us unpack it. First of all, this is a book, not an encyclopedia. What is the difference? An encyclopedia is a massive collection of information, while a book has a theme. Our book certainly does, and that theme is number theory. To be clear, this is not a text book on number theory, but we have decided on which topics to cover with an eye towards number theory, and we have included several sections that show applications of general algebraic ideas to topics in number theory. At the same time, a theme has variations, so we have not strictly restricted ourselves but have covered other topics as well.

This is an introduction, which means that we have presupposed no prior knowledge of abstract algebra. We do, however, assume that you (the student) have had a good course in linear algebra. By a good course we mean one that treats vector spaces and linear transformations in general, not one that is restricted to matrix manipulations (but of course does include that). And naturally, at this point in your mathematical development, you should be comfortable with doing rigorous mathematics, and this is certainly a rigorous book. We prove just about everything we claim or use, except that on occasion we mention a result that goes beyond the bounds of this book for the further edification of the reader.

There is enough material here for a year-long course, but we realize that you (the instructor) may not have the luxury of spending a year on it, so we have tried to write this book in a modular way, so that you may choose which topics to go into, and go into them as far as you like, before moving on to next one, covering what you wish in the course of a semester. Of course, some topics are required for others, so your choice will not be completely free. And for you (the student), if you are in a one-semester course, this book offers you the opportunity to read further in whatever particularly interests you.

Naturally, having written this book, we think highly of it, and think that it would provide an excellent basis for further study in abstract algebra in general. But, given our emphasis, we think that it would provide an ideal basis for further study in algebraic number theory.

The devil is in the details, as the saying goes, so here they are.

We begin, in Chapter 1, with set theory. This is often skipped, or presupposed, but we have decided to begin with it for several reasons. First, you may not be familiar with this material. Second, we treat quotients in many places in the book, so we wanted to present a particularly careful discussion of equivalence relations. And third, we wanted to take the opportunity to present the Schröder–Bernstein theorem (with proof, of course), which you may not be likely to see elsewhere.

In Chapter 2, we turn our attention to group theory. We concentrate on finite groups, but begin by treating groups in general, with examples such as matrix groups, so you can see the widespread appearance of groups throughout mathematics. (Too often, in our opinion, groups are treated purely for their own sake, which is of course appropriate in a specialized text, but is an approach that leads the reader to think of them in isolation rather than being of general interest.) We treat the standard, and essential, topics: homomorphisms, subgroups, quotient groups, etc. We also prove the fundamental structure theorem for finite abelian groups, and for finitely generated abelian groups, something that is not always done in texts at this level. We then have a section on applications to number theory, where we prove Fermat's little theorem and the basic facts on quadratic residues, all from a group-theoretic point of view. We study the actions of groups on sets, in preparation for proving Cauchy's theorem, results on the structure of p-groups, and the Sylow theorems. We briefly treat solvable groups, as we will be studying the solvability of equations by radicals in our chapter on field theory. We conclude this chapter by studying permutations and the symmetric groups.

Chapter 3 deals with ring theory. We begin in complete generality, considering both commutative and noncommutative rings, and rings with and without 1, and we study ideals, both one-sided and two-sided, ring homomorphisms, and quotients. But, in line with our emphasis, we fairly quickly turn our attention to commutative rings with 1, and further to integral domains. We study polynomial rings and prove the Hilbert basis theorem. We concentrate on the issues of divisibility and unique factorization in integral domains, proving the standard results that all Euclidean domains are principal ideal domains and that all principal ideal domains are unique factorization domains, developing Euclid's algorithm in the process. Our approach highlights the role played by the greatest common divisor (GCD), and on our way to our main results we define GCD domains (integral domains in which any set of elements, not all zero, has a gcd) and study their properties. We again have a section on applications to number theory, which has two main results. First, we use the fact that the Gaussian integers are a Euclidean domain (which we have earlier proved) to give Dedekind's proof of Fermat's theorem that every prime congruent to 1 modulo 4 is a sum of two squares in an essentially unique way. Second, we give Zolotarev's proof of Gauss's lemma and the Law of Quadratic Reciprocity by considering signs of permutations. We give, and prove, examples of unique and non-unique factorization, including, in particular, a variety of examples of rings of algebraic integers in quadratic fields. We consider quotient fields and localization, and study polynomial rings in detail. We conclude by studying prime and maximal ideals, proving the standard result that maximal ideals are prime, though not in general conversely, and also the less standard result that an integral domain in which all nonzero prime ideals are maximal is a principal ideal domain if and only if it is a unique factorization domain.

Chapter 4 deals with field theory in general and Galois theory in particular. We feel that our treatment here is quite distinctive.

We begin in a very concrete way, first showing how to make computations in field extensions. We then "front-load" our treatment of Galois theory by giving many examples of field extensions and Galois groups, even before arriving at the fundamental theorem of Galois theory (FTGT). Then we turn to proving the FTGT. As a first step, we show that an extension is Galois if and only if it is normal and separable. We then prove the FTGT *per se*. While our proof follows the spirit of Artin's approach, it is different in detail, and we do not need to use Dedekind's theory of group characters, as Artin does. (We think that our proof is thus more direct and conceptually a bit simple.) Having provide the FTGT, we go on to study further examples of field extensions. We prove the theorem of the primitive element and give a quite extensive study, far more than is usually done, of primitive elements in field extensions. We determine the structure of finite fields, and of cyclotomic fields. We conclude by deriving important consequences of field and Galois theory. We prove Abel's theorem that the general polynomial of degree five is not solvable by radicals. Indeed, we prove this for any degree $d \geq 5$, the proof of this more general result being identical to the proof for $d = 5$. For any prime $p \geq 5$ we give an explicit construction of a polynomial of degree p over the rationals for which this is the case. (Of course, this can be done for any degree at least five, but in light of this being an introductory text, we do not introduce the complications necessary to do so for the general case.) We show that the classical problems of antiquity — trisecting the angle, doubling the cube, and squaring the circle — are impossible to solve by straightedge and compass constructions. (Here we completely prove the first two of these but content ourselves with quoting Lindemann's theorem that π is transcendental in proving the third.) Finally, we give a proof of the fundamental theorem of algebra. Despite its name, this theorem cannot have a purely algebraic proof, as the real and complex numbers cannot be constructed purely algebraically, but we give a proof that uses the irreducible minimum of analytic results — only the theorem that a polynomial of odd degree with real coefficients must have a real roots – but otherwise is entirely algebraic, using Galois theory and group theory. Throughout this chapter we heavily use the viewpoint that an extension $\mathbb{E}$ of a field $\mathbb{F}$ is an $\mathbb{F}$-vector space, and so, as we have said, we are presupposing familiarity with vector spaces in general.

We conclude in Chapter 5 by studying Dedekind rings. Logically speaking, this could be part of Chapter 3, but pedagogically speaking, we feel it would be a mistake to put it there, as at that point we would have no examples to work with. But, having developed field theory in Chapter 4, we have rings of integers in algebraic number fields as examples, and we first prove that these are always Dedekind rings. Then we prove the main result about Dedekind rings, that nonzero ideals have unique factorization as a product of prime ideals, and then we give concrete examples chosen from rings of algebraic integers.

There are two appendices. To get started in Chapter 2 with group theory, we need to know basic properties of the integers (e.g., primes and unique factorization). But these results are part of ring theory, which we do not treat until Chapter 3. So in Appendix A we simply state these results, in order to have them available at the start when we need them. Of course, we do prove them in Chapter 3, and indeed in a more general context. Our proof of the theorem of the primitive element in Chapter 4 uses a result from linear algebra that, although standard, is not always presented, so in Appendix B we provide the statement and a proof of this result in order to have it available as well.

Each chapter concludes with a variety of exercises ranging from the straight-forward to the challenging. Some of these are particular examples that illustrate the theory while others are general results that develop the theory further.

Finally, some remarks on numbering and notation: We use three-level numbering, so that, for example, Theorem 4.10.3 is the third numbered item in Chapter 4, Section 10. We denote the end of proofs by □, as usual. Theorems, etc., are set in italics, so are demarcated by their typeface. Definitions, etc., are not, so we mark their end by ◊. Our mathematical notation is standard, though we want to point out that if A and B are sets, $A \subseteq B$ means that A is a subset of B and $A \subset B$ means that A is a proper subset of B.

We have enjoyed writing this book and we trust that you will enjoy reading it, and thinking deeply about the matter within it, as well.

Steven H. Weintraub
Bethlehem, PA, USA
May 2021

About the Author

 **Steven H. Weintraub** is Professor of Mathematics at Lehigh University. He is an active research mathematician, having been invited to visit and lecture at universities and conferences around the world, an accomplished expositor of mathematics, and serves the mathematical community as an officer of the American Mathematical Society. He has written over 60 research papers in a variety of areas of mathematics, including algebra, number theory, geometry, and topology. This is his 14th book. For more information, see his website https://www.lehigh.edu/~shw2.

Contents

Chapter 1

Set Theory

Set theory is the language of modern mathematics. In fact, this language is so engrained, it is hard to imagine doing mathematics without using it. (Nevertheless, mathematicians did so for literally thousands of years.)

We regard set theory as a tool and will be developing it with a view to using it in our investigations in the subsequent chapters of this book. But there are some quite interesting subtleties, especially when dealing with infinite sets, and we shall present some of these.

1.1 Basic set theory

We begin at the beginning, with the basic definitions of set theory.

Definition 1.1.1. A *set* A is a collection of objects called its *elements*. We write $a \in A$ to mean that the object a is an element of the set A. We write $A = \{a, b, c, \ldots\}$ to mean that $a, b, c, \ldots$ are the elements of A. $\Diamond$

We will often consider that our sets are contained in some "universe" $\mathcal{U}$, consisting of all possible objects (of whatever sort we are considering).

Example 1.1.2. We have the *empty set* $A = \{\}$, the set having no elements. This set is often denoted by ϕ. We also have the set $A = \mathcal{U}$ consisting of all possible elements. $\Diamond$

1

Of course, if these were the only sets we had, set theory would not be very interesting or useful.

Example 1.1.3. We have $A = \{1, 3\}$, the set whose two elements are the integers 1 and 3. $\Diamond$

Instead of listing elements, we may specify a set by giving a property that its elements must satisfy. We let $P(x)$ be a proposition (i.e., a true–false statement) involving the variable x. Then, $\{x \mid P(x)\}$ is the set consisting of those x for which $P(x)$ is true. (The vertical bar in this notation is read as "such that".)

Example 1.1.4. We have $B = \{x \mid x^2 - 4x + 3 = 0\}$. $\Diamond$

Next, we ask when two subsets are equal. If we regard a set as defined by its elements, the answer is forced on us.

Definition 1.1.5. Two sets A and B are *equal*, i.e., $A = B$, if they have the same elements, i.e., if $x \in A \Leftrightarrow x \in B$. $\Diamond$

Example 1.1.6. The sets A and B of Examples 1.1.3 and 1.1.4 are equal. $\Diamond$

Remark 1.1.7. What does it mean to solve an equation? The set B is simply the set of roots of the polynomial $x^2 - 4x + 3$, so solving the equation means explicitly finding the set B of its roots. Mathematicians were solving equations for thousands of years before set theory came along. Indeed, Euclid knew how to solve this equation, while set theory was first developed in the late 19th century by Cantor. True, in this example, we don't need to mention the word "set", as we could simply ask to find the roots of this equation. But, in more complicated situations, the use of set theory is unavoidable. We want to reify (i.e., regard as an object) the roots of equations, for example, to be able to better handle and understand them, and the object we need to introduce is the set of their roots. $\Diamond$

Definition 1.1.8. Let A be a set. Its *complement* A^c is the set

$$A^c = \{x \in \mathcal{U} \mid x \notin A\}.$$ $\Diamond$

Example 1.1.9. Suppose $\mathcal{U} = \{\text{all integers}\}$. If $A = \{x \in \mathcal{U} \mid x \text{ is divisible by 2}\}$, so that $A = \{\text{even integers}\}$, then $A^c = \{\text{odd integers}\}$. $\Diamond$

It is convenient to introduce *Venn diagrams*, which provide a way of visualizing sets and their properties. We have the following Venn diagram of a set A, where the shaded area indicates the elements of A.

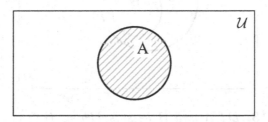

Then, the shaded area in the following diagram indicates the elements of A^c.

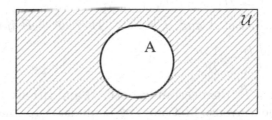

Remark 1.1.10. We easily see that $\phi^c = \mathcal{U}$, $\mathcal{U}^c = \phi$, and that for any set A, $(A^c)^c = A$. ◇

Definition 1.1.11. Let A and B be sets. Then, B is *contained* in A, or is a *subset* of A, if $x \in B \Rightarrow x \in A$. We denote this by $B \subseteq A$. Also, B is *properly contained* in A, or is a *proper subset* of A, if $B \subseteq A$ and $B \neq A$. We denote this by $B \subset A$. ◇

Remark 1.1.12. Many people write, as we do, $B \subseteq A$ to mean that B is a subset of A. But many people denote this by $B \subset A$. We prefer our notation, as containment/proper containment is analogous to the distinction between $x \leq y$ and $x < y$. ◇

The following is the Venn diagram for this situation.

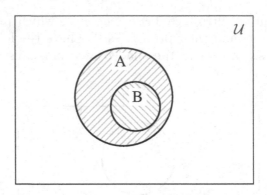

Lemma 1.1.13. *Let A and B be sets. Then, $B \subseteq A$ if and only if $A^c \subseteq B^c$.*

Proof. $B \subseteq A$ means $x \in B \Rightarrow x \in A$. But this implication is logically equivalent to its contrapositive, which is $x \notin A \Rightarrow x \notin B$, which means $A^c \subseteq B^c$. $\qquad\qquad\qquad\qquad\qquad\qquad\qquad\qquad\quad\square$

Lemma 1.1.14. *Let A and B be sets. Then, $A = B$ if and only if*

(i) *$A \subseteq B$ and $B \subseteq A$ or*
(ii) *$A \subseteq B$ and $A^c \subseteq B^c$.*

Proof.   If $A = B$, then certainly both (i) and (ii) are true.

To show the converse, first note, by Lemma 1.1.13, that conditions (i) and (ii) are equivalent. Suppose (i) is true. Since $A \subseteq B, x \in A \Rightarrow x \in B$. Since $B \subseteq A, x \in B \Rightarrow x \in A$. Thus, $x \in A \Leftrightarrow x \in B$, and so $A = B$. $\qquad\qquad\qquad\qquad\qquad\qquad\qquad\qquad\qquad\qquad\qquad\square$

Remark 1.1.15. This innocent looking lemma is the key to proving that two sets are equal. That is, if we want to show that $A = B$, the usual way of doing so is to show that $A \subseteq B$ and that $B \subseteq A$, or to show that $A \subseteq B$ and that $A^c \subseteq B^c$. $\qquad\qquad\qquad\quad\Diamond$

Now, we come to a pair of operations on sets.

Definition 1.1.16. (a) Let A and B be sets. Their *union* $A \cup B$ is the set

$$A \cup B = \{x \mid x \in A \text{ or } x \in B\}.$$

(b) Let A and B be sets. Their *intersection* $A \cap B$ is the set

$$A \cap B = \{x \mid x \in A \text{ and } x \in B\}. \qquad\qquad\qquad\Diamond$$

Example 1.1.17. If $A = \{1, 2\}$ and $B = \{2, 3\}$, then $A \cup B = \{1, 2, 3\}$ and $A \cap B = \{2\}$. ◇

The following is the Venn diagram for this situation, where the union is the hatched region and the intersection is the doubly hatched region.

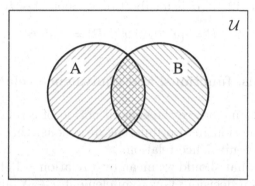

Here are some basic properties of these operations.

Lemma 1.1.18.

(a) *For any set* A, $A \cup A^c = \mathcal{U}$ *and* $A \cap A^c = \phi$.
(b) *For any two sets* A *and* B, $A \cup B = A$ *if and only if* $B \subseteq A$ *and* $A \cap B = A$ *if and only if* $A \subseteq B$.

Proof. Exercise. □

Here are two more properties, which state that each of the operations of union and intersection distributes over the other.

Lemma 1.1.19. *Let* $A, B,$ *and* C *be sets. Then,*

(a) $A \cup (B \cap C) = (A \cup B) \cap (A \cup C)$ *and*
(b) $A \cap (B \cup C) = (A \cap B) \cup (A \cap C)$.

Proof. Exercise. □

Here are two more properties, known as *De Morgan's Laws*.

Lemma 1.1.20. *Let* A *and* B *be sets. Then,*

(a) $(A \cup B)^c = A^c \cap B^c$ *and*
(b) $(A \cap B)^c = A^c \cup B^c$.

Proof. Exercise. □

We introduce a few more definitions.

Definition 1.1.21. Two sets A and B are *disjoint* if $A \cap B = \phi$. If A and B are disjoint, we may write $A \cup B$ as $A \amalg B$ and call this set the *disjoint union* of A and B. $\Diamond$

Definition 1.1.22. If B is a subset of A, then $A - B$ is the set

$$A - B = \{x \in A \mid x \notin B\} = A \cap B^c. \qquad \Diamond$$

1.2 Relations, functions, and equivalence relations

We now want to introduce the general notion of a relation and then specialize this notion in two different directions, that of a function and that of an equivalence relation.

Intuitively, what should we mean by a relation $\sim$ between two sets X and Y? It is a decision: Given an element $x \in X$ and an element $y \in Y$, we decide whether x and y are related, and if so, we write $x \sim y$. For example, we might let $X = \{\text{airlines}\}$ and $Y = \{\text{airports}\}$, and for an airline x_0 (i.e., an element of X) and an airport y_0 (i.e., an element of Y), decide that $x_0 \sim y_0$ if airline x_0 serves airport y_0. But of course a "decision" is not a mathematical construct, and so we must formalize this notion.

First, we need the following construction.

Definition 1.2.1. Let X and Y be sets. Their *product* $X \times Y$ is the set

$$X \times Y = \{(x, y) \mid x \in X, \ y \in Y\}. \qquad \Diamond$$

Example 1.2.2. Let $X = \{1, 2\}$ and $Y = \{a, b, c\}$. Then,

$$X \times Y = \{(1, a), (2, a), (1, b), (2, b), (1, c), (2, c)\}. \qquad \Diamond$$

Now, we define a relation.

Definition 1.2.3. A *relation* R between X and Y is a subset of $X \times Y$. Then, $x \sim_R y$ if $(x, y) \in R$. $\Diamond$

Remark 1.2.4. Strictly speaking, R *is* the relation. But intuitively speaking, we start out by deciding when $x \sim_R y$, and let R be the set of ordered pairs (x, y) for which this is true. Thus, we often call R the *graph* of the relation. $\Diamond$

Definition 1.2.5. Let R be a relation between X and Y.

$$\text{For } x_0 \in X, \ R(x_0) = \{y \in Y \mid (x_0, \ y) \ \in \ R\}.$$
$$\text{For } y_0 \in Y, \ R^{-1}(y_0) = \{x \in X \mid (x, y_0) \ \in \ R\}. \qquad \Diamond$$

We thus see that $R(x_0) = \{y \in Y \mid x_0 \sim_R y\}$ and $R^{-1}(y_0) = \{x \in X \mid x \sim_R y_0\}$, i.e., that $R(x_0)$ consists of those elements of Y that x_0 is related to and that $R^{-1}(y_0)$ consists of those elements of X that are related to y_0.

We may construct new relations from old.

Definition 1.2.6. Let R be a relation between X and Y. Then, its *inverse* R^{-1} is the relation between Y and X given by

$$R^{-1} = \{(y, x) \in Y \times X \mid (x, y) \ \in \ R\}. \qquad \Diamond$$

Definition 1.2.7. Let R be a relation between X and Y, and let S be a relation between Y and Z. Then, their *composition* SR is the relation between X and Z given by

$$SR = \{(x, z) \mid \text{ for some } y \in Y, \ (x, y) \in R \text{ and } (y, z) \ \in \ S\}. \qquad \Diamond$$

Actually, it is not so often in mathematics that we need to consider relations between X and Y in general. It is much more often the case that we want to consider functions from X and Y.

What is a function? Intuitively, a function $f : X \to Y$ is a rule that assigns an element of Y to each element of X. If this rule assigns y_0 to x_0, we write $y_0 = f(x_0)$. Again, we must formalize this notion.

Definition 1.2.8. A *function* (or *mapping*) $f : X \to Y$ is a relation between X and Y with the property that for every $x \in X$, there is exactly one $y \in Y$ with $(x, y) \in f$. In this case, we write $y = f(x)$. $\Diamond$

Remark 1.2.9. The situation here is entirely analogous to (in fact, a special case of) Remark 1.2.4. Strictly speaking, this subset f of $X \times Y$ *is* the function. But intuitively speaking, we start out with the rule $y = f(x)$ and let f be the set of ordered pairs for which this is true. Thus, we often call this set the *graph* of the function.

To see why we use this name, let us look at a very familiar case. Let $X = Y = \{\text{real numbers}\}$, and let $f(x) = x^2$. We identify $X \times Y$ with the plane in the usual way, and then the set $\{(x, x^2) \mid x \in X\}$ is just the graph of this function in the usual sense. $\qquad \Diamond$

Example 1.2.10. Here is a (very simple but) very important function. Let X be any set. Then, the *identity* function id_X on X is the function defined by $\mathrm{id}_X(x) = x$ for every $x \in X$. (We often denote the function simply by id when X is understood.) $\Diamond$

We have some special types of functions.

Definition 1.2.11. Let $f : X \to Y$ be a function. Then:

(a) f is *one-to-one* $(1-1)$, or is an *injection*, if for every $y \in Y$, there is at most one $x \in X$ with $f(x) = y$ (or, equivalently, whenever $x_1 \neq x_2 \in X$, $f(x_1) \neq f(x_2) \in Y$).

(b) f is *onto*, or is a *surjection*, if for every $y \in Y$, there is at least one $x \in X$ with $f(x) = y$.

(c) f is *one-to-one and onto*, or is a *bijection*, if f is both one-to-one and onto, i.e., if for every $y \in Y$, there is exactly one $x \in X$ with $f(x) = y$. $\Diamond$

It is illuminating to view $f : X \to Y$ as a collection of arrows, with an arrow going from $x \in X$ to $f(x) \in Y$.

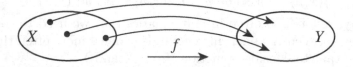

Definition 1.2.12. Let $f = X \to Y$ and $g = Y \to Z$ be functions. Their *composition* $h = gf$ is the function $h = X \to Z$ given by $h(x) = g(f(x))$. $\Diamond$

In the "arrow" representation, we have as follows:

Thus, in this representation, h is the double-length arrows.

Here is the basic fact about composition, which is used throughout mathematics.

Theorem 1.2.13. *Composition of functions is associative, i.e., if* $f : X \to Y$, $g : Y \to Z$, *and* $h : Z \to W$ *are functions, then*

$$h(gf) = (hg)f : X \to W.$$

Proof. On the one hand, for any $x \in X$,

$$(h(gf))(x) = h(gf(x)) = h(g(f(x))),$$

and on the other hand,

$$((hg)f)(x) = hg(f(x)) = h(g(f(x))),$$

and these are equal. $\qquad\square$

In the "arrow" representation, we have

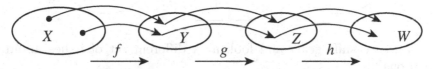

and $h(gf) = (hg)f$ is given by the triple-length arrows.

Definition 1.2.14. Let $f : X \to Y$ be a function. Then, f is *invertible* if there is a function $g : Y \to X$ with $gf = \mathrm{id}_X$ and $fg = \mathrm{id}_Y$. In this case, we call g the *inverse* of f and write $g = f^{-1}$. $\qquad\Diamond$

Remark 1.2.15. We observe that this definition is symmetric in f and g, so if $g = f^{-1}$, then $f = g^{-1}$ (and $(f^{-1})^{-1} = f$, $(g^{-1})^{-1} = g$). $\qquad\Diamond$

There is a very simple criterion for deciding when a function is invertible.

Theorem 1.2.16. *Let $f : X \to Y$ be a function. Then, f is invertible if and only if f is $1 - 1$ and onto.*

Proof. First, suppose f is $1 - 1$ and onto. Then, for every $y \in Y$, there is a unique element $x \in X$ with $f(x) = y$. Thus, setting

$$g(y) = x \text{ if } f(x) = y,$$

we have a well-defined function, and we see that $gf(x) = x$, for every $x \in X$ and $fg(y) = y$, for every $y \in Y$, i.e., $gf = \mathrm{id}_X$ and $fg = \mathrm{id}_Y$, so f is invertible and $f^{-1} = g$.

On the other hand, suppose that f is invertible and let $g = f^{-1}$. First, we show f is onto: Let $y \in Y$ be arbitrary. Set $x = g(y)$. Then,

$$y = \mathrm{id}_Y(y) = fg(y) = f(g(y)) = f(x).$$

Next, we show f is $1 - 1$: Let $y \in Y$ be arbitrary and suppose $y = f(x_1) = f(x_2)$. Then,

$$gf(x_1) = gf(x_2), \ \mathrm{id}_X(x_1) = \mathrm{id}_X(x_2), \ x_1 = x_2. \qquad\square$$

In the "arrow" representation, in case f is invertible, f^{-1} is obtained from f by reversing the direction of the arrows.

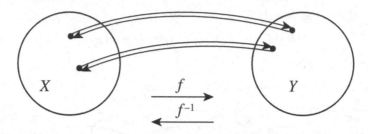

Now, we shift gears and look at a different set of issues about relations.

Definition 1.2.17. Let R be a relation on X (i.e., between X and X). Then:

  (i) R is *reflexive* if $(x, x) \in R$ for every $x \in X$.
 (ii) R is *symmetric* if $(x, y) \in R$ implies $(y, x) \in R$.
(iii) R is *transitive* if $(x, y) \in R$ and $(y, z) \in R$ implies $(x, z) \in R$. ◊

Remark 1.2.18. This is the formal definition. But we usually think of it informally:

  (i) R is reflexive if $x \sim_R x$ for every $x \in X$.
 (ii) R is symmetric if whenever $x \sim_R y$, then $y \sim_R x$.
(iii) R is transitive if whenever $x \sim_R y$ and $y \sim_R z$, then $x \sim_R z$. ◊

We can ask the following three yes/no questions: Is R reflexive? Is R symmetric? Is R transitive? *A priori*, there are eight possibilities for the answers, and in fact they all occur. (See Exercises.) But the case when all these answers are yes is a particularly important one, which we now focus on.

Definition 1.2.19. Let R be a relation on a set X that is reflexive, symmetric, and transitive. Then, R is called an *equivalence relation*. ◊

Example 1.2.20. The archetype of an equivalence relation is equality, i.e., $x \sim_R y$ if $x = y$. This is where the name comes from. But there are many others. ◊

Example 1.2.21. Consider the relation on people: Person $x \sim$ Person y if x and y have the same birthday. This is an equivalence relation. ◇

To state the basic property of an equivalence relation, we need to make two preliminary definitions.

Definition 1.2.22. Let $\sim$ be an equivalence relation on X. For $x_0 \in X$, the *equivalence class* of x is $[x_0] = \{y \in X \mid x_0 \sim y\}$. Any element of an equivalence class is called a *representative* of that equivalence class. ◇

Lemma 1.2.23. *Let $\sim$ be an equivalence relation on X, and let A be a subset of X. Then, A is an equivalence class of $\sim$ if and only if*

(i) *A is nonempty,*
(ii) *$x \sim y$ for every $x, y \in A$, and*
(iii) *$x \nsim z$ for every $x \in A$, $z \notin A$.*

Proof. Suppose A is an equivalence class. Then, by definition, it is $[x_0]$ for some $x_0 \in X$. Since $\sim$ is reflexive, $x_0 \sim x_0$, and so $x_0 \in [x_0]$ and A is nonempty. Now, let $x, y \in A$. By definition, $x_0 \sim x$ and $x_0 \sim y$. Since $\sim$ is symmetric, $x \sim x_0$ and then, since $\sim$ is transitive, $x \sim y$. Also, let $x \in A$ and $z \notin A$. Suppose $x \sim z$. Since $x_0 \sim x$, we have $x_0 \sim z$ and so $z \in A$, a contradiction.

On the other hand, let A be a set satisfying (i), (ii), and (iii). Since A is nonempty, there is some $x_0 \in A$. We claim $A = [x_0]$. First, we show $A \subseteq [x_0]$: Let $y \in A$. Then, $x_0 \sim y$, so $y \in [x_0]$. Next, we show $[x_0] \subseteq A$: Let $z \in [x_0]$. If $z \notin A$, then $x_0 \sim z$ with $z \notin A$, a contradiction. Thus, $A = [x_0]$. □

Corollary 1.2.24. *Let $\sim$ be an equivalence relation on X, and let A and B be equivalence classes of $\sim$. Then, A and B are either identical or disjoint. Equivalently, let A and B be equivalence classes of $\sim$ and suppose there is some $x_0 \in X$ with $x_0 \in A$ and $x_0 \in B$. Then, $A = B$.*

Proof. If A and B are disjoint, there is nothing to prove. Suppose not. Let $x_0 \in A \cap B$. Let $y \in A$ and $y' \in B$.

Then, $x_0 \sim y$ (as $x, y \in A$) and $x_0 \sim y'$ (as $x_0, y' \in B$), so $y \sim y'$. But then, by Lemma 1.2.23, $y' \in A$, so $B \subseteq A$, and similarly $y \in B$, so $A \subseteq B$. Hence, $A = B$. □

Definition 1.2.25. Let X be a set, and let $A_1, A_2, \ldots$ be nonempty subsets of X. Suppose that these subsets are pairwise disjoint, i.e., that $A_i \cap A_j = \phi$ whenever $i \neq j$, and that $X = \underset{i}{\cup} A_i$. Then, $\{A_1, A_2, \ldots\}$ *partitions* X. In this case, we write $X = \underset{i}{\amalg} A_i$. ◇

Theorem 1.2.26. *Let X be a set, and let $\{A_1, A_2, \ldots\}$ be a partition of X. Then, $x \sim y$ if $x \in A_i$ and $y \in A_i$ for some i is an equivalence relation on X.*

Conversely, let $\sim$ be an equivalence relation on X. Then $\{$distinct equivalence classes of $\sim\}$ is a partition of X.

Proof.   Clearly, $x \sim y$ if $x, y \in A_i$ for some i is an equivalence relation on X.

Conversely, let $A_1, A_2, \ldots$ be the distinct equivalence classes of $\sim$. As we have seen, each equivalence class is nonempty. Since for any $x \in X$, $x \in [x]$, we certainly have $X = \underset{x \in X}{\cup} A_i$. But then $X = \underset{i}{\cup} A_i$ as in restricting the union to the distinct equivalence classes, we are merely eliminating duplications. But by Corollary 1.2.24, the distinct equivalence classes are disjoint, so $X = \underset{i}{\amalg} A_i$. □

Definition 1.2.27. Let $\{A_1, A_2, \ldots\}$ be the set of distinct equivalence classes of $\sim$ on X. A set $\{a_1, a_2, \ldots\}$ of elements of X with $a_i \in A_i$, for each i, is called a *complete set of representatives* of the equivalence classes. ◇

Example 1.2.28. Let $X = \{\text{people}\}$, and let $\sim$ be the equivalence relation on X of having the same birthday. Then, the set of distinct equivalence classes is $\{\{\text{people whose birthday is January 1}\}, \{\text{people whose birthday is January 2}\}, \{\text{people whose birthday is January 3}\}, \ldots\}$. If Alice's birthday is January 1, Bob's birthday is January 2, Charlotte's birthday is January 3, ..., then $\{$Alice, Bob, Charlotte, ...$\}$ is a complete set of representatives of $\sim$. But also, if Arthur's birthday is January 1, Brenda's birthday is January 2, Charlie's birthday is January 3, ..., then $\{$Arthur, Brenda, Charlie, ...$\}$ is also a complete set of representatives. The point is that representatives are almost never unique, and that there is no *a priori* reason to prefer one representative to another. ◇

The corresponding Venn diagram (enhanced with representatives) is as follows.

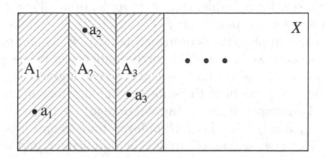

1.3 Cardinality

In this section, we want to investigate the cardinality, or "size", of sets. We start off with a very simple case.

Let S be a finite set, i.e., a set consisting of a finite number of elements. Then, the cardinality of S, $\#(S)$, is the number of elements of S. In other words, if S has a finite number k of elements, then $\#(S) = k$. For example, $\#(\phi) = 0$, $\#(\{\text{banana}\}) = 1$, $\#(\{\text{apple, orange}\}) = 2$, $\#(\{\text{tomato, cucumber}\}) = 2$.

We are now going to make this simple situation a whole lot more complicated. We do this because we have to develop a framework that we can use to investigate the cardinality of arbitrary sets.

We let $\mathbb{N} = \{\text{positive integers}\}$. We use this symbol as the positive integers are sometimes called the *natural numbers*.

For a nonnegative integer k, we let $\mathbb{N}_k = \{i \in \mathbb{N} \mid i \le k\}$. Thus, $\mathbb{N}_0 = \phi$, and for $k > 0$, $\mathbb{N}_k = \{1, 2, \dots, k\}$.

In this section, we call a bijection a *1 – 1 correspondence*.

A moment's thought shows that we can recast our simple notion of size above in the following form.

Definition 1.3.1. Let S be a set. Then, S is a finite set if there is a $1 - 1$ correspondence $f \colon \mathbb{N}_k \to S$ for some nonnegative integer k. In this case, we say the *cardinality* of S, $\#(S) = k$. ◊

Remark 1.3.2. Strictly speaking, in order to know that this definition makes sense, we must know that if j and k are nonnegative integers with $j \neq k$, then there is no bijection between $\mathbb{N}_j$ and $\mathbb{N}_k$. We leave this as an exercise.   ◇

Thus, we see, for example, that $\#(\{\text{apple, orange}\}) = 2$ because we have a $1-1$ correspondence $f \colon \mathbb{N}_2 = \{1, 2\} \to \{\text{apple, orange}\}$ given by $f(1) = \text{apple}$, $f(2) = \text{orange}$. Similarly, $\#(\{\text{tomato, cucumber}\}) = 2$ because we have a $1-1$ correspondence $g \colon \mathbb{N}_2 \to \{\text{tomato, cucumber}\}$ given by $g(1) = \text{tomato}$, $g(2) = \text{cucumber}$.

Now, these two sets have the same cardinality (namely 2), and we have a $1-1$ correspondence $h \colon \{\text{apple, orange}\} \to \{\text{tomato, cucumber}\}$ given by $h = gf^{-1}$. Then, $h(\text{apple}) = \text{tomato}$ and $h(\text{orange}) = \text{cucumber}$. But now, we can eliminate the middleman, and we are led to the following definition.

Definition 1.3.3. Let S and T be sets. Then, S and T have the same cardinality, $\#(S) = \#(T)$, if there is a $1-1$ correspondence $h \colon S \to T$.   ◇

Lemma 1.3.4. *Having the same cardinality is an equivalence relation on sets.*

Proof.   Exercise.   □

Definition 1.3.5. A set S is *countably infinite*, or simply *countable*, if there is a $1-1$ correspondence $f \colon \mathbb{N} \to S$.   ◇

Remark 1.3.6. We use the term because in this situation, we can simply count the elements of $S \colon f(1)$ is the first element of S, $f(2)$ is the second element of S, $f(3)$ is the third element of $S, \ldots$.   ◇

Example 1.3.7. Let $a_1, a_2, a_3, \ldots$ be any infinite sequence all of whose elements are distinct. Then, the set $A = \{a_1, a_2, a_3, \ldots\}$ is countable as we have a $1-1$ correspondence $f \colon \mathbb{N} \to A$ given by $f(i) = a_i$.   ◇

Remark 1.3.8. We use this example in the opposite direction. If we have an infinite set and we can list its elements in order, then we have a $1-1$ correspondence between that set and $\mathbb{N}$, so that the set is countable.   ◇

We think of the cardinality of a set S as its size, so to say that $\#(S) = \#(T)$ is to say that S and T have the same size. But this definition, which is forced on us, has some surprising consequences.

Example 1.3.9. (a) The set $A = \{$even positive integers$\}$ is count-able, as we have the $1 - 1$ correspondence $f\colon \mathbb{N} \to A$ given by $f(n) = 2n$. Thus, even though we might think that $\mathbb{N}$ is twice as big as A, in fact they have the same size.

In light of Remark 1.3.8, we could have shown that A was count-able simply by listing the elements of A in order, $A = \{2, 4, 6, 8, \dots\}$, rather than writing down the function f, and that is what we will do in the remaining parts of this example.

(b) The set $A = \{$nonnegative integers$\}$ is countable as we may list the elements of A as $A = \{0, 1, 2, 3, \dots\}$.

(c) The set $\mathbb{Z} = \{$all integers$\}$ is countable as we may list the elements of $\mathbb{Z}$ as $\{0, 1, -1, 2, -2, 3, -3, \dots\}$.

(d) Consider the set $\mathbb{N} \times \mathbb{N} = \{(i, j) \mid i \in \mathbb{N}, j \in \mathbb{N}\}$. This set contains infinitely many copies of $\mathbb{N}$, the subsets $\{(i, j_0) \mid i \in \mathbb{N}\}$ for any fixed $j \in \mathbb{N}$, so we might suppose its cardinality is greater than that of $\mathbb{N}$. But this set, too, is countable. In the following figure, we let (i, j) denote the point with those coordinates in the plane, and next to each such point, we write the number of that point in a listing of $\mathbb{N} \times \mathbb{N}$.

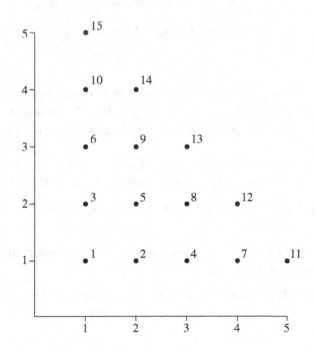

To clarify, this gives the listing of $\mathbb{N} \times \mathbb{N}$ as $\mathbb{N} \times \mathbb{N} = \{(1,1),$ $(2, 1),$ $(1, 2),$ $(3, 1),$ $(2, 2),$ $(1, 3),$ $\ldots\}$, and we see that $\mathbb{N} \times \mathbb{N}$ is countable. $\Diamond$

At this point, you may wonder if there are any sets that are *uncountable* (i.e., not countable). The answer is yes.

Theorem 1.3.10. (*Cantor*) $\mathbb{R} = \{real\ numbers\}$ *is uncountable.*

Proof.   We must show that there is no $1 - 1$ correspondence from $\mathbb{N}$ to $\mathbb{R}$, or equivalently, that no matter how we list the elements of $\mathbb{R}$ in order, our list will not contain all of them.

We can regard real numbers as given by their decimal expansions, and the decimal expansion of a real number is unique, except for one ambiguity: a decimal expansion that ends in an infinite sequence of 9's. (For example, $.999\ldots = 1$.) So, we never use such an expansion.

Thus, suppose we list the real numbers:

$$r_1 = \epsilon_1 \ldots a_3^1\ a_2^1\ a_1^1\ a_0^1 \cdot a_{-1}^1\ a_{-2}^1\ a_{-3}^1\ a_{-4}^1,$$

$$r_2 = \epsilon_2 \ldots a_3^2\ a_2^2\ a_1^2\ a_0^2 \cdot a_{-1}^2\ a_{-2}^2\ a_{-3}^2\ a_{-4}^2,$$

$$r_3 = \epsilon_3 \ldots a_3^3\ a_2^3\ a_1^3\ a_0^3 \cdot a_{-1}^3\ a_{-2}^3\ a_{-3}^3\ a_{-4}^3,$$

$$r_4 = \epsilon_4 \ldots a_3^4\ a_2^4\ a_1^4\ a_0^4 \cdot a_{-1}^4\ a_{-2}^4\ a_{-3}^4\ a_{-4}^4,$$

$$\vdots$$

where ϵ_i is a sign ($\epsilon_i = \pm 1$) and the a_j^i are decimal digits.

Form the real number $s = .b_{-1}b_{-2}b_{-3}b_{-4}\ldots$, where $b_{-i} \neq a_{-i}^i$ and $b_{-i} \neq 9$. (The restriction $b_{-i} \neq 9$ is to guarantee we never end s with $999\ldots$.)

Then, s is not an element of the list $\{r_1, r_2, r_3, \ldots\}$ as s differs from r_i in the ith digit following the decimal point, for every i. $\square$

Remark 1.3.11. The argument in this proof is known as Cantor's diagonal argument. $\Diamond$

We can apply a similar idea in more general circumstances. First, we need a general definition.

Definition 1.3.12. Let S and T be sets. Then, $\#(S) \leq \#(T)$ if there is a $1 - 1$ correspondence from S to a subset of T, and $\#(S) <$

$\# (T)$ if $\# (S) \leq \# (T)$ and $\# (S) \neq \# (T)$, i.e., if there is a 1 – 1 correspondence from S to a subset of T, but there is no 1 – 1 correspondence from S to T. $\Diamond$

Theorem 1.3.13. *Let S be any set, and let $P(S)$ be the power set of S,*

$$P(S) = \{all \ subsets \ of \ S\}.$$

Then, $\# (S) < \# (P(S))$.

Proof. If S is a finite set, this is easy: If $\# (S) = k$, then $\# (P(S)) = 2^k$, and $k < 2^k$ for any nonnegative integer k.

Let S be an arbitrary set. First, we have a 1 – 1 correspondence e from S to a subset of $P(S)$ given by $e(s) = \{s\}$ for $s \in S$.

To show that there is no 1 – 1 correspondence $f : S \to P(S)$, we argue by contradiction. Suppose we have such a correspondence. Let

$$R = \{s \in S \mid s \notin f(s)\}.$$

R is a subset of S, and since a 1 – 1 correspondence is onto, we must have $R = f(s_0)$ for some $s_0 \in S$. We ask whether $s_0 \in R$.

If $s_0 \in R$, then by the defining property of R, $s_0 \notin f(s_0) = R$.
If $s_0 \notin R = f(s_0)$, then by the defining property of R, $s_0 \in R$.
Thus, in either case, we have a contradiction. $\square$

Example 1.3.14. Let us now consider $\mathbb{Q}_+ = \{$positive rational numbers$\}$. Note that we can write every element r of $\mathbb{Q}_+$ as $r = p/q$, a quotient of positive integers, with this fraction in lowest terms. We claim that $\mathbb{Q}_+$ is countable. Let us try to show this.

Before doing so, let us observe that certainly $\# (\mathbb{N}) \leq \# (\mathbb{Q}_+)$ as we have the 1 – 1 function $f : \mathbb{N} \to \mathbb{Q}_+$ given by $f(n) = n/1$.

But let us try to come up with a 1 – 1 correspondence from $\mathbb{Q}_+$ to $\mathbb{N}$. To do so, let us return to Example 1.3.9 (d), where we considered $\mathbb{N} \times \mathbb{N}$. Let us consider $\mathbb{Q}_+$, where we identify $r = p/q$ with the point with coordinates (p, q). Note that we do not get all points of $\mathbb{N} \times \mathbb{N}$. For example, we do not get the point $(2, 2)$ as $2/2$ is not a fraction in lowest terms. Thus, in that example, let us simply erase the points of $\mathbb{N} \times \mathbb{N}$ that we don't get and keep the labels on the points that we do. We have the following figure.

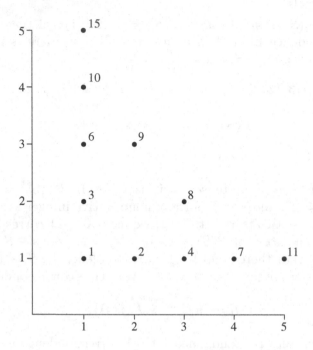

Now, taking a lattice point to its label, we have a $1-1$ function g from $\mathbb{Q}_+$ to $\mathbb{N}$: $g(1,1) = 1$, $g(2,1) = 2$, $g(1,2) = 3$, $g(3,1) = 4$, $g(1,3) = 6, \ldots$

Note that this function g is not onto $\mathbb{N}$ as the set of its values is $\{1, 2, 3, 4, 6, 7, 8, 9, 10, 11, 15, \ldots\}$.

Thus, we have succeeded in showing that $\#\,(\mathbb{Q}_+) \le \#\,(\mathbb{N})$, but we have not shown that $\#\,(\mathbb{Q}_+) = \#\,(\mathbb{N})$. $\qquad\qquad\qquad\diamond$

Now, perhaps we could be more clever and construct a $1-1$ correspondence between $\mathbb{N}$ and $\mathbb{Q}_+$. But this would not help us very much, except in this one example. A theory of cardinalities that required us to be clever each time we wanted to show $\#\,(S) = \#\,(T)$ would not be a very useful theory. So, it is very useful to have the following result. (You can't get something for nothing, so, as you will see, this theorem that avoids the necessity of being clever in particular examples itself has a very clever proof.)

Theorem 1.3.15 (Schröder–Bernstein). *Let S and T be sets with $\#\,(S) \le \#\,(T)$ and $\#\,(T) \le \#\,(S)$. Then, $\#\,(S) = \#\,(T)$.*

Proof (König). We first prove this in the case where S and T are disjoint. This involves all the hard work (and cleverness). Once we have done that, we will see how to easily extend the proof to handle S and T in general.

Thus, let us suppose that S and T are disjoint sets and that we have a $1-1$ correspondence f between S and a subset of T, and a $1-1$ correspondence g between T and a subset of S.

For any element s_0 of S, we may form the "string"

$$s_0 \xrightarrow{f} t_0 \xrightarrow{g} s_1 \xrightarrow{f} t_1 \xrightarrow{g} s_2 \xrightarrow{f} t_2 \longrightarrow \cdots,$$

which is defined by $t_0 = f(s_0), s_1 = g(t_0), t_1 = f(s_1), s_2 = g(t_1), t_2 = f(s_2), \ldots$.

Note that this string extends infinitely far to the right, as we may always keep applying f or g as the case may be.

We now try to extend this string to the left. Consider s_0. Now, g is a $1-1$ function but is not necessarily onto.

If $s_0 \neq g(t)$ for any $t \in T$, then we cannot extend the string past s_0 to the left. But if $s_0 = g(t)$ for some $t \in T$, there is exactly one such t, which we call t_{-1}. Thus, $s_0 = g(t_{-1})$ or $t_{-1} = g^{-1}(s_0)$. Now, consider t_{-1}. By the same logic, either there is no $s \in S$ with $f(s) = t_{-1}$ or there is exactly one such s, which we call s_{-1}. Then, $t_{-1} = f(s_{-1})$ or $s_{-1} = f^{-1}(t_{-1})$.

We keep going with this procedure as long as we can. Perhaps this goes on forever, or perhaps not. So, our string becomes

$$\cdots \longleftarrow s_{-1} \xleftarrow{f^{-1}} t_{-1} \xleftarrow{g^{-1}} s_0 \xrightarrow{f} t_0 \xrightarrow{g} s_1 \xrightarrow{f} t_1 \longrightarrow \cdots,$$

where we do not know if, or where, it stops on the left.

Note that we can rewrite this string as

$$\cdots \longrightarrow s_{-1} \xrightarrow{f} t_{-1} \xrightarrow{g} s_0 \xrightarrow{f} t_0 \xrightarrow{g} s_1 \xrightarrow{f} t_1 \longrightarrow \cdots.$$

Note also that we have simply labeled the arrows for emphasis. A right-pointing arrow must have a label f (resp. g) if it leads away from an element of S (resp. T), and a left-pointing arrow must have a label g^{-1} (resp. f^{-1}) if it leads away from an element of S (resp. T).

We claim that {distinct strings} is a partition of $S \cup T$.

To show this, we must show (i) that every element of $S \cup T$ is in some string and (ii) that distinct strings are disjoint. To show (i), we first show that every element of S is in some string and then show that every element of T is in some string.

Let $s \in S$. Then, we have the string obtained by choosing $s_0 = s$, and that string contains s in position s_0. Let $t \in T$. Set $s = g(t)$. Then, we have the string obtained by choosing $s_0 = s$, and that string contains t in position t_{-1}. This shows (i). Now, we observe that any string is entirely determined by any element of it. Certainly, it is determined by s_0, as from the first way of writing the string, we see that any element of the string is at the head of a sequence of arrows emanating from s_0. But now, note that we can turn the direction of arrows around, as we did to get to the second way of writing the string, so by the same logic the string is entirely determined by any s_i or t_i. From this, we see that any two strings that have an element in common must be identical. (We need not be concerned about the position of that element in each of the strings, as a string always goes infinitely far to the right and always goes as far to the left as possible.) This shows (ii).

Now, let us consider {distinct strings}. We will see that any string is of one of the four types.

How can a string behave? Here is one possibility:

Type I: The string goes infinitely far to the left (and of course to the right) with no repetition of elements.

Here is the next possibility: The string has a repeated element. Choose a pair of repeated elements that are as close together as possible, say $s_{i+k} = s_i$. (As we will see momentarily, the argument is the same if some t element is repeated.) But then, $t_{i+k} = f(s_{i+k}) = f(s_i) = t_i$, and then $s_{i+k+1} = g(t_{i+k}) = g(t_i) = s_{i+1}$, etc., so we have a repetition of both s and t elements to the right. But also, $s_i = s_{i+k} = g(t_{i+k-1})$, so we can continue the string to the left of s_i with $t_{i-1} = t_{i+k-1}$ (as t_{i-1} is defined by the equation $g(t_{i-1}) = s_i$). But then, by the same logic, $s_{i-1} = s_{i+k-1}$, so we have a repetition of both s and t elements to the left. This gives the second possibility:

Type II: For some k, the string contains the substring

$$s_0 \to t_0 \ldots s_{k-1} \to t_{k-1}$$

with no repetition, but $g(t_{k-1}) = s_0$, and the string goes infinitely far to the left (and of course to the right) repeating this substring.

This takes care of the cases in which the string extends infinitely far to the left. Thus, what remains are the cases in which it does not. This gives the remaining two types:

Type III: For some N, a string that begins on the left with s_{-N} (and of course continues infinitely far to the right), i.e., a string of the form

$$s_{-N} \xrightarrow{f} t_{-N} \xrightarrow{g} s_{-N+1} \xrightarrow{f} t_{-N+1} \longrightarrow \cdots .$$

Type IV: For some M, a string that begins on the left with t_{-M-1} (and continues infinitely far to the right), i.e., a string of the form

$$t_{-M-1} \xleftarrow{g^{-1}} s_{-M} \xleftarrow{f^{-1}} t_{-M} \xleftarrow{g^{-1}} s_{-M+1} \xleftarrow{} \cdots .$$

Now, having done all this hard work, we can now easily define a $1 - 1$ correspondence $h : S \to T$. The correspondence is given as follows:

If s is an element of a type-I string, $h(s) = f(s)$.

If $s = s_0, \ldots, s_{k-1}$ in a type-II string, $h(s) = f(s)$.

If s is an element of a type-III string, $h(s) = f(s)$.

If s is an element of a type-IV string, $h(s) = g^{-1}(s)$.

Then, we are done!

It remains to handle the case where S and T are not disjoint. Let $S_0 = S \times \{0\} = \{(s, 0) \mid s \in S\}$ and $T_1 = T \times \{1\} = \{(t, 1) \mid t \in T\}$.

Then, we have a $1 - 1$ correspondence $i : S \to S_0$ given by $i(s) = (s, 0)$ and a $1 - 1$ correspondence $j : T \to T_1$ given by $j(t) = (t, 1)$. Note that S_0 and T_1 are disjoint.

Suppose we have a $1 - 1$ correspondence f between S and a subset of T. This gives us a $1 - 1$ correspondence $\overline{f}$ between S_0 and a subset of T_1 defined by $\overline{f}(s, 0) = (f(s), 1)$. Similarly, a $1 - 1$ correspondence g between T and a subset of S gives a $1 - 1$ correspondence $\overline{g}$ between T_1 and a subset of S_0 defined by $\overline{g}(t, 1) = (g(t), 0)$. Now, we apply the Schröder–Bernstein theorem to the disjoint case (which we have just proved) to obtain a $1 - 1$ correspondence $\overline{h} : S_0 \to T_1$. But then, $h = j^{-1}hi : S \to T$ is a $1 - 1$ correspondence. $\qquad\Box$

Example 1.3.16. Let $S = \mathbb{N}$ and $T = \mathbb{Q}_+$. In Example 1.3.14, we constructed a $1 - 1$ correspondence f from S to a subset of

T and a $1-1$ correspondence g from T to a subset of S. Now, by the Schröder–Bernstein theorem, we can conclude that S and T have the same cardinality, i.e., that $\#\,(\mathbb{Q}_+) = \#\,(\mathbb{N})$ (and so $\mathbb{Q}_+$ is countable). $\lozenge$

Example 1.3.17. Since the proof of the Schröder–Bernstein theorem was so tricky, it is worthwhile to follow the proof in constructing a $1-1$ correspondence from $\mathbb{N}$ to $\mathbb{Q}_+$. It turns out to be more illuminating to proceed differently than in Examples 1.3.14 and 1.3.16.

We let $\mathbb{N} = \{\text{positive integers}\} = \{1, 2, 3, \dots\}$ and $\mathbb{Q}_+ = \{\text{positive rational numbers}\} = \{p/q \mid p, q \text{ positive integers, fraction in lowest terms}\}$. In particular, $\mathbb{Q}_+$ has the elements $1/1$, $2/1$, $3/1$, $\dots$, and we do *not* identify these with 1, 2, 3, $\dots$, so that $\mathbb{N}$ and $\mathbb{Q}_+$ are disjoint.

We let $f\colon \mathbb{N} \to \mathbb{Q}_+$ be the function $f(n) = n/1$, so that f is a $1-1$ correspondence between $\mathbb{N}$ and a subset of $\mathbb{Q}_+$, and we let $g\colon \mathbb{Q}_+ \to \mathbb{N}$ be the function $g(p/q) = 2^p 3^q$, so that g is a $1-1$ correspondence between $\mathbb{Q}_+$ and a subset of $\mathbb{N}$.

We look at the strings in the proof of the Schröder–Bernstein theorem. For convenience, we label the string with $s_0 = n$ by S_n. We have the strings

$$S_1 : 1 \to 1/1 \to 2 \cdot 3 = 6 \to 6/1 \to 2^6 \cdot 3 = 192 \to \cdots,$$

$$S_2 : 2 \to 2/1 \to 4 \cdot 3 = 12 \to 12/1 \to 2^{12} \cdot 3 = 12288 \to \cdots,$$

$$\vdots$$

$$S_5 : 5 \to 5/1 \to 32 \cdot 3 = 96 \to 96/1 \to 2^{96} \cdot 3 \to \cdots$$

and note that these strings do not extend to the left.

Then, we start the next string as

$$6 \to 6/1 \to 2^6 \cdot 3 = 192 \to 192/1 \to 2^{192} \cdot 3 \to \cdots,$$

but now we observe this string *does* extend to the left and we obtain the string

$$S_6 : 1 \to 1/1 \to 6 \to 6/1 \to 2^6 \cdot 3 = 192 \to 192/1 \to 2^{192} \cdot 3 \to \cdots,$$

so we see that $S_6 = S_1$.

Then, $S_7, \ldots, S_{11}$ are all distinct, but we see $S_{12} = S_2$. Then, $S_{13}, \ldots, S_{17}$ are all distinct.

Then, we start the next string as

$$18 \to 18/1 \to 2^{18} \cdot 3 = 786432 \to 786432/1 \to 2^{786432} \cdot 3 \to \cdots,$$

but now we observe that this string *does* extend to the left and we obtain the string

$$S_{18}: \ 1/2 \to 18 \to 18/1 \to 2^{18} \cdot 3 = 786432 \to 786432/1$$
$$\to 2^{786432} \cdot 3 \to \cdots.$$

Note that $S_1, \ldots, S_{17}$ are all type-III strings, while S_{18} is a type-IV string. We continue in this fashion to obtain type-III strings $S_{19}, \ldots, S_{35}$ (with $S_{24} = S_4$) and the next type-IV string is

$$S_{36}: 2/3 \to 36 \to 36/1 \to 2^{36} \cdot 3 \to \cdots.$$

Given these strings, we can now define the $1 - 1$ correspondence $h: \mathbb{N} \to \mathbb{Q}_+$. We give a table of values of $h(n)$ for selected values of n.

n	$h(n)$
1	1/1
2	2/1
17	17/1
18	1/2
19	19/1
35	35/1
36	2/3
192	192/1
12288	12288/1
786432	18/1

$\Diamond$

1.4   Naïve and not-so-naïve set theory

In our development of set theory, one of the ways we have obtained sets is by specifying a property that defines its elements. That is, if $P(x)$ is a proposition (i.e., a true–false statement) involving the variable x, then we have the set $A = \{x \mid P(x) \text{ is true}\}$. For example,

$$A = \{\text{integers } x \mid x \text{ is divisible by } 2\}$$

defines the set of even integers, $A = \{\ldots, -4, -2, 0, 2, 4, \ldots\}$,

$$B = \{\text{integers } x \mid x = y^2 \text{ for some integer } y\}$$

defines the set of perfect squares $B = \{0, 1, 4, 9, 16, \ldots\}$, and if $p(z) = a_n z^n + a_{n-1} z^{n-1} + \cdots + a_1 z + a_0$ is a polynomial with complex coefficients, $C = \{\text{complex numbers } z \mid p(z) = 0\}$ is the set $C = \{\text{roots of } p(z)\}$, even if we cannot find these roots. We will be using this construction throughout this book, and indeed it is used throughout mathematics.

Proceeding in this way is sometimes called naïve set theory because of the (naïve) belief that this always makes sense. Unfortunately, it does *not*.

Example 1.4.1 (Russell's paradox). Is the class of all classes that are not members of themselves a member of itself? If the answer is yes, i.e., if it is a member of itself, then, since this class consists of all classes that are not members of themselves, the answer is no, it is not a member of itself. If the answer is no, i.e., it is not a member of itself, then, since this class consists of all classes that are not members of themselves, the answer is yes, it is a member of itself. Thus, either answer to this question leads to a contradiction. ◊

Thinking about the example, we see that the problem was caused by considering the proposition "$x \notin x$". So, we cannot allow *all* propositions.

Having shown you that there *is* a problem, I am *not* going to show you what the solution is. That would take us very far afield. I am simply going to tell you that there is a solution, which you may look up if you wish. (Most) mathematicians operate under *ZF* set theory, where *ZF* stands for Zermelo–Fraenkel. *ZF* set theory ensures us that $\{x \mid P(x)\}$ makes sense for all the propositions $P(x)$ we normally encounter in doing mathematics.

ZF is a set of axioms for set theory. There is an additional axiom. In the statement of this axiom, I is an arbitrary indexing set.

Axiom of choice: Let $\{X_i\}_{i \in I}$ be pairwise disjoint nonempty sets. Then, there is a set $A = \{A_i\}_{i \in I}$ with $a_i \in X_i$ for each $i \in I$.

This axiom gets its name because we think of obtaining A by choosing one element from each set. If the indexing set I is finite, there is no problem, as we simply choose the elements one at a time. But mathematics is a finite process, so we cannot do that if the indexing set is infinite, so we need an axiom to guarantee the existence of A. It turns out that this axiom is independent of the axioms of ZF. (Most) mathematicians accept this axiom and operate under ZFC, which is ZF plus the axiom of choice. (The axiom of choice has many equivalent formulations. The most useful in constructing mathematical proofs is *Zorn's lemma*, which I am not going to state here but which you may look up if you wish.)

Finally, a few words about cardinalities of sets. Recall that we showed $\#(\mathbb{N}) < \#(\mathbb{R})$. We have the famous

Continuum hypothesis: There is no set S with $\#(\mathbb{N}) < \#(S) < \#(\mathbb{R})$.

It is known that the continuum hypothesis is independent of the axioms of ZFC.

1.5 Exercises

1. Prove Lemma 1.1.18:
   (a) $A \cup A^c = \mathcal{U}, A \cap A^c = \phi$.
   (b) $A \cup B = A \Leftrightarrow B \subseteq A$, $A \cap B = A \Leftrightarrow A \subseteq B$.
2. Prove the distributive laws (Lemma 1.1.19):
   (a) $(A \cup B) \cap C = (A \cap C) \cup (B \cap C)$.
   (b) $(A \cap B) \cup C = (A \cup C) \cap (B \cup C)$.
3. Prove De Morgan's laws (Lemma 1.1.20):
   (a) $(A \cup B)^c = A^c \cap B^c$.
   (b) $(A \cap B)^c = A^c \cup B^c$.
4. Show that $A \cup B = A \amalg (B - A) = B \amalg (A - B)$
   $= (A - B) \amalg (B - A) \amalg (A \cap B)$.
5. The *symmetric difference* of two sets A and B is
   $A * B = (A - B) \cup (B - A)$.

Show that:

(a) $A*B = (A \cap B^c) \cup (A^c \cap B)$.

(b) $(A*B)^c = (A \cap B) \cup (A^c \cap B^c)$.

(c) $(A*B)*C = A*(B*C)$.

6. Let A and B be subsets of X, and let C and D be subsets of Y. Show that:

(a) $(A \cap B) \times (C \cap D) = (A \times C) \cap (B \times D)$.

(b) $(A \cup B) \times (C \cup D) = (A \times C) \cup (A \times D) \cup (B \times C) \cup (B \times D)$.

7. Let $f : X \to Y$ be a mapping. Let A and B be subsets of X, and let C and D be subsets of Y. In each case, prove or find a counterexample:

(a) $f(A \cup B) = f(A) \cup f(B)$.

(b) $f(A \cap B) = f(A) \cap f(B)$.

(c) $f^{-1}(f(A)) = A$.

(a') $f^{-1}(C \cup D) = f^{-1}(C) \cup f^{-1}(D)$.

(b') $f^{-1}(C \cap D) = f^{-1}(C) \cap f^{-1}(D)$.

(c') $f(f^{-1}(C)) = C$.

8. Let $\sim_R$ be a relation on $\mathbb{Z}$. For each of the eight possibilities,

$\sim_R$ reflexive: yes/no,

$\sim_R$ symmetric: yes/no,

$\sim_R$ transitive: yes/no.

Find a relation $\sim_R$ with these properties, except in the case yes/yes/yes, find a relation $\sim_R$ other than equality.

9. (a) Let A and B be finite sets. Show that
$$\#(A \cup B) = \#(A) + \#(B) - \#(A \cap B).$$
(b) Let $A, B,$ and C be finite sets. Show that
$$\#(A \cup B \cup C) = \#(A) + \#(B) + \#(C) - \#(A \cap B) - \#(A \cap C) - \#(B \cap C) + \#(A \cap B \cap C).$$
(c) This generalizes to any numbers of sets: Let $A_1, \ldots, A_n$ be finite sets. Show that
$$\#(A_1 \cup \ldots \cup A_n) = \sum_S (-1)^{\#(S)-1} \#(A_S),$$

when the sum is over nonempty subsets S of $\{1, \ldots, n\}$, and if $S = \{i_1, \ldots i_k\}$ is such a set, $A_S = A_{i1} \cap \cdots \cap A_{ik}$. This is known as the *inclusion–exclusion principle*.

10. (a) Let A and B be finite sets. Show that
$$\#(A \times B) = \#(A) \, \#(B).$$
(b) Let $A_1, \ldots A_n$ be finite sets. Show that
$$\#(A_1 \times \ldots \times A_n) = \#(A_1) \cdots \#(A_n).$$

11. If A is a finite set, $\#(A) = n$. Show that $\#(P(A)) = 2^n$.
12. Prove Lemma 1.3.4: Having the same cardinality is an equivalence relation on sets.
13. In each case, we have a $1-1$ map $f : A \to B$. Find a $1-1$ map $g : B \to A$. (Then, by the Schröder–Bernstein theorem, these two sets here the same cardinality.)

    (a) $A = \mathbb{N}$, $B = \{\text{finite subsets of } \mathbb{N}\}$, $f(n) = \{n\}$.
    (b) $A = \{x \in \mathbb{R} \mid x \geq 0\}$, $B = \mathbb{R}$, $f(x) = x$.
    (c) $A = \{x \in \mathbb{R} \mid 0 \leq x < 1\}$, $B = \mathbb{R}$, $f(x) = x$.
    (d) $A = \mathbb{R}$, $B = \{\text{finite subsets of } \mathbb{R}\}$, $f(x) = \{x\}$.
    (e) $A = \mathbb{R}$, $B = \{\text{countably infinite subsets of } \mathbb{R}\}$, $f(x) = \{x, x+1, x+2, \dots\}$.

Chapter 2

Group Theory

Sets are sets, period. In other words, they have no additional structure. Groups are sets with additional structure, that of an operation (satisfying certain properties). Logically speaking, these are what we should consider next. Afterwards, we will consider rings and fields, which are sets with two operations (again satisfying certain properties). But groups are very important in their own right — indeed, you have certainly already seen them, even if you have not seen them by name — and there is a lot to say about them.

We will consider groups in general. But we will pay particular attention to finite groups, where we can say a lot about their structure. However, we will not only completely determine the structure of finite abelian groups, but also of infinite, though finitely generated, abelian groups.

2.1 Definition, examples, and basic properties

We begin by defining groups.

Definition 2.1.1. A *group* $(G, \cdot)$ is a set of G with an operation (that is, a function) $\cdot$ on $G \times G$ satisfying the following properties:

(0) (Closure) If $a, b \in G$, $a \cdot b \in G$.
(1) (Associativity) If $a, b, c \in G$, $(a \cdot b) \cdot c = a \cdot (b \cdot c)$.
(2) (Identity) There is an *identity element* $e \in G$, i.e., an element e such that $e \cdot a = a \cdot e = a$ for every $a \in G$.

(3) (Inverse) Every $a \in G$ has an *inverse element* $a^{-1} \in G$, i.e., an element a^{-1} such that $a \cdot a^{-1} = a^{-1} \cdot a = e$ for every $a \in G$. $\Diamond$

Definition 2.1.2. An *abelian group* $(G, \cdot)$ is a group $(G, \cdot)$ with the additional property:

(4) (Commutativity) If $a, b \in G$, $a \cdot b = b \cdot a$. $\Diamond$

As a matter of language, when the operation is understood, we will refer to the group G rather than to the group $(G, \cdot)$.

As a matter of notation, when the operation is understood, we write ab instead of $a \cdot b$. In dealing with abelian groups, we will sometimes write ab and sometimes write $a + b$.

Here is the very first invariant of a group.

Definition 2.1.3. Let G be a group. The *order* of G is $|G| = \#(G)$. G is a *finite* or *infinite* group as $|G|$ is finite or infinite. $\Diamond$

Now let us see a bunch of examples.

Example 2.1.4.

(a) We have the *trivial* group $G = \{e\}$, consisting of the single element e (so $|G| = 1$) with the operation $e \cdot e = e$. We also have the following trivial groups: $(\{1\}, \cdot)$ and $(\{0\}, +)$, where $\cdot$ and $+$ are the usual multiplication and addition respectively.

(b) Let $\mathbb{Z}$ denote the integers, $\mathbb{Q}$ the rational numbers, $\mathbb{R}$ the real numbers, and $\mathbb{C}$ the complex numbers. Then if $R = \mathbb{Z}, \mathbb{Q}, \mathbb{R}$, or $\mathbb{C}$, $(R, +)$ is an abelian group. If $\mathbb{R} = \mathbb{Z}$, let $R^* = \{\pm 1\}$. If $R = \mathbb{Q}, \mathbb{R}$, or $\mathbb{C}$, let $R^* = R - \{0\}$. Then $(R^*, \cdot)$ is an abelian group.

(c) Let $S = \{z \in \mathbb{C} | \; |z| = 1\}$. S is the unit circle in the complex plane. Then S is an abelian group under multiplication.

(d) Let n be a positive integer and let $W_n = \{\exp(2\pi i k)/n \, | \, k = 0, \ldots, n-1\}$. Then W_n is the set of complex n-th roots of 1, and W_n is an abelian group under multiplication. Note W_n is a group of order n.

(e) Let R be as in part (b) and let $M_{m,n}(R) = \{m\text{-by-}n \text{ matrices with elements in } R\}$. Then $M_{m,n}(R)$ is an abelian group under matrix addition.

(f) Let R be as above and let $GL_n(R) = \{\text{invertible } n\text{-by-}n \text{ matrices with elements in } R\}$. Then $GL_n(R)$ is a group under multiplication and is not abelian for $n \geq 2$. (There is one subtlety we should point out here: Invertible means that the inverse is also in $GL_n(R)$, i.e., has elements in R. For example, the 1-by-1 matrix $[2]$ is in $GL_1(\mathbb{R})$ but not in $GL_1(\mathbb{Z})$.) ◊

Here are some basic properties of groups.

Lemma 2.1.5. *Let G be a group.*

(a) *The identity element of G is unique.*
(b) *For any $a \in G$, its inverse a^{-1} is unique.*

Proof.

(a) Suppose e and f are two identity elements of G. Then by the properties of the identity, $e = ef = f$.
(b) Suppose b and c are two inverses of a. Then $b = be = b(ac) = (ba)c = ec = c$. □

Lemma 2.1.6. *Let a be an element of a group G. If $ab = e$, or $ba = c$, then $b = a^{-1}$.*

Proof. We do the first case. We have $b = eb = (a^{-1}a)b = a^{-1}(ab) = a^{-1}e = a^{-1}$. □

Remark 2.1.7. Associativity says that if we have any three elements a, b, and c of G, then $(ab)c = a(bc)$. In other words, we may regroup parentheses as we wish without changing the value of the product. This says we may eliminate the parentheses entirely and write the product simply as abc. The same is true no matter how many elements we have. For example, $((ab)c)(de) = a(((bc)d)e)$, and we may write this common product as $abcde$. This is quite tedious to prove in general — in fact, it is even tedious to state it precisely — so rather than a formal statement and proof, we will content ourselves with this remark. ◊

Lemma 2.1.8. *Let G be a group.*

(a) $e^{-1} = e$.
(b) *For any element a of G, $(a^{-1})^{-1} = a$.*
(c) *For any two elements a and b of G, $(ab)^{-1} = b^{-1}a^{-1}$.*

Proof.

(a) $ee = e$.
(b) $(a^{-1})a = a(a^{-1}) = e$.

(c) We simply compute

$$(ab)(b^{-1}a^{-1}) = a(bb^{-1})a^{-1} = aea^{-1} = aa^{-1} = e$$

and

$$(b^{-1}a^{-1})(ab) = b^{-1}(a^{-1}a)b = b^{-1}eb = b^{-1}b = e. \qquad \square$$

We now construct two important examples of groups. These can be done as special cases of more general constructions later, but we construct them "by hand" now so that we will have them available as we proceed.

Example 2.1.9. We fix a positive integer n.

We define a relation on the integers $\mathbb{Z}$ by $x \equiv y \pmod{n}$ if $x - y$ is divisible by n (i.e., $x - y = nq$ for some integer q).

It is straightforward to check that $\equiv \pmod{n}$ is an equivalence relation. It then follows from the division algorithm (see Appendix A) that there are n equivalence classes, and that $\{0, 1, \ldots, n-1\}$ is a complete set of representatives of the equivalence classes. We shall denote the equivalence class of i by $[i]_n$ (for now). We let $\mathbb{Z}_n$ be the set of equivalence classes, $\mathbb{Z}_n = \{[0]_n, [1]_n, \ldots, [n-1]_n\}$. We wish to given $\mathbb{Z}_n$ a group structure, and we define the group operation by the equation

$$[i]_n + [j]_n = [i+j]_n.$$

A priori, this is *not* a valid definition. We defined the sum of two equivalence classes by picking a representative from each equivalence class and taking the equivalence class of the sum of these two representatives. But of course representatives are not unique. What would happen if we were to choose different representatives? In order for this definition to be valid we have to show that we would get the same answer, i.e., that this definition is independent of the choice of representatives.

To do this, suppose $[i']_n = [i]_n$ and $[j']_n = [j]_n$.

Then $i' = i + nq_1$, for some integer q_1, and $j' = j + nq_2$ for some integer q_2. But then $i' + j' = (i+j) + nq_3$ where $q_3 = q_1 + q_2$, so $[i' + j']_n = [i+j]_n$. In other words, we have just shown that if $[i']_n = [i]_n$ and $[j']_n = [j]_n$, then $[i'+j']_n = [i+j]_n$, and so our group operation is well-defined. The rest of the properties of a group are easy to check, and then we find that $\mathbb{Z}_n$ is an abelian group of order n. $\qquad \Diamond$

Example 2.1.10. Again we fix a positive integer n and consider the equivalence relation $x \equiv y \pmod{n}$ on $\mathbb{Z}$. We see (referring to Appendix A again) that if $x \equiv y \pmod{n}$, then $\gcd(x, n) = \gcd(y, n)$. We let $\mathbb{Z}_n^* = \{$equivalence classes $[i]_n$ with i and n relatively prime$\}$, which makes sense (i.e., does not depend on the choice of representative). Then $\{0 \le k < n \mid k$ and n are relatively prime$\}$ is a complete set of coset representatives of those equivalence classes in $\mathbb{Z}_n^*$ (not of all equivalence classes) and so $\mathbb{Z}_n^* = \{[k]_n \mid, 0 \le k \le n - 1$ and $\gcd(k, n) = 1\}$.

We give $\mathbb{Z}_n^*$ a group structure by defining

$$[i]_n [j]_n = [ij]_n.$$

Again we must see that the operation is well-defined, i.e., does not depend on the choice of representatives. So suppose $[i']_n = [i]_n$ and $[j']_n = [j]_n$, so that $i' = i + nq_1$ and $j' = j + nq_2$ for integers q_1 and q_2. Then $i'j' = ij + nq_3$ where $q_3 = iq_1 + jq_2 + nq_1q_2$.

Now the group operations are more interesting to check. We have that $\gcd(k, n) = 1$ if and only if there are integers x and y with $kx + ny = 1$.

Closure: Let i and n be relatively prime, so $iq + nr = 1$ for some q, r. Let j and n be relatively prime, so $js + nt = 1$ for some s, t. Then $(iq + nr)(js + nt) = 1 \cdot 1 = 1$ and $(iq + nr)(js + nt) = (ij)u + nv$ where $u = qs$ and $v = iqt + jrs + nrt$, so ij and n are relatively prime. Thus if $[i]_n \in \mathbb{Z}_n^*$ and $[j]_n \in \mathbb{Z}_n^*$, $[ij]_n \in \mathbb{Z}_n^*$.

Associativity: Easy to check.

Identity: $[1]_n$ is the identity element of $\mathbb{Z}_n^*$.

Inverses: If $[k]_n \in \mathbb{Z}_n^*$, so that k and n are relatively prime, then there are integers x and y with $kx + ny = 1$, i.e., $kx = 1 + n(-y)$, so $kx \equiv 1 \pmod{n}$, and hence $[k]_n[x]_n = [1]_n$, and so $[x]_n = [k]_n^{-1}$.

Commutativity: Easy to check.

Thus we see that $\mathbb{Z}_n^*$ is an abelian group of order

$$\varphi(n) = \#(\{0 \le k \le n - 1 \mid \gcd(k, n) = 1\}).$$

In particular, if $n = p$ is a prime, we see that $\mathbb{Z}_p^* = \{[1]_p, \ldots, [p - 1]_p\}$ has order $p - 1$. $\diamondsuit$

In learning to ride a bicycle, you use training wheels. Then, once you have learned, you remove them. We are just learning group theory, and our notation $[k]_n$ is the analog of training wheels. We shall almost always use this notation here, but your should be aware that experienced mathematicians simply write k in this situation. Thus we have the equations $[6]_9 + [7]_9 = [4]_9$ and $[5]_{11}[8]_{11} = [7]_{11}$, but mathematicians will often simply say $6 + 7 = 4$ in $\mathbb{Z}_9$ and $5 \cdot 8 = 7$ in $\mathbb{Z}_{11}$.

Example 2.1.11. Given a group $(G, \cdot)$, we may write down the *multiplication table* for G, where the entry in row a and column b is the value $a \cdot b$.

We then have, with the abbreviated notation just described, the

multiplication table for $\mathbb{Z}_5$:   and for $\mathbb{Z}_5^*$:

	0	1	2	3	4
0	0	1	2	3	4
1	1	2	3	4	0
2	2	3	4	0	1
3	3	4	0	1	2
4	4	0	1	2	3

	1	2	3	4
1	1	2	3	4
2	2	4	1	3
3	3	1	4	2
4	4	3	2	1

$\Diamond$

One important way that groups often arise is as $\text{Aut}(X)$, the *automorphism group of a structure* X. That is, we have a "structure" X, and we consider the invertible functions from X to itself. We have put the word structure in quotes, and we do not have a precise definition of it, but we will see some examples of it here.

We begin with an example that you have probably seen (at least in disguise).

Example 2.1.12. Let V be a vector space over $\mathbb{F}$, where $\mathbb{F}$ is as in Example 2.1.4. Then $GL(V)$ is the group of all invertible linear transformations $T: V \to V$. Linear transformations are those functions that preserve the vector space structure. If V is finite-dimensional, say $\dim(V) = n$, and we choose a basis B of V, then we can identify T with its matrix $[T]_B$ in the basis B, and in that way we can identify $GL(V)$ with $GL_n(\mathbb{F})$, as in Example 2.1.4. (For purposes of future comparison, we recall that if $S: V \to V$ and $T: V \to V$ are linear transformations, then $ST: V \to V$ is the linear transformation given by $ST(v) = S(T(v))$. We observe several things: Multiplication in the group is composition of functions. In the composition ST, we apply the function on the right (that is, T) first. Associativity in the group follows from the fundamental fact that composition of functions is associative. $\Diamond$

Example 2.1.13. Let X be a set (finite or infinite). (Thus X has no additional structure beyond being a collection of its elements.) Then $\text{Aut}(X)$ consists of the invertible functions $\sigma\colon X \to X$. An invertible function is just a bijection, so we are just rearranging the elements of X. Such a function is called a *permutation* of the elements of X. The group $\text{Aut}(X)$ is called the *symmetric group* on the elements of X.

For example, let $X = \{1, 2, 3, 4, 5\}$. Then we have $\sigma \in \text{Aut}(X)$ when σ is the function $\sigma(1) = 1, \sigma(2) = 3, \sigma(3) = 5, \sigma(4) = 4, \sigma(5) = 2$, and we have $\tau \in \text{Aut}(X)$ where τ is the function $\tau(1) = 2, \tau(2) = 4, \tau(3) = 3, \tau(4) = 5, \tau(5) = 1$. Then $\rho = \sigma\tau \in \text{Aut}(X)$ is the function

$$\rho(1) = \sigma\tau(1) = \sigma(\tau(1)) = \sigma(2) = 3$$

$$\rho(2) = \sigma\tau(2) = \sigma(\tau(2)) = \sigma(4) = 4$$

$$\rho(3) = \sigma\tau(3) = \sigma(\tau(3)) = \sigma(3) = 5$$

$$\rho(4) = \sigma\tau(4) = \sigma(\tau(4)) = \sigma(5) = 2$$

$$\rho(5) = \sigma\tau(5) = \sigma(\tau(5)) = \sigma(1) = 1.$$

You can check that if $X = \{1, 2, \ldots, n\}$, then $\text{Aut}(X)$ is a group of order $n!$. In this case $\text{Aut}(X)$ is called the *symmetric group* on n elements, and is denoted S_n. This is a very important group. While we will say no more about it now, we will be considering it in detail later (see Section 2.10). $\diamond$

Example 2.1.14. Let $n \geq 3$ and let X be a regular n-gon. For the sake of definitiveness we let X have vertices the complex n-th roots of 1, $\exp(2\pi i k/n)$ for $k = 0, \ldots, n - 1$. We label each vertex by the value of k, so that the complex number 1 is labeled as vertex 0 and the numbering increases as we go counterclockwise.

We let $G = \text{Aut}(X)$, the group of symmetries of X. Let us first see how many elements G has. To do this, note that any symmetry of X must take a vertex to a vertex. Thus if we begin with the vertex 0, it can go to any vertex $0, 1, 2, \ldots, n - 1$, so there are n possibilities. Once we choose that 0 goes to some vertex k, we can ask where does 1 go to. The vertex 1 is adjacent to 0, so its image must be adjacent to k. We can choose either of the two possibilities: $k+1$ or $k-1$ (mod n). So there are $2n$ possibilities. But note that once the images of 0 and 1 are fixed, the images of every vertex are determined. Thus,

$|G| = 2n$. This group is denoted D_{2n} and is called the *dihedral group* of order $2n$.

Let us examine this group more closely. Of course we have the identity map on X (i.e., the map which leaves X pointwise fixed), and this is the identity element of D_{2n}, which we denote by e. Next, we have the element α of D_{2n} which rotates X $2\pi/n$ degrees counterclockwise, so that $\alpha(0) = 1, \alpha(1) = 2, \ldots, \alpha(n-2) = n-1, \alpha(n-1) = 0$. And next, we have the element β of D_{2n} which reflects X across the x-axis, so that $\beta(0) = 0, \beta(1) = n - 1, \beta(n - 1) = 1, \beta(2) = n - 2, \beta(n - 2) = 2, \ldots$. Note $\alpha^n = e$ and $\beta^2 = e$. Then as a set D_{2n} has $2n$ elements $\{\alpha^i \beta^j \mid 0 \le i \le n - 1, \ 0 \le j \le 1\}$. Note that $(\alpha^i)^{-1} = \alpha^{n-i}$ and $\beta^{-1} = \beta$. Of course, by $\alpha\beta$ we mean the symmetry of X obtained by first applying β and then applying α. We compute

$$\alpha\beta(0) = \alpha(\beta(0)) = \alpha(0) = 1$$
$$\alpha\beta(1) = \alpha(\beta(1)) = \alpha(n - 1) = 0$$
$$\alpha\beta(2) = \alpha(\beta(2)) = \alpha(n - 2) = n - 1$$
$$\vdots$$
$$\alpha\beta(n - 1) = \alpha(\beta(n - 1)) = \alpha(1) = 2.$$

Let us compute $\beta\alpha$. We find

$$\beta\alpha(0) = \beta(\alpha(0)) = \beta(1) = n - 1$$
$$\beta\alpha(1) = \beta(\alpha(1)) = \beta(2) = n - 2$$
$$\beta\alpha(2) = \beta(\alpha(2)) = \beta(3) = n - 3$$
$$\vdots$$
$$\beta\alpha(n - 1) = \beta(\alpha(n - 1)) = \beta(0) = 0.$$

We observe that $\beta\alpha \ne \alpha\beta$ and so D_{2n} is not abelian. We experiment a little further and find

$$\beta\alpha^{-1}(0) = \beta(\alpha^{-1}(0)) = \beta(n - 1) = 1$$
$$\beta\alpha^{-1}(1) = \beta(\alpha^{-1}(1)) = \beta(0) = 0$$

$$\beta\alpha^{-1}(2) = \beta(\alpha^{-1}(2)) = \beta(1) = n - 1$$

$$\vdots$$

$$\beta\alpha^{-1}(n-1) = \beta(\alpha^{-1}(n-1)) = \beta(n-2) = 2$$

and thus we see that $\beta\alpha^{-1} = \alpha\beta$. Then $\beta\alpha^{-2} = (\beta\alpha^{-1})\alpha^{-1} = (\alpha\beta)\alpha^{-1} = \alpha(\beta\alpha^{-1}) = \alpha(\alpha\beta) = \alpha^2\beta$ and in general we see that $\beta\alpha^{-i} = \alpha^i\beta$. Furthermore, for any i, $(\alpha^i\beta)(\alpha^i\beta) = (\alpha^i\beta)(\beta\alpha^{-i}) = \alpha^i(\beta^2)\alpha^{-i} = \alpha^i(e)\alpha^{-i} = \alpha^i\alpha^{-i} = e$, and more generally $(\alpha^i\beta)(\alpha^j\beta) = (\alpha^i\beta)(\beta\alpha^{-j}) = \alpha^i(\beta^2)\alpha^{-j} = \alpha^i(e)\alpha^{-j} = \alpha^{i-j}$. This enables us to write down the multiplication table for D_{2n}. We do so for $n = 3$.

	e	α	α^2	β	$\alpha\beta$	$\alpha^2\beta$
e	e	α	α^2	β	$\alpha\beta$	$\alpha^2\beta$
α	α	α^2	e	$\alpha\beta$	$\alpha^2\beta$	β
α^2	α^2	e	α	$\alpha^2\beta$	β	$\alpha\beta$
β	β	$\alpha^2\beta$	$\alpha\beta$	e	α^2	α
$\alpha\beta$	$\alpha\beta$	β	$\alpha^2\beta$	α	e	α^2
$\alpha^2\beta$	$\alpha^2\beta$	$\alpha\beta$	β	α^2	α	e

We leave you to check that, geometrically, $\{e, \alpha, \ldots, \alpha^{n-1}\}$ are all rotations and $\{\beta, \alpha\beta, \ldots, \alpha^{n-1}\beta\}$ are all reflections. (Warning: Our notation D_{2n} is the most common one, but some authors denote this group by D_n.) $\Diamond$

If we let $i = \sqrt{-1}$, then it is easy to check that $G = \{1, i, -1, -i\}$ is a group under multiplication. Here is a related, and more interesting, group.

Example 2.1.15. The *quaternion group* Q_8 is the group

$$Q_8 = \{\pm 1, \pm i, \pm j, \pm k\}$$

with $i^2 = j^2 = k^2 = -1$, $ij = k$, and $ji = -k$. These last two relations have the consequences that $jk = i$, $kj = -i$ and $ki = j$, $ik = -j$, as you may check. Here is the multiplication table for Q_8.

	1	i	-1	$-i$	j	$-j$	k	$-k$
1	1	i	-1	$-i$	j	$-j$	k	$-k$
i	i	-1	$-i$	1	k	$-k$	$-j$	j
-1	-1	$-i$	1	i	$-j$	j	$-k$	k
$-i$	$-i$	1	i	-1	$-k$	k	j	$-j$
j	j	k	$-j$	$-k$	-1	1	i	$-i$
$-j$	$-j$	$-k$	j	k	1	-1	$-i$	i
k	k	$-j$	$-k$	j	i	$-i$	-1	1
$-k$	$-k$	j	k	$-j$	$-i$	i	1	-1

$\Diamond$

(By the way, there is no standard way to order the elements of a group, so multiplication tables for the same group may look different. However, although it is not logically necessary to do so, everybody puts the identity element first.)

Here is a way to get new groups from old.

Definition 2.1.16. Let G and H be groups. Their *product* $G \times H$ is the group

$$G \times H = \{(g, h) \mid g \in G, h \in H\}$$

with product defined by $(g_1, h_1) \cdot (g_2, h_2) = (g_1 g_2, h_1 h_2)$. $\Diamond$

We leave it to you to check that $G \times H$ is a group.

Finally, we record a construction for future use.

Definition 2.1.17. Let G be a group and let A and B be subsets of G. Then AB is the subset of G given by $AB = \{ab \mid a \in A, b \in B\}$. If $A = \{a\}$ consists of a single element, we write aB instead of $\{a\}B$, and similarly if B consists of a single element. $\Diamond$

For arbitrary subsets A and B, this construction is not very useful, but we will see particular cases in which it is not only useful, but indeed essential.

We record one property of this construction.

Lemma 2.1.18. *Let G be a group and let A, B, and C be subsets of G. Then $(AB)C = A(BC)$.*

Proof.   This follows directly from the associative law $(ab)c = a(bc)$ for multiplication of individual elements of G. $\square$

2.2 Homomorphisms and isomorphisms

We now want to consider mappings between groups. What should a mapping $\varphi\colon G \to H$ from a group G to a group H be? On the one hand, a group is a collection of elements, so φ should be a function from the set of elements of G to the set of elements of H. But on the other hand, a group is more than just the set of its elements. A group has an operation as well. Thus to be a mapping of groups, φ has to "respect" the operations on G and H. This leads us to the following definition.

Definition 2.2.1. A *homomorphism* $\varphi\colon G \to H$ from a group G to a group H is a function $\varphi\colon G \to H$ with the property that $\varphi(ab) = \varphi(a)\varphi(b)$ for every $a, b \in G$. $\qquad \diamond$

Here are two simple, but basic, properties of homomorphisms.

Lemma 2.2.2. *Let $\varphi\colon G \to H$ be a homomorphism. Then:*

(i) $\varphi(e) = e$; *and*
(ii) *For any $a \in G$, $\varphi(a^{-1}) = \varphi(a)^{-1}$.*

Proof.

(i) $\varphi(e) = \varphi(e \cdot e) = \varphi(e)\varphi(e)$ so $\varphi(e) = e$.
(ii) $e = \varphi(e) = \varphi(aa^{-1}) = \varphi(a)\varphi(a^{-1})$. $\qquad \square$

Example 2.2.3.

(a) For any group G, $\mathrm{id}\colon G \to G$ is a homomorphism.
(b) For any groups G and H, $\varphi\colon G \to H$ defined by $\varphi(g) = e$ for every $g \in G$ is a homomorphism.
(c) Let G and H be any groups. Then we have homomorphisms $i_1\colon G \to G \times H, i_2\colon H \to G \times H, \pi_1\colon G \times H \to G$, and $\pi_2\colon G \times H \to H$ defined by:

$$i_1(g) = (g, e) \quad \text{for every} \quad g \in G$$

$$i_2(h) = (e, h) \quad \text{for every} \quad h \in H$$

$$\pi_1(g, h) = g \quad \text{for every} \quad (g, h) \in G \times H$$

$$\pi_2(g, h) = h \quad \text{for every} \quad (g, h) \in G \times H \qquad \diamond$$

Our next set of examples will be of homomorphisms between abelian groups, and in these examples we will write the groups additively.

Example 2.2.4.

(a) For any fixed integer n, we have a homomorphism $\varphi\colon \mathbb{Z} \to \mathbb{Z}$ defined by $\varphi(i) = ni$. We observe that φ is a homomorphism as $\varphi(i+j) = n(i+j) = ni + nj = \varphi(i) + \varphi(j)$.

(b) For any fixed integer n, we have a homomorphism $\varphi\colon \mathbb{Z} \to \mathbb{Z}_n$ defined by $\varphi(i) = [i]_n$. We observe that φ is a homomorphism as $\varphi(i+j) = [i+j]_n = [i]_n + [j]_n = \varphi(i) + \varphi(j)$ by the definition of addition in $\mathbb{Z}_n$.

(c) For any fixed integers m and n, we have $\varphi\colon \mathbb{Z}_n \to \mathbb{Z}_n$ defined by $\varphi([i]_n) = [mi]_n$. Again $\varphi([i]_n + [j]_n) = \varphi([i+j]_n) = m[i+j]_n = [m(i+j)]_n = [mi + mj]_n = [mi]_n + [mj]_n = m[i]_n + m[j]_n = \varphi(i) + \varphi(j)$.

(d) For any fixed integers m and n *with n a multiple of m*, we have $\varphi\colon \mathbb{Z}_n \to \mathbb{Z}_m$ defined by $\varphi([i]_n) = [i]_m$. Here we need to be careful to ensure that φ is well-defined. The problem is that we have defined $\varphi([i]_n)$ by choosing a representative i of this equivalence class, and we need to know that this value does not depend on our choice. Thus consider i' with $[i']_n = [i]_n$. Then $\varphi([i']_n) = [i']_m$ and so we need to show that $[i']_m = [i]_m$. Now $[i']_m = [i]_m$ is the equation $i' \equiv i \pmod{m}$, or equivalently, that $i' - i$ is divisible by m. However, we chose i' with $[i']_n = [i]_n$, which similarly means that $i' - i$ is divisible by n. But, *since we are requiring that n is a multiple of m*, if $i' - i$ is divisible by n then certainly $i' - i$ is divisible by m, which is what we needed to show. $\Diamond$

Definition 2.2.5. Let G and H be groups. A homomorphism $\varphi\colon G \to H$ is an *isomorphism* if there is an inverse homomorphism $\varphi^{-1}\colon H \to G$. An isomorphism $\varphi\colon G \to G$ is called an *automorphism* of G. If there is an isomorphism $\varphi\colon G \to H$, the groups G and H are *isomorphic*. $\Diamond$

Here is a criterion for a homomorphism to be an isomorphism.

Lemma 2.2.6. *Let $\varphi\colon G \to H$ be a homomorphism. Then φ is an isomorphism if and only if φ is 1-1 and onto (i.e., if and only if φ is a bijection).*

Proof. If $\varphi\colon G \to H$ is an isomorphism, by definition it has an inverse $\varphi^{-1}\colon H \to G$. But φ is a function, and to be invertible as a function it must be 1-1 and onto.

Conversely, suppose $\varphi\colon G \to H$ is 1-1 and onto. Then φ has an inverse function $\varphi^{-1}\colon H \to G$. We must show that φ^{-1} is a homomorphism. To this end, let $x, y \in H$ and let $a = \varphi^{-1}(x), b = \varphi^{-1}(y)$. Then $\varphi(a) = x$ and $\varphi(b) = y$. Since φ is a homomorphism, $\varphi(ab) = \varphi(a)\varphi(b) = xy$, so $\varphi^{-1}(xy) = ab = \varphi^{-1}(x)\varphi^{-1}(y)$, as required. $\square$

Remark 2.2.7. This lemma says that if φ is invertible as a map of sets, it is automatically invertible as a map of groups. But it is not always the case that if φ satisfies an extra condition then φ^{-1} does. Here is an example. Let $X = \{x \in \mathbb{R} \mid 0 \le x < 1 \text{ or } x = 2\}$ and let $Y = \{x \in \mathbb{R} \mid 0 \le x \le 1\}$. Define $f\colon X \to Y$ by $f(x) = x$ if $0 \le x < 1$ and $f(2) = 1$. Then $f\colon X \to Y$ is a continuous function, which is 1-1 and onto, so has an inverse $f^{-1}\colon Y \to X$. in fact, f^{-1} is given by $f^{-1}(x) = x$ if $0 \le x < 1$ and $f^{-1}(1) = 2$. But note that f^{-1} is *not* continuous. $\Diamond$

Here is what, in some sense, is the simplest kind of group.

Definition 2.2.8. A group G is *cyclic* if it consists of the powers (positive and negative) of some element a of G. Such an element is called a *generator* of G. $\Diamond$

Theorem 2.2.9.

(a) *An infinite cyclic group G is isomorphic to $\mathbb{Z}$.*
(b) *A finite cyclic group G is isomorphic to $\mathbb{Z}_n$, where n is the order of G.*

Proof. Let G be cyclic with generator a. Suppose that there is no nonzero value of k for which $a^k = e$. Then we have an isomorphism $\varphi\colon \mathbb{Z} \to G$ given by $\varphi(n) = a^n$. (φ is a homomorphism by the laws of exponents. φ is onto by the definition of a cyclic group. φ is 1-1 as if $n_1 \ne n_2$, $\varphi(n_1) = a^{n_1} \ne \varphi(n_2) = a^{n_2}$, as if $a^{n_1} = a^{n_2}$ then $a^{n_1-n_2} = e$.) This proves part (a).

On the other hand, suppose $a^k = e$ for some nonzero value of k. Since $a^{-k} = (a^k)^{-1}$, then $a^k = e$ for some positive value of k. Let $k = n$ be the smallest such positive value. We first observe that $a^k = e$ if and only if k is a multiple of n. For we may write $k = nq+r$ where r is an integer with $0 \le r < n$.

Then $a^k = a^{nq+r} = (a^n)^q a^r = e^q a^r = a^r$. If $r = 0$ then certainly $a^k = e$. But if $r \neq 0$ then $a^r \neq e$, as r is a positive integer less than n and by definition n is the smallest positive integer with $a^n = e$. This same computation shows that as a set,

$$G = \{a^r \mid 0 \le r < n\} = \{e, a, \dots, a^{n-1}\}$$

and G has order n. We claim we have an isomorphism $\varphi \colon \mathbb{Z}_n \to G$ given by $\varphi([k]_n) = a^k$. The first thing we need to see in that this definition makes sense, as $[k]_n$ is an equivalence class. So suppose $[k]_n = [k']_n$. Then $k' = k + nq$ for some q_1 and then $a^{k'} = a^{k+nq} = a^k a^{nq} = a^k e = a^k$, and so the definition of φ does not depend on our choice of representatives. But given that, φ is an isomorphism. (Again φ is a homomorphism by the laws of exponents, φ is onto by the definition of a cyclic group, and φ is 1-1 as if $[k_1]_n \neq [k_2]_n$, $a^{k_1} \neq a^{k_2}$ as $k_1 - k_2$ is not divisible by n.) This proves part (b). $\square$

Remark 2.2.10. You will notice that in the proof of part (b), we had an extra step where we had to show that φ is well-defined. This will follow as part of a general result we will prove later, but we decided to get started by proving this particular case "by hand". $\Diamond$

Suppose we want to check that φ is onto. In general, there is no shortcut. But suppose we want to check that φ is 1-1. Here we have a shortcut.

Lemma 2.2.11. *Let* $\varphi \colon G \to H$ *be a homomorphism. Then* φ *is 1-1 if* $a \neq e$ *implies* $\varphi(a) \neq e$.

Proof.   To say that φ is 1-1 is to say that for any two elements a and b of G with $a \neq b$, $\varphi(a) \neq \varphi(b)$, so this is just a special case.

On the other hand, suppose that $\varphi(a) \neq e$ for any $a \neq e$ and consider two elements $a \neq b$ of G. Then $ab^{-1} \neq e$, so $\varphi(ab^{-1}) \neq e$. But $\varphi(ab^{-1}) = \varphi(a)\varphi(b)^{-1}$ so $\varphi(a)\varphi(b)^{-1} \neq e$ and hence $\varphi(a) \neq \varphi(b)$. $\square$

Example 2.2.12.

(a) Let $\varphi \colon \mathbb{Z} \to \mathbb{Z}$ be multiplication by n, $\varphi(i) = ni$. Then φ is neither 1-1 nor onto if $n = 0$, φ is both 1-1 and onto if $n = \pm 1$, and φ is 1-1 but not onto if $|n| > 1$.

(b) Let n be a positive integer and let $\varphi \colon \mathbb{Z} \to \mathbb{Z}_n$ by $\varphi(i) = [i]_n$. Then φ is onto but not 1-1.

(c) Let m and n be fixed positive integers and let $\varphi\colon \mathbb{Z} \to \mathbb{Z}_n$ be $\varphi([i]_n) = [mi]_n$. Let $d = gcd(m, n)$ be the greatest common division of m and n. If $d > 1$, then note that $[n/d]_n \neq [0]_n$ but $\varphi([n/d]_n) = [mn/d]_n = [(m/d)n]_n = (m/d)[n]_n = (m/d)[0]_n = [0]_n$. Thus, we see that φ is not 1-1.

Now φ is a function from the finite set $\mathbb{Z}_n$, with n elements, to the finite set $\mathbb{Z}_n$, also with n elements. Thus, since φ is not 1-1, we also have that φ is not onto.

On the other hand, suppose that m and n are relatively prime, i.e., that $d = 1$. Then there are integers x and y with $1 = mx + ny$. Thus for any integer k, $k = mkx + nky$ so $[k]_n = [mkx]_n = m[kx]_n$. In other words, $[k]_n = \varphi([kx]_n)$ and so we see φ is onto. Again we note that φ is a function from the set $\mathbb{Z}_n$, with n elements, to the set $\mathbb{Z}_n$, with n elements, so this time, since φ is onto, φ is also 1-1. ◊

We have said that one way groups arise is as the automorphism groups of various kinds of "structures". We may let that structure itself be a group, and so we are led to the following definition.

Definition 2.2.13. Let G be a group. The *automorphism group* Aut(G) of G is the group of all automorphisms $\varphi\colon G \to G$ (with the group operation being composition). ◊

Lemma 2.2.14. *Aut($\mathbb{Z}_n$) is isomorphic to $\mathbb{Z}_n^*$.*

Proof. Let $\varphi\colon \mathbb{Z}_n \to \mathbb{Z}_n$ be a homomorphism. Then $\varphi([k]_n) = k\varphi([1]_n)$, so φ is determined by $\varphi([1]_n)$. Let $\varphi([1]_n) = [m]_n = m[1]_n$. Then $\varphi([k]_n) = m[k]_n$. Thus φ is one of the homomorphisms considered in Example 2.2.12, and we saw there that φ is an isomorphism exactly when m is relatively prime to n. Now for any m and m' (relatively prime to n or not), if $[m]_n = [m']_n$, and $\varphi([k]_n) = m[k]_n$, $\varphi'([k]_n) = m'[k]_n$, then $\varphi = \varphi'$ (i.e., $\varphi\colon \mathbb{Z}_n \to \mathbb{Z}_n$ is determined by $\varphi([1]_n)$). Thus, we have a 1-1 onto map, i.e., an isomorphism of sets, $\Phi\colon$ Aut($\mathbb{Z}_n$) $\to \mathbb{Z}_n^*$ given by $\Phi(\varphi) = \varphi([1]_n)$. To conclude that Φ is an isomorphism of groups it remains to show that Φ is a homomorphism. So let $\varphi_1([1]_n) = [m_1]_n$ and $\varphi_2([1]_n) = [m_2]_n$. Then $\Phi(\varphi_2\varphi_1) = \varphi_2\varphi_1([1]_n) = \varphi_2(\varphi_1([1]_n)) = \varphi_2([m_1]_n) = m_2[m_1]_n = [m_2 m_1]_n = [m_2]_n[m_1]_n = \Phi(\varphi_2)\Phi(\varphi_1)$ as required. □

We conclude this section with a particularly interesting, and useful, isomorphism, and its consequences.

Theorem 2.2.15. *Let m and n be positive integers. Then $\mathbb{Z}_{mn}$ is isomorphic to $\mathbb{Z}_m \times \mathbb{Z}_n$ if and only if m and n are relatively prime.*

Proof.   The interesting case is when m and n are relatively prime, but let us get the other case out of the way first.

Suppose m and n are not relatively prime. Then m and n have a least common multiple $l < mn$. let $\varphi \colon \mathbb{Z}_{mn} \to \mathbb{Z}_m \times \mathbb{Z}_n$ be any homomorphism. Then $\varphi([1]_{mn}) = ([x]_m, [y]_n)$ for some x, y. But then $[l]_{mn} \neq 0$ while $\varphi([l]_{mn}) = ([lx]_m, [ly]_n) = ([0]_m, [0]_n)$ so φ is not 1-1. Thus, we see there is no possible isomorphism between $\mathbb{Z}_{mn}$ and $\mathbb{Z}_m \times \mathbb{Z}_n$, i.e., these two groups are not isomorphic.

Now suppose m and n are relatively prime. By Example 2.2.4(d), we have homomorphisms $\varphi_1 = \mathbb{Z}_{mn} \to \mathbb{Z}_m$ given by $\varphi_1([k]_{mn}) = [k]_m$ and $\varphi_2 \colon \mathbb{Z}_{mn} \to \mathbb{Z}_n$ given by $\varphi_2([k]_{mn}) = [k]_n$. Then we have a homomorphism $\varphi = (\varphi_1, \varphi_2) \colon \mathbb{Z}_{mn} \to \mathbb{Z}_m \times \mathbb{Z}_n$, i.e., $\varphi([k]_{mn}) = (\varphi_1([k]_{mn}), \varphi_2([k]_{mn})) = ([k]_m, [k]_n)$. (In particular, $\varphi([1]_{mn}) = ([1]_m, [1]_n)$.) We claim that φ is an isomorphism. Suppose $\varphi([k]_{mn}) = ([0]_m, [0]_n)$. Then $[k]_m = [0]_m$ and $[k]_n = [0]_n$, i.e., k is divisible by m and k is divisible by n. Since m and n are relatively prime, that implies k is divisible by their product mn, i.e., $[k]_{mn} = [0]_{mn}$. Thus φ is 1-1. Since $\mathbb{Z}_{mn}$ has mn elements, as does $\mathbb{Z}_m \times \mathbb{Z}_n$, that implies that φ is onto as well, so φ is an isomorphism, and these two groups are isomorphic. $\square$

Corollary 2.2.16. *Let $m_1, \ldots, m_k$ be pairwise relatively prime positive integers (i.e., m_i and m_j are relatively prime whenever $i \neq j$.) Let $M = m_1 \ldots m_k$. Then $\mathbb{Z}_M$ is isomorphic to $\mathbb{Z}_{m_1} \times \cdots \times \mathbb{Z}_{m_k}$, with an isomorphism φ given by $\varphi([1]_M) = ([1]_{m_1}, \ldots, [1]_{m_k})$.*

Proof.   This follows from Theorem 2.2.15 by induction on k. $\square$

Corollary 2.2.17 (Chinese remainder theorem). *Let $m_1, \ldots,$ m_k be pairwise relatively prime positive integers and set $M = m_1 \ldots m_k$. Let $a_1, \ldots, a_k$ be arbitrary integers. Then there is an integer x satisfying the congruences*

$$x \equiv a_1 (mod\ m_1)$$

$$\vdots$$

$$x \equiv a_k (mod\ m_k)$$

and x is unique $(mod\ M)$.

Proof. By Corollary 2.2.16, we have an isomorphism $\varphi\colon \mathbb{Z}_M \to \mathbb{Z}_{m_1} \times \cdots \times \mathbb{Z}_{m_k}$. Since φ is an isomorphism, it has an inverse φ^{-1}. Then x is given by

$$[x]_M = \varphi^{-1}([a_1]_{m_1}, \ldots, [a_k]_{m_k}).$$ □

Remark 2.2.18. The Chinese Remainder Theorem has a generalization, which is best understood in the context of ring theory. But, as we have just seen, Corollary 2.2.17, the original case of this theorem, can be understood as a theorem in group theory/number theory. ◇

2.3 Subgroups

The basic idea of a subgroup is a very simple one, but subgroups play an important role.

In this section we will usually write groups multiplicatively, as we will usually be considering general (i.e., not necessarily abelian) groups.

Definition 2.3.1. Let G be a group. H is a *subgroup* of G if $H \subseteq G$ (i.e., H is a subset of G) and H is a group (with the same operation as in G). H is a *proper subgroup* of G if $H \subset G$. ◇

First, we have a very simple criterion for a subset H of G to be a subgroup of G.

Lemma 2.3.2. *A subset H of G is a subgroup of G if and only if:*

 (i) $e \in H$.
 (ii) *If $a, b \in H$ then $ab \in H$.*
 (iii) *If $a \in H$ then $a^{-1} \in H$.*

Proof. These are just the group axioms for H, except for associativity, which is automatic, as it holds in G. □

Remark 2.3.3. If G is abelian then any subgroup H of G is abelian. ◇

Remark 2.3.4. Every group G has the trivial subgroup $\{e\}$, and G itself is a subgroup of G. ◇

Example 2.3.5.

(a) We have subgroups

$$(\mathbb{Z}, +) \subseteq (\mathbb{Q}, +) \subseteq (\mathbb{R}, +) \subseteq (\mathbb{C}, +).$$

(b) We have subgroups

$$(\{\pm 1\}, \cdot) \subseteq (\mathbb{Q} - \{0\}, \cdot) \subseteq (\mathbb{R} - \{0\}, \cdot) \subseteq (\mathbb{C} - \{0\}, \cdot). \qquad \Diamond$$

Example 2.3.6.

(a) Let $R = \mathbb{Z}, \mathbb{Q}, \mathbb{R}$, or $\mathbb{C}$ and let n be a positive integer.
We have (among others) the following subgroups of $GL_n(R)$, the group of invertible n-by-n matrices with entries in R. (We do not bother to repeat n and R, for simplicity.)

$$\{\text{identity matrix } I\}$$
$$\subseteq \{\text{invertible scalar matrices}\}$$
$$\subseteq \{\text{invertible diagonal matrices}\}$$
$$\subseteq \{\text{invertible upper triangular matrices}\}$$
$$\subseteq GL_n(R).$$

(b) $GL_n(R)$ also has the subgroup

$$SL_n(R) = \{\text{matrices of determinant } 1\}. \qquad \Diamond$$

Example 2.3.7. Let G_1 and G_2 be groups, and let $G = G_1 \times G_2$. Then $G_1 \times \{e\}$ and $\{e\} \times G_2$ are both subgroups of G. $\qquad \Diamond$

Lemma 2.3.8. *The distinct subgroups of $\mathbb{Z}$ are $n\mathbb{Z} = \{ni \,|\, i \in \mathbb{Z}\}$ for some nonnegative integer n.*

Proof. Let H be a subgroup of $\mathbb{Z}$. If $H = \{0\}$, then $H = 0\mathbb{Z}$. Assume $H \neq \{0\}$. Then H contains a nonzero integer k. Since H is a subgroup, if $k \in H$ then $-k \in H$, so we may assume k is positive. Thus, we see that H contains a positive integer k. Let n be the smallest positive integer in H. We claim $H = n\mathbb{Z}$.

Certainly $n\mathbb{Z} \subseteq H$, by closure. To show $H \subseteq n\mathbb{Z}$, let $m \in H$. Then we may write $m = nq + r$ for integers q and r with $0 \leq r < n$. If $r \neq 0$, then $r = m - nq = m + (-n)q \in H$ with $1 \leq r < n$, impossible as n is the smallest positive integer in H. Hence $r = 0$, $m = nq$, and $H \subseteq n\mathbb{Z}$.

From our construction we see that these subgroups are all distinct. □

We now consider the situation of a group G and a subgroup H, and define a pair of (closely related, but in general *not* the same) equivalence relations on the elements of G, which will lead us to the important notions of left and right cosets of H in G.

Definition 2.3.9. Let G be a group and let H be a subgroup of G. We define the equivalence relations $x \equiv_l y \pmod{H}$ and $x \equiv_r y \pmod{H}$ by:

$$x \equiv_l y \pmod{H} \quad \text{if} \quad x^{-1}y \in H,$$
$$x \equiv_r y \pmod{H} \quad \text{if} \quad xy^{-1} \in H. \qquad \Diamond$$

Remark 2.3.10. If G is abelian then $x^{-1}y = y^{-1}x$ so these two equivalence relations are the same, and we abbreviate them to $x \equiv y \pmod{H}$. $\Diamond$

While in general distinct, these two equivalence relations are similar, so we will henceforth state results for both, but only prove them for $x \equiv_l y \pmod{H}$.

Lemma 2.3.11. *The relations* $x \equiv_l y \pmod{H}$ *and* $x \equiv_r y \pmod{H}$ *are equivalence relations.*

Proof. We must check the three properties of an equivalence relation.

Reflexive: $x \equiv_l x \pmod{H}$, as $x^{-1}x = e \in H$, since H is a subgroup.

Symmetric: Suppose $x \equiv_l y \pmod{H}$, so $x^{-1}y \in H$. Since H is a subgroup, $(x^{-1}y)^{-1} \in H$. But $(x^{-1}y)^{-1} = y^{-1}x$, so $y^{-1}x \in H$ and $y \equiv_l x \pmod{H}$.

Transitive: Suppose $x \equiv_l y \pmod{H}$ and $y \equiv_l z \pmod{H}$. Then $x^{-1}y \in H$ and $y^{-1}z \in H$. Since H is a subgroup $(x^{-1}y)(y^{-1}z) \in H$. But $(x^{-1}y)(y^{-1}z) = x^{-1}z$, so $x^{-1}z \in H$ and $x \equiv_l z \pmod{H}$. □

Definition 2.3.12. The equivalence classes of elements of G under the relation $x \equiv_l y \pmod{H}$ are the *left cosets* of H in G, and the equivalence classes of elements of G under the relation $x \equiv_r y \pmod{H}$ are the *right cosets* of H in G. $\Diamond$

Lemma 2.3.13.

(a) *Consider the relation $x \equiv_l y \pmod{H}$. Then*

  (i) *The equivalence class of e is H, i.e., the left coset of H containing e is H.*

  (ii) *More generally, the equivalence class of the element x of G is xH, i.e., the left coset of H containing x is xH.*

  (iii) *Any two left cosets of H are either identical or disjoint.*

  (iv) *The left cosets of H partition G.*

(b) *Consider the relation $x \equiv_r y \pmod{H}$. Then*

  (i) *The equivalence class of e is H, i.e., the right coset of H containing e is H.*

  (ii) *More generally, the equivalence class of the element x of G is Hx, i.e., the right coset of H containing x is Hx.*

  (iii) *Any two right cosets of H are either identical or disjoint.*

  (iv) *The right cosets of H partition G.*

Proof.

  (i) The equivalence class of e under $x \equiv_l y \pmod{H}$ is $\{y \in G \,|\, e \equiv_l y \pmod{H}\} = \{y \in G \,|\, e^{-1}y \in H\} = \{y \in G \,|\, ey \in H\} = \{y \in G \,|\, y \in H\} = H$.

  (ii) First we observe that $zH = H$ if and only if $z \in H$. To see this, note that if $z \in H$, then $zh \in H$ for any $h \in H$, so $zH \subseteq H$, and for any $h \in H$, $h = z(z^{-1}h)$ so $H \subseteq zH$, and so they are equal. On the other hand, if $z \notin H$, then $ze = z \in zH$ so $zH \neq H$.
  Then $x \equiv_l y \pmod{H} \Leftrightarrow x^{-1}y \in H \Leftrightarrow x^{-1}yH = H \Leftrightarrow yH = xH$ (where the last $\Leftrightarrow$ is Lemma 2.1.18) and, similarly, $yH = xH \Leftrightarrow y \in xH$.

  (iii) and (iv) are properties of equivalence classes in general (Corollary 1.2.24 and Theorem 1.2.26). $\qquad\square$

Lemma 2.3.14. *Let H be a subgroup of G. Then for any $x \in G$. $|xH| = |H|$ and $|Hx| = |H|$.*

Proof.   We have a 1-1 correspondence $f \colon H \to xH$ given by $f(h) = xh$. $\qquad\square$

Lemma 2.3.15. *Let H be a subgroup of G and let $\{x_i\}$ be a complete set of left coset representatives of H in G. Then $\{x_i^{-1}\}$ is a complete set of right coset representatives of H in G.*

Proof. We have a 1-1 correspondence $f\colon G \to G$ given by $f(g) = g^{-1}$. Then $G = \amalg x_i H$ so applying this correspondence, $f(G) = \amalg f(x_i H)$. But $f(G) = G$ and

$$f(x_i H) = \{f(x_i h) \mid h \in H\} = \{(x_i h)^{-1} \mid h \in H\}$$
$$= \{h^{-1} x_i^{-1} \mid h \in H\} = H x_i^{-1}. \qquad \square$$

Definition 2.3.16. Let H be a subgroup of G. The *index* $[G : H]$ of H in G is the number of left (or right) cosets of H in G. $\qquad \diamond$

Theorem 2.3.17. *Let G be a finite group and let H be a subgroup of G. Then $|G| = |H|[G : H]$.*

Proof. Let $|G| = n, |H| = m$, and $[G : H] = k$. Let $\{x_1, \ldots, x_k\}$ be a complete set of left coset representatives of H in G. Then

$$G = \coprod_{i=1}^{k} x_i H \quad \text{so} \quad |G| = \sum_{i=1}^{k} |x_i H|, \quad n = \sum_{i=1}^{k} |x_i H|.$$

But $|x_i H| = |H| = m$ for every i, by Lemma 2.3.14. $\qquad \square$

Corollary 2.3.18 (Lagrange's theorem). *Let G be a finite group and let H be a subgroup of G. Then $|H|$ divides $|G|$.*

Proof. Immediate from Theorem 2.3.17. $\qquad \square$

Definition 2.3.19. Let G be a group and let $a \in G$. The *order* of a is m if m is the smallest positive integer such that $a^m = e$. The element a has infinite order if no such m exists. $\qquad \diamond$

Corollary 2.3.20. *Let G be a finite group and let $a \in G$. Then the order of a divides the order of G.*

Proof. Let $|G| = n$. If a has order m, then $H = \{e, a, \ldots, a^{m-1}\}$ is a subgroup of G, and so $|H| = m$ divides $|G| = n$ by Lagrange's Theorem. $\qquad \square$

Corollary 2.3.21. *Let G be a group of prime order p. Then G is cyclic, generated by any element $a \neq e$ of G.*

Proof.   By Corollary 2.3.20, the order of a must divide p. Since $a \neq e$, the order of a is not 1. Hence a has order p, and so $H = \{e, a, \ldots, a^{p-1}\} = G$. $\square$

We now want to look carefully at the (potential) difference between left and right cosets of a subgroup H of a group G. They can only differ in case G is nonabelian, so we look at a nonabelian group G.

Example 2.3.22.   We have introduced the dihedral groups D_{2n} in Example 2.1.14.

We begin by considering $G = D_6$. Then, in the notation of that example, $G = \{e, \alpha, \alpha^2, \beta, \alpha\beta, \alpha^2\beta\}$. We let $N = \{e, \alpha, \alpha^2\}$, $H_0 = \{e, \beta\}$, $H_1 = \{e, \alpha\beta\}$, $H_2 = \{e, \alpha^2\beta\}$, and we observe that N, H_0, H_1, and H_2 are all subgroups of G, of orders 3, 2, 2, 2 respectively.

First let us consider N. Since N has order 3, and G has order 6, $[G : N] = 2$, so N has two left cosets, and two right cosets. A little computation shows:

   {left cosets of N} $= \{\{e, \alpha, \alpha^2\}, \{\beta, \alpha\beta, \alpha^2\beta\}\}$,
   {right cosets of N} $= \{\{e, \alpha, \alpha^2\}, \{\beta, \alpha\beta, \alpha^2\beta\}\}$.

Thus in this case the left cosets of N and the right cosets of N *agree*.

On the other hand, let us consider H_0. Since H_0 has order 2, and G has order 6, $[G : H_0] = 3$, so H_0 has three left cosets, and three right cosets. A little computation shows:

   {left cosets of H_0} $= \{\{e, \beta\}, \{\alpha, \alpha\beta\}, \{\alpha^2, \alpha^2\beta\}\}$,
   {right cosets of H_0} $= \{\{e, \beta\}, \{\alpha, \alpha^2\beta\}, \{\alpha^2, \alpha\beta\}\}$.

Thus in this case the left cosets of H_0 and the right cosets of H_0 *disagree*.

Similarly we compute:

   {left cosets of H_1} $= \{\{e, \alpha\beta\}, \{\alpha, \alpha^2\beta\}, \{\alpha^2, \beta\}\}$,
   {right cosets of H_1} $= \{\{e, \alpha\beta\}, \{\alpha, \beta\}, \{\alpha^2, \alpha^2\beta\}\}$,

and

   {left cosets of H_2} $= \{\{e, \alpha^2\beta\}, \{\alpha, \beta\}, \{\alpha^2, \alpha\beta\}\}$,
   {right cosets of H_2} $= \{\{e, \alpha^2\beta\}, \{\alpha, \alpha\beta\}, \{\alpha^2, \beta\}\}$,

so the left cosets of H_1 and the right cosets of H_1, and the left cosets of H_2 and the right cosets of H_2, also *disagree*. $\Diamond$

Remark 2.3.23.   Before going any further we want to emphasize that when we write a left coset of a subgroup H (for example) as aH, we are making a *choice* of the coset representative (i.e., the representative of the equivalence class), and that we could equally will make a *different choice* and write the coset as $a'H$. For example, referring to our computations in Example 2.3.22, the two left cosets of N (in the order written) are $eN = \alpha N = \alpha^2 N$, and $\beta N = \alpha\beta N = \alpha^2\beta N$, while the two right cosets are $Ne = N\alpha = N\alpha^2$, and $N\beta = N\alpha\beta = N\alpha^2\beta$. For H_0, the three left cosets are $eH_0 = \beta H_0$, $\alpha H_0 = \alpha\beta H_0$, and $\alpha^2 H_0 = \alpha^2\beta H_0$, while the three right cosets are $H_0 e = H_0\beta$, $H_0\alpha = H_0\alpha^2\beta$, and $H_0\alpha^2 = H_0\alpha\beta$ (and similarly for H_1 and H_2). In principle, all choices of coset representatives are equally good. (In practice, we always choose the representative of the left or right coset H of H to be the identity element e, but that is a choice, and we could equally well make a different one.) $\Diamond$

We now want to look at subgroups whose left and right cosets agree (such as the subgroup N of D_6 in Example 2.3.22.) While this may seem to be a curiosity, it turns out to be an extremely important property.

Definition 2.3.24. Let G be a group, and let N be a subgroup of G. Then N is a *normal* subgroup of G, written $N \triangleleft G$, if every left coset of N in G is a right coset, and vice-versa. $\Diamond$

Lemma 2.3.25. *Let G be a group and let N be a subgroup of G. The following are equivalent:*

(i) *N is a normal subgroup of G.*
(ii) *$aN = Na$ for every $a \in G$.*
(iii(a)) *$a^{-1}Na = N$ for every $a \in G$.*
(iii(b)) *$aNa^{-1} = N$ for every $a \in G$.*
(iv(a)) *$a^{-1}Na \subseteq N$ for every $a \in G$.*
(iv(b)) *$aNa^{-1} \subseteq N$ for every $a \in G$.*

Proof. First observe that (iii(a)) and (iii(b)) are equivalent, and that (iv(a)) and (iv(b)) are equivalent, since each is obtained from the other simply by replacing a by a^{-1}.

Suppose (i) is true. Then for any $a \in G$, $aN = Nb$ for some b. But note $a = ae \in aN$ and so $a \in Nb$. But $a = ea \in Na$. Since two right cosets are either identical or disjoint, we must have $Na = Nb$ and so $aN = Na$. On the other hand, if (ii) is true then (i) is certainly true, as (ii) is *a* more specific statement than (i). Thus (i) and (ii) are equivalent.

Now $aN = Na \Leftrightarrow a^{-1}(aN) = a^{-1}(Na) \Leftrightarrow N = a^{-1}Na$ by Lemma 2.1.18, so (ii) and (iii(a)) are equivalent.

Certainly (iii(a)) implies (iv(a)), as (iii(a)) is a stronger statement than (iv(a)). On the other hand, suppose (iv(a)) is true. Then, as we have observed, (iv(b)) is true. But also $a^{-1}Na \subseteq N \Leftrightarrow a(a^{-1}Na)a^{-1} \subseteq aNa^{-1} \Leftrightarrow N \subseteq aNa^{-1}$, again by Lemma 2.1.18, so (iii(b)) is true, and hence so is (iii(a)).   □

Lemma 2.3.26. *Let G be an abelian group. Then every subgroup of G is normal.*

Proof.   For every $a \in G$, and every $x \in G$, $a^{-1}xa = x$.   □

Example 2.3.27. The converse of Lemma 2.3.26 is false. Let $G = Q_8$ be the quaternion group of order 8 as in Example 2.1.15. Then G is a nonabelian group, but it is straightforward to check that every subgroup of G is normal.   ◇

Now let us relate subgroups and homomorphisms.

Definition 2.3.28. Let $\varphi\colon G \to H$ be a group homomorphism. The *image* of φ is $\mathrm{Im}(\varphi) = \{h \in H \mid h = \varphi(g) \text{ for some } g \in G\}$. The *kernel* of φ is $\mathrm{Ker}(\varphi) = \{g \in G \mid \varphi(g) = e\}$.   ◇

Much of the importance of normal subgroups comes from part (b) of the following lemma.

Lemma 2.3.29. *Let $\varphi\colon G \to H$ be a homomorphism.*

(a) *$\mathrm{Im}(\varphi)$ is a subgroup of H.*
(b) *$\mathrm{Ker}(\varphi)$ is a normal subgroup of G.*

Proof.

(a) We have to verify the three properties of a subgroup. By Lemma 2.2.2, $\varphi(e) = e$ so $e \in \mathrm{Im}(\varphi)$. If $h_1, h_2 \in \mathrm{Im}(\varphi)$, then $h_1 = \varphi(g_1)$ for some $g_1 \in G$, and $h_2 = \varphi(g_2)$ for some $g_2 \in G$, so, since φ is a homomorphism, $h_1h_2 = \varphi(g_1)\varphi(g_2) = \varphi(g_1g_2)$ and $h_1h_2 \in \mathrm{Im}(\varphi)$. Finally, if $h_1 \in \mathrm{Im}(\varphi)$, so $h_1 = \varphi(g_1)$ for some $g_1 \in G$, then $h_1^{-1} = \varphi(g_1^{-1})$ by Lemma 2.2.2 again, so $h_1^{-1} \in \mathrm{Im}(\varphi)$.

(b) First we verify the three properties of a subgroup. By Lemma 2.2.2, $\varphi(e) = e$ so $e \in \mathrm{Ker}(\varphi)$. If $g_1, g_2 \in \mathrm{Ker}(\varphi)$, then $\varphi(g_1) = e$, $\varphi(g_2) = e$, so, since φ is a homomorphism,

$\varphi(g_1 g_2) = \varphi(g_1)\varphi(g_2) = ee = e$, so $g_1 g_2 \in \text{Ker}(\varphi)$. Finally, if $g_1 \in \text{Ker}(\varphi)$ then $\varphi(g_1) = e$, so, by Lemma 2.2.2 again, $\varphi(g_1^{-1}) = e^{-1} = e$, so $g_1^{-1} \in \text{Ker}(\varphi)$.

Now we must show that $\text{Ker}(\varphi)$ is in fact a normal subgroup of G. We do so by verifying condition (iv(a)) of Lemma 2.3.25. Let $g_1 \in \text{Ker}(\varphi)$, so that $\varphi(g_1) = e$, and let a be an arbitrary element of G. Then $\varphi(a^{-1}g_1 a) = \varphi(a^{-1})\varphi(g_1)\varphi(a) = \varphi(a)^{-1}\varphi(g_1)\varphi(a) = \varphi(a)^{-1}e\varphi(a) = \varphi(a)^{-1}\varphi(a) = e$, so $a^{-1}g_1 a \in \text{Ker}(\varphi)$, and hence $\text{Ker}(\varphi)$ is a normal subgroup of G. $\qquad\square$

It is natural to ask whether *every* subgroup of H is the image of some homomorphism (from some group), and whether *every* normal subgroup of G is the kernel of some homomorphism (to some group).

Here is the easy positive answer to the first of these two questions.

Lemma 2.3.30. *Let K be a subgroup of H. Then K is the image of a homomorphism.*

Proof. Let $i: K \to H$ be the inclusion, given by $i(k) = k$ for every $k \in K$. Then $\text{Im}(i) = K$. $\qquad\square$

The answer to the second question is also yes, but it takes much more work to show it, and we do so in the following section.

But let us conclude this section with a few more examples.

Definition 2.3.31. Two elements g_1 and g_2 of a group g are *conjugate* if $g_2 = a^{-1}g_1 a$ for some $a \in G$. Two subgroups H_1 and H_2 of G are *conjugate* if $H_2 = a^{-1}H_1 a$ for some $a \in G$. $\qquad\diamond$

Lemma 2.3.32. *Being conjugate is an equivalence relation on $\{elements\ of\ G\}$ and on $\{subgroups\ of\ G\}$.*

Proof. Straightforward. $\qquad\square$

Remark 2.3.33. We observe that H is a normal subgroup of G if and only if its only conjugate is itself. $\qquad\diamond$

Example 2.3.34.

(a) Let $G = GL_n(R)$ be the group of Example 2.3.6. Let $\varphi: G \to H$ (where $H = \{\pm 1\}$ if $R = \mathbb{Z}$, and $H = \mathbb{Q}-\{0\}, \mathbb{R}-\{0\}$, or $\mathbb{C}-\{0\}$ if $R = \mathbb{Q}, \mathbb{R}$, or $\mathbb{C}$) by $\varphi(A) = \det(A)$, where $\det(A)$ denotes the determinant of the matrix A. φ is a homomorphism by a basic property of determinants: $\varphi(AB) = \det(AB) = \det(A)\det(B) = \varphi(A)\varphi(B)$. Then $\mathrm{Ker}(\varphi) = SL_n(R)$ (as in Example 2.3.6(b)) is a normal subgroup of G.

(b) Let $n \geq 2$ and let U and L be the subgroups of G is given by:

$$U = \{\text{upper triangular matrices in } G\}$$
$$L = \{\text{lower triangular matrices in } G\}$$

Then U and L are conjugate subgroups of G, so neither U nor L is a normal subgroup of G. We do the computations to show this in case $n = 2$, and leave the general case to the reader:

$$\begin{bmatrix} 0 & 1 \\ 1 & 0 \end{bmatrix}^{-1} \begin{bmatrix} a & b \\ 0 & c \end{bmatrix} \begin{bmatrix} 0 & 1 \\ 1 & 0 \end{bmatrix} = \begin{bmatrix} c & 0 \\ b & a \end{bmatrix} \quad \text{for every } a, b, c \in R. \quad \Diamond$$

2.4   Quotient groups

Definition 2.4.1. Let G be a group and let N be a subgroup of G. Then G/N is the set of left cosets of N in G. $\qquad \Diamond$

In general, G/N is a set — nothing more. But if N is a normal subgroup of G, then G/N has the structure of a group. We prove this in stages.

Lemma 2.4.2. *Let G be a group and let N be a subgroup of G. The following are equivalent:*

(i) *N is a normal subgroup of G.*
(ii) *For any two left cosets L_1 and L_2 of N in G, their product $L_1 L_2$ (in the sense of Definition 2.1.17) is a left coset of N in G.*

Proof.   First suppose (i) is true. Let L_1 and L_2 be left cosets of N in G. Choosing left coset representatives, we have $L_1 = aH$ and $L_2 = bH$ for some $a, b \in G$. Observe that, for any subgroup N of G, $NN = N$. Then, using the fact that N is normal, and using Lemma 2.1.18, we have that

$$L_1 L_2 = (aN)(bN) = (aN)(Nb) = aNNb = aNb = abN$$

so $L_1 L_2$ is a left coset of N in G.

Next suppose (ii) is true. Let a be any element of G. Let L_1 be the left coset $L_1 = aN$ and let L_2 be the left coset $L_2 = a^{-1}N$. Then $L_1 L_2$ is a left coset of N in G. Which left coset can it be?

Since $e \in N$, the left coset $L_1 L_2 = aNa^{-1}N$ contains the element $aea^{-1}e = e$. But $e \in N$ as N is a subgroup of G. Recalling that any two left cosets of N in G are either identical or disjoint (Lemma 2.3.13), we must have $L_1 L_2 = N$. That is, $aNa^{-1}N = N$, and so $aNa^{-1} \subseteq N$, and then by Lemma 2.3.25, N is a normal subgroup of G. □

Theorem 2.4.3. *Let N be a normal subgroup of G. Then G/N is a group. Furthermore, $|G/N| = [G: N]$.*

Proof. G/N is closed under multiplication by Lemma 2.4.2. Multiplication is associative by Lemma 2.1.18. The identity element is the left coset $N = eN$, as for any left coset $L = aN$, $LN = (aN)(eN) = aeNN = aeN = aN = L$ and $NL = eNaN = eaNN = eaN = aN = L$. The inverse of the left coset $L_1 = aN$ is the coset $L_2 = a^{-1}N$ as $L_1 L_2 = (aN)(a^{-1}N) = aa^{-1}NN = aa^{-1}N = eN = N$ and $L_2 L_1 = (a^{-1}N)(aN) = a^{-1}aNN = a^{-1}aN = eN = N$. Thus G/N is a group.

By definition, $|G/N|$ is the number of elements of G/N. But the elements of G/N are the left cosets of N in G, and by definition the number of these is $[G: N]$. □

Definition 2.4.4. The group G/N of Theorem 2.4.3 is the *quotient* of G by the normal subgroup N. ◊

Example 2.4.5. Let $G = \mathbb{Z}$ and let n be a positive integer. Let $N = n\mathbb{Z}$. Then the quotient $G/N = \mathbb{Z}/n\mathbb{Z}$ is the group $\mathbb{Z}_n$ of Example 2.1.9. (If you go back to that example, and look at it closely, you will see that our construction of $\mathbb{Z}_n$ "by hand" there agrees with our construction of the quotient group $\mathbb{Z}/n\mathbb{Z}$ in Theorem 2.4.3.) ◊

Remark 2.4.6. This illustrates a common theme in mathematics. We look at a particular situation and make a particular construction. Later on we look at a more general situation, and see that our particular construction is a special case of a more general construction, or, viewed the other way, we see that our particular construction can be generalized to a (much) wider one. ◊

Remark 2.4.7. It is impossible to overemphasize the distinction between a subgroup of a group G and a quotient group of a group G. These are two completely different animals. The elements of a subgroup of G are some of the elements of G, while the elements of a quotient group of G are equivalence classes of elements of G, and these are two different things. $\Diamond$

Remark 2.4.8. Again to point out the difference between subgroups and quotient groups, note that in Example 2.4.5, for every positive integer n, the subgroup $n\mathbb{Z}$ of $\mathbb{Z}$ is an infinite group, while the quotient group $\mathbb{Z}_n$ is a finite group of order n. $\Diamond$

Remark 2.4.9. As we have seen, left cosets are equivalence classes. If L is a left coset of N, then choosing any representative of this equivalence class, i.e., any element a of L, we have that $L = aN$, and we have been using that notation in this chapter. But we have the notation $[a]$ for the equivalence class of a, that we used in Chapter 1, and we will feel free to use that notation here as well, as it is sometimes helpful to emphasize the fact that we are dealing with equivalence classes when we are considering elements of G/N. $\Diamond$

Lemma 2.4.10. *Let G be a group and let N be a normal subgroup of G. Let $\pi\colon G \to G/N$ be defined by $\pi(a) = [a]$. Then π is a homomorphism.*

Proof.   Let $a, b \in G$. Then $\pi(a)\,\pi(b) = [a][b] = (aH)(bH) = abH = [ab] = \pi(ab)$. $\square$

Definition 2.4.11. The map $\pi\colon G \to G/N$ of Lemma 2.4.10 is the *quotient map* or the *canonical projection* of G onto G/N. $\Diamond$

We now give some more examples of quotient groups.

Example 2.4.11. Let $G = D_6$ as in Example 2.3.22 and let $N = \{e, \alpha, \alpha^2\}$ as in that example. Then $N \lhd G$ and $[G\colon N] = 2$, so G/N is a group of order 2. Indeed, as we saw there, N has two left cosets: N itself, and $\{\beta, \alpha\beta, \alpha^2\beta\}$. Then G/N has two elements, $[e]$ and $[\beta]$, and $\pi\colon G \to G/N$ is given by $\pi(e) = \pi(\alpha) = \pi(\alpha^2) = [e]$, and $\pi(\beta) = \pi(\alpha\beta) = \pi(\alpha^2\beta) = [\beta]$. Note that $[\beta]^2 = [e]$ as $\beta^2 = e \in N$.

You may well ask why did we choose $[e]$ and $[\beta]$ as the elements of G/N. The answer is simply that we had to make a choice, and, once again, any choice is as good as any other. We follow convention by choosing e as the left coset representative of N. But we could equally well choose $[\alpha\beta]$ instead of $[\beta]$ as the other representative, so G/N would have elements $[e]$ and $[\alpha\beta]$, and $[\alpha\beta]^2 = [e]$ as $(\alpha\beta)^2 = e \in$

N. Similarly we could have chosen $[\alpha^2\beta]$ as the other representative, and $[\alpha^2\beta]^2 = [e]$ as $(\alpha^2\beta)^2 = e \in N$.

In fact, this generalizes without change to D_{2n}. Let $N = \{e, \alpha, \ldots, \alpha^{n-1}\}$. Then $N \lhd G$ and $[G: N] = 2$, so G/N is a group of order 2, $G/N = \{[e], [\beta]\}$ with $[\beta]^2 = [e]$ (and again we could replace $[\beta]$ by $[\alpha^i\beta]$ for any i). If $n = p$ is prime, N is the only nontrivial proper normal subgroup of G. If n is composite, there are others. $\Diamond$

Example 2.4.12. Let $G = Q_8$ be the quaternion group of order 8 as in Example 2.1.15. G has the following four nontrivial proper normal subgroups:

$$Z = \{\pm 1\}$$
$$C_i = \{1, i, -1, -i\}$$
$$C_j = \{1, j, -1, -j\}$$
$$C_k = \{1, k, -1, -k\}.$$

First let us consider G/C_i. Since $[G: C_i] = 2$, G/C_i is a group of order 2. We choose coset representatives so that $G/C_i = \{[e], [j]\}$ and we observe that $[j]^2 = [e]$ as $j^2 = -1 \in C_i$. Again we remark that this is simply a choice and we could have chosen different coset representatives if we had wanted to. Similarly, $G/C_j = \{[e], [k]\}$ with $[k]^2 = [e]$ and $G/C_k = \{[e], [i]\}$ with $[i]^2 = [e]$. Note that as abstract groups these are all isomorphic, and indeed are all isomorphic to $\mathbb{Z}_2$ (compare Corollary 2.3.21).

Now let us consider G/Z. Since $[G: Z] = 4$, G/Z is a group of order 4. We note that $G/Z = \{\{\pm 1\}, \{\pm i\}, \{\pm j\}, \{\pm k\}\}$, so, choosing coset representatives, we write $G/Z = \{[1], [i], [j], [k]\}$. Now $[i]^2 = [i^2] = [-1] = [1]$ and similarly $[j]^2 = [1]$, $[k]^2 = [1]$. Moreover $[i][j] = [ij] = [k]$ and $[j][i] = [ji] = [-k] = [k]$. Similarly $[j][k] = [k][j] = [i]$ and $[k][i] = [i][k] = [j]$. Thus G/Z has "multiplication table"

	$[1]$	$[i]$	$[j]$	$[k]$
$[1]$	$[1]$	$[i]$	$[j]$	$[k]$
$[i]$	$[i]$	$[1]$	$[k]$	$[j]$
$[j]$	$[j]$	$[k]$	$[1]$	$[i]$
$[k]$	$[k]$	$[j]$	$[i]$	$[1]$

and as an abstract group G/Z is isomorphic to $\mathbb{Z}_2 \times \mathbb{Z}_2$. $\Diamond$

Example 2.4.13. Let G_1 and G_2 be arbitrary groups and let $G = G_1 \times G_2$. Let $N_1 = G_1 \times \{e\}$. Then N_1 is a normal subgroup of G, and so we have the quotient group G/N_1. Note that $\{(e, g_2) \mid g_2 \in G_2\}$ is a complete set of left coset representatives of N_1 in G, and so $G/N_1 = \{[(e, g_2)] \mid g_2 \in G_2\}$. It is straightforward to check that the group operation in G/N_1 is given by $[(e, g_2)] [(e, g_2')] = [(e, g_2 g_2')]$. An entirely analogous situation holds if we let $N_2 = \{e\} \times G_2$ and consider the quotient group G/N_2. $\Diamond$

Here is a result we promised the reader in the last section.

Lemma 2.4.14. *Let N be a normal subgroup of G. Then N is the kernel of a homomorphism.*

Proof. N is the kernel of the quotient map $\pi\colon G \to G/N$. $\square$

Remark 2.4.15. Let G be a group and let N be a subgroup of G. As we have seen, the multiplication on G/N is given by $(aH)(bH) = (ab)H$, or equivalently $[a][b] = [ab]$. But it cannot be emphasized strongly enough that this does *not* imply $(a)(b) = ab$. Indeed, if a' is *any* element of aH, i.e., if $[a'] = [a]$, and if b' is *any* element of bH, i.e., $[b'] = [b]$, and if c' is *any* element of $(ab)H$, i.e., $[c'] = [ab]$, then $[a'][b'] = [c']$ but not necessarily $(a')(b') = c'$.

As a concrete example of this, let us take $G = \mathbb{Z}$, $N = 10\,\mathbb{Z}$, so $G/N = \mathbb{Z}_{10} = \{[0], [1], \ldots, [9]\}$. Then, sure enough, $[3] + [4] = [7]$ and $3 + 4 = 7$, but $[6] + [7] = [3]$, even though $6 + 7 \neq 3$. $\Diamond$

Remark 2.4.16. This having been said, it is *sometimes* possible to choose left coset representatives so that we do have $(a)(b) = ab$ whenever $[a][b] = [ab]$. With some thought, we can see that this will be the case if and only if there is a subgroup H of G with $\pi\colon H \to G/N$ an isomorphism. (We would be choosing our coset representatives to be elements of H.) This is *not* possible in the situation of Example 2.4.5 or Example 2.4.12. But it is possible in the situation of Example 2.4.11. We could choose $H = \{e, \beta\}$, or, as we mentioned, $H = \{e, \alpha\beta\}$ or $H = \{e, \alpha^2\beta\}$. It is also possible in the situation of Example 2.4.13. In the case of $N = G_1 \times \{e\}$, we choose $H = \{e\} \times G_2$, and in the case of $N = \{e\} \times G_2$, we choose $H = G_1 \times \{e\}$.

Another place in which you may have encountered a very analogous situation is that of vector spaces. Let V be a vector space and let U be a subspace of V. Let W be any complement of U. Then $\pi\colon W \to V/U$ is an isomorphism. So this situation always occurs in the case of vector spaces, but is an exceptional situation in the case of groups. ◊

We now proceed to investigate this situation, but on the way we will obtain some results interesting in their own right.

Lemma 2.4.17. *Let G be a group and let H_1 and H_2 be subgroups of G. Then*

$$|H_1 H_2| = \frac{|H_1||H_2|}{|H_1 \cap H_2|}.$$

Proof. Let $K = H_1 \times H_2 = \{(h_1, h_2) \mid h_1 \in H_1, h_2 \in H_2\}$ and note $|K| = |H_1| \, |H_2|$. Let $\sim$ be the relation on K given by $(h_1, h_2) \sim (h_1', h_2')$ if $h_1 h_2 = h_1' h_2'$. It is easy to check that $\sim$ is an equivalence relation, and the equivalence classes of K under $\sim$ are in $1 - 1$ correspondence with the elements of $H_1 H_2$ (where the equivalence class of (h_1, h_2) corresponds to $h_1 h_2 \in H_1 H_2$), so the number of equivalence classes is $|H_1 H_2|$. Now for any element h_0 of $H_1 \cap H_2$, $(h_1', h_2') = (h_1 h_0, h_0^{-1} h_2) \sim (h_1, h_2)$ as $(h_1 h_0)(h_0^{-1} h_2) = h_1 h_2$. Furthermore, every (h_1', h_2') in the equivalence class of (h_1, h_2) arises in this way, as we see from the following computation. Suppose $h_1' h_2' = h_1 h_2$. Then $h_1^{-1} h_1' = h_2 (h_2')^{-1}$. Call this common value h_0. Since H_1 is a subgroup, $h_0 = h_1^{-1} h_1' \in H_1$, and since H_2 is a subgroup, $h_0 = h_2 (h_2')^{-1} \in H_2$. Thus $h_0 \in H_1 \cap H_2$. But $h_0 = h_1^{-1} h_1'$ gives $h_1' = h_1 h_0$ and $h_0 = h_2 (h_2')^{-1}$ gives $h_2' = h_0^{-1} h_2$. Thus every equivalence class has $|H_1 \cap H_2|$ elements and so we see that $|H_1| \, |H_2| = |H_1 H_2| \, |H_1 \cap H_2|$, yielding the result. □

Lemma 2.4.18. *Let H_1 and H_2 be subgroups of G with $H_1 H_2 = H_2 H_1$. Then $H_1 H_2$ is a subgroup of G. In particular, if at least one of H_1 and H_2 is normal, then $H_1 H_2$ is a subgroup of G.*

Proof. We must verify the properties of a subgroup.
(Closure) Let $h_1 h_2 \in H_1 H_2$ and $h_1' h_2' \in H_1 H_2$. Then

$$(h_1 h_2)(h_1' h_2') = h_1 (h_2 h_1') h_2' = h_1 (h_1'' h_2'') h_2' = (h_1 h_1'')(h_2'' h_2') \in H_1 H_2$$

where $h_2 h_1' = h_1'' h_2''$ for some $h_1'' \in H_1$, $h_2'' \in H_2$ as $H_1 H_2 = H_2 H_1$.
(Identity) $e = ee \in H_1 H_2$.
(Inverses) Let $h_1 h_2 \in H_1 H_2$. Then $(h_1 h_2)^{-1} = h_2^{-1} h_1^{-1} \in H_1 H_2$ as $H_1 H_2 = H_2 H_1$.

Now suppose that at least one of H_1 and H_2 is normal. We give the argument in case H_2 is normal; the argument for H_1 normal is entirely analogous. Since H_2 is normal in G, $h_1 H_2 = H_2 h_1$ for every $h_1 \in H_1$ (in fact, for every $h_1 \in G$, but we don't need that here), by Lemma 2.3.25, and so $H_1 H_2 = H_2 H_1$. □

Example 2.4.19. If neither H_1 nor H_2 is normal, then $H_1 H_2$ may not be a subgroup of G. For example, let $G = D_6$. Then $\{e, \alpha\}\{e, \alpha\beta\} = \{e, \alpha, \alpha\beta, \alpha^2\beta\}$ is not a subgroup of G. ◇

Lemma 2.4.20. *Let N_1 and N_2 be normal subgroups of G with $N_1 \cap N_2 = \{e\}$. Then $n_1 n_2 = n_2 n_1$ for every $n_1 \in N_1$, $n_2 \in N_2$.*

Proof. Consider $n_1 n_2 n_1^{-1} n_2^{-1}$. On the one hand, $n_1 n_2 n_1^{-1} n_2^{-1} = n_1 (n_2 n_1^{-1} n_2^{-1}) \in N_1$ as $N_1 \lhd G$. On the other hand, $n_1 n_2 n_1^{-1} n_2^{-1} = (n_1 n_2 n_1^{-1}) n_2^{-1} \in N_2$ as $N_2 \lhd G$. Hence $n_1 n_2 n_1^{-1} n_2^{-1} \in N_1 \cap N_2 = \{e\}$, i.e., $n_1 n_2 n_1^{-1} n_2^{-1} = e$, $(n_1 n_2 n_1^{-1} n_2^{-1}) (n_2 n_1) = e (n_2 n_1)$, $n_1 n_2 = n_2 n_1$ as claimed. □

Definition 2.4.21. Let N_1 and N_2 be normal subgroups of G with $G = N_1 N_2$ and $N_1 \cap N_2 = \{e\}$. Then G is the *direct product*, $G = N_1 \times N_2$ of N_1 and N_2. ◇

Example 2.4.22. Let G_1 and G_2 be groups. Then $G = G_1 \times G_2$ has normal subgroups $N_1 = G_1 \times \{e\}$ and $N_2 = \{e\} \times G_2$ with $N_1 N_2 = G$ and $N_1 \cap N_2 = \{e\}$. Then we have an isomorphism $\varphi \colon G_1 \times G_2 \to G = N_1 N_2$ given by $\varphi(g_1, g_2) = (g_1, e)(e, g_2)$. ◇

Example 2.4.23. Let m and n be positive integers. Suppose that m and n are relatively prime. Let $G = \mathbb{Z}_{mn}$. Let $N_1 = \{[0], [n], [2n], \ldots, [(m-1)n]\}$ and let $N_2 = \{[0], [m], [2m], \ldots, [(n-1)m]\}$. Note that N_1 is isomorphic to $\mathbb{Z}_m$ and N_2 is isomorphic to $\mathbb{Z}_n$. Then G is the direct product of N_1 and N_2. ◇

Definition 2.4.24. Let H be a subgroup of G and N be a normal subgroup of G with $G = HN$ and $H \cap N = \{e\}$. Then G is the *semidirect product*, $G = H \ltimes N$, of H and N. (We also write $G = N \rtimes H$.) ◇

Example 2.4.25. If G is the direct product of N_1 and N_2 (as in Definition 2.4.21) then G is the semidirect product of N_1 and N_2. ◇

Here is a lemma that makes it easier to check that G is a semidirect product.

Lemma 2.4.26. *Let G be a finite group. Let H be a subgroup of G and N a normal subgroup of G with $|G| = |H|\,|N|$. If $G = HN$ or $H \cap N = \{e\}$, then G is the semidirect product of H and N.*

Proof. This follows directly from Lemma 2.4.17. $\qquad\qquad\square$

Example 2.4.27. (a) Let $G = D_6$, let $H = \{e, \beta\}$, and let $N = \{e, \alpha, \alpha^2\}$. Then $G = H \ltimes N$.
(b) More generally, let $G = D_{2n}$, let $H = \{e, \beta\}$ and let $N = \{e, \alpha, \alpha^2, \ldots, \alpha^{n-1}\}$. Then $G = H \ltimes N$.
(c) Let n be a positive integer. We have the following subgroups of $GL_n(R)$:
$U = \{\text{invertible upper triangular matrices}\}$
$U_0 = \{\text{upper triangular matrices with all diagonal entries equal to } 1\} \lhd U$
$D = \{\text{invertible diagonal matrices}\} \subseteq U$.
Then $U = D \ltimes U_0$. $\qquad\qquad\Diamond$

Here is the most general construction of semidirect products.

Lemma 2.4.28. *Let H and N be groups and let $\Phi\colon H \to \operatorname{Aut}(N)$ be a homomorphism. For $h \in H$, let $\varphi_h = \Phi(h)$, so that $\varphi_h\colon N \to N$ is an automorphism of N. Let $G = H \times N$ with the group operation on G being given by*

$$(h_1, n_1)(h_2, n_2) = (h_1 h_2, \varphi_{h_2^{-1}}(n_1)n_2).$$

Then G is a group and G is the semidirect product of the subgroup $H \times \{e\}$ of G and the normal subgroup $\{e\} \times N$ of G.

Proof. First we must check that G is a group. Because of the (complicated) definition of the group operation, this will require careful consideration.
(Closure) Clear from the definition.
(Associativity) We compute

$$[(h_1, n_1)(h_2, n_2)](h_3, n_3) = (h_1 h_2, \varphi_{h_2^{-1}}(n_1)n_2)(h_3, n_3)$$

$$= (h_1 h_2 h_3, \varphi_{h_3^{-1}}(\varphi_{h_2^{-1}}(n_1)n_2)n_3)$$

$$= (h_1 h_2 h_3, \varphi_{h_3^{-1}}(\varphi_{h_2^{-1}}(n_1))\varphi_{h_3^{-1}}(n_2)n_3)$$

$$= (h_1 h_2 h_3, \varphi_{(h_2 h_3)^{-1}}(n_1)\varphi_{h_3^{-1}}(n_2)n_3)$$

$$(h_1, n_1)[(h_2, n_2)](h_3, n_3)] = (h_1, n_1)(h_2h_3, \varphi_{h_3^{-1}}(n_2)n_3)$$
$$= (h_1h_2h_3, \varphi_{(h_2h_3)^{-1}}(n_1)\varphi_{h_3^{-1}}(n_2)n_3)$$

and these are equal.

(Identity) The identity element of G is (e, e).

(Inverses) The inverse of the element (h, n) of G is $(h^{-1}, \varphi_h(n)^{-1})$ as

$$(h, n)(h^{-1}, \varphi_h(n)^{-1}) = (hh^{-1}, \varphi_h(n)\varphi_h(n)^{-1}) = (e, e),$$
$$(h^{-1}, \varphi_h(n)^{-1})(h, n) = (h^{-1}h, \varphi_{h^{-1}}(\varphi_h(n)^{-1})n)$$
$$= (h^{-1}h, \varphi_{h^{-1}}(\varphi_h(n)^{-1})n)$$
$$= (h^{-1}h, n^{-1}n) = (e, e).$$

Now that we know G is a group we see that $(h, e)(e, n) = (h, n)$ and also that $H \times \{e\} \cap \{e\} \times N = \{(e, e)\}$.

It remains to show that $\{e\} \times N$ is a normal subgroup of G. We compute

$$(h_0, n_0)(e, n)(h_0, n_0)^{-1} = (h_0, n_0)(e, n)(h_0^{-1}, n_1) \quad \text{where } n_1 \in N$$
$$= (h_0, n_0)(h_0^{-1}, n_2) \quad\quad\quad \text{where } n_2 \in N$$
$$= (e, n_3) \quad\quad\quad\quad\quad\quad\quad \text{where } n_3 \in N$$

as required. □

We isolate some of the salient computations.

Corollary 2.4.29. *In the situation of Lemma 2.4.28,*

(a) $(h_1, e)(e, n_2) = (h_1, n_2)$

(b) $(e, n_1)(h_2, e) = (h_2, \varphi_{h_2^{-1}}(n_1))$

(c) $(h, e)(e, n)(h, e)^{-1} = (e, \varphi_h(n))$

Proof.   These are special cases of the above computations.   □

Corollary 2.4.30. *Let G be a semidirect product $G = H \ltimes N$ as in Definition 2.4.24. Then G can be obtained from the construction of Lemma 2.4.28.*

Proof.   Define $\Phi \colon H \to \operatorname{Aut}(N)$ by $\Phi(h) = \varphi_h$ where $\varphi_h(n) = hnh^{-1}$.   □

Example 2.4.31. Let N be the cyclic group $\mathbb{Z}_n = \{[0], \ldots, [n-1]\}$ and let $H = \mathbb{Z}_2 = \{[0], [1]\}$. Define Φ by $\varphi_{[1]}([k]) = [-k]$. Then $G = HN$ is isomorphic to the dihedral group D_{2n}. (Here we have written the groups additively. Switching to multiplicative notation, $N = \{e, \alpha, \ldots, \alpha^{n-1}\}$, $H = \{e, \beta\}$, and (using Corollary 2.4.30) $\varphi_\beta(\alpha) = \beta\alpha\beta^{-1} = \alpha^{-1}$, giving us D_{2n}).

Corollary 2.4.32. *In the situation of Lemma 2.4.28,*

$$\pi \colon H \times \{e\} \to G/\{e\} \times N$$

defined by $\pi(h, e) = [(h, e)] = (h, e)(\{e\} \times N)$ *is an isomorphism.*

Proof. From the definition of multiplication in G, we see that π is a homomorphism. Since $\{(h, e) \mid h \in H\}$ is a complete set of left coset representatives of $\{e\} \times N$, π is onto. Also, $\pi((h, e)) = [(e, e)]$ implies $h = e$, so $\operatorname{Ker}(\pi) = \{(e, e)\}$ and hence π is $1-1$. Thus π is an isomorphism. □

We now give a different perspective on semidirect products.

Lemma 2.4.33. *Let G be a group and let $\varphi \colon G \to Q$ be a homomorphism from G onto the group Q. Suppose there is a subgroup H of G such that $\varphi|_H \colon H \to Q$ is an isomorphism. Then G is the semidirect product of H and the normal subgroup $N = \operatorname{Ker}(\varphi)$.*

Proof. We must show that $G = HN$ and that $H \cap N = \{e\}$.

$G = HN$: Let $g \in G$ and let $q = \varphi(g)$. Since $\varphi|_H \colon H \to Q$ is an isomorphism, in particular it is onto, and so there is an element $h \in H$ with $\varphi(h) = q$. Then $\varphi(h^{-1}g) = \varphi(h^{-1})\varphi(g) = \varphi(h)^{-1}\varphi(g) = q^{-1}q = e$, so $n = h^{-1}g \in \operatorname{Ker}(\varphi) = N$. But then $g = hn$.

$H \cap N = \{e\}$: Let $g \in H \cap N$. Since $g \in N = \operatorname{Ker}(\varphi)$, $\varphi(g) = e$. Since $\varphi|_H \colon H \to Q$ is an isomorphism, in particular it is $1-1$, and so, since $g \in H$ with $\varphi(g) = e$, we must have $g = e$. □

There is another way of looking at this situation.

Definition 2.4.34. Let $\varphi \colon G \to Q$ be a homomorphism. A homomorphism $\lambda \colon Q \to G$ is a *splitting* of φ if $\varphi\lambda \colon Q \to Q$ is the identity. If φ has a splitting, then φ *splits*. ◇

Lemma 2.4.35.

(a) *Let $G = H \ltimes N$ be the semidirect product of the subgroup H and the normal subgroup N. Then the quotient map $\pi\colon\ G \to G/N$ splits.*

(b) *Suppose that $\varphi\colon G \to Q$ splits. Let λ be a splitting of φ. Then G is the semidirect product $G = H \ltimes N$ where $H = \lambda(Q)$ and $N = \operatorname{Ker}(\varphi)$.*

Proof.

(a) We observe that $\pi|_H$, the restriction of π to H, is an isomorphism $\pi|_H\colon H \to G/N$. Let $\lambda = (\pi|_H)^{-1}$.

(b) We know that $N \lhd G$ as N is the Kernel of a homomorphism. We must show $G = HN$ and $H \cap N = \{e\}$.

$G = HN$: Let $g \in G$ and let $q = \varphi(g)$. Let $h = \lambda(q)$ and set $n = h^{-1}g$, so $g = hn$. Certainly $h \in H$. Also, $\varphi(n) = \varphi(h^{-1}g) = \varphi(h^{-1})\varphi(g) = \varphi(h)^{-1}\varphi(g) = \varphi(i(q))^{-1}\varphi(g) = q^{-1}q = e$ so $n \in N$.

$H \cap N = \{e\}$: Let $g \in H \cap N$ and let $q = \varphi(g)$. On the one hand, $g \in N$ so $\varphi(g) = e$. On the other hand, $g \in H$ and $\varphi\lambda\colon Q \to Q$ is the identity, so in particular $\varphi|_H$ (the restriction of φ to H) is $1 - 1$. Since $\varphi(g) = e$, we must have $g = e$.

Thus G is the semidirect product of H and N, as claimed.   □

Example 2.4.36.

(a) For $R = \mathbb{Z}, \mathbb{Q}$, or $\mathbb{R}$, let $GL_n^+(R) = \{A \in GL_n(R) \mid \det(A) > 0\}$. Then $GL_n^+(R)$ is a normal subgroup of $GL_n(R)$ of index 2.
For $r \in R$, $r \neq 0$, let $\operatorname{sign}(r) = 1$ if $r > 0$ and $\operatorname{sign}(r) = -1$ if $r < 0$. Then we have a homomorphism $\varphi\colon GL_n(R) \to \{\pm 1\}$ given by $\varphi(A) = \operatorname{sign}(\det(A))$, and $\operatorname{Ker}(\varphi) = GL_n^+(R)$. This homomorphism φ has a splitting given by

$$\lambda(\varepsilon) = \begin{bmatrix} \varepsilon & & & \\ & 1 & & \\ & & \ddots & \\ & & & 1 \end{bmatrix} \quad \text{for} \quad \varepsilon = \pm 1.$$

If we let $H = \{\lambda(1), \lambda(-1)\}$ then H is a subgroup of $GL_n(R)$ of order 2 and $GL_n(R) = H \ltimes GL_n^+(R)$.

(b) In this situation, suppose now that n is *odd*. Then φ has another splitting given by

$$\lambda'(\varepsilon) = \begin{bmatrix} \varepsilon & & & \\ & \varepsilon & & \\ & & \ddots & \\ & & & \varepsilon \end{bmatrix} \quad \text{for} \quad \varepsilon = \pm 1.$$

Now let $H' = \{\lambda'(1), \lambda'(-1)\}$. Then we can check that H' is a normal subgroup of $GL_n(R)$, or that every element of H' commutes with every element of $GL_n^+(R)$. Either of these facts tells us that for n odd, $GL_n(R)$ is the direct product $GL_n(R) = H' \times GL_n^+(R)$. $\Diamond$

We close with an observation.

Remark 2.4.37. If G is abelian and G is the semidirect product of H and N, then G is automatically the direct product of H and N. $\Diamond$

2.5 The Noether isomorphism theorems

In this section, we will prove some basic, and widely applicable, theorems about isomorphisms of groups.

Definition 2.5.1. Let $\sigma\colon G \to H$ be a homomorphism from G onto H. A homomorphism $\varphi\colon G \to K$ *factors* through σ if there is a homomorphism $\tau\colon H \to K$ with $\varphi = \tau\sigma$.

It is illuminating to consider the following diagram:

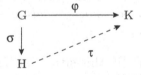

Then φ factors through σ if we can fill in the dotted arrow to a solid arrow. $\Diamond$

Lemma 2.5.2. *In the situation of Definition 2.5.1, φ factors through σ if and only if $Ker(\sigma) \subseteq Ker(\varphi)$.*

Proof. First suppose $Ker(\sigma) \subseteq k(\varphi)$. We define τ as follows: Let $h \in H$. Then $h = \sigma(g)$ for some $g \in G$. Let $\tau(h) = \varphi(g)$.

We have to see that this definition makes sense, i.e., that it only depends on h and not on the element g we have chosen. So suppose $h = \sigma(g')$ for some $g' \in G$. We need to show that $\varphi(g') = \varphi(g)$. Now $e = hh^{-1} = \sigma(g')\sigma(g)^{-1} = \sigma(g')\sigma(g^{-1}) = \sigma(g'g^{-1})$ so $g'g^{-1} \in \mathrm{Ker}(\sigma) \subseteq \mathrm{Ker}(\varphi)$. Thus $g'g^{-1} \in \mathrm{Ker}(\varphi)$, i.e., $\varphi(g'g^{-1}) = e$. But $e = \varphi(g'g^{-1}) = \varphi(g')\varphi(g^{-1}) = \varphi(g')\varphi(g)^{-1}$ so $\varphi(g') = \varphi(g)$ as required.

Next suppose that φ factors through σ and let $\varphi = \tau\sigma$. Let $g \in \mathrm{Ker}(\sigma)$. Then $\varphi(g) = \tau\sigma(g) = \tau(\sigma(g)) = \tau(e) = e$ so $g \in \mathrm{Ker}(\varphi)$. Thus $\mathrm{Ker}(\sigma) \subseteq \mathrm{Ker}(\varphi)$. $\qquad\square$

Theorem 2.5.3. (First isomorphism theorem) *Let* $\varphi\colon G \to H$ *be a homomorphism. Then* $Im(\varphi)$ *is isomorphic to* $G/Ker(\varphi)$.

Proof.   Let $\pi\colon G \to G/\mathrm{Ker}(\varphi)$ be the quotient map. Then $\mathrm{Ker}(\pi) = \mathrm{Ker}(\varphi)$ so by Lemma 2.5.2 φ factors as $\varphi = \bar{\varphi}\pi$.

First we claim that $\mathrm{Im}(\varphi) = \mathrm{Im}(\bar{\varphi})$. If $h \in \mathrm{Im}(\varphi)$, so $h = \varphi(g)$ for some $g \in G$, and then $h = \bar{\varphi}\pi(g) = \bar{\varphi}(\pi(g))$ so $h \in \mathrm{Im}(\bar{\varphi})$. On the other hand, suppose $h \in \mathrm{Im}(\bar{\varphi})$, so $h = \bar{\varphi}(\bar{g})$ for some $\bar{g} \in G/\mathrm{Ker}(\pi)$. But π is onto, so $\bar{g} = \pi(g)$ for some $g \in G$, so $h = \bar{\varphi}(\pi(g)) = \varphi(g)$ so $h \in \mathrm{Im}(\varphi)$. Thus, if we let $H_0 = \mathrm{Im}(\varphi)$, then $\bar{\varphi}\colon G/\mathrm{Ker}(\varphi) \to H_0$ is onto.

Next we claim that $\bar{\varphi}\colon G/\mathrm{Ker}(\varphi) \to H_0$ is 1-1. Let $\bar{g} \in G/\mathrm{Ker}(\varphi)$. Then by the definition of $\bar{\varphi}$, $\bar{\varphi}(\bar{g}) = \varphi(g)$ where g is an element of G with $\pi(g) = \bar{g}$. Thus $e = \bar{\varphi}(\bar{g}) \Leftrightarrow e = \varphi(g) \Leftrightarrow g \in \mathrm{Ker}(\varphi) \Leftrightarrow \pi(g) = e$.

Hence $\bar{\varphi}\colon G/\mathrm{Ker}(\varphi) \to H_0$ is both 1-1 and onto, i.e., is an isomorphism. In particular, $G/\mathrm{Ker}(\varphi)$ and $H_0 = \mathrm{Im}(\varphi)$ are isomorphic. $\qquad\square$

Theorem 2.5.4 (Second isomorphism theorem). *Let* H *be a subgroup of* G *and let* N *be a normal subgroup of* G. *Then* $H/H \cap N$ *is isomorphic to* HN/N.

Proof.   First recall that in this situation, HN is a subgroup of G. Then, since N is a normal subgroup of G, it is certainly a normal subgroup of HN. Also, it is straightforward to verify that if N is a normal subgroup of G, then $H \cap N$ is a normal subgroup of H. Thus both $H/H \cap N$ and HN/N are groups.

Let $\pi\colon G \to G/N$ be the quotient map and let π_0 be the restriction of π to H. From the fact that $\mathrm{Ker}(\pi) = N$ it is straightforward to check that $\mathrm{Ker}(\pi_0) = H \cap N$. Thus from the first isomorphism theorem $H/H \cap N = H/\mathrm{Ker}(\pi_0)$ is isomorphic to $\mathrm{Im}(\pi_0)$. Now $\pi\colon G \to G/N$ is defined by $\pi(g) = gN$, so $\mathrm{Im}(\pi_0) = \{hN \mid h \in H\}$. But $\{hN \mid h \in H\}$ is just HN/N, yielding the theorem. $\square$

Theorem 2.5.5 (Third isomorphism theorem). *Let H and K be normal subgroups of G with $K \subseteq H$. Then H/K is a normal subgroup of G/K and G/H is isomorphic to $(G/K)/(H/K)$.*

Proof. Let π_K be the quotient map $\pi_K\colon G \to G/K$ and let π_H be the quotient map $\pi_H\colon G \to G/H$. Then $\mathrm{Ker}(\pi_K) = K$ and $\mathrm{Ker}(\pi_H) = H$ so $\mathrm{Ker}(\pi_K) \subseteq \mathrm{Ker}(\pi_H)$. Thus by Lemma 2.5.2, π_H factors through π_K, i.e., $\pi_H = \pi'\pi_K$ for some homomorphism $\pi'\colon G/K \to G/H$. (In other words, we have the diagram

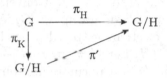

where we have filled in the dotted arrow.)

From the first isomorphism theorem, we know that $\mathrm{Im}(\pi')$ is isomorphic to $(G/K)/\mathrm{Ker}(\pi')$. Thus we must identify these two subgroups.

$\mathrm{Im}(\pi')$: We claim $\mathrm{Im}(\pi') = G/H$. We know that π_H is onto, and in the situation of a factorization, that implies π' is onto. (Let $\bar{g} \in G/H$. Then $\bar{g} = \pi(g)$ for some element of G, i.e., $\bar{g} = \pi'\pi_K(g) = \pi'(\pi_K(g))$ and so $\bar{g} \in \mathrm{Im}(\pi')$.)

$\mathrm{Ker}(\pi')$: We claim $\mathrm{Ker}(\pi') = H/K$. (Note this implies that H/K is a normal subgroup of G/K.) By the definition of the quotient maps, $\pi_H(g) = gH$ and $\pi_K(g) = gK$. Then, by the definition of π', $\pi'(gK) = gH$. Remember that the identity element of G/H is the left coset H. Thus $gK \in \mathrm{Ker}(\pi') \Leftrightarrow gH = H \Leftrightarrow g \in H$. Thus $\mathrm{Ker}(\pi') = \{hK \mid h \in H\}$. But $\{hK \mid h \in H\}$ is just H/K, as claimed. $\square$

Theorem 2.5.6 (Correspondence theorem). *Let N be a normal subgroup of G and let $\pi\colon G \to G/N$ be the quotient map. For any subgroup H of G with $N \subseteq H$ we have the restriction $\pi|_H\colon H \to H/N$. This gives a 1-1 correspondence between $S_1 = \{$subgroups of G containing $N\}$ and $S_2 =$*

$\{subgroups\ of\ G/N\}$. *Furthermore, this restricts to a 1-1 correspondence between* $T_1 = \{normal\ subgroups\ of\ G\ containing\ N\}$ *and* $T_2 = \{normal\ subgroups\ of\ G/N\}$.

Proof. Clearly if $H \in S_1$ then $\pi|_H(H) \in S_2$. Also $H_1 \subseteq H_2 \Leftrightarrow \pi|_{H_1}(H_1) \subseteq \pi|_{H_2}(H_2)$ and similarly $H_2 \subseteq H_1 \Leftrightarrow \pi|_{H_2}(H_2) \subseteq \pi|_{H_1}(H_1)$ so $H_1 = H_2 \Leftrightarrow \pi|_{H_1}(H_1) = \pi|_{H_2}(H_2)$ and so the map from S_1 to S_2 is 1-1. Also, if $\bar{H} \in S_2$ then, setting $H = \pi^{-1}(\bar{H})$, we have $\pi|_H(H) = \bar{H}$ so the map from S_1 to S_2 is onto, and hence we have a 1-1 correspondence between S_1 and S_2.

Now suppose $H \in T_1$, so that H is a normal subgroup of G. Then for any $a \in G$, $aHa^{-1} = H$. Thus for any $aN \in G/N$, $(a/N)(H/N)(aN)^{-1} = (aHa^{-1})/N = H/N$ so H/N is a normal subgroup of G/H, i.e., $H/N \in T_2$. Conversely, if $H/N \in T_2$, so that H/N is a normal subgroup of G/N, let $\pi_1 : G/N \to (G/N)/(H/N)$ be the quotient map. Also, we have an isomorphism $i : (G/N)/(H/N) \to G/H$. Let $\pi' = i\pi_1\pi : G \to G/H$ be the composition. Then $H = \mathrm{Ker}(\pi')$ is a normal subgroup of G, i.e., $H \in T_1$. Thus we also have a 1-1 correspondence between T_1 and T_2.   □

2.6   The structure of finite, and finitely generated, Abelian groups

In this section, we first completely determine the structure of finite abelian groups, and then completely determine the structure of infinite, but finitely generated, abelian groups. Along the way, we shall prove some results of interest in themselves.

Since we will only be dealing with abelian groups in this section, we will write all groups additively. Also, if $G = G_1 \times G_2$ is the direct product of G_1 and G_2, we will write $G = G_1 \oplus G_2$ and call G the *direct sum* of G_1 and G_2.

We begin with finite groups.

Theorem 2.6.1 (Cauchy's theorem for abelian groups). *Let G be a finite abelian group of order n, and let p be a prime dividing n. Then G has an element of order p.*

Proof. Let $n = pk$. We prove the theorem by complete induction on k. If $k = 1$, then $n = p$. We have proved (Corollary 2.3.21) that in this case G is cyclic. Let a be any element of G except $a = e$. Then a has order p.

Now for the inductive step. Let G have order $n = pk$ with $k > 1$. Choose any element a of G other than $a = e$. Then a has order m for some positive integer m with $m > 1$ and m dividing pk (by Corollary 2.3.20).

There are two possibilities:

(i) p divides m. In that case, let $b = (m/p)a$. Then b is an element of G of order p.

(ii) p does not divide m. In that case, let H be the subgroup generated by a. Then H has order m. Since G is abelian, H is normal, so G/H is a group. Let $\pi\colon G \to G/H$ be the quotient map. Then $|G/H| = n/m$ is still divisible by p, so $|G/H| = pj$ for some $j < k$. By the inductive hypothesis, G/H has an element b of order p. Let c be any element of G with $\pi(c) = b$. Then the order of b divides the order of c, so c has order pq for some q. Let $d = qc$. Then d is an element of G of order p. $\square$

Remark 2.6.2. Cauchy's theorem is true for arbitrary finite groups, not just abelian ones. We have proved it here in the abelian case because this case is simpler. The proof in the general case requires new ideas. But we will prove it in general later (as Theorem 2.9.1). $\lozenge$

Lemma 2.6.3. *Let G be an abelian group and let a and b be elements of G of orders m and n respectively, with m and n relatively prime. Then $c = a + b$ is an element of G of order mn.*

Proof. First observe that $(mn)c = (mn)(a+b) = (mn)a + (mn)b = n(ma) + m(nb) = n(0) + m(0) = 0$. Thus c has order k for some k dividing mn. Now $0 = kc = k(a+b) = ka + kb$ so $kb = k(-a)$.

Now the order of $k(-a)$ divides the order of $-a$, which is the order of a, which is m, and the order of kb divides the order of b, which is n. Hence the order of $kb = k(-a)$ divides $\gcd(m,n)$. But m and n are assumed to be relatively prime, so $\gcd(m,n) = 1$, and $kb = k(-a)$ has order 1, i.e., $kb = k(-a) = 0$. But then, since $k(-a) = 0$, m divides k, and since $kb = 0$, n divides k. Again, m and n are assumed to be relatively prime, so mn divides k. Hence $k = mn$, as claimed. $\square$

Remark 2.6.4. This lemma is really Theorem 2.2.15 "in disguise" but rather than showing how it is equivalent to that theorem, we decided to prove it directly, for the sake of simplicity.   ◇

Theorem 2.6.5. *Let G be a finite abelian group and let e be the exponent of G, i.e., the least common multiple of the orders of the elements of G. Then G has an element of order e.*

Proof.   Factor e as $e = p_1^{d_1} p_2^{d_2} \ldots p_k^{d_k}$ where $p_1, \ldots, p_k$ are distinct primes. Then G has an element a_i of order $p_i^{d_i} q_i$, for some q_i, for each $i = 1, \ldots, k$. Then $q_i a_i$ has order $p_i^{d_i}$, for each i. Since $p_1^{d_1}, \ldots, p_k^{d_k}$ are pairwise relatively prime, we see, applying Lemma 2.6.3 inductively, that $b = q_1 a_1 + q_2 a_2 + \cdots + q_k a_k$ is an element of G of order e.   □

Corollary 2.6.6. *Let G be an abelian group of order n. The following are equivalent:*

   (i) *G is cyclic.*
   (ii) *For any integer d dividing n, G has exactly d elements whose order divides d.*
   (iii) *For any integer d dividing n, G has at most d elements whose order divides d.*

Proof.   Let e be the least common multiple of the orders of the elements of G. From Corollary 2.3.20, we see that e divides n. There are two possibilities:

   $e = n$: In this case, by Theorem 2.6.5, G has an element of order n, so G is a cyclic group of order n, i.e., G is isomorphic to $\mathbb{Z}_n$. Thus (i) is true. It is straightforward to check that all the subgroups of $\mathbb{Z}_n$ are as follows: Let d be any integer dividing n. Then there is a unique subgroup of $\mathbb{Z}_n$ of order d, whose elements are $\{[0], [n/d], [2(n/d)], \ldots, [(d-1)n/d]\}$, from which (ii) and (iii) follow.

   $e \neq n$: In this case G is not cyclic (as if G were cyclic, it would have an element of order n) which would imply that $e = n$. Thus (i) is false. But $ea = 0$ for *every* $a \in G$, so G has $n > e$ elements whose order divides e, and (ii) and (iii) are false as well.   □

We will now investigate the structure of $\mathbb{Z}_m \oplus \mathbb{Z}_n$ in general.

Theorem 2.6.7. *Let* $G = \mathbb{Z}_m \oplus \mathbb{Z}_n$ *with* m *and* n *positive integers. Let* $g = gcd(m, n)$ *and* $l = lcm(m, n)$. *Then* G *is isomorphic to* $\mathbb{Z}_l \oplus \mathbb{Z}_g$.

Proof.   Denote the element $(1, 0)$ of $\mathbb{Z}_m \oplus \mathbb{Z}_n$ by a and the element $(0, 1)$ of $\mathbb{Z}_m \oplus \mathbb{Z}_n$ by b. Let $y = a + b$. If $ky = 0$, since G is the direct sum of the subgroup generated by a and the subgroup generated by b, we must have $ka = 0$ and $kb = 0$, and this will be true if and only if both m and n divide k, i.e., if and only if l divides k. Thus y is an element of G of order l, so y generates a cyclic subgroup N_y of G of order l, i.e., isomorphic to $\mathbb{Z}_l$.

Write $m = gm'$ and $n = gn'$. Then m' and n' are relatively prime, so there are integers s and t with $m's + n't = 1$. Let $z = -m'sa + n'tb$.

You may well ask where does this choice of z come from? Note that we have a symbolic matrix equation

$$\begin{bmatrix} y \\ z \end{bmatrix} = \begin{bmatrix} 1 & 1 \\ -m's & n't \end{bmatrix} \begin{bmatrix} a \\ b \end{bmatrix}$$

and if we let A be this 2-by-2 matrix, A has determinant 1. This implies that A is invertible over the integers, i.e., that A^{-1} is a matrix with integer entries. We will crucially use this fact soon.

We claim that z is an element of G of order g. To see that, first observe that $gz = g(-m'sa) + g(n'tb) = -s(gm'a) + t(gn'b) = -s(ma) + t(nb) = -s(0) + t(0) = 0$, as a has order m and b has order n. Thus if z has order r, then r divides g. But suppose that $rz = 0$. Then $r(-m'sa) + r(n'tb) = 0$. Since G is the direct sum of the subgroup generated by a and the subgroup generated by b, we must have $r(-m'sa) = 0$ and $r(n'tb) = 0$, or $(m'rs)a = 0$ and $(n'rt)b = 0$. Since a has order m and b has order n, we must have that $m = m'g$ divide $m'rs$ and that $n = n'g$ divides $n'rt$, and so g divides rs and g divides rt. But, since $m's + n't = 1$, s and t are relatively prime, so that implies that g divides r. Hence $r = g$. Thus z generates a cyclic subgroup N_z of order g, i.e., isomorphic to $\mathbb{Z}_g$.

We claim that G is the semidirect product of N_y and N_z. Since G is abelian, this semidirect product is automatically the direct product (Remark 2.4.36), or direct sum, as we are calling it here, $G = N_y \oplus N_z$, isomorphic to $\mathbb{Z}_l \oplus \mathbb{Z}_g$.

To see this we must verify that $G = N_y + N_z$ (we are writing the group additively) and that $N_y \cap N_z = \{0\}$.

$G = N_y + N_z$: Referring to the above symbolic matrix computation, we see that

$$\begin{bmatrix} a \\ b \end{bmatrix} = A^{-1} \begin{bmatrix} y \\ z \end{bmatrix} = \begin{bmatrix} n't & -1 \\ m's & 1 \end{bmatrix} \begin{bmatrix} y \\ z \end{bmatrix},$$

i.e., $a = n'ty - z$ and $b = m'sy + z$.

Thus a and b are both in $N_y + N_z$. Since $N_y + N_z$ is a subgroup of G, and a and b generate the group G, we must have $G = N_y + N_z$.

Actually, at this point we are done. Note $|G| = mn = lg = |N_y| \, |N_z|$ so by Lemma 2.4.26, G is the semidirect product of N_y and N_z, as claimed.

But, although it is not logically necessary, it is illuminating to verify independently that $N_y \cap N_z = \{0\}$, so we shall do so.

$N_y \cap N_z = \{0\}$: Let $x \in N_y \cap N_z$. Then $x = jy$ for some j and $x = kz$ for some k. Thus $jy = kz$ and so $j(a + b) = k(-m'sa + n'tb)$. Once again, since G is the direct sum $G = \mathbb{Z}_m \oplus \mathbb{Z}_n$ of the subgroup generated by a and the subgroup generated by b, we must have $ja = k(-m'sa)$ and $jb = k(n'tb)$, i.e., $(j + km's)a = 0$ and $(j - kn't)b = 0$. Since a has order m and b has order n, we must have that m divides $j + km's$ and that n divides $j - kn't$. Since g divides both m and n, we have that g divides both $j + km's$ and $j - kn't$, and hence that g divides their difference $(j + km's) - (j - kn't) = k(m's + n't) = k(1) = k$. Since g divides k, and z has order g, we must have $kz = 0$. Thus $x = 0$ (and $jy = 0$ so l divides j) as claimed.   □

Remark 2.6.8. Theorem 2.6.7, and its proof, are valid for any m and n. But in case m and n are relatively prime, then $l = mn$ and $g = 1$. In this case the first paragraph of the proof (again) yields that $\mathbb{Z}_m \oplus \mathbb{Z}_n$ is isomorphic to $\mathbb{Z}_{mn}$. But if we go through the whole proof, we find that $\mathbb{Z}_m \oplus \mathbb{Z}_n$ is isomorphic to $\mathbb{Z}_{mn} \oplus \mathbb{Z}_1$. But $\mathbb{Z}_1$ is the trivial group, so $\mathbb{Z}_{mn} \oplus \mathbb{Z}_1$ is isomorphic to $\mathbb{Z}_{mn}$.   ◊

Example 2.6.9.

(a) Let N be a positive integer. Then any factorization of N as $N = mn$ with m and n relatively prime gives an isomorphism of $\mathbb{Z}_N$ with $\mathbb{Z}_m \oplus \mathbb{Z}_n$. For example, if $N = 30$, we have that $\mathbb{Z}_{30}$

is isomorphic to $\mathbb{Z}_{15} \oplus \mathbb{Z}_2$, $\mathbb{Z}_{10} \oplus \mathbb{Z}_3$, and $\mathbb{Z}_6 \oplus \mathbb{Z}_5$. ($\mathbb{Z}_{30}$ is also isomorphic to $\mathbb{Z}_{30} \oplus \mathbb{Z}_1$, but again that is trivial since $\mathbb{Z}_1$ is the trivial group.)

(b) Let N be a positive integer and let $N = mn$. Then any factorization of N as $N = m'n'$ with gcd $(m', n') = \gcd(m, n)$ and $\operatorname{lcm}(m', n') = \operatorname{lcm}(m, n)$ (in fact, these two conditions are equivalent) gives an isomorphism of $\mathbb{Z}_m \oplus \mathbb{Z}_n$ with $\mathbb{Z}_{m'} \oplus \mathbb{Z}_{n'}$. For example, if $N = 4320 = 180 \cdot 24 = 120 \cdot 36 = 360 \cdot 12$, $\mathbb{Z}_{180} \oplus \mathbb{Z}_{24}$ is isomorphic to $\mathbb{Z}_{120} \oplus \mathbb{Z}_{36}$ and $\mathbb{Z}_{360} \oplus \mathbb{Z}_{12}$. $\Diamond$

In preparation for proving the first main theorem of this section, we derive a result that we will use in that proof.

We begin with a definition and a result that should remind you very much of vector spaces — indeed, the situation here is entirely analogous.

Let $\mathbb{Z}^n = \mathbb{Z} \times \cdots \times \mathbb{Z}$ where there are n factors. Let $e_1, \ldots, e_n$ be the elements of $\mathbb{Z}^n$ given by $e_1 = (1, 0, \ldots, 0)$, $e_2 = (0, 1, \ldots, 0), \ldots$, $e_n = (0, 0, \ldots, 1)$. Note that every element $a = (a_1, \ldots, a_n)$ of $\mathbb{Z}^n$ can be written as $a = a_1 e_1 + \cdots + a_n e_n$ and this expression is unique.

Definition 2.6.10. An abelian group G is *free abelian group of rank* n if G is isomorphic to $\mathbb{Z}^n$. If $\alpha\colon G \to \mathbb{Z}^n$ is any isomorphism, let $b_1 = \alpha^{-1}(e_1)$, $b_2 = \alpha^{-1}(e_2), \ldots, b_n = \alpha^{-1}(e_n)$. Then $\{b_1, \ldots, b_n\}$ is a *basis* of G. $\Diamond$

We thus see that any element $g \in G$ can be written uniquely as $g = a_1 b_1 + \cdots + a_n b_n$ for integers $a_1, \ldots, a_n$. Indeed if $\{b_1, \ldots, b_n\}$ is any set of elements of G with this property, we have an isomorphism $\alpha\colon G \to \mathbb{Z}^n$ given by

$$\alpha(a_1 b_1 + \cdots + a_n b_n) = (a_1, \ldots, a_n).$$

Now recall that if V is a vector space with basis $\{v_1, \ldots, v_n\}$ and $\{w_1, \ldots, w_n\}$ are any elements of a vector space W, there is a unique linear transformation $T\colon V \to W$ with $T(v_i) = w_i$ for each i. Here is the analog for abelian groups.

Lemma 2.6.11. *Let G be a free abelian group of rank n and let $\{b_1, \ldots, b_n\}$ be a basis of G. Let H be an arbitrary abelian group and let $\{h_1, \ldots, h_n\}$ be elements of H. Then there is a unique group homomorphism $\varphi\colon G \to H$ with $\varphi(b_i) = h_i$ for each i.*

Proof.   First let us consider $G_0 = \mathbb{Z}^n$.

We start out by defining $\varphi_0 \colon G_0 \to H$ "coordinatewise". That is, we let $\varphi_0(e_1) = h_1, \ldots, \varphi_0(e_n) = h_n$. Then, if $a \in G_0$, $a = (a_1, \ldots, a_n)$, then $a = a_1 e_1 + \cdots + a_n e_n$, so we let

$$\varphi_0(a) = a_1 \varphi_0(e_1) + \cdots + a_n \varphi_0(e_n) = a_1 h_1 + \cdots + a_n h_n.$$

Then $\varphi_0 \colon G_0 \to H$ is well-defined, as the expression for a in terms of $\{e_1, \ldots, e_n\}$ is unique, and it is straightforward to check that φ_0 is a homomorphism.

Now let G be a free group of rank n, let $\alpha \colon G \to G_0$ be an isomorphism, so that $b_i = \alpha^{-1}(e_i)$ for each $i = 1, \ldots, n$. Then we let $\varphi = \varphi_0 \alpha$. $\square$

Our proof of the lemma was a little indirect. We could have directly worked with G without going through G_0. But we have done things this way to more directly lead to our next result.

We now consider the group $\mathbb{Z}_{m_1} \times \cdots \times \mathbb{Z}_{m_n}$, and we let $\sigma \colon \mathbb{Z}^n \to \mathbb{Z}_{m_1} \times \cdots \times \mathbb{Z}_{m_n}$ be the homomorphism

$$\sigma(a_1, \ldots, a_n) = ([a_1]_{m_1}, \ldots, [a_n]_{m_n}).$$

We let
$$\bar{e}_1 = \sigma(e_1) = ([1]_{m_1}, [0]_{m_2}, \ldots, [0]_{m_n}), \ \ldots$$
$$\bar{e}_n = \sigma(e_n) = ([0]_{m_1}, [0]_{m2}, \ldots, [1]_{m_n}).$$

We observe that $\mathrm{Ker}(\sigma) = (m_1 \mathbb{Z}) \times \cdots \times (m_n \mathbb{Z})$.

Lemma 2.6.12. *Let G be a group that is isomorphic to $\bar{G}_0 = \mathbb{Z}_{m_1} \times \cdots \times \mathbb{Z}_{m_n}$ and let $\alpha : G \to \bar{G}_0$ be an isomorphism. For each $i = 1, \ldots, n$, let $b_i = \alpha^{-1}(\bar{e}_i)$. Let H be an arbitrary abelian group and let $\{h_1, \ldots, h_n\}$ be elements of H. Suppose that the order of h_i divides m_i for each $i = 1, \ldots, n$. Then there is a unique group homomorphism $\varphi \colon G \to H$ with $\varphi(b_i) = h_i$ for each i.*

Proof.   By Lemma 2.6.11, if $G_0 = \mathbb{Z}^n$, we have a unique group homomorphism $\varphi_0 \colon G_0 \to H$. Now the condition that the order of h_i divides m_i for each i gives us that $\varphi_0(m_i e_i) = m_i \varphi_0(e_i) = m_i h_i = 0$ for each i, so $\mathrm{Ker}(\varphi_0) \supseteq (m_1 \mathbb{Z}) \times \cdots \times (m_n \mathbb{Z}) = \mathrm{Ker}(\sigma)$. Then by Lemma 2.5.2, φ_0 factors through σ, i.e., $\varphi_0 = \varphi \sigma$ for some homomorphism $\varphi \colon G \to H$, and this is the desired homomorphism φ.

Now we have $\varphi(b_i) = h_i$ for each i, and since every element of G can be expressed as $a_1 b_1 + \cdots + a_n b_n$ for some $a_1, \ldots, a_n$, this determines φ uniquely. $\qquad\qquad\qquad\qquad\qquad\qquad\qquad\square$

In thinking about the proof of Lemma 2.6.12, you may wonder why we didn't simply define $\varphi\colon G \to H$ by the formula $\varphi(a_1 b_1 + \cdots + a_n b_n) = a_1 h_1 + \cdots + a_n h_n$ as we did in Lemma 2.6.11. The point is that in Lemma 2.6.11, this expression is unique, giving us an unambiguous definition of φ, but in Lemma 2.6.12 it is *not*. So we had to have some method of showing that the potential ambiguity in fact doesn't matter. We set things up so that we could use Lemma 2.5.2 to do so.

Now we come to the first main result of this section, the structure theorem for finite abelian groups. This tells us that every finite abelian group G is isomorphic to the direct sum of cyclic group $Z_{e_1} \oplus Z_{e_2} \oplus \cdots \oplus Z_{e_k}$. We would like to conclude that these integers $e_1, e_2, \ldots, e_k$ are unique, giving us a classification of finite abelian groups, but we have already seen that they are not (compare Example 2.6.9). We can readily see one potential source of ambiguity: We do not want to allow any e_i to be 1, as that would give us a trivial $\mathbb{Z}_1$ summand, which we could eliminate without changing the isomorphism class of G. (You can think of this as being analogous to factoring positive integers. We would not want to consider $2 \cdot 3$, $2 \cdot 3 \cdot 1$, $2 \cdot 3 \cdot 1 \cdot 1, \ldots$ to be distinct factorization of 6.) But even without that, Example 2.6.9 shows there is a lot of other ambiguity. However, if in Example 2.6.9 we exclude $e_i = 1$ then in (a) we have the unique expression $\mathbb{Z}_{30}$. If we also require in (b) that e_2 divides e_1 then we have the unique expression $\mathbb{Z}_{360} \oplus \mathbb{Z}_{12}$. It turns out that the analog (or, more precisely, the generalization) of this condition is enough to give us uniqueness.

But this is really step 2 of our main result. Step 1 is to prove existence, that G *is* isomorphic to the direct sum of cyclic groups, and afterwards we will prove uniqueness. However, this is one of the (not uncommon) situations in mathematics where it turns out to be easier to prove a stronger (in our case, more specific) result. Rather than trying to show that G is isomorphic to some direct sum of cyclic groups, it is easier to show that G is isomorphic to a direct sum of cyclic groups whose orders satisfy our condition. So that is what we shall do.

Theorem 2.6.13   (Structure theorem for finite abelian groups).   *Let G be a finite abelian group. If G is the trivial group, then G is isomorphic to $\mathbb{Z}_{e_1}$ for $e_1 = 1$. Suppose that G is nontrivial. Then there are positive integers $e_1, \ldots, e_k$ (for some k) with $e_k \neq 1$, and, if $k > 1$, with e_{i+1} dividing e_i for $i = 1, \ldots, k-1$, such that G is isomorphic to $\mathbb{Z}_{e_1} \oplus \cdots \oplus \mathbb{Z}_{e_k}$. Furthermore, the integers $e_1, \ldots, e_k$ are unique.*

Proof.   We first prove existence. We proceed by complete induction on $n = |G|$.

If $n = 1$ then G is isomorphic to $\mathbb{Z}_1$ and there is nothing further to show. Assume now that the theorem is true for all abelian groups of order less than n and let G have order n.

Let e be the exponent of G. If $e = n$, then by Theorem 2.6.5, G has an element a of order n, so in this case G is cyclic of order n, i.e., G is isomorphic to $\mathbb{Z}_n$, and we are done.

Suppose that $e < n$. Set $e_1 = e$. By Theorem 2.6.5, G has an element a_1 of order e_1. Let H_1 be the subgroup generated by a_1. Then H_1 is isomorphic to $\mathbb{Z}_{e_1}$.

Now consider the group $Q = G/H_1$. This is an abelian group of order n/e_1, so by the inductive hypothesis there is an isomorphism $\varphi \colon Q \to \mathbb{Z}_{e_2} \oplus \cdots \oplus \mathbb{Z}_{e_k}$ where $e_k \neq 1$ and if $k \geq 3$, e_{i+1} divides e_i for $i = 2, \ldots, k-1$. Let $b_2 \in Q$ with $\varphi(b_2) = (1, 0, \ldots, 0), \ldots, b_k \in Q$ with $\varphi(b_k) = (0, 0, \ldots, 1)$. Now the quotient map $\pi \colon G \to Q$ is onto, so there are elements $a_2', \ldots, a_k'$ of G with $\pi(a_2') = b_2, \ldots, \pi(a_k') = b_k$.

Now if a_2' had order $e_2, \ldots, a_k'$ had order e_k, we would be in business, as we shall see below. But there is no reason to believe that is the case, and it may not be, so we will have to modify these elements.

Fix $i \geq 2$ and consider a_i'. $\pi(a_i') = b_i$ and b_i has order e_i. Then $\pi(e_i a_i') = e_i \pi(a_i') = e_i b_i = 0$. But $\pi(e_i a_i') = 0$ means $e_i a_i' \in H_1$. Thus $e_i a_i' = k_i a_1$ for some k_i. Now by the definition of $e_1 = e$, e_i divides e_1. Then $(e_1/e_i)(e_i a_i') = (e_1/e_i)k_i a_1$, i.e., $e_1 a_i' = (e_1/e_i)k_i a_1$. Now the order of any element of G, and in particular of a_i', divides e_1, so $e_1 a_i' = 0$. Thus we see that $(e_1/e_i)(k_i a_1) = 0$, i.e., that $k_i a_1$ is an element of H_1 whose order divides e_1/e_i. But H_1 is a cyclic group of order e_1, generated by a_1, and we know exactly what those elements are. They are the elements of the subgroup generated by $e_i a_1$. Thus we must have that $k_i = e_i j_i$ for some integer j_i.

Assembling this, we see that $e_i a_i' = e_i j_i a_1$, or that $e_i(a_i' - j_i a_1) = 0$. Thus if we set $a_i = a_i' - j_i a_1$, we see that $e_i a_i = 0$, so the order of a_i divides e_i. But $\pi(a_i) = \pi(a_i') = b_i$ has order e_i, so e_i divides the order of a_i. Hence a_i has order e_i, and this is true for every $i = 2, \ldots, k$.

Now let H_2 be the subgroup of G generated by $a_2, \ldots, a_k$. By Lemma 2.6.12. there is a homomorphism $\lambda: Q \to H_2$ with $\lambda(b_i) = a_i$, $i = 2, \ldots, k$. Then $\pi\lambda(b_i) = b_i$, $i = 2, \ldots, k$, so $\pi\lambda: Q \to Q$ is the identity. In the language of Definition 2.4.34, λ is a splitting of π. Hence, by Lemma 2.4.35, G is the semidirect product of H_1 and H_2. Since G is abelian, G is then the direct product of H_1 and H_2 (Remark 2.4.37), or, as we are calling it here, the direct sum $G = H_1 \oplus H_2$. But then, since H_1 is isomorphic to Z_{e_1}, and H_2 is isomorphic to $Z_{e_2} \oplus \cdots \oplus Z_{e_k}$, G is isomorphic to $Z_{e_1} \oplus Z_{e_2} \oplus \cdots \oplus Z_{e_k}$.

Hence, by induction, we have the existence of a direct sum decomposition for every finite abelian group G.

Now we must show that, under our conditions, the integers $e_1, \ldots, e_k$ are unique. Thus, suppose G is isomorphic to $Z_{e_1} \oplus \cdots \oplus Z_{e_k}$ and also to $Z_{f_1} \oplus \cdots \oplus Z_{f_l}$, with both $e_1, \ldots, e_k$ and $f_1, \ldots, f_l$ satisfying the conditions of the theorem. We must show that $l = k$, and then that $f_i = e_i$ for $i = 1, \ldots, k$.

We begin with an observation that we will use crucially. Let s and t be positive integers. Then the number of elements of the cyclic group Z_s whose order divides t is equal to $\gcd(s, t)$. In particular, if s divides t the number of these elements is s and if t divides s the number of these elements is t. Also, if $t = p$ is prime, the number of these elements is p if p divides s and is 1 if p does not.

With this observation in hand, we go to work. Since k and l are positive integers, we many as well assume that $l \leq k$. (Otherwise switch the e's and the f's.) Now $e_k > 1$, so is divisible by some prime p. Then, since e_k divides e_i for every $i \leq k$, by the above observation we see that G has p^k elements whose order divides p. But we can also compute this number from the second decomposition, and we see it is $p^{l'}$ where l' is the number of $f_1, \ldots, f_l$ that are divisible by p. Certainly $l' \leq l$. Counting the number of elements of G whose order divides p in two different ways must yield the same result, so $p^{l'} = p^k$ and hence $l' = k$. Thus $k \leq l$, so we must have $l = k$.

Now we must show that $e_i = f_i$ for each $i = 1, \ldots, k$. We proceed by downward induction on i. We begin with $i = k$. Let us count the number of elements of G whose order divides e_k. From the first decomposition we see it is e_k^k. From the second decomposition we see that it is the product $\gcd(f_1, e_k)\gcd(f_2, e_k) \ldots \gcd(f_k, e_k)$. Thus, since again the two counts must yield the same result, we must have that $\gcd(f_k, e_k) = e_k$, i.e., that e_k divides f_k (and then automatically $\gcd(f_i, e_k) = e_k$ for every $i = 1, \ldots, k$). Reversing the roles of f_k and e_k, we see that $\gcd(e_k, f_k) = f_k$, i.e., that f_k divides e_k. Hence $f_k = e_k$.

Now for the inductive step. Suppose that $f_k = e_k$, $f_{k-1} = e_{k-1}, \ldots, f_{i+1} = e_{i+1}$ and consider e_i and f_i. Let us count the number of elements of G whose order divides e_i. From the first decomposition we see that this number is the product $e_i^i \, e_{i+1} \ldots e_k$ while from the second decomposition (using the inductive hypothesis) it is the product $\gcd(f_1, e_i)\gcd(f_2, e_i)\ldots\gcd(f_i, e_i)e_{i+1} \ldots e_k$. Again these two answers must be the same, so in particular we must have $\gcd(f_i, e_i) = e_i$, i.e., that e_i divides f_i. And again we can reverse the roles of e_i and f_i to conclude that $\gcd(e_i, f_i) = f_i$, i.e., that f_i divides e_i. Hence $f_i = e_i$.

Then by induction this is true for every $i = k, \ldots, 1$, and we are done. $\square$

Definition 2.6.14. Let G be a finite abelian group.

The integers $e_1, \ldots, e_k$ as in the conclusion of Theorem 2.6.13 are called the *elementary divisors* of G. $\Diamond$

Before proceeding further, we should observe an important point about the proof of Theorem 2.6.13. The best way to make this observation is by looking at an example.

Let $G = \mathbb{Z}_4 \oplus \mathbb{Z}_2$. Of course, this group is given to us as a direct sum of cyclic groups, but suppose we didn't notice that and wanted to prove that it was. So we follow the strategy of the proof of Theorem 2.6.13. This group has exponent 4, so we begin by choosing an element a_1 of order 4, say $a_1 = (1, 0)$. Now we look at the quotient of G by the subgroup generated by a_1, which is isomorphic to $\mathbb{Z}_2$, a cyclic group of order 2, and look for an element a_2 of G of order 2 that projects to the element of order 2 in this quotient. Such an element is $a_2 = (0, 1)$, and we're done. But all we know is that we can find an element whose projection has order 2, so we might instead have chosen $a_2' = (1, 1)$, which is an element of G of order 4, which doesn't work. But the proof of Theorem 2.6.13 shows that we can "fix" this element by letting $a_2 = a_2' - a_1 = (0, 1)$, which does work. Note that what was crucial in being able to fix a_2' was that a_1 was an element of maximal order.

But suppose we hadn't done things this way, and just had started out picking any element. If we had started by choosing $a_1 = (0, 1)$, of order 2, then we could have chosen $a_2 = (1, 0)$, of order 4, and our proof would have gone through. But instead, suppose we had started by choosing $a_1 = (2, 0)$, also an element of order 2. Then we would have been stuck. There would have been no way of choosing a_2 of order 4 so that G would be the direct sum of the subgroup generated by a_1 and a_2, as if a_2 is *any* element of G of order 4, a_1 is an element of the subgroup of G generated by a_2. (We leave this for you to check.) Thus our strategy of choosing an element of maximal order is what guarantees that our proof indeed does work.

By the way, although the choices of $a_1 = (1, 0)$ and $a_2 = (0, 1)$ are the "obvious" choices for $G = \mathbb{Z}_4 \oplus \mathbb{Z}_2$, these are not the only choices. Indeed, our proof shows that we can make any choices for a_1 and a_2, providing that these elements have the right orders (4 and 2 respectively). Thus, for example, we could have *equally well* chosen $a_1 = (1, 1)$, or $a_2 = (2, 1)$.

There is a second kind of decomposition of a finite abelian group G as a direct sum of cyclic groups that is often more useful.

We begin with the case of G cyclic.

Theorem 2.6.15. *Let G be a cyclic group of order $n > 1$. Let $n = p_1^{d_1} \ldots p_k^{d_k}$ be the prime factorization of N, with $p_1, \ldots, p_k$ distinct primes. Then G is isomorphic to the direct sum $\mathbb{Z}_{p_1^{d_1}} \oplus \cdots \oplus \mathbb{Z}_{p_k^{d_k}}$.*

Proof. By induction on the number of distinct prime factors k of n.

If $k = 1$ there is nothing to prove.

The crucial case is $k = 2$. Suppose $n = p_1^{d_1} p_2^{d_2}$. Since $p_1^{d_1}$ and $p_2^{d_2}$ are relatively prime, we know that $\mathbb{Z}_n$ is isomorphic to $\mathbb{Z}_{p_1^{d_1}} \oplus \mathbb{Z}_{p_2^{d_2}}$.

Now suppose the theorem is true for $k - 1$ and let $n = p_1^{d_1} \ldots p_k^{d_k}$. Then $p_1^{d_1}$ and the product $p_2^{d_2} \ldots p_k^{d_k}$ are relatively prime, so by the $k = 2$ case $\mathbb{Z}_n$ is isomorphic to $\mathbb{Z}_{p_1^{d_1}} \oplus \mathbb{Z}_{p_2^{d_2} \ldots p_k^{d_k}}$. But then by the $k - 1$ case, $\mathbb{Z}_{p_2^{d_2} \ldots p_k^{d_k}}$ is isomorphic to $\mathbb{Z}_{p_2^{d_2}} \oplus \cdots \oplus \mathbb{Z}_{p_k^{d_k}}$, so G is isomorphic to $\mathbb{Z}_{p_1^{d_1}} \oplus \cdots \oplus \mathbb{Z}_{p_k^{d_k}}$ and by induction we are done. $\square$

Before proving our next general theorem, let us return to Example 2.6.9 and reexamine it from this perspective.

Example 2.6.16.

(a) Let $G = \mathbb{Z}_{30}$. Since $30 = 2 \cdot 3 \cdot 5$, we see that $\mathbb{Z}_{30}$ is isomorphic to $\mathbb{Z}_2 \oplus \mathbb{Z}_3 \oplus \mathbb{Z}_5$. But we could regroup the right hand side as $(\mathbb{Z}_3 \oplus \mathbb{Z}_5) \oplus \mathbb{Z}_2$, so $\mathbb{Z}_{30}$ is isomorphic to $\mathbb{Z}_{15} \oplus \mathbb{Z}_2$, or as $(\mathbb{Z}_2 \oplus \mathbb{Z}_5) \oplus \mathbb{Z}_3$, so $\mathbb{Z}_{30}$ is isomorphic to $\mathbb{Z}_{10} \oplus \mathbb{Z}_3$, or as $(\mathbb{Z}_2 \oplus \mathbb{Z}_3) \oplus \mathbb{Z}_5$, so that $\mathbb{Z}_{30}$ is isomorphic to $\mathbb{Z}_6 \oplus \mathbb{Z}_5$.

(b) Let $G = \mathbb{Z}_{180} \oplus \mathbb{Z}_{24}$. Since $180 = 4 \cdot 9 \cdot 5$, $\mathbb{Z}_{180}$ is isomorphic to $\mathbb{Z}_4 \oplus \mathbb{Z}_9 \oplus \mathbb{Z}_5$, and since $24 = 8 \cdot 3$, $\mathbb{Z}_{24}$ is isomorphic to $\mathbb{Z}_8 \oplus \mathbb{Z}_3$. Thus G is isomorphic to $(\mathbb{Z}_4 \oplus \mathbb{Z}_9 \oplus \mathbb{Z}_5) \oplus (\mathbb{Z}_8 \oplus \mathbb{Z}_3)$. Regrouping terms on the right hand side as $(\mathbb{Z}_8 \oplus \mathbb{Z}_3 \oplus \mathbb{Z}_5) \oplus (\mathbb{Z}_4 \oplus \mathbb{Z}_9)$, we see that G is isomorphic to $\mathbb{Z}_{120} \oplus \mathbb{Z}_{36}$, and regrouping them as $(\mathbb{Z}_8 \oplus \mathbb{Z}_9 \oplus \mathbb{Z}_5) \oplus (\mathbb{Z}_4 \oplus \mathbb{Z}_3)$, we see that G is isomorphic to $\mathbb{Z}_{360} \oplus \mathbb{Z}_{12}$. $\diamondsuit$

Here is a more general decomposition theorem.

Theorem 2.6.17. *Let G be an abelian group of order mn, with m and n relatively prime. Let $\varphi_m \colon G \to G$ by $\varphi_m(g) = mg$, and let $\varphi_n \colon G \to G$ by $\varphi_n(g) = ng$. Let $H_m = \{g \in G \mid \text{order of } g \text{ divides } m\}$ and $H_n = \{g \in G \mid \text{the order of } g \text{ divides } n\}$. Then*

(a) $H_m = \mathrm{Ker}(\varphi_m) = \mathrm{Im}(\varphi_n)$ and $H_n = \mathrm{Ker}(\varphi_n) = \mathrm{Im}(\varphi_m)$.
(b) $G = H_m \oplus H_n$.

Proof.

(a) By definition, $H_m = \mathrm{Ker}(\varphi_m)$. We must show $\mathrm{Ker}(\varphi_m) = \mathrm{Im}(\varphi_n)$. We show this by showing that each is contained in the other.
$\mathrm{Im}(\varphi_n) \subseteq \mathrm{Ker}(\varphi_m)$: Let $h \in \mathrm{Im}(\varphi_m)$. Then $h = mg$ for some $g \in G$. But then $\varphi_n(h) = \varphi_n(mg) = (mn)g = 0$ as the order of any element g of G divides $|G| = mn$.
$\mathrm{Ker}(\varphi_m) \subseteq \mathrm{Im}(\varphi_n)$: Since m and n are relatively prime, there are integers a and b with $am + bn = 1$. Now let $g \in \mathrm{Ker}(\varphi_m)$, so that $mg = 0$. Then $g = 1g = (am + bn)g = a(mg) + n(bg) = a(0) + n(bg) = \varphi_n(bg)$.
The second claim in (a) is proved the same way, switching the roles of m and n.

(b) We must show $H_m \cap H_n = \{0\}$ and $G = H_m + H_n$. Again choose integers a and b with $am + bn = 1$.
$H_m \cap H_n = \{0\}$: Let $g \in H_m \cap H_n = \mathrm{Ker}(\varphi_n) \cap \mathrm{Ker}(\varphi_m)$. Then $g = 1g = (am + bn)g = a(mg) + b(ng) = a(0) + b(0) = 0$.

$H_m + H_n = G$: Let $g \in G$. Then $g = 1g = (am + bn)g = m(ag) + n(bg) = \varphi_m(ag) + \varphi_n(bg)$. □

Example 2.6.18.

(a) Returning to Example 2.6.16(a), we see that example is an illustration here as well.

(b) Returning to Example 2.6.16(b), we see that $G = \mathbb{Z}_{180} \oplus \mathbb{Z}_{24}$ is a group of order $4320 = 2^5 \cdot 3^3 \cdot 5$. Writing $4320 = 2^5 \cdot (3^3 \cdot 5)$, we obtain that G is isomorphic to $(\mathbb{Z}_8 \oplus \mathbb{Z}_4) \oplus (\mathbb{Z}_9 \oplus \mathbb{Z}_3 \oplus \mathbb{Z}_5)$; writing $4320 = 3^3 \cdot (2^5 \cdot 5)$, we obtain that G is isomorphic to $(\mathbb{Z}_9 \oplus \mathbb{Z}_3) \oplus (\mathbb{Z}_8 \oplus \mathbb{Z}_4 \oplus \mathbb{Z}_5)$; writing $4320 = 5 \cdot (2^5 \cdot 3^3)$, we obtain that G is isomorphic to $\mathbb{Z}_5 \oplus (\mathbb{Z}_8 \oplus \mathbb{Z}_4 \oplus \mathbb{Z}_9 \oplus \mathbb{Z}_3)$. ◇

Remark 2.6.19. In Example 2.6.18, we know the structure of G exactly. But the point of Theorem 2.6.17 is that we don't have to. For example, let G be a group of order $96 = 32 \cdot 3$ whose structure we don't know. Then we can conclude that $G = H_{32} \oplus H_3$, where H_{32} is the subgroup of G consisting of those elements of G whose order divides 32 and H_3 is the subgroup of G consisting of those elements of G whose order divides 3. Similarly, if G is any group of order $240 = 16 \cdot 15$, there we can conclude that $G = H_{16} \oplus H_{15}$, where H_{16} is the subgroup of G consisting of those elements of G whose order divides 16 and H_{15} is the subgroup of G consisting of those elements of G whose order divides 15.

Also, we have put Theorem 2.6.17 here as it serves as a lead-in to our next result. By we wish to observe that this theorem does not depend on Theorem 2.6.13. Its proof is completely independent of the proof of Theorem 2.6.13, and in fact much easier. ◇

Here is our next general result. (To preclude confusion, let us specifically state that we allow the set S in Theorem 2.6.20 to have repeated entries.)

Theorem 2.6.20 (Alternate structure theorem for finite abelian groups). *Let G be a finite abelian group. If G is the trivial group, then G is isomorphic to $\mathbb{Z}_1$. Suppose that G is nontrivial. Then there is a set $S = \{q_1, \ldots, q_t\}$ of positive integers, each of which is a prime power, such that G is isomorphic to $\mathbb{Z}_{q_1} \oplus \cdots \oplus \mathbb{Z}_{q_t}$. Furthermore, the set S is unique.*

Proof.   If G is the trivial group, the theorem is certainly true. Suppose not.

By Theorem 2.6.13, G is isomorphic to $\mathbb{Z}_{e_1} \oplus \cdots \oplus \mathbb{Z}_{e_k}$ with all $e_i > 1$. Factor each e_i into a product of powers of distinct primes:

$$e_1 = p_{11}^{d_{11}} \cdots p_{1j1}^{d_{1j1}}$$

$$\vdots$$

$$e_k = p_{k1}^{d_{k1}} \cdots p_{kjk}^{d_{kjk}}.$$

Then by Theorem 2.6.15,

$$\mathbb{Z}_{e_1} \text{ is isomorphic to } \mathbb{Z}_{p_{11}^{d_{11}}} \oplus \cdots \oplus \mathbb{Z}_{p_{1j1}^{d_{1j1}}}$$

$$\vdots$$

$$\mathbb{Z}_{e_k} \text{ is isomorphic to } \mathbb{Z}_{p_{k1}^{d_{k1}}} \oplus \cdots \oplus \mathbb{Z}_{p_{kjk}^{d_{kjk}}}$$

and so G is isomorphic to

$$\left(\mathbb{Z}_{p_{11}^{d_{11}}} \oplus \cdots \oplus \mathbb{Z}_{p_{1j1}^{d_{1j1}}} \right) \oplus \cdots \oplus \left(\mathbb{Z}_{p_{k1}^{d_{k1}}} \oplus \cdots \oplus \mathbb{Z}_{p_{kjk}^{d_{kjk}}} \right),$$

and $S = \left\{ p_{11}^{d_{11}}, \ldots, p_{1j1}^{d_{1j1}}, \ldots, p_{k1}^{d_{k1}}, \ldots, p_{kjk}^{d_{kjk}} \right\}$.

Now we must show that S is unique.

We see from the above construction that the elementary divisors $e_1, \ldots, e_k$ determine S. We shall show that, conversely, the set S determines the elementary divisors. Then, since the elementary divisors are unique, S is unique.

Here is an algorithm for recovering the elementary divisors from S. Set $S_1 = S$. Let $i = 1$.

For any value of i, let R_i be the subset of S_i consisting of, for every prime, an element of S_i of the highest power of that prime. Let e_i be the product of the elements of R_i, and let $S_{i+1} = S_i - R_i$. Change i to $i+1$ and iterate this process. (Note that this process will stop as S is finite and we remove at least one element from S_i at every step. In fact, it will stop at step k, yielding e_k, when k is the largest integer such that, for some prime p, there are k powers of p in S.)□

Definition 2.6.21. Let S be the set in the conclusion of Theorem 2.6.20. In case G is the trivial group, let S be the empty set. Then the elements of S are the *invariant factors* of G.   ◇

We have just proved the structure theorem for finite abelian groups.

Now we wish to generalize this to finitely generated abelian groups, which we first define.

Definition 2.6.22. Let G be an abelian group. A set $A = \{x_i\}$ of elements of G *generates* G if every element g of G can be expressed as $g = \Sigma_i n_i x_i$ for some integers $\{n_i\}$. If there is a finite set of generators for G, then G is called *finitely generated*. The minimum number of elements in a generating set for G is called the *rank* of G. $\Diamond$

Example 2.6.23.

(a) Any finite abelian group is finitely generated, as we may let $A = G$.

(b) If $G = \{0\}$, G has rank 0. (By definition, the value of the empty sum is 0.)

(c) If G is a nontrivial cyclic group, then G has rank 1.

(d) If G is a finite abelian group with elementary divisors $e_1, \ldots, e_k$ as in Definition 2.6.14, G has rank k.

(e) If $G = \mathbb{Z}^n$, then G is finitely generated. We see right away that G has rank at most n. With considerably more work we will see that G has rank exactly n. $\Diamond$

Actually, our statement that we will generalize the structure theorem from finite abelian groups to finitely generated abelian groups, while true, is in a way misleading, as the key to our analysis will be to consider the diametrically opposite case, finitely generated abelian groups that have *no* elements of finite order other than the identity. Once we have done that, it will be easy to combine the two situations.

A lot of what we will be doing should very much remind you of linear algebra, but, as you will see, the situation is much more involved here. Indeed, the proofs here turn out to be rather subtle.

Let us introduce some standard terminology.

Definition 2.6.24. An abelian group G is *torsion-free* if it has no elements of finite order other than the identity. $\Diamond$

Now let us make some definitions that should remind you a lot of linear algebra.

Definition 2.6.25. Let G be an abelian group and let $A = \{x_i\}$ be a set of elements of G.

(a) A is *linearly independent* if the equation $\Sigma_i n_i x_i = 0$ only has the solution $n_i = 0$ for all i.

(b) A *spans* G if the equation $\Sigma_i n_i x_i = g$ has a solution for every $g \in G$.

(c) A is a *basis* of G if A is linearly independent and spans G. $\Diamond$

Note that "spans" is a synonym for "generates". The word spans is usually used in linear algebra, while the word generates is usually used in group theory.

Of course, we know that every vector space has a basis. We will prove, with a lot of work, that every finitely generated torsion-free abelian group has a basis.

We do linear algebra over fields, e.g., the field $\mathbb{Q}$. Here we will be doing analogous arguments over the integers $\mathbb{Z}$. But you should not think that everything just goes through the same way — *it does not.* Here is a simple example that already illustrates the difference.

Example 2.6.26. Recall the following theorem from linear algebra.

Theorem. Let V be a vector space and let B be a set of elements of V. The following are equivalent:

(i) B is a basis of V (i.e., B is linearly independent and spans V).

(ii) B is a maximal linearly independent set of vectors in V.

(iii) B is a minimal spanning set of vectors in V.

Now let us think about $G = \mathbb{Z}$. We could choose $B = \{1\}$. Then, indeed, B is a basis of G, and is a maximal linearly independent set of elements of G, and is also a minimal spanning set of elements of G.

But if we choose $B = \{2\}$, then B is a maximal linearly independent set of elements of G, but B does *not* span G (and so is *not* a basis of G). And if we choose $B = \{2, 3\}$, then B is a minimal spanning set for G, but is *not* linearly independent (and so is *not* a basis of G). $\Diamond$

Definition 2.6.27. An abelian group G is *free* if it has a basis. $\Diamond$

Note that if G has a basis $\{x_1, \ldots, x_k\}$ then we have an isomorphism $\alpha\colon G \to \mathbb{Z}^k$ defined as follows: Let $g \in G$ and write

$g = \Sigma_i n_i x_i$. (Since $\{x_1, \ldots, x_k\}$ is a basis of G, every element g of G can be written in the form in a unique way.) Then $\alpha(g) = (n_1, \ldots, n_k)$. Thus we see that our definition here of a free abelian group agrees with our previous definition, Definition 2.6.10.

Remark 2.6.28. Note that no element x of G of finite order can be linearly independent, as for such an element x we have $nx = 0$ for some $n \neq 0$. Similarly, if G has any elements of finite order other than the identity, then G cannot possibly have a basis, so G cannot be free. $\Diamond$

Here is a result that is exactly analogous to the situation in vector spaces.

Lemma 2.6.29. *Let G be an abelian group of rank k. Then any set S of more than k elements of G is linearly dependent.*

Proof. By induction on k. This is trivial if $k = 0$, so assume $k > 0$. Since G has rank k, by definition it is spanned by some set $A = \{x_1, \ldots, x_k\}$ of k elements of G.

Let $S = \{y_1, y_2, \ldots\}$ be a set of more than k elements of G.

The case $k = 1$: If $y_1 = 0$ or $y_2 = 0$ then S is linearly dependent. Otherwise, since $\{x_1\}$ spans G, $y_1 = ax_1$ for some $a \in \mathbb{Z}$, $a \neq 0$, and $y_2 = bx_1$, for some $b \in \mathbb{Z}$, $b \neq 0$. But then $by_1 - ay_2 = 0$.

Now suppose the result is true for $k - 1$. Let H be the subgroup of G spanned by $\{x_1, \ldots, x_{k-1}\}$.

If $S \subseteq H$, we are done by induction. Suppose not. For each i, let

$$y_i = \tilde{y}_i + r_i x_k \quad \text{where } \tilde{y}_i \in H \text{ and } r_i \text{ is an integer.}$$

(This expression for y_i may not be unique. Choose any one.) At least one $r_i \neq 0$. Assume it is r_1.

Suppose $S = \{y_1, y_2, \ldots\}$ is linearly independent. Then $\{y_1, r_1 y_2, r_1 y_3, \ldots\}$ is also linearly independent. (If $n_1 y_1 + n_2(r_1 y_2) + n_3(r_1 y_3) + \cdots = 0$, then $n_1 y_1 + (n_2 r_1)y_2 + (n_3 r_1)y_3 + \cdots = 0$.) Then $\{y_1, r_1 y_2 - r_2 y_1, r_1 y_3 - r_3 y_1 \ldots\}$ is linearly independent. (If $n_1 y_1 + n_2(r_1 y_2 - r_2 y_1) + n_3(r_1 y_3 - r_3 y_1) + \cdots = 0$, then $(n_1 - n_2 r_2 - n_3 r_3 \ldots)y_1 + (n_2 r_1)y_2 + (n_3 r_1)y_3 + \cdots = 0$.) Then $T = \{r_1 y_2 - r_2 y_2, r_1 y_3 - r_3 y_1, \ldots\}$ is linearly independent (being a subset of a linearly independent set). But for $i > 1$,

$$r_1 y_i - r_i y_1 = r_1(\tilde{y}_i + r_i x_k) - r_i(\tilde{y}_1 + r_1 x_k) = r_1 \tilde{y}_i - r_i \tilde{y}_1 \in H,$$

as $\tilde{y}_i \in H$ and $\tilde{y}_1 \in H$. Thus T is a linearly independent set of more than $k - 1$ vectors in H; contradiction. $\qquad\square$

This lemma has a consequence that is also analogous to the situation in vector spaces.

Corollary 2.6.30. *Let G be a free abelian group with a basis B consisting of k elements. Then $k = \mathrm{rank}(G)$. In particular, any basis of G consists of $\mathrm{rank}(G)$ elements.*

Proof.   By definition, the rank of G is the smallest number of elements in any generating set B for G. Since B is a generating set, we must have $\mathrm{rank}(G) \leq k$. But we cannot have $\mathrm{rank}(G) < k$, as then, by Lemma 2.6.29, B would be linearly dependent. $\qquad\square$

Now we have a technical lemma we will use later.

Lemma 2.6.31. *Let G be an abelian group and let $S = \{x_1, \ldots, x_k\}$ be a finite set of elements of G. Let H be the subgroup of G spanned by S. Let A be any k-by-k matrix that is invertible over $\mathbb{Z}$, and let*

$$
\begin{bmatrix} y_1 \\ \vdots \\ y_k \end{bmatrix} = A \begin{bmatrix} x_1 \\ \vdots \\ x_k \end{bmatrix}.
$$

Let H' be the subgroup of G generated by $S' = \{y_1, \ldots, y_k\}$. Then $H' = H$. Furthermore, S is linearly independent if and only if S' is linearly independent.

Proof.   Since H' consists of those elements of G that can be expressed as linear combinations of the elements of S', and every element of S' can be expressed as a linear combination of the elements of S, we see that every element of H' can be expressed as a linear combination of the elements of S, and so $H' \subseteq H$.
But also

$$
\begin{bmatrix} x_1 \\ \vdots \\ x_k \end{bmatrix} = A^{-1} \begin{bmatrix} y_1 \\ \vdots \\ y_k \end{bmatrix}
$$

so by the same logic $H \subseteq H'$, and hence $H' = H$.

Now suppose S is linearly independent. Consider the equation $\Sigma_{i=1}^{k} b_i y_i = 0$. We can write this equation as

$$[b_1 \cdots b_k] \begin{bmatrix} y_1 \\ \vdots \\ y_k \end{bmatrix} = 0$$

and then

$$[b_1 \cdots b_k] \left[A \begin{bmatrix} x_1 \\ \vdots \\ x_k \end{bmatrix} \right] = 0$$

$$[[b_1 \cdots b_k] A] \begin{bmatrix} x_1 \\ \vdots \\ x_k \end{bmatrix} = 0$$

Since S is linearly independent, we must have $[b_1 \ldots b_k] A = 0$. But A is invertible, so we must have $[b_1 \ldots b_k] = [0 \ldots 0]$, and thus we see that S' is linearly independent. By the same logic, if S' is linearly independent then so is S. □

Lemma 2.6.32. *Let G be an abelian group of finite rank, and let $A = \{x_i\}$ be any finite generating set of G. Then a subgroup H of G has finite index in G if and only if there is an integer $N \neq 0$ such that $N x_i \in H$ for every $x_i \in A$.*

Proof. First of all, note that the condition in the lemma is equivalent to the condition that for each $x_i \in A$ there is an integer $N_i \neq 0$ such that $N_i x_i \in H$. To see this, let $A = \{x_1, \ldots, x_k\}$. If for some x_i there is no nonzero integer N_i with $N_i x_i \in H$, then there is certainly no nonzero integer N with $N x_i \in H$ for every i. On the other hand, if there is such an integer N_i for every x_i, we could take N to be any nonzero integer that is divisible by $N_1, \ldots, N_k$, e.g., $N = \mathrm{lcm}(N_1, \ldots, N_k)$. (In fact, it is easy to check that any integer N that is not divisible by $\mathrm{lcm}(N_1, \ldots, N_k)$ will not have this property.)

Now suppose H has finite index d in G. Then for each x_i, we may consider the elements $\{0, x_i, 2x_i, \ldots, dx_i\}$. These cannot be in distinct cosets of H in G, as there are $d + 1$ such elements. Thus we must have two distinct elements of this set, say mx_i and nx_i, in the same coset, in which case, if $N_i = (n - m)$, $N_i x_i = (n - m)x_i \in H$.

On the other hand, suppose there is such an integer N. Then any expression $\Sigma_{i=1}^{k} n_i x_i$ with each n_i divisible by N gives an element of H. We thus see that we have representatives of all of the cosets of H in G given by

$$\left\{ \sum_{i=1}^{k} n_i x_i \, \middle| \, 0 \le n_i < N \text{ for each } i \right\}.$$

These representations may contain duplications, but whether or not they do the number of element in this set is N^k, so H has at most N^k cosets in G, and hence is a subgroup of finite index (in fact, index at most N^k) in G.   □

Theorem 2.6.33. *Let G be a free abelian group of rank k.*

(a) *Let $C = \{y_1, \ldots, y_k\}$ be a set of k element of G that spans G. Then C is linearly independent, and hence C is a basis of G.*

(b) *Let $C = \{y_1, \ldots, y_k\}$ be a set of k element of G that is linearly independent. Let H be the subgroup of G spanned by C. Then H is a subgroup of G of finite index, and hence C is a basis of a subgroup of G of finite index.*

Proof.   Let $B = \{x_1, \ldots, x_k\}$ be a basis of G.
(a) Since B spans G, we have

$$y_i = \sum_{j=1}^{k} p_{ij} x_j \quad \text{for some integers } \{p_{ij}\}, \text{ for each } j = 1, \ldots, k.$$

Similarly, since C spans G, we have

$$x_i = \sum_{j=1}^{k} q_{ij} y_j \quad \text{for some integers } \{q_{ij}\}, \text{ for each } j = 1, \ldots, k.$$

Let $P = (p_{ij})$, the k-by-k the matrix whose entry in position (i_{ij}) is p_{ij}, and similarly let $Q = (q_{ij})$. Then

$$\begin{bmatrix} y_1 \\ \vdots \\ y_k \end{bmatrix} = P \begin{bmatrix} x_1 \\ \vdots \\ x_k \end{bmatrix} \quad \text{and} \quad \begin{bmatrix} x_1 \\ \vdots \\ x_k \end{bmatrix} = Q \begin{bmatrix} y_1 \\ \vdots \\ y_k \end{bmatrix},$$

so

$$\begin{bmatrix} x_1 \\ \vdots \\ x_k \end{bmatrix} = QP \begin{bmatrix} x_1 \\ \vdots \\ x_k \end{bmatrix}.$$

But B is a basis of G, which means that every element of G can be expressed in a unique way as a linear combination of the elements of B, so we must have $QP = I$, the identity matrix. In particular, the matrices P and Q are invertible.

Since B is a basis, it is linearly independent, and then we have that C is linearly independent by Lemma 2.6.31.

(b) Again let us write

$$y_i = \sum_{j=1}^{k} p_{ij} r_j, \quad \text{for each } i = 1, \ldots, k,$$

and form the matrix $P = (p_{ij})$ as in part (a).

Now B is a basis of G, so every element of G can be written uniquely as

$$\sum_{j=1}^{k} r_i x_i \quad \text{for integer } r_1, \ldots, r_k.$$

We define a new group $G_{\mathbb{Q}}$ by

$$G_{\mathbb{Q}} = \left\{ \sum_{i=1}^{k} f_i x_i \mid f_i \in \mathbb{Q} \right\}.$$

Since these expressions for the elements of G were unique, this gives us unambiguous expressions for the elements of $G_{\mathbb{Q}}$. Then $G_{\mathbb{Q}}$ is an abelian group containing G. Not only that, but we can multiply any element of $G_{\mathbb{Q}}$ by any rational number f. Thus, we see that $G_{\mathbb{Q}}$ has the structure of a vector space over $\mathbb{Q}$.

We claim that B is a vector space basis for $G_{\mathbb{Q}}$. It certainly spans $G_{\mathbb{Q}}$, so we must show it is linearly independent over $\mathbb{Q}$. So suppose we have an expression

$$\sum_{i=1}^{k} f_i x_i = 0 \quad \text{with } f_i \in \mathbb{Q}.$$

Write f_i as a quotient of two integers, $f_i = r_i/s_i$, for each $i = 1, \dots, k$. Then

$$\sum_{i=1}^{k} (r_i/s_i) \, x_i = 0.$$

Let s be the product $s = s_1 \dots, s_k$. Then, multiplying this equation by s, we obtain the equation

$$\sum_{i=1}^{k} (r_i \, (s/s_i)) \, x_i = 0.$$

Now s_i divides s, for each i, so every coefficient $r_i(s/s_i)$ is an integer. By hypothesis B is linearly independent over $\mathbb{Z}$, so each of these coefficients must be 0, so each r_i must be zero, and so each f_i must be zero, as required.

Thus we see that $G_{\mathbb{Q}}$ is a $\mathbb{Q}$–vector space which has $B = \{x_1, \dots, x_k\}$ as a vector space basis, so in particular it has dimension k.

Now consider the set C. We are assuming that C is linearly independent over $\mathbb{Z}$, so, by the same argument as we gave above for B, we can conclude that C is linearly independent over $\mathbb{Q}$.

Thus, C is a linearly independent set of k elements in the vector space $G_{\mathbb{Q}}$ of dimension k, so C is a vector space basis of $G_{\mathbb{Q}}$. Thus C spans $G_{\mathbb{Q}}$ and so we have

$$x_i = \sum_{j=1}^{k} q_{ij} y_j \text{ for some rational numbers} \{q_{ij}\}, \quad \text{for each } i = 1, \dots, k.$$

Again we let $Q = (q_{ij})$. We note that P is a k-by-k matrix with entries in $\mathbb{Z}$, and Q is a k-by-k matrix with entries in $\mathbb{Q}$. We argue as before:

$$
\begin{bmatrix} y_1 \\ \vdots \\ y_k \end{bmatrix} = P \begin{bmatrix} x_1 \\ \vdots \\ x_k \end{bmatrix}, \quad \begin{bmatrix} x_1 \\ \vdots \\ x_k \end{bmatrix} = Q \begin{bmatrix} y_1 \\ \vdots \\ y_k \end{bmatrix}
$$

so

$$
\begin{bmatrix} x_1 \\ \vdots \\ x_k \end{bmatrix} = QP \begin{bmatrix} x_1 \\ \vdots \\ x_k \end{bmatrix}
$$

and once again B is a basis, so $QP = I$, and P and Q are invertible matrices over $\mathbb{Q}$. In particular, $Q = P^{-1}$. Now Q is a matrix with rational coefficients, but we can write any such matrix as $Q = (1/s_0) Q'$ where Q' has entries in $\mathbb{Z}$ and s_0 is some nonzero integer: Just take s_0 to be the least common multiple of the denominators of all of the entries of Q.

Then

$$
\begin{bmatrix} x_1 \\ \vdots \\ x_k \end{bmatrix} = Q \begin{bmatrix} y_1 \\ \vdots \\ y_k \end{bmatrix} = (1/s_0) Q' \begin{bmatrix} y_1 \\ \vdots \\ y_k \end{bmatrix}
$$

so

$$
s_0 \begin{bmatrix} x_1 \\ \vdots \\ x_k \end{bmatrix} = Q' \begin{bmatrix} y_1 \\ \vdots \\ y_k \end{bmatrix}, \quad \begin{bmatrix} s_0 x_1 \\ \vdots \\ s_0 x_k \end{bmatrix} = Q' \begin{bmatrix} y_1 \\ \vdots \\ y_k \end{bmatrix}
$$

In other words, if H is the subgroup of G spanned by $C = \{y_1, \ldots, y_k\}$, $s_0 x_i \in H$ for each $i = 1, \ldots, k$. But then by Lemma 2.6.32, H is a subgroup of finite index in G. $\square$

Theorem 2.6.34. *Let G be a free abelian group of rank k. Then any subgroup H of G is a free abelian group.*

Proof. This is certainly true if $H = \{0\}$, so assume $H \neq \{0\}$. We prove this by induction on k.

If $k = 1$, G is isomorphic to $\mathbb{Z}$, i.e., G is an infinite cyclic group, and we have already shown, in Lemma 2.3.8 that a nontrivial subgroup H of an infinite cyclic group is itself infinite cyclic. (If G is generated by an element x_1, then H is generated by nx_1 for some nonzero integer n.)

Now suppose the theorem is true for any free abelian group of rank $k - 1$, and let G have rank k. Let $B = \{x_1, \dots, x_k\}$ be a basis of G.

Let J be the subgroup of G generated by $\{x_1, \dots, x_{k-1}\}$ and let K be the subgroup of G generated by $\{x_k\}$. Note that J is free abelian of rank $k - 1$, K is free abelian of rank 1, and $G = J \oplus K$.

We have a group homomorphism $\pi \colon G \to G$ given by

$$\pi(n_1 x_1 + \cdots + n_{k-1} x_{k-1} + n_k x_k) = n_1 x_1 + \cdots + n_{k-1} x_{k-1}$$

and we observe that $\mathrm{Ker}(\pi) = K$ and $\mathrm{Im}(\pi) = J$. Let φ be the restriction of π to the subgroup H, so that φ is given by the same formula, but is only defined for elements of H. Let $J_1 = \mathrm{Im}(\varphi)$ and $K_1 = \mathrm{Ker}(\varphi)$. Then $J_1 \subseteq J$ and $K_1 \subseteq K$ (more precisely, $K_1 = H \cap K$).

Since $J_1 \subseteq J$, and J is free abelian of rank $k - 1$, we have, by the inductive hypothesis, that J_1 is a free abelian group (of some rank $j \leq k - 1$ by Lemma 2.6.29).

If $K_1 = \mathrm{Ker}(\varphi) = \{0\}$, then $\varphi \colon H \to J_1$ is an isomorphism and hence H is free abelian.

Suppose $K_1 \neq \{0\}$. Let $\{z_1, \dots, z_j\}$ be a basis of J_1 and let $\{y_1, \dots, y_j\}$ be elements of H with $\varphi(y_i) = z_i$, $i = 1, \dots, j$. Then, by Lemma 2.6.11, there is a homomorphism $\lambda \colon J_1 \to H$ with $\lambda(z_i) = y_i$, $i = 1, \dots, j$. In the language of Definition 2.4.34, λ is a splitting of φ, and so by Lemma 2.4.35, H is isomorphic to $J_1 \oplus K_1$. (Note our argument here is very similar to the proof of Theorem 2.6.13.) But $K_1 \subseteq K$ is free abelian by the $k = 1$ case. Thus H is isomorphic to $\mathbb{Z}^j \oplus \mathbb{Z}$, i.e., H is free abelian of rank $j + 1$. $\square$

Now we come to our second main result.

Theorem 2.6.35. *Let G be a finitely generated torsion-free abelian group. Then G is free.*

Proof.   First we shall show that G contains a subgroup of finite index that is a free abelian group, and then we will use a "trick" to show that G itself is a free abelian group.

Let $A = \{x_1, \ldots, x_k\}$ be any finite set that generates G. Choose a maximal linearly independent subset B of A. Renumbering the elements, if necessary, we may assume that $B = \{x_1, \ldots, x_j\}$. Let H be the subgroup generated by B. Then H is free abelian (as B is a basis of H). We claim that H is a subgroup of G of finite index.

If $j = k$, then $H = G$. Suppose not. Then for each $i = j+1, \ldots, k$, the set $\{x_1, \ldots, x_j, x_i\}$ is not linearly independent, so we have a relation

$$\left(\sum_{n=1}^{j} b_n^i x_n \right) + a_i x_i = 0 \text{ for some integers } b_1^i, \ldots, b_j^i$$
$$\text{and some integer } a_i \text{ with } a_i \neq 0.$$

But that gives us that $a_i x_i \in H$ for each $i = j+1, \ldots, k$. Of course, $1 x_i = x_i \in H$ for each $i = 1, \ldots, j$. But then, from Lemma 2.6.32 (see in particular the first paragraph of the proof), H is a subgroup of G of finite index.

Now for the trick: Again by Lemma 2.6.32 there is some nonzero integer N such that $N x_i \in H$ for every $x_i \in A$, and since every element x of G can be written as $x = n_1 x_1 + \cdots + n_x x_k$, we see that $N x \in H$ for every $x \in G$. In other words, if we define $\varphi: G \to G$ by $\varphi(x) = Nx$, then $\mathrm{Im}(\varphi) \subseteq H$. Now φ is 1-1 precisely because G is torsion-free: $Nx = 0$ implies $x = 0$. Thus φ is an isomorphism onto its image $G_1 = \varphi(G)$. But G_1 is a subgroup of the free abelian group H (and H has finite rank; indeed we must have that the rank of H is at most k, again by Lemma 2.6.29), so, by Theorem 2.6.34, G_1 is a free abelian group; since G is isomorphic to G_1, G is free abelian as well. $\quad\square$

We can simply assemble some of the results we have proved to state a slightly sharper result.

Corollary 2.6.36. *Let G be a free abelian group of rank k. Then any subgroup H of G is free abelian of rank $j \leq k$. Furthermore, $j = k$ if and only if H is a subgroup of G of finite index.*

Proof. By Theorem 2.6.35 we know that H is free. By Lemma 2.6.29 we know that $j \leq k$.

If $j = k$, we know from Theorem 2.6.33(b) that H is a subgroup of finite index.

Suppose that H is a subgroup of G of finite index. We apply the "trick" in the proof of Theorem 2.6.35: G is isomorphic to a subgroup G_1 of H. But then G_1 also has rank k, and by Lemma 2.6.29 again we know that $k \le j$, so $j = k$. $\qquad\square$

Remark 2.6.37. Note the hypothesis in Theorem 2.6.35 that G be finitely generated. This theorem is false in general without this hypothesis. For example, $G = \mathbb{Q}$ is a torsion-free abelian group that is not free. $\qquad\Diamond$

We have previously handled finite abelian groups, and we have just handled finitely generated torsion-free abelian groups. It is easy to put these together to obtain a final structure theorem.

Definition 2.6.38. Let G be an abelian group. Its *torsion subgroup* G_{tor} is the subgroup of G defined by

$$G_{tor} = \{x \in G \mid x \text{ is an element of finite order}\}. \qquad\Diamond$$

It is easy to check that G_{tor} is a subgroup of G. Note, however, that if G has both elements of infinite order and nontrivial elements of finite order, $\{x \in G \mid x \text{ is an element of infinite order}\} \cup \{\text{the identity}\}$ is *never* a subgroup of G. For example, if $G = \mathbb{Z} \oplus \mathbb{Z}_2$, $(1, 0)$ is an element of infinite order, $(-1, 1)$ is an element of infinite order, but their sum $(1, 0) + (-1, 1) = (0, 1)$ is an element of finite order (order 2).

Theorem 2.6.39 (Structure theorem for finitely generated abelian groups). *Let G be a finitely generated abelian group. Then G is the direct sum $T \oplus F$, where T is a finite abelian group and F is a finitely generated free abelian group. More precisely, T is the subgroup G_{tor} of G and F is isomorphic to the quotient group G/G_{tor}.*

Proof. Let π be the quotient map $\pi\colon G \to G/G_{tor} = Q$. Observe that $T = \operatorname{Ker}(\pi) = G_{tor}$. Then Q is torsion-free, and is finitely generated (as any set that generates G generates Q). Thus, by Theorem 2.6.35, Q is free. But then, just as in the proof of Theorem 2.6.34, π has a splitting $\lambda\colon Q \to G$. Let $F = \lambda(Q)$. Then $G = T \oplus F$. $\qquad\square$

Corollary 2.6.40. *Let G and G' be finitely generated abelian groups (possibly $G' = G$). If G is isomorphic to $T \oplus F$, T finite and F free abelian, and G' is isomorphic to $T' \oplus F'$, T' finite and F' free abelian,*

*then G and G′ are isomorphic if and only if T and T′ are isomorphic
and F and F′ are isomorphic.*

Proof. Any isomorphism $\varphi\colon G \to G'$ must restrict to an isomorphism between G_{tor} and G'_{tor}, and hence give an isomorphism
between G/G_{tor} and G'/G'_{tor}.

On the other hand, if T and T' are isomorphic, and F and F' are
isomorphic, then G and G' are certainly isomorphic. $\qquad\square$

Corollary 2.6.41. *A finitely generated abelian group G is determined up to isomorphism by the elementary divisors, or invariant
factors, of G_{tor} and the rank of G/G_{tor}.*

Proof. This data determines G_{tor} and G/G_{tor} up to
isomorphism. $\qquad\square$

2.7 Applications to number theory

In this section we apply our knowledge of finite groups to prove some
results in number theory. These results are interesting and important
in themselves, and we will also be using them later (in Section 3.8).

Throughout this section p will denote a prime.

Recall that $\mathbb{Z}_p$ is the quotient group $\mathbb{Z}_p = \mathbb{Z}/p\mathbb{Z}$ and, following
our previous notation, if $x, y \in \mathbb{Z}$ then $[x]_p = [y]_p$ if and only if $x \equiv
y \pmod{p\mathbb{Z}}$. We will follow universal practice and write this relation
as $x \equiv y \pmod{p}$. We recall that $\mathbb{Z}_p = \{[0]_p, [1]_p, \ldots, [p-1]_p\}$ and
that $p - 1 \equiv -1 \pmod{p}$.

We let $G = \mathbb{Z}_p^* = \{[1]_p, \ldots, [p-1]_p\}$ with group operation
$[m]_p[n]_p = [mn]_p$. Note we are writing G multiplicatively. Since we
will exclusively be considering elements and subgroups of G in this
section, we will write all our groups multiplicatively.

Also, we will abbreviate $[i]_p$ to $[i]$ throughout this section.

Theorem 2.7.1 (Fermat's little theorem). *Let p be a prime,
and let a be an integer relatively prime to p. Then $a^{p-1} \equiv 1 \pmod{p}$.*

Proof. Consider $[a] \in G$. Since G is a group of order $p-1$, the order
of $[a]$ divides $p-1$. Thus $[a]^{p-1} = [1]$ in G. But $[a]^{p-1} = [a^{p-1}]$, so
$[a^{p-1}] = [1]$ in G, i.e., $a^{p-1} \equiv 1 \pmod{p}$. $\qquad\square$

The following easy number-theoretic lemma turns out to play a key role.

Lemma 2.7.2. *Let x and y be integers. Then $x^2 \equiv y^2 \ (mod \ p)$ if and only if $x \equiv \pm y \ (mod \ p)$. In particular, $x^2 \equiv 1 \ (mod \ p)$ if and only if $x \equiv \pm 1 \ (mod \ p)$.*

Proof. Certainly if $x \equiv y \ (mod \ p)$ then $x^2 \equiv y^2 \ (mod \ p)$, and if $x \equiv -y \ (mod \ p)$ then $x^2 \equiv (-y)^2 = y^2 \ (mod \ p)$.

Conversely, suppose that $x^2 \equiv y^2 \ (mod \ p)$. Then p divides $x^2 - y^2 = (x-y)(x+y)$. Since p is a prime, p must divide one of the factors. If p divides $x - y$, then $x \equiv y \ (mod \ p)$, and if p divides $x + y$, then $x \equiv -y \ (mod \ p)$.

The second conclusion of the lemma is the special case $y = 1$. $\square$

Theorem 2.7.3 (Wilson's theorem). *Let p be a prime. Then*

$$(p-1)! = \prod_{n=1}^{p-1} n \equiv -1 \ (mod \, p).$$

Proof. If $p = 2$, then $1! = 1 \equiv -1 \ (mod \ 2)$.

Suppose that $p > 2$. Then

$$[(p-1)!] = \prod_{n=1}^{p-1}[n] = [1]\left(\prod_{n=2}^{p-2}[n]\right)[p-1] \in G.$$

Let us examine the product inside the parentheses. We make two observations. Let $[n] = [j]$, $2 \le j \le p-2$. Since G is a group, $[j]$ has an inverse $[j]^{-1} = [k]$ for some k. We observe:

(1) Since $[1]^{-1} = [1]$ and $[p-1]^{-1} = [p-1]$ (since $[1]^2 = [1]$ and $[p-1]^2 = [1]$), $[k] \neq [1]$ and $[k] \neq [p-1]$.
(2) $[k] \neq [j]$. For if $[k] = [j]$, then we would have $[1] = [j][k] = [j]^2$, i.e., $j^2 \equiv 1 \ (mod \ p)$, and that cannot happen by Lemma 2.7.2.

Thus we may group the $p-3$ terms in this product into $(p-3)/2$ pairs $\{[j], [j]^{-1}\}$. Now the product of terms in each pair is $[1]$, so this product is $[1]^{(p-3)/2} = [1]$, and then the entire right-hand side is $[1][1][p-1]$. Thus $[(p-1)!] = [p-1]$ in G, i.e.,

$$(p-1)! \equiv p-1 \equiv -1 \ (mod \, p),$$

as claimed. $\square$

Our goal in the remainder of this section is to investigate quadratic residues and nonresidues (mod p). These are defined as follows.

Definition 2.7.4. Let a be an integer relatively prime to p. Then a is a *quadratic residue* (mod p) if there is an integer b with $b^2 \equiv a$ (mod p). Otherwise, a is a *quadratic nonresidue* (mod p).

Equivalently, let a be an integer relatively prime to p. Then a is a *quadratic residue* (mod p) if $[a] = [b]^2$ for some $[b] \in G$. Otherwise, a is a *quadratic nonresidue* (mod p).

We call $[a] \in G$ a quadratic residue/nonresidue according as a is a quadratic residue/nonresidue (mod p). ◊

At this point we are faced with a choice. Our goal is to prove Euler's theorem (Theorem 2.7.12). There are two ways to reach this goal.

The first way is purely by group theory. We will show, just using group theory, that a certain subgroup of G is cyclic, and that is enough for us to be able to derive Euler's theorem.

But in fact, the group G is cyclic. We can't prove this just using group theory, but we will be able to prove it using some (easier) group theory and some ring theory, and that will easily give us Euler's theorem (see Section 3.8).

So we have a choice. You can read the group theory proof now, or the group/ring theory proof later. But as usual in mathematics, "or" is inclusive — if you wish, you can do both!

Corollary 2.7.5. *Let p be a prime, and let a be an integer relatively prime to p. Then $a^{(p-1)/2} \equiv \pm 1 \pmod{p}$.*

Proof. Let $b = a^{(p-1)/2}$. Then $b^2 = (a^{(p-1)/2})^2 = a^{p-1} \equiv 1 \pmod{p}$ by Fermat's little theorem. But then $b = \pm 1 \pmod{p}$ by Lemma 2.7.2. □

Lemma 2.7.6. *For any odd prime p, there are $(p-1)/2$ quadratic residues and $(p-1)/2$ quadratic nonresidues in G.*

Proof. Let $\varphi\colon G \to G$ be the homomorphism defined by $\varphi([k]) = [k]^2 = [k^2]$. By definition, $R_+ = \{\text{quadratic residues in } G\} = \operatorname{Im}(\varphi)$.

By the first isomorphism theorem (Theorem 2.5.3), R_+ is isomorphic to $G/\operatorname{Ker}(\varphi)$. Now $\operatorname{Ker}(\varphi) = \{[k] \in G \mid [k]^2 = [1]\}$. By Lemma 2.7.2, $\operatorname{Ker}(\varphi) = \{[1], [p-1]\}$, a subgroup of G of order 2. Thus

$|R_+| = [G \colon \mathrm{Ker}(\varphi)] = |G| / |\mathrm{Ker}(\varphi)| = (p-1)/2$. If $R_- = \{$quadratic nonresidues in $G\}$, then $R_- = G - R_+$, so $\#(R_-) = \#(G) - \#(R_+) = (p-1) - (p-1)/2 = (p-1)/2$. $\qquad\square$

Remark 2.7.7. Note that R_+ is a subgroup of G and that R_- is a coset of R_+. $\qquad\Diamond$

Now, in the notation of the proof of Lemma 2.7.6, let us consider the quotient group G/R_+. This is a group of order $|G|/|R_+| = (p-1)/((p-1)/2) = 2$, so must be isomorphic to $\mathbb{Z}_2$. Following common mathematical practice here, we write this group multiplicatively, as $\{\pm 1\}$.

Definition 2.7.8. With the above identification, we let $\chi_p \colon G \to \{\pm 1\}$ be the quotient map, so that $\chi_p([a]) = 1$ if $[a]$ is a quadratic residue in G and $\chi_p([a]) = -1$ if not. The homomorphism χ_p is called the *quadratic residue character*.

For an integer a relatively prime to p, we define the *Legendre symbol* $\left(\frac{a}{p}\right)$ to be $\left(\frac{a}{p}\right) = \chi_p([a])$, so that $\left(\frac{a}{p}\right) = 1$ if a is a quadratic residue (mod p) and $\left(\frac{a}{p}\right) = -1$ if a is a quadratic nonresidue (mod p). $\qquad\Diamond$

Remark 2.7.9. We are doing violence to mathematical history here. In the development of mathematics (at least on planet earth) number theory preceded group theory, so the Legendre symbol was defined long before the quadratic residue character was.

But in a way this is part of our point here. Looking back on number theory from the point of view of group theory, we can see how group theory both enriches and clarifies number theory. $\qquad\Diamond$

Lemma 2.7.10. *Let a and b integers relatively prime to p. Then*

$$\left(\frac{a}{p}\right)\left(\frac{b}{p}\right) = \left(\frac{ab}{p}\right).$$

In other words:

(i) If a and b are both quadratic residues (mod p), then ab is a quadratic residue (mod p).

(ii) If one of a and b is a quadratic residue (mod p) and the other one is a quadratic nonresidue (mod p), then ab is a quadratic nonresidue (mod p).

(iii) If a and b are both quadratic nonresidues (mod p), then ab is a quadratic residue (mod p).

Proof. This is just the statement that $\chi_p\colon G{\to}\{\pm 1\}$ is a homomorphism. □

In the statement of the next lemma, we will write $G = H_{2^k} \times H_n$ rather that $H_{2^k} \oplus H_n$ as we are writing G multiplicatively, and similarly in its proof. Thus, in particular, we will write the elements of a cyclic group as powers of a generator, and denote the identity element of any group by e.

Lemma 2.7.11. *Let $p - 1 = 2^k n$ where n is odd, and write $G = H_{2^k} \times H_n$ as in Theorem 2.6.14. Then the subgroup H_{2^k} of G is cyclic.*

Proof. Suppose not. Then by either of our two structure theorems (Theorem 2.6.13 or Theorem 2.6.20), H_{2^k} is isomorphic to a direct product of at least two cyclic groups, each of whose orders is a power of 2. So in particular H_{2^k} has a subgroup K isomorphic to $\mathbb{Z}_{2^e} \times \mathbb{Z}_{2^f}$ with $e, f \geq 1$. Let $\varphi\colon K \to \mathbb{Z}_{2^e} \times \mathbb{Z}_{2^f}$ be an isomorphism.

Let a be a generator of $\mathbb{Z}_{2^e}$ and let b be a generator of $\mathbb{Z}_{2^f}$. Of course, (e, e) is the identity in $\mathbb{Z}_{2^e} \times \mathbb{Z}_{2^f}$, so has order 1. But notice that $(e, e)^2 = (e, e)$, $(a^{2^{e-1}}, e)^2 = (e, e)$, $(e, b^{2^{f-1}})^2 = (e, e)$, and $(a^{2^{e-1}}, b^{2^{f-1}})^2 = (e, e)$. Thus we see that $\mathbb{Z}_{2^e} \times \mathbb{Z}_{2^f}$ has at least (and in fact exactly) 4 elements of order dividing 2.

But then H_{2^k}, and hence G, has at least 4 elements of order dividing 2, namely $\varphi^{-1}((e,e)), \varphi^{-1}((a^{2^{e-1}}, e)), \varphi^{-1}((e, b^{2^{f-1}}))$, and $\varphi^{-1}((a^{2^{e-1}}, b^{2^{f-1}}))$. But this is impossible by Lemma 2.7.2. □

Now we arrive at our goal.

Theorem 2.7.12 (Euler). *Let p be an odd prime. For any integer a relatively prime to p, $a^{(p-1)/2} \equiv \left(\dfrac{a}{p}\right) \pmod{p}$.*

Proof. Write $G = H_{2^k} \times H_n$ as in Lemma 2.7.11. Choose a generator x of H_{2^k}. Then $G = \{(x^i, y) \mid 0 \leq i \leq 2^k - 1, \ y \in H_n\}$. Let $\varphi\colon G{\to}G$ by $\varphi([m]) = [m^2]$. Note that $\varphi(H_{2^k}) \subseteq H_{2^k}$ and $\varphi(H_n) \subseteq H_n$. Also note that $\varphi|H_n\colon H_n \to H_n$ is an isomorphism, since n is odd. (The argument for this is very much like the argument in the proof of Theorem 2.6.17.)

As we have observed, $\text{Im}(\varphi) = R_+ = \{\text{quadratic residues in } G\}$. We claim $\text{Im}(\varphi) = S = \{(x^i, y) \mid 0 \leq i \leq 2^k - 1 \text{ with } i \text{ even}, y \in H_n\}$. Consider such an element (x^i, y) with i even. Since $\varphi|H_n$ is an isomorphism, $y = \varphi(z) = z^2$ for some $z \in H_n$. But then

$$\varphi\left(x^{i/2}, z\right) = \left(\left(x^{i/2}\right)^2, z^2\right) = (x^i, y).$$

Thus $S \subseteq \text{Im}(\varphi)$. On the other hand, consider any element (x^j, y) of G. Then $\varphi(x^j, y) = (x^{2j}, y^2)$. To be precise, this exponent is only defined $(\text{mod } 2^k)$. But since 2^k is even, any i with $i \equiv 2j \ (\text{mod } 2^k)$ must also be even. Thus $\text{Im}(\varphi) \subseteq S$ and these two sets are equal, i.e., $S = R_+$.

Now let $g = [a] \in G$. We have two possibilities:

(1) $g \in R_+$. Then on the one hand, $\left(\frac{a}{p}\right) = 1$ by the definition of the Legendre symbol. On the other hand, $g = \varphi(f)$ for some $f \in G$ and then $g^{(p-1)/2} = \left(f^2\right)^{(p-1)/2} = f^{p-1} = e \in G$, or in other words

$$[a]^{(p-1)/2} = \left[a^{(p-1)/2}\right] = [1] \in G, \text{ i.e., } a^{(p-1)/2} \equiv 1 \ (\text{mod } p),$$

and the theorem is true in this case.

(2) $g \in R_-$. Then on the one hand, $\left(\frac{a}{p}\right) = -1$ by the definition of the Legendre symbol. On the other hand, $g = (x^i, y)$ with i odd. (Again, this exponent is only defined $(\text{mod } 2^k)$, but since 2^k is even, any i' with $i' \equiv i \ (\text{mod } 2^k)$ must also be odd.) Now $(p-1)/2 = 2^{k-1}n$ so $g^{(p-1)/2} = \left(x^{2^{k-1}in}, y^{2^{k-1}n}\right)$. But $y^n = 1$ so $g^{(p-1)/2} = \left(x^{2^{k-1}in}, e\right)$. Now x has order 2^k and $2^{k-1}in$ is not a multiple of 2^k (as i and n are both odd) so $x^{2^{k-1}in} \neq e$ and hence $g^{(p-1)/2} \neq e$. But then by Corollary 2.7.5

$$[a]^{(p-1)/2} = \left[a^{(p-1)/2}\right] = [-1] \in G, \text{ i.e., } a^{(p-1)/2} \equiv -1 \ (\text{mod } p),$$

and the theorem is true in this case as well. $\square$

Corollary 2.7.13. *Let p be an odd prime.*

(a) *If $p \equiv 1 \ (\text{mod } 4)$, then -1 is a quadratic residue $(\text{mod } p)$.*
(b) *If $p \equiv 3 \ (\text{mod } 4)$, then -1 is a quadratic nonresidue $(\text{mod } p)$.*

Proof. By Euler's theorem, $\left(\frac{-1}{p}\right) \equiv (-1)^{(p-1)/2}$ (mod p). If $p \equiv 1$ (mod 4), $(p-1)/2$ is even, and if $p \equiv 3$ (mod 4), $(p-1)/2$ is odd. $\qquad\square$

This tells us the quadratic character of -1. Next we want to find the quadratic character of 2. We first prove a general result, and then apply it to easily determine that.

Lemma 2.7.14 (Gauss's lemma). *Let p be an odd prime and let a be an integer that is relatively prime to p. Let*

$$S = \{i \mid 1 \leq i \leq (p-1)/2 \text{ and } [ai] = [k] \text{ for some } k$$
$$\text{with } 1 \leq k \leq (p-1)/2\},$$

$$T = \{i \mid 1 \leq i \leq (p-1)/2 \text{ and } [ai] = [k] \text{ for some } k$$
$$\text{with } (p+1)/2 \leq k \leq p-1\}.$$

Let $s = \#(S)$ and $t = \#(T)$. Then a is a quadratic residue (mod p) if t is even, and a is a quadratic nonresidue (mod p) if t is odd.

Proof. Write $[m_i] = [ai]$ if $i \in S$ and $[n_i] = [ai]$ if $i \in T$.

Of course, $S \cup T = \{i \mid 1 \leq i \leq (p-1)/2\}$ and $S \cap T = \emptyset$. That is, $\{S, T\}$ is a partition of the set $\{1, \ldots, (p-1)/2\}$.

We first compute

$$\prod_{i \in S} [m_i] \prod_{i \in T} [n_i] = \prod_{i \in S} [ai] \prod_{i \in T} [ai]$$

$$= \prod_{i \in S \cup T} [ai] = \prod_{i=1}^{(p-1)/2} [ai]$$

$$= [a]^{(p-1)/2} \prod_{i=1}^{(p-1)/2} [i] = [a^{(p-1)/2}][((p-1)/2)!].$$

Next, we claim $\{[m_i]\} \cup \{[p - n_i]\} = \{1, \ldots, (p-1)/2\}$. Since the two sets on the left-hand side have $s + t = (p-1)/2$ elements, if suffices to show they are disjoint. Suppose not. Then $[m_i] = [p - n_j] = [-n_j]$ for some i and j, i.e., $[ai] = -[aj]$, or $ai + aj = a(i + j)$ is divisible by p. Since p is a prime, it must divide one of the factors. Now p does not divide a, by hypothesis, so p must divide $i + j$. But this is impossible as $1 \leq i, j \leq (p-1)/2$, so $i + j$ is between 1 and $p-1$.

Thus

$$[((p-1)/2)!] = \prod_{i=1}^{(p-1)/2} [i] = \prod_{i \in S}[m_i]\prod_{i \in T}[-n_i]$$

$$= \prod_{i \in S}[m_i]\prod_{i \in T}[-1][n_i]$$

$$= [(-1)^t]\prod_{i \in S}[m_i]\prod_{i \in T}[n_i]$$

so

$$\prod_{i \in S}[m_i]\prod_{i \in T}[n_i] = [(-1)^t][((p-1)/2)!].$$

Comparing these two computations, and using Euler's theorem, we see

$$\left[\left(\frac{a}{p}\right)\right] = \left[a^{(p-1)/2}\right] = [(-1)^t],$$

yielding the result. $\qquad\qquad\qquad\qquad\qquad\qquad\qquad\qquad\quad\square$

Corollary 2.7.15. *Let p be an odd prime.*

(a) *If $p \equiv 1$ or 7 (mod 8), then 2 is a quadratic residue (mod p).*
(b) *If $p \equiv 3$ or 5 (mod 8), then 2 is a quadratic nonresidue (mod p).*

Proof.   Let $p = 8k + r$ with $r = 1, 3, 5,$ or 7. Let $a = 2$ in Gauss's Lemma.
If $r = 1$, $T = \{2k+1, \ldots, 4k\}$ has $2k$ elements.
If $r = 3$, $T = \{2k+1, \ldots, 4k+1\}$ has $2k+1$ elements.
If $r = 5$, $T = \{2k+2, \ldots, 4k+2\}$ has $2k+1$ elements.
If $r = 7$, $T = \{2k+2, \ldots 4k+3\}$ has $2k+2$ elements. $\qquad\qquad\square$

Corollary 2.7.16. *Let p be an odd prime.*

(a) *If $p \equiv 1$ or 3 (mod 8), then -2 is a quadratic residue (mod p).*
(b) *If $p \equiv 5$ or 7 (mod 8), then -2 is a quadratic nonresidue (mod p).*

Proof. This follows immediately from Corollary 2.7.10, 2.7.13 and 2.7.15. □

What about $\left(\frac{a}{p}\right)$ for other values of a? This is the subject of the famous Law of Quadratic Reciprocity, which we prove in Section 3.8.

2.8 Actions of groups on sets

In this section, we discuss the action of groups (both abelian and nonabelian) on sets, an important and useful topic in its own right, and one which will play a crucial role in proving the results of the next section.

Let X be set. Recall from Example 2.1.13 that $\mathrm{Aut}(X) = \{\text{bijections } \sigma\colon X \to X\}$ is a group under the operation of composition of functions. We often denote this group by S_X, and call it the *symmetric group* on the elements of X. It is easy to check that if we have a bijection, i.e., a 1-1 correspondence, $f\colon X \to Y$, then S_X and S_Y are isomorphic; indeed $\varphi\colon S_X \to S_Y$ by $\varphi(\upsilon) = f\sigma f^{-1}$ is an isomorphism. Thus if X is a nonempty finite set, with n elements, then X is isomorphic to the symmetric group on $\{1, 2, \ldots, n\}$, we denote this group by S_n. We recall that S_n is a group of order $n!$.

Here is the basic definition.

Definition 2.8.1. An *action* of a group G on a set X is a homomorphism $\Phi\colon G \to \mathrm{Aut}(X)$. ◇

This is a cryptic definition, so let us expand on it to see what it means. Let $\varphi_g = \Phi(g)$. Then $\mathrm{Im}(\Phi) = \{\varphi_g \mid g \in G\}$ is a group of automorphisms of X (as it is a subgroup of $\mathrm{Aut}(X)$, which is the group consisting of all automorphisms of X), i.e., $\{\varphi_g \mid g \in G\}$ is a set of automorphisms of X satisfying:

(i) $\varphi_e = \mathrm{id}\colon X \to X$,
(ii) $\varphi_{gh} = \varphi_g\varphi_h$ for any $g, h \in G$,
(iii) $\varphi_{g^{-1}} = (\varphi_g)^{-1}$ for any $g \in G$.

We will see some examples — in fact, we have already seen some examples, without giving them this name — but we make another definition first.

Definition 2.8.2. An action of a group G on a set X as in Definition 2.8.1 is *effective* if $\mathrm{Ker}(\Phi) = \{e\}$. ◇

In other words, an action is effective if the only element g of G for which $\varphi_g = \mathrm{id}$ is $g = e$ (the identity element of G). As usual, Φ is effective $\Leftrightarrow \varphi_g = \varphi_h$ if and only if $g = h$.

Then, by Lemma 2.5.2, any action Φ of G on X factors through $\mathrm{Ker}(\Phi)$, and gives an effective action of the quotient group $G/\mathrm{Ker}(\Phi)$ on X.

Example 2.8.3.

(a) If G is a subgroup of $\mathrm{Aut}(X)$ (or $G = \mathrm{Aut}(X)$) then the inclusion of G into $\mathrm{Aut}(X)$ (or the identity map from G to $\mathrm{Aut}(X)$) gives an action of G on X. This may not sound like it gives us anything new, but we should point out that some of our examples of groups arose exactly from this construction. For example, we constructed D_{2n} as the automorphism group of a regular n-gon. From our point of view here, D_{2n} is a subgroup of $\mathrm{Aut}(\{\text{vertices of a regular } n\text{-gon}\})$. Similarly, although we did not construct $\mathbb{Z}_n^*$ in this way, we observed that $\mathbb{Z}_n^* = \mathrm{Aut}(\mathbb{Z}_n)$, i.e., $\mathbb{Z}_n^*$ is a subgroup of the automorphism group of the set $\{[0]_n, [1]_n, \ldots, [n-1]_n\}$ of the elements of $\mathbb{Z}_n$, consisting of the automorphisms of this set that preserve the group structure. (For $[i] \in \mathbb{Z}_n^*, \varphi_{[i]} : \mathbb{Z}_n \to \mathbb{Z}_n$ is the homomorphism $\varphi_{[i]}([k]) = [ik]$.)

(b) We constructed semidirect products in Lemma 2.4.28 precisely by using group actions, though we did not use that term there. In particular, we constructed the dihedral group D_{2n} in that way in Example 2.4.31. $\qquad\qquad\diamond$

Our next family of examples of group actions is a well-known theorem.

Theorem 2.8.4 (Cayley's theorem). *Let G be a group. Then G is isomorphic to a subgroup of S_G. In particular, if G is a finite group order n, then G is isomorphic to a subgroup of S_n.*

Proof.   For any element g_0 of G, let $\varphi_{g_0} : G \to G$ be the map $\varphi_{g_0}(g) = g_0 g$. We leave it to the reader to check that this is an effective group action of G on the set of elements of G. $\qquad\qquad\square$

We can generalize this example, but first we need a lemma, and a definition.

Lemma 2.8.5. *Let H be a subgroup of G. Then there is a largest normal subgroup of G contained in H, i.e., there is a subgroup N of H such that*:

(i) *N is a normal subgroup of G; and*
(ii) *If N′ is any subgroup of H that is a normal subgroup of G, then N′ is contained in N.*

Proof. We let N be the product of all of the normal subgroup of G contained in H. □

Definition 2.8.6. The subgroup N of Lemma 2.8.5 is the *normal core* of H, written $\mathrm{Core}_G(H)$ (or $\mathrm{Core}(H)$, when there is no possibility of confusion). ◊

Remark 2.8.7. We observe that if H is a normal subgroup of G, then $\mathrm{Core}(H) = H$. Otherwise $\mathrm{Core}(H)$ is a proper subgroup of H. ◊

Theorem 2.8.8. *Let H be a subgroup of G. Then there is an action Φ of G on G/H, the set of left cosets of H in G, given by $\varphi_{g_0}(L) = g_0 L$ for every $g_0 \in G$ and every left coset L of H in G. The kernel $\mathrm{Ker}(\Phi)$ of this action is the normal core $\mathrm{Core}(H)$ of H.*

Proof. We leave it to the reader to check that Φ is a group action. We determine $\mathrm{Ker}(\Phi)$. We begin with two observations:

(i) $\mathrm{Ker}(\Phi) \subseteq H$: Let $g_0 \in G$, $g_0 \notin H$. Then $\varphi_{g_0}(H) = g_0 H \neq H$, so φ_{g_0} is not the identity on G/H.
(ii) $\mathrm{Ker}(\Phi) \triangleleft G$: $\mathrm{Ker}(\Phi)$ is the kernel of a homomorphism.

With these observations in hand, we complete the proof by showing that if N' is any subgroup of H that is a normal subgroup of G, then $\varphi_n(L) = L$ for every $n \in N'$ and every $L \in G/H$.

By choosing a coset representative, we can write $L = gH$ for same $g \in G$. Then

$$\varphi_n(gH) = (ng)H = \left(g\left(g^{-1}ng\right)\right)H = g\left(\left(g^{-1}ng\right)H\right) = gH,$$

where the last equality is first of all because $g^{-1}ng \in N'$, as N' is a normal subgroup of G, and second of all because N' is a subgroup of H. □

Here is an application of these ideas.

Corollary 2.8.9. *Let H be a subgroup of the finite group G. Let $k = [G: H]$, and suppose that $|G|$ does not divide $k!$ Then $\text{Core}(H) \neq \{e\}$. In particular, in this case H contains a nontrivial normal subgroup of G.*

Proof.   Let $N = \text{Core}(H)$, so that $N = \text{Ker}(\Phi)$ where Φ is as in Lemma 2.8.9. We know that G/N is isomorphic to $\text{Im}(\Phi)$, a subgroup of $S_{G/H}$. But $S_{G/H}$ is isomorphic to S_k. Hence $|G/N| = |\text{Im}(\Phi)|$ divides $|S_k| = k!$. By hypothesis, $|G|$ does not divide $k!$, so we must have $|N| > 1$. $\qquad\square$

Corollary 2.8.10. *Let G be a finite group and let p be the smallest prime dividing $|G|$. Then any subgroup H of G of index p is normal.*

Proof.   Let $N = \text{Core}(H)$. We shall show that $N = H$.

By the proof of Corollary 2.8.9, $|G/N|$ divides $p! = 1 \cdot 2 \cdots (p-1)p$. But, since $|G/N|$ divides $|G|$, and p is the smallest prime dividing $|G|$, this is only possible if $|G/N| = 1$ or p, i.e., $[G:N] = 1$ or p. But $N \subseteq H$ and $[G:H] = p$, so we must have $[G:N] = p = [G:H]$, in which case $N = H$. $\qquad\square$

We now return to considering general group actions.

Definition 2.8.11. Let $\Phi: G \to \text{Aut}(X)$ be an action of the group G on the set X.

(a) For $x \in X$, the *orbit* $\text{Orbit}(x) = \{y \in X \mid y = \varphi_g(x) \text{ for some } x \in X\}$.

(b) For $x \in X$, the *stabilizer* $\text{Stab}(x) = \{g \in G \mid \varphi_g(x) = x\}$

(Note that $\text{Orbit}(x)$ is a subset of X while $\text{Stab}(x)$ is a subgroup of G). $\qquad\qquad\Diamond$

We observe:

Lemma 2.8.12. *The relation $\sim$ as on X defined by $x \sim y$ if $y \in \text{Orbit}(x)$ is an equivalence relation on X. Consequently, if $\{X_i\}_{i \in I}$ is the set of equivalence classes of elements of X under $\sim$, $\{X_i\}_{i \in I}$ is a partition of X.*

Proof.   We leave this as an exercise for the reader. $\qquad\square$

Definition 2.8.13. If there is only one orbit of X under the action of Φ of G (or, equivalently, if for any $x, y \in X$ there is a $g \in G$ with $\varphi_g(x) = y$), this action is called *transitive*. $\qquad\qquad\Diamond$

Lemma 2.8.14. *Let* $\Phi: G \to \mathrm{Aut}(X)$ *be an action of the group G on the set X.*

(a) *Let $x \in X$. Then there is a 1–1 correspondence*

$$\mathrm{Orbit}\,(x) \leftrightarrow G/\mathrm{Stab}\,(x)$$

given by

$$y \in \mathrm{Orbit}(x) \leftrightarrow g\,\mathrm{Stab}(x) \text{ where } g \in G \text{ with } y = \varphi_g(x).$$

(b) *Let $x \in X$. Let K be any subgroup of G that is conjugate to $\mathrm{Stab}(x)$. Then $K = \mathrm{Stab}(y)$ for some y in the orbit of x in this action of G on X.*

Proof.

(a) Note that the given correspondence is obtained by a choice of element $g \in G$, so our first job is to show that this correspondence is well defined, i.e., independent of the choice. So suppose we have elements g and h of G with $y = \varphi_g(x) = \varphi_h(x)$. Then $x = \varphi_g^{-1}(\varphi_h(x)) = \varphi_{g^{-1}h}(x)$ so $g^{-1}h \in \mathrm{Stab}(x)$ and hence $h \in g\,\mathrm{Stab}(x)$, and similarly $g \subset h\,\mathrm{Stab}(x)$, so $g\,\mathrm{Stab}(x) = h\,\mathrm{Stab}(x)$.
Now we must show this is a 1-1 correspondence. Note that any element g' of $g\,\mathrm{Stab}(x)$ is $g' = gg_0$ for some $g_0 \in \mathrm{Stab}(x)$ and then $\varphi_{g'}(x) = \varphi_{gg_0}(x) = \varphi_g(\varphi_{g_0}(x)) = \varphi_g(x)$. Thus if $y_1 \neq y_2$ are two elements of $\mathrm{Orbit}(x)$, we cannot have $y_1 \in g\,\mathrm{Stab}(x)$ and $y_2 \in g\,\mathrm{Stab}(x)$ (for any g) so this correspondence is 1-1. And also, if we choose y to be the element $y = \varphi_g(x)$, an element of X in the orbit of x, then y corresponds to $g\,\mathrm{Stab}(x)$, for any $g \in G$, so this correspondence is onto.

(b) Suppose that x and y are in the same orbit, so that $y = \varphi_{g_o}(x)$ for some $g_o \in G$. We claim $\mathrm{Stab}(y) = g_o\,\mathrm{Stab}(x)\,g_o^{-1}$. To see this, let g be any element of $\mathrm{Stab}(x)$. Then

$$\varphi_{g_0 g g_0^{-1}}(y) = \varphi_{g_0}\left(\varphi_g\left(\varphi_{g_0^{-1}}(y)\right)\right) = \varphi_{g_0}\left(\varphi_g\left(\varphi_{g_0^{-1}}(y)\right)\right)$$

$$= \varphi_{g_0}(\varphi_g(x)) = \varphi_{g_0}(x)$$

so that $g_0\,\mathrm{Stab}(x)\,g_0^{-1} \subseteq \mathrm{Stab}(y)$. By the same logic, $g_0^{-1}\,\mathrm{Stab}(y)\,g_0 \subseteq \mathrm{Stab}(x)$, and so $\mathrm{Stab}(y) \subseteq g_0\,\mathrm{Stab}(x)\,g_0^{-1}$, and hence these two subgroups of G are equal.

On the other hand, suppose that K and $\text{Stab}(x)$ are conjugate, so that $K = g_0 \, \text{Stab}(x) \, g_0^{-1}$ for some $g_0 \in G$. Let $y = \varphi_{g_0}(x)$, an element in the orbit of x. Then the above argument shows that $K = \text{Stab}(y)$.

□

Corollary 2.8.15.

(a) *Let $x \in X$. Then $\#(Orbit(x)) = [G\colon Stab(x)]$.*
(b) *Let $\{x_i\}_{i \in I}$ be a complete a set of representatives of the equivalence classes $\{X_i\}_{i \in I}$ as in Lemma 2.8.12. Then*

$$\#(X) = \sum_{i \in I} [G : \text{Stab}(x_i)].$$

Proof.   (a) is immediate from Lemma 2.8.14(a). As for (b), note that, since $\{X_i\}_{i \in I}$ is a partition of X, where $X_i = \text{Orbit}(x_i)$, $\#(X) = \Sigma_{i \in I} \#(X_i)$, so this follows directly from (a).

Now let us apply these general considerations to a particular situation. First, a definition.

□

Definition 2.8.16. The *center* $Z(G)$ of a group G is the set of elements of G that commute with every element of G,

$$Z(G) = \{z \in G \mid zg = gz \quad \text{for every } g \in G\}. \qquad \Diamond$$

Remark 2.8.17. $Z(G)$ is an abelian normal subgroup of G, as is any subgroup of $Z(G)$. $\Diamond$

We now consider another action of the group G on the set of elements of G, this time not by left multiplication (as in Cayley's theorem) but rather by *conjugation*.

That is, we let $\Phi\colon G \to \text{Aut}(\{\text{elements of } G\})$ by $\varphi_g(h) = ghg^{-1}$. We leave it to the reader to check that this is a group action. Then two elements of G are in the same orbit if and only if they are conjugate (Definition 2.3.31). We call the orbits of G under conjunction the *conjugacy classes* of G. For $x \in G$, $\text{Stab}(x) = \{g \in G \mid gxg^{-1} = x\} = \{g \in G \mid gx = xg\}$ consists of the elements of G that commute with x; we call this the *centralizer* of x and denote it by $C(x)$.

Corollary 2.8.18 (The class equation for G). *Let G be a finite group. Then*

$$|G| = |Z(G)| + \sum_{i \in I'} [G : C(x_i)]$$

where the sum is taken over a complete set of representatives of the conjugacy classes of elements of G not in the center of G.

Proof. Let $\{X_i\}_{i \in I}$ be the conjugacy classes of G. Then we know that these partition G, and so

$$|G| = \sum_{i \in I} \#(X_i) = \sum_{i \in I} [G : C(x_i)]$$

by Corollary 2.8.15.

Now if $x \in Z(G)$ then $C(x_i) = G$ (and conversely), so in this case $[G : C(x_i)] = 1$, and also in this case the orbit of x consists of x alone (and conversely). Thus we may rewrite this sum as

$$|G| = \sum_{x \in Z(G)} 1 + \sum_{x \in I'} [G : C(x_i)] = |Z(G)| + \sum_{x_i \in I'} [G : C(x_i)]$$

as claimed. □

Remark 2.8.18. We have been very careful here. Given an action $\Phi : G \to \text{Aut}(X)$, we have let $\varphi_g = \Phi(g)$ so for each $g \in G$, and each $x \in X$ we have the result $\varphi_g(x)$ of the action of the element g on x. When Φ is understood, $\varphi_g(x)$ is often (indeed usually) abbreviated to $g(x)$. ◇

2.9 Structural results on Nonabelian groups

As opposed to the case of finite abelian groups, where we fully described their structure in Section 2.6, there is no way to fully describe the structure of finite nonabelian groups. But there are important things we can say, and we say some of the most important ones in this section. (Actually, our results here are true whether or not the group is abelian, but in the abelian case they tell us nothing new.)

Theorem 2.9.1 (Cauchy's theorem). *Let G be a finite group of order n, and let p be a prime dividing n. Then G has an element of order p.*

Proof. (McKay) Let $G^p = G \times \cdots \times G$, where there are p factors, and let

$$X = \left\{ (g_1, g_2, \ldots, g_p) \in G^p \,\middle|\, \prod_{i=1}^{p} g_i = e \right\}.$$

Note that $\#(X) = |G|^{p-1}$ as we may choose $g_1, \ldots, g_{p-1}$ arbitrary and then $g_p = (g_1 \cdots g_{p-1})^{-1}$.

We now define an action of the group $\mathbb{Z}_p$ on X by "rotation": If $\mathbb{Z}_p = \{[0], [1], \ldots, [p-1]\}$ then

$$\varphi_{[1]}(g_1, g_2, \ldots, g_p) = (g_p, g_1, g_2, \ldots, g_{p-1})$$

and $\varphi_{[k]} = \varphi_{[1]}^k$. Note this makes sense as $\varphi_{[p]} = \varphi_{[1]}^p = \mathrm{id}$. Clearly this is an action of $\mathbb{Z}_p$ on G^p but it is also an action of $\mathbb{Z}_p$ on X, since if $(g_1, \ldots, g_p) \in X$, $\varphi_{[1]}(g_1, \ldots, g_p) \in X$ as we see from the following argument: Suppose $g_1 \cdots g_p = e$. Then

$$g_p g_1 \cdots g_{p-1} = (g_p g_1 \cdots g_{p-1})\left(g_p g_p^{-1}\right) = g_p\left(g_1 \cdots g_{p-1} g_p\right) g_p^{-1}$$
$$= g_p e g_p^{-1} = g_p g_p^{-1} = e.$$

Let $\{X_i\}_{i \in I}$ be the distinct orbits of X under this action of G, and let $\{x_i\}_{i \in I}$ be a complete set of coset representatives of the orbits. Then by Corollary 2.8.15,

$$|X| = \sum_{i \in I} [\mathbb{Z}_p \colon \mathrm{Stab}(x_i)]$$

Now $\mathrm{Stab}(x_i)$ is a subgroup of $\mathbb{Z}_p$, and p is prime, so is either $\{[0]\}$ or $\mathbb{Z}_p$ itself. Let n_1 be the number of x_i for which this index is 1 (i.e., for which $\mathrm{Stab}(x_i) = \mathbb{Z}_p$) and n_p be the number of x_i for which this index is p (i.e., for which $\mathrm{Stab}(x_i) = \{[0]\}$). Then

$$|X| = n_1 \cdot 1 + n_p \cdot p$$

Now $|X| = |G|^{p-1}$ is divisible by p, so n_1 is divisible by p. Let $x = (g_1, \ldots, g_p)$ be an element of X with $\mathrm{Stab}(x) = \mathbb{Z}_p$. Then $\varphi_{[1]}(x) = x$, i.e.,

$$(g_1, g_2, \ldots, g_p) = (g_p, g_1, g_2, \ldots, g_{p-1}),$$

i.e., $g_1 = g_p$, $g_2 = g_1, \ldots$, $g_p = g_{p-1}$, which implies $g_1 = g_2 = \cdots = g_{p-1} = g_p$. Call this common value g. Then $x = (g, \ldots, g)$ and, since $x \in X$, $g^p = e$. Conversely, any element x of X of this form has $\mathrm{Stab}(x) = \mathbb{Z}_p$. Thus we see that if $S = \{g \in G \mid g^p = e\}$, then $n_1 = \#(S)$.

As we have just seen, n_1 is divisible by p. But we cannot have $n_1 = 0$, as $e \in S$. Thus $n_1 \geq p$. Choosing g_0 to be any element of S other than $g_0 = e$, we see that g_0 is an element of G of order dividing p, and hence, since p is a prime, of order (exactly) p. $\square$

Now we turn to the study of groups of prime power order.

Definition 2.9.2. A group G is a *p-group* if the order of G is p^n, for some prime p and some positive integer n. ◊

Lemma 2.9.3. *Let G be a p-group. Then the center $Z(G)$ of G is nontrivial.*

Proof. If G is abelian, then $Z(G) = G$ is certainly nontrivial. Suppose not. Consider the equation class equation for G (Corollary 2.8.18):

$$|G| = |Z(G)| + \sum_{i \in I'} [G\colon C(x_i)].$$

Now $|G|$ is divisible by p. In the summation, each $C(x_i)$ is a proper subgroup of G, so its index $[G\colon C(x_i)]$ is greater than 1, and divides $|G|$, so is also divisible by p. Thus every term on the right-hand side, except possibly $|Z(G)|$, is divisible by p, so $|Z(G)|$ is divisible by p as well; in particular, $Z(G)$ is nontrivial. □

If G has order p, then we know that G is cyclic, so in particular, it is abelian. What if G has order p^2?

Corollary 2.9.4. *Let G be a group of order p^2. Then G is isomorphic to $\mathbb{Z}_{p^2}$ or $\mathbb{Z}_p \times \mathbb{Z}_p$. In particular, G is abelian.*

Proof. Since a priori G may not be abelian, we will write G multiplicatively.

Let $|G| = p^2$. We ask whether G has an element α of order p^2. If so, $G = \{e, \alpha, \alpha^2, \ldots, \alpha^{p^2-1}\}$ is cyclic of order p^2, so is isomorphic to $\mathbb{Z}_{p^2}$.

Suppose not. By Lemma 2.9.3, $Z(G)$ is nontrivial. Let $\alpha \in Z(G)$, $\alpha \neq e$. Then α has order p, so α generates a subgroup $H_1 = \{e, \alpha, \ldots, \alpha^{p-1}\}$ of G of order p. Now let $\beta \in G$, $\beta \notin H_1$. Then β generates a subgroup $H_2 = \{e, \beta, \ldots, \beta^{p-1}\}$ of G of order p. Then $H_1 \cap H_2 = \{e\}$, as $H_1 \cap H_2$ is a proper subgroup of H_1 (and of H_2). Since α is in the center of G, α commutes with every element of G; in particular α commutes with β, and consequently every element of H_1 commutes with every element of H_2. This readily implies that $H_1 H_2$ is a subgroup of G. (Alternatively, since H_1 is a normal subgroup of G, $H_1 H_2$ is a subgroup of G by Lemma 2.4.18.)

Also, $|H_1 H_2| = p^2 = |G|$ by Lemma 2.4.17, so $G = H_1 H_2$. But again, since every element of H_1 commutes with every element of H_2, G is a product $H_1 \times H_2$ so is isomorphic to $\mathbb{Z}_p \times \mathbb{Z}_p$.

(For use in our next result, let us make the isomorphism explicit. Let $\mathbb{Z}_p = \{e, a, \ldots, a^{p-1}\}$ and also $\mathbb{Z}_p = \{e, b, \ldots, b^{p-1}\}$. Then we have an isomorphism $\varphi \colon \mathbb{Z}_p \times \mathbb{Z}_p \to G$ given by $\varphi((a, e)) = \alpha$, $\varphi((e, b)) = \beta$, and in general $\varphi\left((a^i, b^j)\right) = \alpha^i \beta^j$. □

What about groups of order p^3? We show that there is a non-abelian group of order p^3 by explicitly constructing it as a semidirect product of a subgroup isomorphic to $\mathbb{Z}_p$ and a normal subgroup isomorphic to $\mathbb{Z}_p \times \mathbb{Z}_p$.

Example 2.9.5. Let $N_0 = \mathbb{Z}_p \times \mathbb{Z}_p$, which we write multiplicatively, $N_0 = \{(a^i, b^j) \mid 0 \leq i \leq p-1, 0 \leq j \leq p-1\}$, where $a^p = e$ and $b^p = e$, as in the proof of Corollary 2.9.4. Then N_0 has an automorphism $\varphi \colon N_0 \to N_0$ given by $\varphi((a, e)) = (a, b)$ and $\varphi((e, b)) = (e, b)$; more generally $\varphi((a^i, b^j)) = (a^i, b^{i+j})$. It is straightforward to check that φ is indeed an automorphism of N_0, and moreover that $\varphi^p = $ id$\colon N_0 \to N_0$. Thus, letting H_0 be another copy of $\mathbb{Z}_p$, which we also write multiplicatively, $H_0 = \{e, c, \ldots, c^{p-1}\}$ with $c^p = e$, we have a homomorphism $\Phi \colon H_0 \to \mathrm{Aut}(N_0)$ defined by $\Phi(c) = \varphi$ and, more generally, $\Phi(c^k) = \varphi^k$. Then the semidirect product $H_0 \ltimes N_0$ defined as in Lemma 2.4.28 is a nonabelian group of order p^3.

Referring to the calculations in Corollary 2.4.29, we see that if G is the group $G = \{\gamma^k \alpha^i \beta^j \mid 0 \leq k \leq p-1, 0 \leq i \leq p-1, 0 \leq j \leq p-1\}$ with multiplication given by

$$\gamma^p = e, \quad \alpha^p = e, \quad \beta^p = e, \quad \alpha\beta = \beta\alpha, \gamma\beta\gamma^{-1} = \beta, \quad \gamma\alpha\gamma^{-1} = \alpha\beta$$

then we have an isomorphism $\psi \colon H_0 \ltimes N_0 \to G$, given by

$$\psi(c, (e, e)) = \gamma, \quad \psi(e, (a, e)) = \alpha, \quad \psi(e, (e, b)) = \beta,$$

and in general $\psi\left(c^k, (a^i, b^j)\right) = \gamma^k \alpha^i \beta^j$.   ◊

Here is a general result about p-groups.

Lemma 2.9.6. *Let G be group of order p^k, $k \geq 1$. Then there are normal subgroups $\{e\} = N_0 \subset N_1 \subset \cdots \subset N_k = G$ of G with $|N_i| = p^i$, $i = 0, \ldots, k$.*

Proof.   We prove this by induction on k.

In case $k = 1$, we certainly have $\{e\} = N_0 \subset N_1 = G$.

Now suppose the lemma is true for all groups of order p^{k-1} and let G have order p^k. By Lemma 2.9.3, $Z(G)$ is nontrivial. Then $Z(G)$ is a p-group, so in particular its order is divisible by p. Then by Cauchy's theorem (for abelian groups, the easy case of this theorem), Theorem 2.6.1, $Z(G)$ has an element of order p. Let H be the cyclic subgroup generated by this element, let $Q = G/H$ be the quotient group, and let $\pi\colon G \to Q$ be the quotient map. Then Q is a group of order p^{k-1}. By the inductive hypothesis Q has normal subgroups $M_0 \subset M_1 \cdots \subset M_{k-1}$. Let $N_0 = \{e\}$ and for $i = 1, \ldots, k$, let $N_i = \pi^{-1}(M_{i-1})$. Then $N_0 \subset N_1 \subset \cdots \subset N_k$ is as claimed. $\square$

By Theorem 2.6.4, a finite abelian group G is the direct product of its subgroups of prime power order. There is no corresponding result for finite groups in general, but, nevertheless, the prime-power-order subgroups of a general group G play an important role in determining, and analyzing, its structure. We now prove the fundamental theorem about these subgroups.

Definition 2.9.7. Let G be a group of order $n = p^k m$ where $k \geq 1$ and m is relatively prime to p.

(a) A subgroup H of G is a *p-subgroup* if H has order p^j for some j with $1 \leq j \leq k$.
(b) A subgroup H of G is a *p-Sylow subgroup* if H has order p^k. $\Diamond$

Theorem 2.9.8 (Sylow). *Let G be a group of order $n = p^k m$, where $k \geq 1$ and m is relatively prime to p.*

(a) *G has a p-Sylow subgroup, and every p-subgroup of G is contained in some p-Sylow subgroup.*
(b) *The p-Sylow subgroups of G are all mutually conjugate.*
(c) *The number of p-Sylow subgroups of G is congruent to 1 modulo p and divides the order of G (or, equivalently, divides m).*

Proof. We first show that G has a p-Sylow subgroup.

We prove this by complete induction on $n = |G|$.

If $n = p^k$ then G itself is a p-Sylow subgroup. Suppose that $n > p^k$. Consider the class equation for G (Corollary 2.8.18),

$$|G| = |Z(G)| + \sum_{i \in I'} [G\colon C(x_i)].$$

If G is abelian then $G = Z(G)$ and G has a p-Sylow subgroup by Theorem 2.6.14. Suppose not.

There are two possibilities:

(1) For some x_i, $[G\colon C(x_i)] = j$ is not divisible by p. Then $|C(x_i)| = p^k m'$ with $m' = m/j < m$. Then by the inductive hypothesis $C(x_i)$ has a p-Sylow subgroup H, i.e., a subgroup of order p^k, and then H is a p-Sylow subgroup of G.

(2) $[G\colon C(x_i)]$ is divisible by p for every i. Then, since $|G|$ is divisible by p, $|Z(G)|$ is divisible by p as well. Now we argue as in the proof of Lemma 2.9.6. $Z(G)$ has an element of order p, hence a (cyclic) subgroup of H of order p, which is a normal subgroup of G. Let $Q = G/H$ and let $\pi\colon G \to Q$ be the quotient map. Then Q is a group of order $p^{k-1}m < n$, so by the inductive hypothesis Q has p-Sylow subgroup, i.e., a subgroup K of order p^{k-1}. Then $H = \pi^{-1}(K)$ is a subgroup of G of order p^k, i.e., a p-Sylow subgroup of G.

Thus, by induction, every finite group has a p-Sylow subgroup. Now we establish the more precise claims of the theorem.

(a) Let P be a p-Sylow subgroup of G. Let

$$X = \{\text{subgroups of } G \text{ conjugate to } P\}.$$

Since every element of X is a subgroup of G conjugate to P, and conjugate subgroups have the same order, every element of X is a p-Sylow subgroup of G. (Part (b) of the theorem tells us that in fact X consists of all of the p-Sylow subgroups of G, but we don't know that yet.)

Now G acts on X by conjugation, and by the definition of X, there is only one orbit, i.e., this action is transitive. Then by Lemma 2.8.14, $\#(X) = [G\colon \mathrm{Stab}(P)]$. But we certainly have that $P \subseteq \mathrm{Stab}(P)$, so $[G\colon \mathrm{Stab}(P)]$ divides $[G\colon P] = |G|/|P| = m$. In particular, $\#(X)$ divides $|G|$ and is relatively prime to p.

Now let H be any p-subgroup of G. We show $H \subseteq P'$ for some $P' \in X$. To that end, consider the action of H on X by conjugation. Then, if $\{x_i\}$ is a complete set of representatives of orbits of H on X, we have by Lemma 2.8.15,

$$\#(X) = \sum_{i \in I} [H \colon \mathrm{Stab}(x_i)].$$

We have established that $\#(X)$ is not divisible by p, so some term on the right hand side is not divisible by p. But $[H \colon \mathrm{Stab}(x_i)]$ divides $|H|$, which is a power of p, so that term must be equal to 1. Thus there is some p-Sylow subgroup P' of G with $hP'h^{-1} = P'$ for every $h \in H$, i.e., with $HP' = P'H$. Then by Lemma 2.4.18, HP' is a subgroup of G, and then by Lemma 2.4.17, $|HP'| = |H| \, |P'| \, / \, |H \cap P'| = |P'| (|H| \, / \, |H \cap P'|) = p^k \, [H \colon H \cap P']$. Now on the one hand HP' is a subgroup of G, so $|HP'|$ must divide $|G| = p^k m$. On the other hand $[H \colon H \cap P']$ must divide $|H|$, and, since H is a p-group, $[H \colon H \cap P']$ must be a power of p. But p^k is the highest power of p dividing, $|G|$, so we must have $[H \colon H \cap P'] = 1$, i.e., $H \cap P' = H$, i.e., $H \subseteq P'$, establishing (a).

(b) Note that the p-Sylow subgroup P' is an element of X, i.e., is a p-Sylow subgroup conjugate to P. So let $H = P''$ be any p-Sylow subgroup of G. Then $P'' \subseteq P'$ by (a). But $|P''| = |P'|$ so $P'' = P'$. Thus $P'' \in X$, i.e., P'' is conjugate to P.

(c) From (b), we know that the set X consists of all of the p-Sylow subgroups of G. We have already observed that $\#(X)$ divides $|G|$. Now consider the action of P on X by conjugation. Again we have

$$\#(X) = \sum_{j \in J} [P \colon \mathrm{Stab}(x_j)]$$

where $\{x_j\}$ is a complete set of representatives of the orbits of P on X. Consider $x \in X$. There are two possibilities:

(1) $x = P$. Then P is stabilized by all P, i.e., $\mathrm{Stab}(P) = P$, so $[P \colon \mathrm{Stab}(P)] = 1$.

(2) $x = P' \neq P$. Let $H = \mathrm{Stab}(P')$, a subgroup of P. The argument here parallels the argument for (a). $HP' = P'H$, so HP' is a subgroup of G, and then, as in (a), $H \subseteq P'$. But $H \subseteq P$, so $H \subseteq P' \cap P$. Now $|P| = |P'|$ and $P' \neq P$ so $P' \cap P$ is a proper subgroup of P. Thus H is a proper subgroup of P, so in particular

we have $[P: H] = [P: Stab\,(P')] > 1$, and since P is a p-group, we must have $[P: \mathrm{Stab}\,(P')]$ divisible by p. Thus in the sum on the right-hand side, we have one term equal to 1 and every other term divisible by p, so the total, which is $\#(X)$, is congruent to 1 (mod p). $\qquad\qquad\square$

Theorem 2.9.8 had three parts. These are often called the three Sylow theorems.

Corollary 2.9.9. *Let G be finite group and let p be a prime dividing the order of G. Then G has a unique p-Sylow subgroup if and only if that subgroup is normal.*

Proof.   Immediate from Theorem 2.9.8(b). $\qquad\qquad\square$

Lemma 2.9.10. *Let p and q be distinct primes with $p < q$. Let G be group of order pq. Then the q-Sylow subgroup S_q of G is normal.*

Proof.   The number m of q-Sylow subgroups must be congruent to 1 (mod q), so m is either 1 or greater than q. It must also divide $|G|$, so m must be 1, p, or q. But m cannot equal q, and m cannot equal p as we are assuming $p < q$. Hence, $m = 1$. $\qquad\qquad\square$

Corollary 2.9.11.

(a) *Let p and q be distinct primes with $p < q$. If $q \not\equiv 1\,(mod\ p)$ then any group of order pq is isomorphic to the cyclic group $\mathbb{Z}_{pq}$. In particular, any group of order pq is abelian.*

(b) *Let p and q be distinct primes with $p \not\equiv \pm 1\,(mod\ q)$ and $q \not\equiv 1\,(mod\ p)$. Then any group of order $p^2 q$ is abelian.*

(c) *Let p and q be distinct primes with $p \not\equiv \pm 1\,(mod\ q)$ and $q \not\equiv \pm 1\,(mod\ p)$. Then any group of order $p^2 q^2$ is abelian.*

Proof.   First note that in case (a), since $p < q$ we have $p \not\equiv 1\,(mod\ q)$.

Let S_p be a p-Sylow group of G and S_q be a q-Sylow subgroup of G. By Corollary 2.3.21, or by Corollary 2.9.4, S_p and S_q are both abelian. By the Sylow theorems, the number of p-Sylow subgroups is not divisible by p and must divide $|G|$, so must be a power of q. Similarly the number of q-Sylow subgroups must be a power of p. But also the number of p-Sylow subgroups must be congruent to 1 (mod p) and the number of q-Sylow subgroups must be congruent to 1 (mod q). Under the given conditions p and q, the number of each these subgroups must be 1, i.e., they must each be normal (Corollary 2.9.9). But observe that $|G| = |S_p||S_q|$ and $S_p \cap S_q = \{e\}$ (as $S_p \cap S_q$, is a subgroup of S_p, and of S_q, so its order must divide both p and q and hence must be 1). Then, by Lemma 2.4.17, $G = S_p S_q$ and then, referring to Definition 2.4.21, we see that $G = S_p \times S_q$ is the direct product of S_p and S_q. Finally, in case (a) we conclude from Theorem 2.2.15 that G is cyclic. □

In the cases where p and q do not satisfy the conditions of Corollary 2.9.11, we will construct examples where G is not abelian.

Example 2.9.12.

(a) Suppose that $q \equiv 1 \pmod{p}$, so that p divides $q - 1$. We know that $\mathrm{Aut}(\mathbb{Z}_q)$ is isomorphic to $\mathbb{Z}_q^*$, of order $q - 1$, so by Cauchy's theorem (for abelian groups), $\mathbb{Z}_q^*$ has an element of order p, i.e., there is an automorphism $\varphi \colon \mathbb{Z}_q \to \mathbb{Z}_q$ of order p. Then if $\Phi \colon \mathbb{Z}_p \to \mathrm{Aut}(\mathbb{Z}_q)$ is defined by $\Phi([k]) = \varphi^k$, from Lemma 2.4.28 we obtain a nonabelian group of order pq as the semidirect product of a subgroup isomorphic to $\mathbb{Z}_p$ and a normal subgroup isomorphic to $\mathbb{Z}_q$.

(b) If $p \equiv 1 \pmod{q}$ or $q \equiv 1 \pmod{p}$, then (a) gives us a nonabelian group of order pq, and taking the direct product of this group with $\mathbb{Z}_p$ gives us a nonabelian group of order $p^2 q$.

This leaves the (harder and more interesting) case $p \equiv -1 \pmod{q}$. Note in this case $p^2 \equiv -1 \pmod{q}$. We claim that $\mathrm{Aut}(\mathbb{Z}_p \times \mathbb{Z}_p)$ is a group whose order is divisible by q. Then we proceed as in part (a): There is an element φ of $\mathrm{Aut}(\mathbb{Z}_p \times \mathbb{Z}_p)$ of order q, and if $\Phi \colon \mathbb{Z}_q \to \mathrm{Aut}(\mathbb{Z}_p \times \mathbb{Z}_p)$ is defined by $\Phi([k]) = \varphi^k$, then again from Lemma 2.4.28 we obtain a nonabelian group of order $p^2 q$ as the semidirect product of a subgroup isomorphic to $\mathbb{Z}_q$ and a normal subgroup isomorphic to $\mathbb{Z}_p \times \mathbb{Z}_p$.

That it remains to prove our claim, and to do so we must carefully investigate the group $\mathrm{Aut}(\mathbb{Z}_p \times \mathbb{Z}_p)$.

As in the proof of Corollary 2.9.4 we write

$$\mathbb{Z}_p \times \mathbb{Z}_p = \left\{ (a^i, b^j) \mid 0 \le i \le p - 1, 0 \le j \le p - 1 \right\}.$$

Recall we have the automorphism of $\mathbb{Z}_p$ given by taking any element of $\mathbb{Z}_p$ to its k-th power, for any k relatively prime to p. Clearly we may perform this automorphism on both factors, and we may choose the powers independently. Thus we have an automorphism given by $\psi_{m,n} : \mathbb{Z}_p \times \mathbb{Z}_p \to \mathbb{Z}_p \times \mathbb{Z}_p$ given by $\psi_{m,n}((a,b)) = (a^m, b^n)$ and hence $\psi_{m,n}((a^i, b^j)) = (a^{mi}, b^{jn})$ for any integers m and n, both relatively prime to p. We also have the automorphism we used in Example 2.9.5, which here we denote by $\sigma: \mathbb{Z}_p \times \mathbb{Z}_p \to \mathbb{Z}_p \times \mathbb{Z}_p$, given by $\sigma((a,e)) = (a,b)$, $\sigma((e,b)) = (e,b)$ and hence $\sigma((a^i, b^j)) = (a^i, b^{i+j})$. Finally, we have an automorphism $\tau: \mathbb{Z}_p \times \mathbb{Z}_p \to \mathbb{Z}_p \times \mathbb{Z}_p$ given by $\tau((a,e)) = (e,b)$, $\tau((e,b)) = (a,e)$ and hence $\tau((a^i, b^j)) = (a^j, b^i)$. With these in hand we go to work.

Any automorphism of $\mathbb{Z}_p \times \mathbb{Z}_p$ must take the identity element (e, e) to itself, so we have an orbit $\{(e, e)\}$ consisting of the identity element alone in the action of $\mathrm{Aut}(\mathbb{Z}_p \times \mathbb{Z}_p)$ on $\mathbb{Z}_p \times \mathbb{Z}_p$. Consider any element of $\mathbb{Z}_p \times \mathbb{Z}_p$ other then the identity. We distinguish three cases:

  (i) An element (a^m, e). Then $(a^m, e) = \psi_{m,1}((a, e))$.
  (ii) An element (e, b^n). Then $(e, b^n) = \tau(\psi_{n,1}((a, e)))$.
  (iii) An element (a^m, b^n) with both m and n not divisible by p. Then $(a^m, b^n) = \psi_{m,n}(\sigma((a, e)))$.

Thus we see that all elements of $\mathbb{Z}_p \times \mathbb{Z}_p$ other than the identity are in the orbit of the element (a, e) of $\mathbb{Z}_p \times \mathbb{Z}_p$ in the action of $\mathrm{Aut}(\mathbb{Z}_p \times \mathbb{Z}_p)$ on $\mathbb{Z}_p \times \mathbb{Z}_p$. In other words, $\mathbb{Z}_p \times \mathbb{Z}_p - \{(e,e)\}$ is a single orbit in this action. (Otherwise said, $\mathrm{Aut}(\mathbb{Z}_p \times \mathbb{Z}_p)$ acts transitively on $\mathbb{Z}_p \times \mathbb{Z}_p - \{(e,e)\}$). Now $\mathbb{Z}_p \times \mathbb{Z}_p - \{(e,e)\}$ has $p^2 - 1$ elements, so by Lemma 2.8.14,

$$p^2 - 1 = \#(\mathbb{Z}_p \times \mathbb{Z}_p - \{(e,e)\}) = [\mathrm{Aut}(\mathbb{Z}_p \times \mathbb{Z}_p) : \mathrm{Stab}((a,e))]$$

so in particular $p^2 - 1$ divides $|\mathrm{Aut}(\mathbb{Z}_p \times \mathbb{Z}_p)|$, as claimed.

(c) In this case we have a nonabelian group of order $p^2 q$, or of order pq^2 (or both), by part (b), and then taking the direct product of this group with $\mathbb{Z}_q$, or with $\mathbb{Z}_p$, we obtain a nonabelian group of order $p^2 q^2$.   ◊

Remark 2.9.13. In Example 2.9.12(b), the p-Sylow subgroup of G was normal and hence the group G, of order $p^2 q$ was the semidirect product of its p-Sylow subgroup and its normal p-Sylow subgroup. This is no surprise. Suppose that $p \equiv -1 \pmod{q}$. Then, except in the case $p = 2$, $q = 3$, $p > q$. Thus q is the smallest prime dividing $|G|$, and its p-Sylow subgroup, of order p^2, has index q, so this subgroup must be normal by Corollary 2.8.10.

In case $p = 2$, $q = 3$ this construction still works, so the p-Sylow subgroup may still be normal, but need not be: take $G = D_6 \times \mathbb{Z}_2$. ◊

One of the basic theorems of group theory is Lagrange's theorem (Corollary 2.3.18): The order of the subgroup of a finite group must divide the order of the group. In other words, if G has order n and H is a subgroup of G of order d, then d divides n. It does *not* say that if d is any positive integer dividing n, then G has subgroup of order d. In fact, that is false, and the group constructed in Example 2.9.12(b) provides a counter example.

Example 2.9.14. Let $p \equiv -1 \pmod{q}$ and let G be the group of order $p^2 q$ constructed in Example 2.9.12(b). We shall show that G does not have a subgroup of order pq. Let S_p be a p-Sylow subgroup of G and S_q be a q-Sylow subgroup of G. Then S_p is a normal subgroup of G isomorphic to $\mathbb{Z}_p \times \mathbb{Z}_p$ and S_q is a group of order of q so is isomorphic to $\mathbb{Z}_q$. Let $X = \{$cyclic subgroups of S_p other than $\{e\}\}$. Note that S_p has $p^2 - 1$ elements of order p, a cyclic subgroup has $p - 1$ elements of order p, and two distinct cyclic subgroups of S_p only intersect in the identity element. Thus $\#(X) = (p^2 - 1)/(p - 1) = p + 1$. Also, since S_p is a normal subgroup of G, it is the unique p-Sylow subgroup of G (Corollary 2.9.9), so by the Sylow theorem, Theorem 2.9.8(a), X is in fact the set of all subgroups of G of order p. Since conjugate subgroups have the same order, we see that S_q acts on X by conjugation. Then, if $\{X_i\}_{i \in I}$ is the partition of X into the orbits of S_q, and x_i is a representative of the orbit X_i, we have

$$p + 1 = \#(X) = \sum_{i \in I} \#(X_i) = \sum_{i \in I} [S_q : \mathrm{Stab}(x_i)].$$

Now every orbit has size dividing $|S_q| = q$, i.e., has size 1 or q. Note that q divides $p + 1$ (as we began with $p \equiv -1 \pmod{q}$) so if there is an orbit of size 1, there must be at least q of them; in particular there must be at least 2 of them. Suppose this is the case, and let H_1 and H_2 be two subgroups of S_p of order p with $S_q H_1 S_q^{-1} = H_1$ and $S_q H_2 S_q^{-1} = H_2$. Then $S_q H_1 = H_1 S_q$ and $S_q H_2 = H_2 S_q$ so by Lemma 2.4.18 both of these are subgroups of G, of order pq, and then by Corollary 2.9.11(a) these subgroups are abelian. (Note that the roles of p and q here are reversed compared to those in that Corollary: Here q is the smaller prime and $p \equiv -1 \pmod{q}$ so $p \not\equiv 1 \pmod{q}$.)

Thus every element of S_q commutes with every element of H_1, and with every element of H_2, so with at least $2p-1$ elements of S_p. But it is easy to check that $K = \{g \in S_p \mid gh = hg \text{ for every } h \in S_q\}$ is a subgroup of S_p. Since any subgroup of must have order 1, p, or p^2, we conclude that in fact $K = S_p$, i.e., that every element of S_q commutes with every element of S_p, or that conjugation by any element of S_q is the identity automorphism of S_p. But this is impossible, as we constructed G as a semidirect product beginning with an automorphism φ of S_p of order q. (Otherwise said, any two elements of S_p commute with each other, and if every element of S_p commuted with every element of S_q, G would be abelian, and it is not.) Thus we see that this is impossible, and hence that every orbit X_i must have q elements, and so $\mathrm{Stab}(H) = \{e\}$ for every $H \in X$. Now the subgroup S_q may not be unique, but note this argument holds for *any* q-Sylow subgroup of G. (In fact, S_q cannot be unique, as if it were, it would be normal, and then $G = S_p \times S_q$ would be abelian, which it is not.)

Now suppose G has a subgroup F of order pq. Then certainly $F = HS_q$ for some subgroup H, cyclic of order p, of G, and some subgroup S_q, cyclic of order q, of G. Let r be a generator of S_q. Then, on the one hand, $rHr^{-1} = H$ as F is abelian, while on the other hand, $rHr^{-1} = H' \neq H$ as $r \notin \mathrm{Stab}(H) = \{e\}$; contradiction. $\Diamond$

To conclude this section we introduce, and study the basic properties, of a class of group known as solvable groups. The reason for this name is that they originally arose out of the question of being able to solve polynomial equations, and that is our interest in them here. We will be considering the question of solvability of polynomial equations in Chapter 4, on field theory, and we introduce solvable groups in this chapter on group theory in order to have them available when we need them.

It turns out that solvable groups are important in themselves, through we will not be studying them for their own sake. This once again illustrates the unity of mathematics.

Definition 2.9.15. Let G be a group.

(a) A *subnormal series* is a series of subgroups of G

$$G = G_0 \supset G_1 \supset \cdots \supset G_k = \{1\}$$

with $G_i \lhd G_{i-1}$ (i.e., with G_i a normal subgroup of G_{i-1}) for each $i \geq 1$.

(b) A *composition series* is a subnormal series with G_i a maximal proper normal subgroup of G_{i-1} (i.e., there is no proper normal subgroup H of G_{i-1} with $G_i \subset H \subset G_{i-1}$) for each $i \geq$ i. ◇

Definition 2.9.16. A group G is *solvable* if it has a subnormal series with the quotient groups G_{i-1}/G_i abelian for each $i \geq 1$. ◇

Example 2.9.17.

(a) Any abelian group is solvable as the abelian group G has the subnormal series $G \supset \{1\}$ with $G/\{1\}$ abelian.

(b) Let G be a nonabelian group of order pq, with p and q distinct primes. Let $p < q$. Then, by Lemma 2.9.10, G has a normal subgroup G_1 isomorphic to $\mathbb{Z}_q$. Then, we have the subnormal series $G \supset G_1 \supset \{1\}$ with G/G_1 isomorphic to $\mathbb{Z}_p$ and $G_1/\{1\}$ isomorphic to $\mathbb{Z}_q$, both of which are abelian.

(c) Let G be a p-group. Then Lemma 2.9.6 gives a composition series for G, so G is solvable. ◇

Lemma 2.9.18. *A finite group G is solvable if and only if it has a composition series with G_{i-1}/G_i cyclic of prime order for all $i \geq 1$.*

Proof. Certainly if G has such a composition series it is solvable.

Conversely, suppose that G is solvable and consider a subnormal series $G = G_0 \supset G_1 \supset \cdots \supset G_k = \{1\}$ with each G_i/G_{i-1} abelian. Suppose that for some i, $H = G_i/G_{i-1}$ is not cyclic of prime order. Let $\pi \colon G_i \to H$ be the quotient map.

Let p be a prime dividing $|H|$. Then by Cauchy's theorem (for abelian groups — the easy case) H has an element of order p. Let K be the subgroup generated by this element, so $|K| = p$. Since H is abelian, $K \lhd H$. Let $\tilde{G}_{i-1} = \pi^{-1}(K)$. Then we have

$$G_i \supset \tilde{G}_{i-1} \supset G_{i-1}$$

with $\tilde{G}_{i-1}/G_i$ isomorphic to K and $G_i/\tilde{G}_{i-1}$ isomorphic to H/K, a group of smaller order than H. If H/K is of prime order, there is nothing more to do. If not, repeat this process. It eventually stops as we are decreasing the order of the group at every stage.

Then "refine" the original subnormal series by inserting all of these intermediate groups to obtain a composition series for G as claimed. $\square$

Lemma 2.9.19. *Let G and H be solvable groups. Then $G \times H$ is solvable.*

Proof. Let $G = G_0 \supset G_1 \supset \cdots \supset G_k = \{1\}$ be a subnormal series for G and $H = H_0 \supset H_1 \supset \cdots \supset H_l = \{1\}$ be a subnormal series for H. Then

$$G \times H \supset G_1 \times H \supset \cdots \supset \{1\} \times H \supset \{1\} \times H_1 \supset \cdots \supset \{1\} \times \{1\}$$

is a composition series for $G \times H$. $\square$

Theorem 2.9.20.

(a) *Let G be a solvable group and let H be a subgroup of G. Then H is solvable.*

(b) *Let G be a solvable group and let $Q = G/N$ be a quotient of G. Then Q is solvable.*

(c) *Let G be a group and let N be a normal subgroup of G. If N and $Q = G/N$ are solvable, then G is solvable.*

Proof.

(a) Consider a subnormal series $G = G_0 \supset G_1 \supset \cdots \supset G_k = \{1\}$. Let $H_i = H \cap G_i$. Then $H = H_0 \supseteq H_1 \supseteq \cdots \supseteq H_k = \{1\}$. We may eliminate the terms in which $H_i = H_{i-1}$ to obtain a subnormal series. But then

$$H_{i-1}/H_i = (H \cap G_{i-1})/(H \cap G_i) = (H \cap G_{i-1})/(H \cap G_{i-1}) \cap G_i$$
$$\cong (H \cap G_{i-1})G_i/G_i \subseteq G_{i-1}/G_i$$

is isomorphic to a subgroup of an abelian group and hence is abelian.

(b) Consider a subnormal series $G = G_0 \supset G_1 \supset \cdots \supset G_k = \{1\}$. Let $\pi \colon G \to Q$ be the quotient map and let $Q_i = \pi(G_i)$. Then $Q = Q_0 \supseteq Q_1 \supseteq \cdots \supseteq Q_k = \{1\}$. We may eliminate the terms, in which $Q_i = Q_{i-1}$ to obtain a subnormal series. But then

$$Q_i = G_i N/N \cong G_i/G_i \cap N, \text{ so that}$$
$$Q_{i-1}/Q_i = (G_{i-1}N/N)/(G_i N/N)$$
$$\cong (G_{i-1}/G_{i-1} \cap N)/(G_i/G_i \cap N)$$

is isomorphic to a quotient of the abelian group G_{i-1}/G_i and hence is abelian.

(c) Let $\pi\colon G \to Q$ be the quotient map. Consider subnormal series $Q = Q_0 \supset \cdots \supset Q_k = \{1\}$ and $N = N_0 \supset \cdots \supset N_l = \{1\}$. Then we have the subnormal series for G.

$$G = \pi^{-1}(Q_0) \supset \pi^{-1}(Q_1) \supset \cdots \pi^{-1}(Q_k)$$
$$= N \supset N_1 \supset \cdots \supset N_l = \{1\}$$

and so G is solvable. $\qquad\qquad\qquad\qquad\qquad\qquad\qquad\qquad\square$

Remark 2.9.21. At this point you may wonder if there are any groups that are not solvable. There are. We will see in the next section that the symmetric groups S_n and the alternating groups A_n, both for $n \geq 5$, are not solvable. In fact, it is known that A_5, of order 60, is the nonsolvable group of smallest order. $\qquad\qquad\Diamond$

2.10 The symmetric groups

In this section, we will be considering the symmetric groups S_n, n a positive integer. These are very specific groups and we will be proving most of our results by doing specific computations.

Recall that

$$S_n = \mathrm{Aut}(\{1,\ldots,n\}) = \{\text{bijections } \sigma\colon \{1,\ldots,n\} \to \{1,\ldots,n\}\}).$$

We will abbreviate $\{1,\ldots,n\}$ to $\mathbb{N}_n$, and we will call the action of S_n on $\mathbb{N}_n$ given by $\sigma \in S_n$ takes $k \in \mathbb{N}_n$ to $\sigma(k)$ the *canonical action*.

Lemma 2.10.1. *The canonical action of S_n on $\mathbb{N}_n$ is transitive.*

Proof. If $n = 1$ this is trivial. Suppose $n > 1$. Then for any j with $2 \leq j \leq n$ we have $\sigma \in S_n$ given by

$$\sigma(1) = j, \sigma(j) = 1, \sigma(i) = i \quad \text{for } i = 1,\ldots,n, \ i \neq 1, i \neq j.$$

Thus the orbit of 1 is $\mathbb{N}_n$ and the action is transitive. $\qquad\qquad\square$

We have observed that $|S_n| = n!$ We now give a proof of this from the viewpoint of the canonical action.

Lemma 2.10.2. *For any positive integer n, S_n is a group of order $n!$*

Proof.   By induction on n.

If $n = 1$, there is only a single bijection $\sigma: \mathbb{N}_1 \to \mathbb{N}_1$, namely the bijection given by $\sigma(1) = 1$, so $|S_1| = 1$.

Assume true for $n - 1$ and consider the canonical action of S_n on $\mathbb{N}_n$. By Lemma 2.8.14, we know that

$$\# \, (\text{Orbit}(n)) = [S_n : \text{Stab}(n)].$$

By Lemma 2.10.1, $\text{Orbit}(n) = \mathbb{N}_n$, so $\#(\text{Orbit}(n)) = n$. Now $\text{Stab}(n) = \{\sigma \in S_n \,|\, \sigma(n) = n\}$ and we have an isomorphism $\varphi: S_{n-1} \to \text{Stab}(n)$ given as follows:

Let $\sigma_0 \in S_{n-1}$, so that $\sigma_0 : \mathbb{N}_{n-1} \to \mathbb{N}_{n-1}$ is a bijection. Then $\sigma = \varphi(\sigma_0)$ is the bijection $\sigma: \mathbb{N}_n \to \mathbb{N}_n$ given by $\sigma(i) = \sigma_0(i)$ for $1 \le i \le n - 1$ and $\sigma(n) = n$.

Then by the induction hypothesis $|S_{n-1}| = (n - 1)!$ and then $|S_n| = n(n - 1)! = n!$, and by induction we are done. $\square$

Remark 2.10.3.   A bijection from a set X to itself is often called a *permutation* of X. Thus in this language S_n is the group of permutations of $\mathbb{N}_n$. $\Diamond$

We now introduce two notations for permutations. The first one is logically simpler, and is unambiguous. The second is more complicated, but is also more illuminating, and is the notation most commonly used.

The first is the "two-line" notation. An element σ of S_n is a function $\sigma: \mathbb{N} \to \mathbb{N}_n$, and the two-line notation for σ is essentially a table of values:

$$\sigma = \begin{pmatrix} 1 & 2 & & n \\ & & \cdots & \\ \sigma(1) & \sigma(2) & & \sigma(n) \end{pmatrix}.$$

For example

$$\sigma_1 = \begin{pmatrix} 1 & 2 & 3 & 4 & 5 \\ 2 & 3 & 1 & 5 & 4 \end{pmatrix}$$

is the function $\sigma: \mathbb{N}_5 \to \mathbb{N}_5$ given by $\sigma_1(1) = 2$, $\sigma_1(2) = 3$, $\sigma_1(3) = 1$, $\sigma_1(4) = 5$, $\sigma_1(5) = 4$.

This notation makes it easy to compose functions. (Remember that composition is the group operation in S_n.) Note that $\sigma\sigma'(i) = \sigma(\sigma'(i))$ so that we apply the permutation *on the right* first.

For example, if σ_1 is the permutation above and σ_2 is the permutation

$$\sigma_2 = \begin{pmatrix} 1 & 2 & 3 & 4 & 5 \\ 2 & 3 & 1 & 5 & 4 \end{pmatrix},$$

then $\sigma_1\sigma_2$ is

$$\sigma_1\sigma_2 = \begin{pmatrix} 1 & 2 & 3 & 4 & 5 \\ 2 & 3 & 1 & 5 & 4 \end{pmatrix} \begin{pmatrix} 1 & 2 & 3 & 4 & 5 \\ 2 & 4 & 3 & 5 & 1 \end{pmatrix}$$

$$= \begin{pmatrix} 1 & 2 & 3 & 4 & 5 \\ 3 & 5 & 1 & 4 & 2 \end{pmatrix}$$

as $\sigma_1\sigma_2(1) = \sigma_1(\sigma_2(1)) = \sigma_1(2) = 3$, $\sigma_1\sigma_2(2) = \sigma_1(\sigma_2(2)) = \sigma_1(4) = 5$, etc. We may perform this computation simply by following the arrows in the following diagram, where we just indicate how to compute $\sigma_1\sigma_2(1)$:

$$\begin{pmatrix} 1 & 2 & 3 & 4 & 5 \\ 2 & 3 & 1 & 5 & 4 \end{pmatrix} \begin{pmatrix} 1 & 2 & 3 & 4 & 5 \\ 2 & 4 & 3 & 5 & 1 \end{pmatrix},$$

Composition of functions is not in general commutative, and so there is no reason to expect that $\sigma_1\sigma_2 = \sigma_2\sigma_1$, and in fact in this example it is not:

$$\sigma_2\sigma_1 = \begin{pmatrix} 1 & 2 & 3 & 4 & 5 \\ 2 & 4 & 3 & 5 & 1 \end{pmatrix} \begin{pmatrix} 1 & 2 & 3 & 4 & 5 \\ 2 & 3 & 1 & 5 & 4 \end{pmatrix}$$

$$= \begin{pmatrix} 1 & 2 & 3 & 4 & 5 \\ 4 & 3 & 2 & 1 & 5 \end{pmatrix}.$$

The identity element of S_n is just the identity function $\sigma\colon \mathbb{N}_n \to \mathbb{N}_n$ given by $\sigma(i) = i$, for each i, so, for example, the identity element of S_5 is

$$\begin{pmatrix} 1 & 2 & 3 & 4 & 5 \\ 1 & 2 & 3 & 4 & 5 \end{pmatrix}.$$

Also, in this notation it is easy to find inverses. If $\sigma\colon \mathbb{N}_n \to \mathbb{N}_n$ is a bijection, then $\sigma^{-1}(\sigma(i)) = i$ for each i, i.e., if the value of σ on i

is $\sigma(i)$, then the value of σ^{-1} on $\sigma(i)$ is i. So, looking at the two-line notation, σ^{-1} should be given by

$$\sigma^{-1} = \begin{pmatrix} \sigma(1) & \sigma(2) & & \sigma(n) \\ & & \cdots & \\ 1 & 2 & & n \end{pmatrix},$$

except that the top lines is out of order, so rearranging it gives,

$$\sigma^{-1} = \begin{pmatrix} 1 & 2 & & n \\ \sigma^{-1}(1) & \sigma^{-1}(2) & \cdots & \sigma^{-1}(n) \end{pmatrix}$$

For example, if σ_1 is as above then

$$\sigma_1^{-1} = \begin{pmatrix} 2 & 3 & 1 & 5 & 4 \\ 1 & 2 & 3 & 4 & 5 \end{pmatrix} = \begin{pmatrix} 1 & 2 & 3 & 4 & 5 \\ 3 & 1 & 2 & 5 & 4 \end{pmatrix}$$

and if σ_2 is as above then

$$\sigma_2^{-1} = \begin{pmatrix} 2 & 4 & 3 & 5 & 1 \\ 1 & 2 & 3 & 4 & 5 \end{pmatrix} = \begin{pmatrix} 1 & 2 & 3 & 4 & 5 \\ 5 & 1 & 3 & 2 & 4 \end{pmatrix}.$$

We now shift gears and develop the notion of a cycle. On the one hand, the "cycle structure" of a permutation gives us essential information about it, and on the other hand, this will lead us to our second notation for a permutation, the "cycle notation". In fact, the reason this notation is preferred is because it displays the cycle structure.

Definition 2.10.4. A permutation $\sigma \in S_n$ is an *r-cycle* if there is a subset $C = \{i_1, \ldots, i_r\}$ of $\mathbb{N}_n$ of cardinality r with

$$\sigma(i_1) = i_2, \sigma(i_2) = \sigma_3, \ldots, \sigma(i_{r-1}) = i_{r_1}, \sigma(i_r) = i_1$$

$$\text{and} \quad \sigma(i) = i \quad \text{for } i \notin C.$$

We denote this *r*-cycle by $(i_1 \ i_2 \ \ldots \ i_r)$. We say that $(i_1, \ldots, i_r)$ is the cyclic order on C given by σ and that in this order i_1 precedes $i_2, \ldots, i_{r-1}$ precedes i_r, and i_r precedes i_1. $\diamond$

Remark 2.10.5.

(a) Note that any 1-cycle is the identity.
(b) Note that this notation is not well-defined, as

$$(i_1 i_2 \ldots i_{r-1} i_r) = (i_2 i_3 \ldots i_r i_1) = (i_3 i_4 \ldots i_1 i_2) = \cdots$$

$$= (i_r i_1 \ldots i_{r-2} i_{r-1})$$

and all of these give the same cyclic order on C. $\diamond$

Remark 2.10.6. Observe that an r-cycle is an element of S_n of order r. $\diamond$

Definition 2.10.7. Two cycles $\sigma_1 = (i_1, \ldots, i_r)$ and $\sigma_2 = (j_1, \ldots, j_s)$ are disjoint if $\{i_1, \ldots, i_r\} \cap \{j_1, \ldots, j_s\} = \emptyset$. $\diamond$

Lemma 2.10.8. *Let $\sigma_1 = (i_1, \ldots, i_r)$ and $\sigma_2 = (j_1, \ldots, j_s)$ be disjoint cycles, and let $\sigma_3 = \sigma_2 \sigma_1$. Then σ_3 is given by*

$$\sigma_3(k) = \sigma_1(k) \quad \text{if } k \in \{i_1, \ldots, i_r\}, \quad \sigma_3(k) = \sigma_2(k) \quad \text{if } k \in \{j_1, \ldots, j_s\},$$

$\sigma_3(k) = \sigma_1(k) = \sigma_2(k) = k$ *otherwise. In particular, disjoint cycles commute.*

Proof. Direct computation. $\square$

Lemma 2.10.9. *Every $\sigma \in S_n$ can be written as a product of disjoint cycles, with the only ambiguity being given by Remark 2.10.5 and Lemma 2.10.8.*

Proof. Let H be the cyclic subgroup generated by σ. Then, in the canonical action of S_n on $\mathbb{N}_n$, if the orbits of H on $\mathbb{N}_n$ are $C_1, \ldots, C_t$, these sets are the cycles in a decomposition of σ into a product of cycles, and the action of σ on each of $C_1, \ldots, C_t$ gives the cyclic order on that cycle. Furthermore, products of disjoint cycles that differ by more than in the statement of the lemma give distinct elements of S_n. $\square$

For example, referring to the elements σ_1 and σ_2 above, we may write

$$\sigma_1 = (1\,2\,3)(4\,5) = (2\,3\,1)(4\,5) = (3\,1\,2)(4\,5)$$

$$= (1\,2\,3)(5\,4) = (2\,3\,1)(5\,4) = (3\,1\,2)(5\,4)$$

$$= (4\,5)(1\,2\,3) = (4\,5)(2\,3\,1) = (4\,1)(3\,1\,2)$$

$$= (5\,4)(1\,2\,3) = (5\,4)(2\,3\,1) = (5\,4)(3\,1\,2)$$

and

$$\begin{aligned} \sigma_2 &= (1\,2\,4\,5) = (2\,4\,5\,1) = (4\,5\,1\,2) = (5\,1\,2\,4) \\ &= (1\,2\,4\,5)(3) = (2\,4\,5\,1)(3) = (4\,5\,1\,2)(3) = (5\,1\,2\,4)(3) \\ &= (3)(1\,2\,4\,5) = (3)(2\,4\,5\,1) = (3)(4\,5\,1\,2) = (3)(5\,1\,2\,4). \end{aligned}$$

By convention, we do not write cycles of length 1, so if some integer i does not appear in a decomposition of a permutation σ into a product of disjoint cycles, we understand that i lies in a 1-cycle (i), i.e., that $\sigma(i) = i$.

Again we can multiply permutations, remembering that cycle notation is encoding the values of a function, and that multiplication is composition, again *from right to left*. For example, given σ_1 and σ_2 as above, we wish to compute

$$\sigma_3 = \sigma_2\sigma_1 = (1\,2\,4\,5)(1\,2\,3)(4\,5).$$

Again we perform the computation by following arrows. We first compute $\sigma_3(1)$:

$$(1\ 2 \to 4\ 5)\ (1 \to 2\ 3)\ (4\ 5)$$

so we see $\sigma_3(1) = 4$ and we have a partial cycle (14. Since we wish to express our result as a product of disjoint cycles, we next compute $\sigma_3(4)$:

$$(1\ 2\ 4\ 5)\ (1\ 2\ 3)\ (4 \to 5)$$

Thus, $\sigma_3(4) = 1$ and we have the cycle (14) in σ_3. So next we compute $\sigma_3(2)$:

$$(1\,2\,4\,5)(1\ 2 \to 3)(4\,5).$$

Thus $\sigma_3(2) = 3$. Then we compute $\sigma(3)$:

$$(1 \to 2\ 4\ 5)\ (1\ 2\ 3)\ (4\ 5)$$

Thus $\sigma_3(3) = 2$ and we have another cycle (23) in σ_3. Thus so far we have the product (14)(23) in σ_3. Now in this case we only have a single element, 5, left over, and σ_3 is a bijection, so we must have $\sigma_3(5) = 5$, but we compute it anyway:

$$(1\ 2\ 4 \to 5)\ (1\ 2\ 3)\ (4\ 5)$$

Thus, recalling our conventional of not writing 1-cycles, we see $\sigma_3 = (14)(23)$, and you can check that this agrees with our previous computation. Similarly, $\sigma_1\sigma_2 = (13)(25)$, also agreeing with our previous computation.

In this notation, it is easy to find the inverse of a cycle — we just reverse the cyclic order, so that we rewrite the cycle from right to left. That is, $(i_1 \ldots i_r)^{-1} = (i_r \ldots i_1)$. In our examples we have

$$\sigma_1^{-1} = (5\ 4)(3\ 2\ 1) = (1\ 3\ 2)(4\ 5)$$
$$\sigma_2^{-1} = (5\ 4\ 2\ 1) = (1\ 5\ 4\ 2).$$

Finally, we note that, under our convention of not writing 1-cycles, the identity element of S_n is (blank space), the empty product.

Now let us carefully investigate S_n.

Definition 2.10.10. Let $\sigma \in S_n$. The *cycle structure* of σ is the set (repetitions allowed) of lengths of the cycles in a decomposition of σ into a product of disjoint cycles. ◇

For example, σ_1 has cycle structure $\{3, 2\}$, σ_2 has cycle structure $\{4, 1\}$, and $\sigma_3 = \sigma_2\sigma_1$ has cycle structure $\{2, 2, 1\}$.

Lemma 2.10.11. *Two elements σ and σ' of S_n are conjugate if and only if they have the same cycle structure.*

Proof. Let α be any element of S_n. If

$$\sigma = (i_{11} \ldots i_{1r_1})(i_{21} \ldots i_{2r_2}) \ldots$$

then

$$\alpha\sigma\alpha^{-1} = (\alpha(i_{11}) \ldots \alpha(i_{1r_1}))(\alpha(i_{21}) \ldots \alpha(i_{2r_2})) \ldots$$

has the same cycle structure as σ. And if

$$\sigma' = (j_{11} \ldots j_{1r_1})(j_{21} \ldots j_{2r_2}) \ldots$$

has the same cycle structure as σ, then $\sigma' = \alpha\sigma\alpha^{-1}$ where α is defined by

$$\alpha(i_{11}) = j_{11}, \ldots, \alpha(i_{1r_1}) = j_{1r_1}, \alpha(i_{21})$$

$$= j_{21}, \ldots, \alpha(i_{2r_2}) = j_{2r_2}, \ldots. \qquad \square$$

Corollary 2.10.12.

(a) *For any fixed value of r, any two r-cycles in S_n are conjugate.*
(b) *If σ is an r-cycle and m is any integer relatively prime to r, then σ and σ^m are conjugate.*
(c) *Any element of S_n is conjugate to its inverse.*

Proof.   This follows directly from Lemma 2.10.12, noting in particular that that lemma states that two elements of S_n with the same cycle structure are conjugate, regardless of the cyclic order in each of the cycles. $\qquad \square$

Definition 2.10.13. A 2-cycle $\tau = (i_1 i_2)$ in S_n is called a *transposition*. $\qquad \Diamond$

Lemma 2.10.14.

(a) *An r-cycle σ in S_n can be written as the product of $r-1$ transpositions.*
(b) *Any element of S_n can be written as a product of transpositions.*

Proof.

(a) By induction on r. If $r = 1$, σ is the identity and is the empty product. If $r = 2$, σ is itself a transposition. Now assume the theorem is true for $r-1$, and let σ be an r-cycle, $\sigma = (i_1 \ldots i_r)$. Direct computation shows

$$\sigma = (i_1 \ldots i_r) = (i_1 i_2)(i_2 \ldots i_r).$$

Now $(i_2 \ldots i_r)$ is an $r-1$ cycle, so is a product of $r-2$ transpositions, so σ is a product of $r-1$ transpositions. Then by induction we are done.
(a) By Lemma 2.10.9 any $\sigma \in S_n$ is a product of cycles, and by (a) any cycle is a product of transpositions. $\qquad \square$

Lemma 2.10.15. *Let p be a prime. Let G be a subgroup of S_p such that*

   (a) *the canonical action of G on $\mathbb{N}_p$ is transitive; or*

(a') *G contains an element of order p;*
and (b) *G contains a transposition.*
Then $G = S_p$.

Proof. First we show that (a) and (a') are equivalent.

To say that the action of G on $\mathbb{N}_p$ is transitive is to say that $\mathbb{N}_p$ is an orbit of G. Now $\mathbb{N}_p$ has p elements, so this means that G has a subgroup of index p and hence that $|G|$ is divisible by p. Thus G has an element of order p.

Now observe that an element of S_p of order p must be a p-cycle, and that the subgroup of S_p generated by a p-cycle acts transitively on $\mathbb{N}_p$.

Thus G contains a p-cycle σ_0 and a transposition τ.

Let us suppose, for the sake of simplicity, that $\tau = (1\,2)$. Now for some k, $\sigma(1) = \sigma_0^k(1) = 2$. Again, for the sake of simplicity, let us suppose that $\sigma = (1\,2\,\ldots\,p)$. Now direct computation shows that

$$\sigma^j \tau \sigma^{-j} = (1\,2\,\ldots\,p)^j (1\,2)(1\,2\,\ldots\,p)^{-j} = ((j+1)(j+2))$$

for every $j = 0, \ldots, p-2$. Thus G contains the transpositions $(1\,2)$, $(2\,3)$, $(3\,4)\,\ldots,\,(p-1\,p)$. But then G also contains the transpositions

$$(2\,3)(1\,2)(2\,3) = (1\,3)$$

$$(3\,4)(1\,3)(3\,4) = (1\,4)$$

$$\vdots$$

$$(p-1\,p)(1\,p-1)(p-1\,p) = (1\,p)$$

i.e., G contains the transpositions $(1\,2), (1\,3), \ldots, (1\,p)$.

But then, for any j, k, G contains the transposition

$$(1\,k)(1\,j)(1\,k) = (j\,k).$$

In other words, G contains *every* transposition. But then, by Lemma 2.10.14(b), G contains every element of S_p, i.e., $G = S_p$. $\square$

We observed in Lemma 2.10.14(b), and just used, that every element of S_n can be written as a product of transpositions. But we did not claim, and it is certainly not true, that any element of S_n can be written as a product of transpositions in a unique (or anything like a unique) way. For example, the identity element is the empty product, but also, since a transposition is an element of order 2, the identity is equal to τ^2 for any transposition τ. More interestingly, we have (as we also used) that $(1\,3)$ is a transposition, but it is also the product of three transpositions $(1\,3) = (2\,3)(1\,2)(2\,3)$. Thus we see it is not even the case that when we write an element of S_n as a product of transpositions, the number of transpositions in the product is well-defined. As we shall soon see, what is well-defined is the parity of this number. In order to see this, we introduce another quantity.

Definition 2.10.16. Let $\sigma \in S_n$ be a permutation. The number of *inversions* in σ is the number of pairs of integers (i, j) such that $i < j$ but $\sigma(i) > \sigma(j)$. $\qquad\qquad\Diamond$

This number is easiest to read off from the two-line notation for σ. In the two-line notation, every time we see

$$\begin{pmatrix} \cdots & & \cdots & & \cdots \\ \cdots & a & \cdots & b & \cdots \end{pmatrix}$$

with $a > b$, that is an inversion.

For example, in σ_1 above, 2 is to the left of 1, 3 is to the left of 1, and 5 is to the left of 4, so there are 3 inversions in σ_1.

In σ_2 above, 2 is to the left of 1, 4 is to the left of 3 and 1, 3 is to the left of 1, and 5 is to the left of 1, so there are 5 inversions in σ_2.

Lemma 2.10.17. *Let $\sigma \in S_n$ be a permutation and suppose that σ is the product of t transpositions. Then t is even (resp. odd) according as the number of inversions in σ is even (resp. odd).*

Proof. Of course, the identity element is the product of zero transpositions and has zero inversions, so the lemma is true for it. We prove the lemma in general by showing that every time we multiply a permutation σ_0 by a transposition τ on the left, the number of inversions in $\sigma = \tau\sigma_0$ differs from the number of inversions in σ_0 by an odd number.

Let $\tau = (a\,b)$ and suppose that a is in position i and b is in position j in σ_0 (i.e., that $\sigma_0(i) = a, \sigma_0(j) = b$), with $i < j$. Let k be the number of positions between positions i and j for which the corresponding entry c is between a and b. We obtain σ from σ_0 by interchanging a and b. Let us see the effect of this on the number of inversions.

First suppose that $a < b$. Note that in this case we have k new inversions, since each c is greater than a, another k new inversions, since b is greater than each c, and one more new inversion, since b is greater than a. Thus in this case the number of inversions in σ is $2k + 1$ more than the number of inversions in σ_0. On the other hand, if $a > b$ then a similar argument shows that the number of inversions in σ is $2k + 1$ less than the number of inversions in σ_0. Thus in any case the number of inversions changes by an odd number, as claimed. $\qquad\qquad\square$

Given this lemma, the following definition makes sense.

Definition 2.10.18. Let $\sigma \in S_n$ be a permutation and suppose that σ is the product of t transpositions. Then σ is even (resp. odd) as t is even (resp. odd). Furthermore, $\mathrm{sign}(\sigma)$ is defined by $\mathrm{sign}(\sigma) = (-1)^t$. $\qquad\qquad\Diamond$

Remark 2.10.19. We observe that $\mathrm{sign}\colon S_n \to \{\pm 1\}$ is a homomorphism. $\qquad\qquad\Diamond$

Up until now, everything we have said is valid for S_n for any positive integer n, including $n = 1$. Now S_n is the trivial group, and there is really nothing more to be said about it, so in this case, everything we have said, though true, is superfluous. But henceforth we need $n \geq 2$ so that S_n *has a transposition.*

Lemma 2.10.20. *For $n \geq 2$ there are exactly two homomorphisms $\varphi\colon S_n \to \{\pm 1\}$. These are the trivial homomorphism $\varphi(\sigma) = 1$ for every $\sigma \in S_n$, and the homomorphism $\varphi(\sigma) = \mathrm{sign}(\sigma)$.*

Proof. First we observe that, since $\{\pm 1\}$ is an abelian group, if σ and σ' are conjugate elements of S_n, then $\varphi(\sigma) = \varphi(\sigma')$.

Let $\tau_0 = (1\,2)$, a transposition, and set $\epsilon = \varphi(\tau_0)$, $\epsilon = \pm 1$. As s special case of Lemma 2.10.11, all transpositions are conjugate in S_n, so $\varphi(\tau) = \epsilon$ for every transposition τ. Now let $\sigma \in S_n$ be arbitrary. Write σ as a product of transpositions $\sigma = \tau_1 \ldots \tau_t$. Then $\varphi(\sigma) = \varphi(\tau_1 \ldots \tau_t) = \varphi(\tau_1) \ldots \varphi(\tau_t) = \epsilon^t$.

Then we see there are exactly two possibilities:

(1) $\epsilon = 1$, in which case $\varphi(\sigma) = 1$ for every $\sigma \in S_n$; or
(2) $\epsilon = -1$, in which case $\varphi(\sigma) = (-1)^t = \mathrm{sign}(\sigma)$. $\qquad\qquad\square$

Definition 2.10.21. For $n \geq 2$ the *alternating group* A_n is the subgroup of S_n defined by

$$A_n = \{\text{even permutations in } S_n\}. \qquad \Diamond$$

Lemma 2.10.22. *For $n \geq 2$ A_n is a normal subgroup of S_n of index 2.*

Proof. A_n is the kernel of the homomorphism sign: $S_n \to \{\pm 1\}$, and this homomorphism is onto. $\qquad\qquad\qquad\qquad\qquad\qquad\qquad \square$

Lemma 2.10.23. *Let $\sigma \in S_n$ be an r-cycle. Then σ is even if r is odd and σ is odd if r is even. In particular, $\sigma \in A_n$ if and only if r is odd.*

Proof.   Immediate from Lemma 2.10.14(a). $\qquad\qquad\qquad\qquad\quad \square$

We now want to investigate the structure of A_n.

Lemma 2.10.24. *For any n, A_n is generated by 3-cycles.*

Proof.   Since, by definition, any element of A_n is a product of pairs of 2-cycles, it suffices to show that any product of pairs of 2-cycles is a product of 3-cycles. There are only two nontrivial cases:

$$(ab)(ac) = (acb),$$
$$(ab)(cd) = (acb)(acd). \qquad\qquad\qquad \square$$

Lemma 2.10.25. *Let G be any normal subgroup of A_n that contains a 3-cycle. Then $G = A_n$.*

Proof.   We shall show that G contains every 3-cycle. Then the lemma follows immediately from Lemma 2.10.24.

If $n = 2$, $A_2 = \{\text{id}\}$ is the trivial group. If $n = 3$, $A_3 = \{\text{id}, (1\,2\,3), (1\,3\,2)\}$.

Suppose $n = 4$. For simplicity suppose that G contains the 3-cycle $\sigma_0 = (1\,2\,3)$. Up to renumbering, any 3-cycle σ must be one of $(1\,2\,3)$, $(1\,3\,2)$, $(1\,2\,4)$ or $(1\,4\,2)$. Now $(1\,2\,3) = \sigma_0$ and $(1\,3\,2) = \sigma_0^2$. Also $(1\,2\,4) = (1\,4\,2)^2$. Thus, we need only show that G contains $(1\,4\,2)$. But that is true as

$$((1\,2)(3\,4))(1\,2\,3)((1\,2)(3\,4))^{-1} = (1\,4\,2).$$

Suppose $n \geq 5$. Here we shall show that every 3-cycle is conjugate to $\sigma_0 = (1\,2\,3)$. Let $\sigma = (a\,b\,c)$ be any 3-cycle in S_n. Note by

Lemma 2.10.11 that σ is conjugate to σ_0 in S_n (i.e., there is an element of α of S_n with $\sigma = \alpha\sigma_0\alpha^{-1}$) but that is not good enough–we want to show that σ is conjugate to σ_0 in A_n (i.e., there is an element α of A_n with $\sigma = \alpha\sigma_0\alpha^{-1}$). But the proof of Lemma 2.10.11 tells us how to proceed.

Let α_0 be any element of S_n with $\alpha_0(1) = a$, $\alpha_0(2) = b$, $\alpha_0(3) = c$. Note that $n \geq 5$, so let $d = \alpha_0(4)$ and $e = \alpha_0(5)$. Let α_1 be the element of S_n with $\alpha_1(1) = a$, $\alpha_1(2) = b$, $\alpha_1(3) = c$, $\alpha_1(4) = e$, $\alpha_1(5) = d$, and $\alpha_1(i) = \alpha_0(i)$, for $i > 5$. Note that $\alpha_1 = (d\,e)\alpha_0$ (the product of the transposition $(d\,e)$ with the permutation α_0). Now if we write α_0 as the product of t_0 transpositions, we see that we can write α_1 as the product of $t_1 = t_0 + 1$ transpositions. Hence, α_0 and α_1 have opposite parity. Let α be whichever of α_0 and α_1 is even. Then $\sigma = \alpha\sigma_0\alpha^{-1}$ is conjugate to σ_0 in A_n. $\square$

Definition 2.10.26. A group G is *simple* if it has no nontrivial proper normal subgroups. $\Diamond$

Theorem 2.10.27. *For* $n \geq 5$, A_n *is simple.*

Proof. Let G be a nontrivial normal subgroup of A_n. We shall show that G contains a 3-cycle. Then the theorem follows immediately from Lemma 2.10.25.

Since G is nontrivial, it has an element σ of order $k > 1$. We prove the theorem by induction on k.

Let $k = 2$. Since σ has order 2, when written as a product of disjoint cycles, all the cycles have length 2, i.e., they are all transpositions. Since $\sigma \in A_n$, there are an even number of them. For simplicity, let

$$\sigma = (1\,2)(3\,4)\sigma'$$

where σ' is the product of the remaining transpositions, if any.

Case 1: $\sigma = (1\,2)(3\,4)$. Then (and here we crucially use the fact that $n \geq 5$), σ is conjugate to

$$(1\,2)(3\,5)((1\,2)(3\,4))((1\,2)(3\,5))^{-1} = (1\,2)(4\,5)$$

and then the product

$$((1\,2)(3\,4))((1\,2)(4\,5)) = (3\,4\,5)$$

is a 3-cycle in G.

Case 2: σ' is not the identity. In that case, σ is conjugate in A_n to

$$(1\,2\,3)((1\,2)(3\,4)\sigma')(1\,2\,3)^{-1} = (1\,4)(2\,3)\sigma'$$

and then the product

$$((1\,2)(3\,4)\sigma')((1\,4)(2\,3)\sigma') = (1\,3)(2\,4)$$

is an element of G, so we may proceed as in Case 1.

Next let $k = 3$. Then, when σ is written as a product of disjoint cycles, σ is a product of 3-cycles. If there is only one of them, we are done. Assume there are at least two. For simplicity, let

$$\sigma = (1\,2\,3)(4\,5\,6)\sigma'.$$

(In this case we must have $n \geq 6$.) Then

$$\sigma^2 = (1\,3\,2)(4\,6\,5)(\sigma')^2 \text{ is in } G,$$

and also

$$(1\,2\,4)((1\,2\,3)(4\,5\,6)\sigma')(1\,2\,4)^{-1} = (1\,5\,6)(2\,4\,3)\sigma' \text{ is in } G,$$

so their product

$$((1\,3\,2)(4\,6\,5)(\sigma')^2)((1\,5\,6)(2\,4\,3)\sigma') = (1\,4\,2\,6\,3) \text{ is in } G.$$

But then

$$(1\,2)(3\,5)(1\,4\,2\,6\,3)\,((1\,2)(3\,5))^{-1} = (1\,6\,5\,2\,4) \text{ is in } G,$$

so their product

$$(1\,4\,2\,6\,3)(1\,6\,5\,2\,4) = (1\,3)(5\,6) \text{ is in } G,$$

and we are back in the $k = 2$ case.

Next let $k \geq 4$. If k is composite, and σ is an element of G of order k, let p be a prime dividing k. Then $\sigma^{k/p}$ is an element of G of order $p < k$, so we are done by induction. Thus it suffices to consider the case $k = p$, a prime. Note $p \geq 5$; in particular p is odd. Note in this case that σ must be a product of p-cycles.

For simplicity let

$$\sigma = (1\,2\,\ldots\,p)\sigma' = \sigma_1\sigma_2\,\ldots\,\sigma_j,$$

i.e., $\sigma_1 = (1\,2\,\ldots\,p)$. We concentrate on σ_1.

Now σ_1 is conjugate in A_n to

$$((1\,2)(p{-}1\,p))(1\,2\,\ldots\,p)((1\,2)(p{-}1\,p))^{-1} = (1\,3\,4\,5\,6\,\ldots\,p{-}2\,p\,p{-}1\,2)$$

and then

$$(1\,3\,4\,5\,6\,\ldots\,p-2\,p\,p-1\,2)(1\,2\,3\,\ldots\,p-1\,p)$$

$$= (2\,4\,6\,\ldots\,p-3\,p\,3\,5\,\ldots\,p-4\,p-2)$$

is a $p-2$ cycle. Performing the analogous operation on $\sigma_2,\ldots,\sigma_j$, we find that we have an element of G that is a product of $(p-2)$-cycles, and by induction we are done. $\qquad\square$

Corollary 2.10.28. *Let G be a normal subgroup of S_n, $n \geq 5$. Then $G = \{e\}$, A_n, or S_n.*

Proof. Suppose $G \subseteq A_n$. Then G is a normal subgroup of A_n, so $G = \{e\}$ or A_n by Theorem 2.10.27.

Suppose $G \not\subseteq A_n$. Let $\sigma \in G$, $\sigma \notin A_n$. Then $\sigma^2 \in A_n$. Thus if $\sigma^2 \neq e$, $G \cap A_n \neq \{e\}$. But the intersection of normal subgroups of S_n is a normal subgroup of S_n, and hence of A_n. Thus $G \cap A_n = A_n$. Thus $A_n \subset G$. But A_n is a subgroup of index 2 of S_n, so $G = S_n$.

Now suppose $\sigma^2 = e$. Then, when written as a product of disjoint cycles, they must all be transpositions. Suppose $\sigma = \tau_1\tau_2\ldots\tau_j$ is such a product, with $j > 1$. For simplicity we may suppose

$$\sigma = (1\,2)(3\,4)\tau_3\ldots\tau_j.$$

But then

$$(1\,3)((1\,2)(3\,4)\tau_3\ldots\tau_j)(1\,3)^{-1} = (1\,4)(2\,3)\tau_3\ldots\tau_j \in G.$$

and then

$$((1\,2)(3\,4)\tau_3\ldots\tau_j)((1\,4)(2\,3)\tau_3\ldots\tau_j) = (1\,3)(2\,4) \in G,$$

so again $G \cap A_n \neq \{e\}$ and, as above, $G = S_n$.

This leaves the case $j = 1$, i.e., $\sigma = \tau_1$ is a transposition. But all transpositions are conjugate in S_n (Corollary 2.10.12) and every element of S_n is a product of transpositions (Lemma 2.10.14(b)), so in this case $G = S_n$ as well. $\qquad\square$

2.11   Exercises

1. We have written down the "multiplication tables" for the groups $\mathbb{Z}_5$ and $\mathbb{Z}_5^*$ in Example 2.1.11, for the group D_6 in Example 2.1.14, and for the group Q_8 in Example 2.1.15. Write down the multiplication tables for each of the following groups. While there is no preferred order for the elements in a group, in each case use the given order.

(a) $(\mathbb{Z}_4, +)$ $\mathbb{Z}_4 = \{[0]_4, [1]_4, [2]_4, [3]_4\}$

(b) $(\mathbb{Z}_6, +)$ $\mathbb{Z}_6 = \{[0]_6, [1]_6, [2]_6, [3]_6, [4]_6, [5]_6\}$

(c) $(\mathbb{Z}_2 \times \mathbb{Z}_2, +)$ $\mathbb{Z}_2 \times \mathbb{Z}_2 = \{([0]_2, [0]_2), ([0]_2, [1]_2),$
$([1]_2, [0]_2), ([1]_2, [1]_2)\}$

(d) $(\mathbb{Z}_2 \times \mathbb{Z}_3, +)$ $\mathbb{Z}_2 \times \mathbb{Z}_3 = \{([0]_2, [0]_3), ([0]_2, [1]_3), ([0]_2, [2]_3),$
$([1]_2, [0]_3), ([1]_2, [1]_3), ([1]_2, [2]_3)\}$

(e) $(\mathbb{Z}_7^*, \cdot)$ $\mathbb{Z}_7^* = \{[1]_7, [2]_7, [3]_7, [4]_7, [5]_7, [6]_7\}$

(f) $(\mathbb{Z}_8^*, \cdot)$ $\mathbb{Z}_8^* = \{[1]_8, [3]_8, [5]_8, [7]_8\}$

(g) $(\mathbb{Z}_9^*, \cdot)$ $\mathbb{Z}_9^* = \{[1]_9, [2]_9, [4]_9, [5]_9, [7]_9, [8]_9\}$

(h) $(P(\{a, b\}), *)$ $P(\{a, b\}) = \{\{\ \}, \{a\}, \{b\}, \{a, b\}\}$

(i) $D_8 =< \alpha, \beta \,|\, \alpha^4 = 1, \beta^2 = 1, \alpha\beta = \beta\alpha^{-1} >= \{e, \alpha, \alpha^2, \alpha^3, \beta, \alpha\beta, \alpha^2\beta, \alpha^3\beta\}$.

2. (a) Show that the groups $(\mathbb{Z}_4, +)$ and $(\mathbb{Z}_5^*, \cdot)$ are isomorphic.
   (b) Show that the groups $(\mathbb{Z}_2 \times \mathbb{Z}_2, +)$, $(\mathbb{Z}_8^*, \cdot)$, and $(P(\{a, b\}), *)$ are isomorphic.
   (c) Show that the groups in (a) and the groups in (b) are not isomorphic.
   (d) Show that the groups $(\mathbb{Z}_6, +)$, $(\mathbb{Z}_2 \times \mathbb{Z}_3, +)$, $(\mathbb{Z}_7^*, \cdot)$, and $(\mathbb{Z}_9^*, \cdot)$ are isomorphic.
   (e) Show that the groups D_8 and Q_8 are not isomorphic.

3. (a) For any $n \geq 3$, let $\zeta = \exp(2\pi i/n)$, and let G be the subgroup of $M_2(\mathbb{C})$ generated by the matrices $\begin{bmatrix} \zeta & 0 \\ 0 & \zeta^{-1} \end{bmatrix}$ and $\begin{bmatrix} 0 & 1 \\ 1 & 0 \end{bmatrix}$.
   Show that G is isomorphic to the dihedral group D_{2n}.
   (b) Let G be the subgroup of $M_2(\mathbb{C})$ generated by the matrices $\begin{bmatrix} 0 & i \\ i & 0 \end{bmatrix}$ and $\begin{bmatrix} 0 & 1 \\ -1 & 0 \end{bmatrix}$. Show that G is isomorphic to the quaternion group Q_8.

4. Let G be a group and let X be a set. Let

$$G^X = \{\text{functions } f\colon X \to G\}.$$

Define an operation $*$ on G^X by

$$(f * g)(x) = f(x)g(x) \quad \text{for every } x \in X.$$

Show that $(G, *)$ is a group.
(Note that if $|G| = n$ and $\#(X) = k$, then $|G^X| = n^k$.)

5. As a special case of problem 4, let $G = (\mathbb{Z}_2, +) = \{[0]_2, [1]_2\}$. As we have seen, we can identify G^X with $P(X) = \{\text{subsets of } X\}$ by $f \longleftrightarrow f^{-1}([1]_2)$.

   (a) Under this identification, if $f \longleftrightarrow A$ and $g \longleftrightarrow B$, show that $h = f * g \longleftrightarrow C = A * B$. In this way we see that $(P(X), *)$ is a group.

   (b) What is the identity element of this group? If $A \in P(X)$, what is the inverse of A in this group?

6. Let G be a group and let g be a fixed element of G. Define a new operation $*$ on G by $a * b = agb$.

   (a) Show that $(G, *)$ is a group. What is the identity element of $(G, *)$? What is the inverse of $a \in G$ in $(G, *)$?

   (b) Show that the group $(G, *)$ is isomorphic to G.

7. Let X be a set and let $f\colon X \to G$ be a bijection from X to a group G. Define an operation $*$ on X by

$$x * y = f^{-1}(f(x)f(y)).$$

Show that $(X, *)$ is a group. What is the identity element of this group? What is the inverse of $x \in X$ in this group?
(Note that problem 6 is the special case of this problem where $f\colon G \to G$ by $f(a) = ag$, or by $f(a) = ga$.)

8. Let $f\colon P(X) \to P(X)$ be the bijection $f(A) = A^c$. Then by problem 7 we may use this bijection to define a new group structure which we shall denote by $(P(X), *^c)$.

   (a) Show that $A *^c B = (A * B)^c$.

   (b) What is the identity element of $(P(X), *^c)$? What is the inverse of A in $(P(X), *^c)$?

(c) Show that $\varphi\colon (P(X), *) \to (P(X), *^c)$ given by $\varphi(A) = A^c$ is a group isomorphism.

9. A semigroup $(S, *)$ is a set S which is closed under an associative binary operation $*$. It is a cancellation semigroup if $a * b = a * c$ implies $b = c$ for all $a, b, c \in S$, and also $b * a = c * a$ implies $b = c$ for all $a, b, c \in S$. Show that a finite cancellation semigroup is a group.

10. Let $(H, *)$ be an abelian semigroup (i.e., $a * b = b * a$ for all $a, b \in H$). Then $(H \times H, * \times *)$ is a semigroup. Define a relation $\sim$ on $H \times H$ by $(a, b) \sim (c, d)$ if $a * d \equiv b * c$.

    (a) Show that $\sim$ is an equivalence relation on $H \times H$.
    (b) Show that $G = H \times H/\sim$ is a group. (Note that if $(H, *) = (\mathbb{N}, +)$ then $G = (\mathbb{Z}, +)$ and if $(H, +) = (\mathbb{N}, \cdot)$ then $G = (\mathbb{Q}^+, \cdot)$, where $\mathbb{Q}^+$ denotes the positive rational numbers.)

11. Let G be a group and let a and b be elements of G. Show that ab and ba are conjugate.

12. Let G be a group and let a and b be conjugate elements of G. Show that a and b have the same order.

13. Let H be a subgroup of G of finite index k.

    (a) If H is a normal subgroup of G, show that $g^k \in H$ for every $g \in G$.
    (b) Give an example to show that (a) may not be true if H is not a normal subgroup of G.

14. Let G be a group of odd order. Let $g \in G$, $g \neq e$. Show that g is not conjugate to g^{-1}.

15. Let G be a group of order n and let k be an integer relatively prime to n. Let $\varphi\colon G \to G$ be the map $\varphi(g) \equiv g^k$.

    (a) If G is abelian, show that φ is an isomorphism.
    (b) If G is any finite group, show that φ is a bijection.

16. Let $\varphi\colon G \to G$ by $\varphi(g) = g^{-1}$. Show that φ is a homomorphism if and only if G is abelian.

17. Let G be a finite group and let H be a proper subgroup of G. Show that G is not a union of conjugates of H.

18. Let G be a group and let H be a subgroup of G of index 2. Show that H is a normal subgroup of G.

19. Let m_0, n_0, and q_0 be fixed nonzero integers. Let S be the subset of $M_3(\mathbb{Z})$ defined by

$$S = \left\{ \begin{bmatrix} 1 & m & n \\ 0 & 1 & q \\ 0 & 0 & 1 \end{bmatrix} \;\middle|\; \begin{array}{l} m \text{ is divisible by } m_0, \\ n \text{ is divisible by } n_0, \\ q \text{ is divisible by } q_0. \end{array} \right\}.$$

Under what conditions on m_0, n_0 and q_0 is S a group under matrix multiplication?

20. Show that D_{4n} is isomorphic to $D_{2n} \times \mathbb{Z}_2$ if and only if n is odd.

21. For $r = 1, 3, 5, 7$ let G_r be the group

$$G_r = \langle \alpha, \beta \,|\, \alpha^8 = 1, \beta^2 = 1, \beta\alpha\beta = \alpha^r \rangle.$$

Show that these are four pairwise non-isomorphic groups of order 16.

22. Let $\varphi\colon G_1 \to G_2$ be a group homomorphism.

    (a) If H_1 is a subgroup of G_1, show that $\varphi(H_1)$ is a subgroup of G_2.

    (b) If N_1 is a normal subgroup of G_1 and φ is onto, show that $\varphi(N_1)$ is a normal subgroup of G_2.

    (c) If H_2 is a subgroup of G_2, show that $\varphi^{-1}(H_2)$ is a subgroup of G_1.

    (d) If N_2 is a normal subgroup of G_2, show that $\varphi^{-1}(N_2)$ is a normal subgroup of G_1.

23. Let $\varphi\colon G_1 \to G_2$ be a homomorphism. Let $N = \mathrm{Ker}(\varphi)$. If H_1 is a subgroup of G_1, let $H_2 = \varphi(H_1)$, a subgroup of G_2. Show that H_2 is isomorphic to $H_1/N \cap H_1$.

24. Let $\varphi\colon G_1 \to G_2$ be a homomorphism. Let $N = \mathrm{Ker}(\varphi)$. If $g \in G_2$ is an element of $\mathrm{Im}(\varphi)$, show that $\varphi^{-1}(g)$ is a coset of N in G_1.

25. Let $\varphi\colon G_1 \to G_2$ be a homomorphism. Let H_2 be a subgroup of $\mathrm{Im}(\varphi)$. Show that $K_2 \longleftrightarrow \varphi^{-1}(K_2)$ is a 1-1 correspondence between (left or right) cosets K_2 of H_2 in $\mathrm{Im}(\varphi)$ and (left or right) cosets of $H_1 = \varphi^{-1}(H_2)$ in G_1.

26. Let H and K be subgroups of a group G. An (H, K)-double coset of G is a subset of G of the form

$$H a K = \{hak \,|\, h \in H, k \in K\}$$

for some element a of G.

(a) Show that $\{(H, K)\text{-double cosets}\}$ is a partition of G.
(b) Give an example to show that $\{(H, K)\text{-double cosets}\}$ may not all have the same size.

27. Let G be a group. Let Δ be the diagonal of $G \times G$,

$$\Delta = \{(g, g) \in G \times G\}.$$

(a) Show that Δ is a subgroup of $G \times G$.
(b) Show that Δ is a normal subgroup of $G \times G$ if and only if G is abelian.
(c) Show that in this case the quotient $G \times G/\Delta$ is isomorphic to G.

28. Let $Z(G) = \{g \in G \,|\, gx = xg \text{ for every } x \in G\}$. $Z(G)$ is called the *center* of G.

(a) Show that $Z(G)$ is a normal subgroup of G.
(b) More generally, let H be any subgroup of $Z(G)$. Show that H is a normal subgroup of G.

29. Let G and H be groups. Show that $Z(G \times H) = Z(G) \times Z(H)$.
30. Let G be a group and let $Z(G)$ be the center of G.

(a) If G has a normal subgroup N such that $Z(N) = \{e\}$ and $Z(G/N) = \{e\}$, show that $Z(G) = \{e\}$.
(b) More generally, show that $|Z(G)|$ divides $|Z(N)|\,|Z(G/N)|$.
(c) Give an example with N a nontrivial proper subgroup of G where we have equality in (b), and on example where we do not.

31. Show that if G is not abelian, then $G/Z(G)$ is not cyclic.
32. Let G and H be groups and let $\varphi \colon G \to H$ be a homomorphism. Let $\Delta_1(\varphi) = \{(g, \varphi(g)) \,|\, g \in G\} \subseteq G \times H$.

(a) Show that $\Delta_1(\varphi)$ is a group and that $\Phi_1 \colon G \to \Delta_1(\varphi)$ defined by $\Phi_1(g) = (g, \varphi(g))$ is an isomorphism.
(b) Show that $\Delta_1(\varphi)$ is a subgroup of $G \times H$.
(c) Show that $\Delta_1(\varphi)$ is a normal subgroup of $G \times H$ if and only if $\mathrm{Im}(\varphi) \subseteq Z(H)$.
     In an entirely analogous way, given a homomorphism $\psi \colon H \to G$ we may define $\Delta_2(\psi) = \{(\psi(h), h) \,|\, h \in H\} \subseteq G \times H$ and $\Phi_2 \colon H \to \Delta_2(\psi)$ by $\Phi_2(h) = (\psi(h), h)$.

(d) Determine under what conditions on φ and ψ is $G \times H$ isomorphic to a semidirect product $\Delta_1(\varphi) \rtimes \Delta_2(\psi)$, or, analogously, to $\Delta_1(\varphi) \ltimes \Delta_2(\psi)$, and under what conditions on φ and ψ is $G \times H$ isomorphic to the direct product $\Delta_1(\varphi) \times \Delta_2(\psi)$.

33. For a positive integer n, let

$$GL_n(\mathbb{R}) = \{A \in M_n(\mathbb{R}) \,|\, \det(A) \neq 0\}$$
$$GL_n^+(\mathbb{R}) = \{A \in M_n(\mathbb{R}) \,|\, \det(A) > 0\}$$
$$SL_n(\mathbb{R}) = \{A \in M_n(\mathbb{R}) \,|\, \det(A) = 1\}.$$

(a) Find $Z(GL_n(\mathbb{R}))$.
(b) Show that $GL_n(\mathbb{R})$ is a semidirect product

$$GL_n(\mathbb{R}) = GL_n^+(\mathbb{R}) \rtimes H$$

with H a subgroup isomorphic to $\mathbb{Z}_2$, for every n.
(c) Show that $GL_n(\mathbb{R})$ is a direct product

$$GL_n(\mathbb{R}) = GL_n^+(\mathbb{R}) \times H$$

with H a subgroup isomorphic to $\mathbb{Z}_2$, if and only if n is even.
(d) Show that $GL_n^+(\mathbb{R})$ is a direct product

$$GL_n^+(\mathbb{R}) = SL_n(\mathbb{R}) \times K$$

with K a subgroup isomorphic to $\mathbb{R}^+ = \{\text{positive real numbers}\}$, for every n.

34. Let $G = M_2(\mathbb{C})$ and let U be the subgroup of G,

$$U = \left\{ \begin{bmatrix} u & v \\ 0 & w \end{bmatrix} \right\} \subseteq G.$$

Show that G is the union of conjugates of U.
(Compare problem 17.)

35. Let H be a subgroup of G. The *normalizer* $N_G(H)$ is

$$N_G(H) = \{g \in G \,|\, gHg^{-1} = H\}$$

(a) Show that $N_G(H)$ is a subgroup of G. (Observe that $H \subseteq N_G(H)$ and furthermore that $H \triangleleft N_G(H)$.)

(b) Show that there is a 1-1 correspondence between conjugates of H and left cosets of $N_G(H)$ in G. Thus # (conjugates of H) = $[G: N_G(H)]$.

36. The *commutator* $[a, b]$ of two elements of a group G is the element $[a, b] = aba^{-1}b^{-1}$ of G. The commutator subgroup of G is the subgroup G' generated by all commutators of G, i.e.,

$$G' = \left\{ \prod [a_i, b_i] \right\}$$

(a) Show that G' is a subgroup of G.
(b) Show that G' is a normal subgroup of G.
(c) Show that the quotient G/G' is abelian.
(d) Let H be a subgroup of G with $G' \subseteq H$. Show that H is a normal subgroup of G.
(e) Let N be a normal subgroup of G. Show that G/N is abelian if and only if $G' \subseteq N$.

37. For a group G, let $\Phi \colon G \to \mathrm{Aut}(G)$ be the map $\Phi(g) = \varphi_g$, where $\varphi_g \colon G \to G$ by $\varphi_g(x) = gxg^{-1}$ for $x \in G$. Such an automorphism φ_g is called an *inner* automorphism of G.

(a) Show that Φ is a homomorphism. Hence $\mathrm{Inn}(G) = \mathrm{Im}(\Phi)$ is a subgroup of $\mathrm{Aut}(G)$.
(b) Show that $\mathrm{Inn}(G)$ is a normal subgroup of $\mathrm{Aut}(G)$. The quotient $\mathrm{Out}(G) = \mathrm{Aut}(G)/\mathrm{Inn}(G)$ is called the *outer* automorphism group of G.

38. Find $\mathrm{Inn}(G)$, $\mathrm{Aut}(G)$, and $\mathrm{Out}(G)$ for $G = D_6, D_8, Q_8$.

39. For a positive integer n, show that

$$\mathrm{Aut}(\mathbb{Z}^n) = GL_n(\mathbb{Z}) = \{\text{invertible } n\text{-by-}n \text{ matrices}$$

$$\text{with entries in } \mathbb{Z}\}.$$

(Here invertible means that the inverse must also have entries in $\mathbb{Z}$.)

40. (a) Let G be an abelian group with elementary divisors 2, 6, 6, 12, 60, 300, 2100. Find the invariant factors of G.
(b) Let G be an abelian group with invariant factors 2, 4, 8, 8, 3, 27, 125. Find the elementary divisors of G.

41. Let $M = \mathbb{Z}^2$ (written as column vectors) and let N be the subgroup of M generated by $\{z_1, z_2\}$ with

$$z_1 = \begin{bmatrix} 36 \\ 42 \end{bmatrix}, z_2 = \begin{bmatrix} 40 \\ 130 \end{bmatrix}.$$

Find a basis $\{x_1, x_2\}$ of M and positive integers s_1 and s_2 with s_1 dividing s_2 such that N has basis $\{s_1 x_1, s_2 x_2\}$. (Then M/N is isomorphic to $(\mathbb{Z}/s_1\mathbb{Z}) \oplus (\mathbb{Z}/s_2\mathbb{Z})$).

42. Let $M = \mathbb{Z}^k$ (written as column vectors) and let $N = \{v_1, \ldots, v_k\}$ be a set of k elements of M. Let A be the k-by-k matrix whose ith column is v_i, $i = 1, \ldots, k$.

   (a) If $\det(A) = 0$ show that M/N is infinite.
   (b) If $\det(A) \neq 0$ show that M/N is a finite group of order $|\det(A)|$.

43. Let G be a finite abelian group.

   (a) If G has odd order, show that the sum of the elements of G is equal to the identity.
   (b) If G is a cyclic group of even order, show that the sum of the elements of G is equal to the unique element of G of order 2.
   (c) Let G be a group of even order, and write $G = H \oplus K$, where $|H|$ is a power of 2 and $|K|$ is odd. Show that the sum of the elements of G is equal to the identity if and only if H is not cyclic.

44. Let G be a finite abelian group. Let p be a prime.

   (a) Show that the number of elementary divisors of G that are divisible by p is equal to the number of invariant factors of G that are divisible by p.
   (b) If this common value is m, show that G has exactly $p^m - 1$ elements of order p and that these elements, together with the identity, form a subgroup H of G that is isomorphic to $\mathbb{Z}_p \oplus \cdots \oplus \mathbb{Z}_p$, where there are m summands.

45. Let G act on a set X. For an element $x \in X$, let $G_x = \{g \in G \mid g(x) = x\}$.

   (a) Prove Lemma 2.8.12: The relation $x \sim y$ if $y \in \text{Orbit}(x)$ is an equivalence relation on X.
   (b) Show that G_x and G_y are conjugate if and only if $x \sim y$.

(c) Show that, for any $x \in X$, there is a bijection $f \colon G/G_x \to$ Orbit(x).

(d) Conclude that if any two of $|G|, |G_x|$, and #(Orbit(x)) are finite, so is the third, and $|G| = |G_x| \#(\text{Orbit}(x))$.

46. (a) Let $G = \{\text{symmetries of a regular tetrahedron}\}$. Show $|G| = 24$.

    (b) Let $G = \{\text{symmetries of a cube}\}$. Show $|G| = 48$.

47. Let p be a prime. Let $V = (\mathbb{Z}_p)^n$, written as column vectors. Then $GL_n(\mathbb{Z}_p)$ acts on V by multiplying a matrix times a vector.

    (a) Show that $GL_n(\mathbb{Z}_p)$ acts transitively on {nonzero vectors in V}.

    (b) Let e_1 be the vector in V with first entry 1 and all other entries 0. For $n > 1$, show that G_{e_1} is the semidirect product of a normal subgroup isomorphic to $(\mathbb{Z}_p)^{n-1}$ and a subgroup isomorphic to $GL_{n-1}(\mathbb{Z}_p)$. (For $n = 1$, G_{e_1} is trivial.)

    (c) Use (b), and induction, to derive a formula for $|GL_n(\mathbb{Z}_p)|$.

48. Prove Cauchy's Theorem (Theorem 2.9.1) as follows: Let G have order n. Proceed by complete induction on n.

    If $|Z(G)|$ is divisible by p, $Z(G)$ has an element of order p by the abelian case of Cauchy's Theorem (Theorem 2.6.1).

    Suppose that $|Z(G)|$ is not divisible by p. Use the class equation (Corollary 2.8.18) to show that G has a proper subgroup H of order divisible by p, so that H has an element of order p by the inductive hypothesis.

49. Let p and q be distinct primes with $p < q$.

    (a) If G is an abelian group of order pq, count the number of elements of G of order 1, p, q, and pq.

    (b) If G is a nonabelian group of order pq, count the number of elements of G of order 1, p, q, and pq.

50. Let G be a finite group of order divisible by a prime p. For $i \geq 1$ let s_i be the number of elements of G of order p^i. Show that $s_1 \equiv -1 \pmod{p}$ and $s_i \equiv 0 \pmod{p}$ for $i \geq 2$.

51. Let G be a finite group and let p be a prime dividing $|G|$. Let P be a p-Sylow subgroup of G.

    (a) Let N be a normal subgroup of G with $|N|$ divisible by p. Show that $P \cap N$ is a p-Sylow subgroup of N.

(b) Let H be a subgroup of G with $|H|$ divisible by p. Show that $gPg^{-1} \cap H$ is a p-Sylow subgroup of H for some element g of G.

(c) Let N be a normal subgroup of G with $[G : N]$ not divisible by p. Show that N contains every p-Sylow subgroup of G.

(d) Let N be a normal subgroup of G with $|N|$ a power of p. Show that N is contained in every p-Sylow subgroup of G.

52. (a) Show that a group of order 56 cannot be simple.

(b) Show that a group of order 312 cannot be simple.

(c) Show that a group of order 600 cannot be simple.

(d) Show that a group of order 1176 cannot be simple.

53. Let G be a group of order $p^k m$, p a prime, $k \geq 1$, p not dividing m. If p^k does not divide $(m - 1)!$, show that G is not simple.

54. (a) Let G be a p-group. Let $g \in G$, $g \neq e$. Show that g is not congruent to g^k for any k with $k \not\equiv 1 \pmod{p}$.

(b) More generally, let G have order n and let k be any integer with $k^n - 1$ relatively prime to n. Let $g \in G$, $g \neq e$. Show that g is not congruent to g^k.

55. Consider the following elements of S_9, written in two-line form:

$$\alpha_1 = \begin{pmatrix} 1\,2\,3\,4\,5\,6\,7\,8\,9 \\ 3\,8\,4\,9\,5\,2\,1\,6\,7 \end{pmatrix}$$

$$\alpha_2 = \begin{pmatrix} 1\,2\,3\,4\,5\,6\,7\,8\,9 \\ 7\,8\,9\,6\,2\,3\,1\,5\,4 \end{pmatrix}$$

$$\alpha_3 = \begin{pmatrix} 1\,2\,3\,4\,5\,6\,7\,8\,9 \\ 2\,4\,8\,7\,1\,9\,5\,6\,3 \end{pmatrix}$$

(a) Find $\alpha_1\alpha_2$, $\alpha_2\alpha_1$, $\alpha_1\alpha_3$, $\alpha_3\alpha_1$, $\alpha_2\alpha_3$, $\alpha_3\alpha_2$. Write your answers in two-line form.

(b) By counting inversions, find $\text{sign}(\alpha_1)$, $\text{sign}(\alpha_2)$, $\text{sign}(\alpha_3)$.

(c) Write $\alpha_1, \alpha_2, \alpha_3$ as products of disjoint cycles.

56. Consider the following elements of S_9, written as products of disjoint cycles:

$$\beta_1 = (1\,5\,3\,8\,9\,2)(4\,7)$$

$$\beta_2 = (1\,3\,8\,7)(2\,5\,9)(4\,6)$$

$$\beta_3 = (1\,4\,8)(2\,3\,7)(5\,6\,9)$$

(a) Find $\beta_1\beta_2$, $\beta_2\beta_1$, $\beta_1\beta_3$, $\beta_3\beta_1$, $\beta_2\beta_3$, $\beta_3\beta_1$. Write your answers as products of disjoint cycles.

(b) From the cycle structure, find $\text{sign}(\beta_1)$, $\text{sign}(\beta_2)$, $\text{sign}(\beta_3)$.

(c) Write β_1, β_2, β_3 in two-line form.

57. (a) Find all possible orders of elements of S_n, for each $n = 1, \ldots, 12$.

   (b) Find all possible orders of elements of A_n, for each $n = 1, \ldots, 12$.

58. (a) Find all subgroups of S_4, up to conjugacy. (That is, find one representative of each conjugacy class.) Which are normal?

   (b) Find all subgroups of A_4, up to conjugacy. (That is, find one representative of each conjugacy class.) Which are normal?

59. Let $G = S_n$, considered as the group of permutations of $\mathbb{N}_n = \{1, \ldots, n\}$. Let M_k be a subgroup of S_n with $\#(M_k) = k$. Let

$$H_1(M_k) = \{\sigma \in S_n \,|\, \sigma(m) = m \text{ for every } m \in M_k\}$$

$$H_2(M_k) = \{\sigma \in S_n \,|\, \sigma(M_k) \subseteq M_k\}.$$

   Show that $H_1(M_k)$ is isomorphic to S_{n-k} and that $H_2(M_k)$ is isomorphic to $S_{n-k} \times S_k$.

60. Show that two elements of S_n are conjugate if and only if the cyclic subgroups of S_n they generate are conjugate.

61. Let p be a prime. Show that S_p has $(p-2)!$ p-Sylow subgroups.

62. Let p be a prime. Describe the p-Sylow subgroups of S_{p^2}.

63. Let H be a subgroup of S_n. Suppose that H contains an n-cycle, an $(n-1)$-cycle, and a 2-cycle. Show that $H = S_n$.

64. Let $G = S_n$, considered as the group of permutations of $\mathbb{N}_n = \{1, \ldots, n\}$.

   Let H be a subgroup of S_n that acts transitively on $\mathbb{N}_n$. If $|H| > n$, show that H is not abelian.

Chapter 3

Ring Theory

Having dealt with groups, we now move on to rings. Groups have one operation, while rings have two related operations, "addition" and "multiplication". We will begin by studying rings in general, but will quickly turn our attention to "commutative rings with 1". Among these is one of the most familiar and important objects in mathematics, the integers $\mathbb{Z}$. On the one hand, we will be proving some of the most important properties of $\mathbb{Z}$ from a ring-theoretic viewpoint, and on the other hand, one of our principal concerns will be to see whether, and in what degree, these properties generalize.

The study of the (positive) integers is a domain known as number theory. Number theory is a deep and beautiful subject in mathematics whose study, historically, long preceded that of ring theory. But as an application of our ideas, we will be proving a couple of the (justly) most famous theorems of number theory by using results from group theory and ring theory. This not only gives a different perspective on these results, but demonstrates the essential unity of mathematics.

3.1 Definition, examples, and basic properties

We begin by defining rings.

Definition 3.1.1. A *ring* R is a set with two operations, addition $(+)$ and multiplication $(\cdot)$, such that $(R, +)$ is an abelian group and in addition

(1) R is closed under multiplication, i.e., $a \cdot b \in R$ for every $a, b \in R$.

(2) Multiplication is associative, i.e., $a \cdot (b \cdot c) = (a \cdot b) \cdot c$ for every $a, b, c \in R$.

(3) Multiplication distributes over addition, i.e., $a \cdot (b+c) = a \cdot b + a \cdot c$ and $(b + c) \cdot a = b \cdot a + c \cdot a$ for every $a, b, c \in R$.

R is a *commutative ring* if R is a ring and in addition

(4) Multiplication is commutative, i.e., $a \cdot b = b \cdot a$ for every $a, b \in R$.

R is a *ring with 1* (or *ring with identity*) if R is a ring and in addition

(5) There is an element $1 \neq 0$ in R such that $a \cdot 1 = 1 \cdot a = a$ for every $a \in R$.

R is a *commutative ring with 1* (or *commutative ring with identity*) if R is a ring and in addition both properties (4) and (5) bold for R.   ◊

The *trivial* ring is $R = \{0\}$ with $0 + 0 = 0$ and $0 \cdot 0 = 0$. Any other ring is nontrivial.

Henceforth, we write ab for $a \cdot b$.

Here are some basic properties of arithmetic in a ring.

Lemma 3.1.2. *Let R be a ring.*

(a) $-(-a) = a$ *for every* $a \in R$.
(b) $0a = a0 = 0$ *for every* $a \in R$.
(c) $(-a)b = a(-b) = -(ab)$ *for every* $a, b \in R$.
(d) *If R is a ring with 1,* $(-1)a = a(-1) = -a$ *for every* $a \in R$.
(e) $(-a)(-b) = ab$ *for every* $a, b \in R$.

Proof.

(a) $(R, +)$ is a group.
(b) $0a + 0 = 0a = (0 + 0)a = 0a + 0a$ so $0a = 0$ by cancellation in $(R, +)$.
(c) $0 = 0b = (a + (-a))b = ab + (-a)b$ so $(-a)b = -(ab)$ and similarly $a(-b) = -(ab)$.
(d) If R has 1, this is just the special case of (c) with $b = 1$.
(e) $(-a)(-b) = -(a(-b)) = -(-ab) = ab$.   □

Remark 3.1.3. It is easy to check (from Lemma 3.1.2(b)) that if R is a nontrivial ring, then an identity element for multiplication in R must be nonzero. Thus, the condition $1 \neq 0$ in Definition 3.1.1 is simply because we do *not* wish to consider the trivial ring $R = \{0\}$ to be a ring with 1.   ◊

We now give a whole bunch of examples.

Example 3.1.4.

(a) The integers $\mathbb{Z}$ is a commutative ring with 1.
(b) $0\mathbb{Z} = \{0\}$ is a trivial ring and $1\mathbb{Z} = \mathbb{Z}$.
For $n > 1$, $n\mathbb{Z}$ is a commutative ring without 1.
(c) For $n > 1$, $\mathbb{Z}_n$, the integers (mod n), is a commutative ring with 1. ($\mathbb{Z}_1$ is a trivial ring.) Note that there is something to check here, as the elements of $\mathbb{Z}_n$ are equivalence classes. We showed in Chapter 1 that $(\mathbb{Z}_n, +)$ is a group. We need to show that multiplication in $\mathbb{Z}_n$ satisfies the properties in Definition 3.1.1. We leave this to the reader.
(d) For p a prime, $\mathbb{Z}_p$, the integers (mod p), is a commutative ring with 1. (Of course, this is just a special case of (c).) Also, the rationals $\mathbb{Q}$, the real number $\mathbb{R}$, and the complex numbers $\mathbb{C}$ are all commutative rings with 1.
(e) Let R be any commutative ring with 1. Then

$$R[x] = \{\text{polynomials in } x \text{ with coefficients in } R\}$$
$$= \{a_0 + a_1 x + \cdots + a_n x^n \mid a_0, a_1, \ldots, a_n \in R\}$$

with the usual operations of polynomial addition and multiplication is a commutative ring with 1.
(f) Let R be any commutative ring with 1. Then

$$R[[x]] = \{\text{formal power series on } x \text{ with coefficients in } R\}$$

where a formal power series is an expression $\Sigma_{n=0}^{\infty} a_n x^n$. We call these "formal" power series because (as opposed to the case of polynomials) they do not define functions but rather are just expressions. The operations of addition and multiplication are the usual ones with power series. $R[[x]]$ is a commutative ring with 1.
(g) Let R be any commutative ring with 1. For any positive integer n,

$$M_n(R) = \{n\text{-by-}n \text{ matrices with entries in } R\}$$

with the usual operations of matrix addition and multiplication is a ring with 1. It is not commutative for $n > 1$.

(h) Let R be a commutative ring with 1.

$$\{\text{diagonal matrices in } M_n(R)\}$$

is a commutative ring with 1.

$$\{\text{upper triangular matrices in } M_n(R)\}$$

is a commutative ring with 1 if $n = 1$ and a noncommutative ring with 1 if $n > 1$.

$$\{\text{strictly upper triangular matrices in } M_n(R)\}$$

is a trivial ring if $n = 1$, a commutative ring without 1 if $n = 2$, and a noncommutative ring without 1 if $n > 2$.
(A strictly upper triangular matrix is an upper triangular matrix all of whose diagonal entries are zero.)

(i) Let G be an abelian group. An *endomorphism* of G is a group homomorphism $\varphi \colon G \to G$. Then

$$\text{End}(G) = \{\text{endomorphism of } G\}$$

is a noncommutative ring with 1 with operations

$$(\varphi + \psi)(g) = \varphi(g) + \psi(g), \qquad (\varphi\psi)(g) = \varphi(\psi(g)).$$

(j) Let R be a ring with 1, and let X be any set. Then

$$\{\text{functions } f \colon X \to R\}$$

is a ring with 1, with operations

$$(f + g)(x) = f(x) + g(x)$$
$$(fg)(x) = f(x)g(x).$$

(k) For a ring R, let us denote by 0_R the zero element of R, and, if R has 1, let us denote by 1_R the identity element of R.
Let R and S be rings. Their *product* $T = R \times S$ is the ring

$$T = \{(r, s) \mid r \in R, s \in S\}$$

with operations defined componentwise, i.e.,

$$(r_1, s_1) + (r_2, s_2) = (r_1 + r_2, s_1 + s_2),$$
$$(r_1, s_1)(r_2, s_2) = (r_1 r_2, s_1 s_2).$$

Then T is a ring with $0_T = (0_R, 0_S)$. If R and S are both rings with 1, then T is a ring with 1 with $1_T = (1_R, 1_S)$. ◊

Let us now look at some conditions on elements of a ring, and with those in hand have another look at the above examples, as well as introducing new ones.

Definition 3.1.5. A nonzero element r of a ring R is a *zero divisor* if for some nonzero element s of R, $rs = 0$ or $sr = 0$. ◊

Definition 3.1.6. An *integral domain* is a commutative ring R with 1 that has no zero divisors. ◊

Example 3.1.7.

(a) The integers $\mathbb{Z}$ is an integral domain.
(b) The ring $n\mathbb{Z}$, for $n \geq 2$, has no zero divisors, but is not an integral domain, as it does not have 1.
(c) If p is a prime, then $\mathbb{Z}_p$ is an integral domain. Also, $\mathbb{Q}, \mathbb{R}$, and $\mathbb{C}$ are integral domains.
(d) Suppose that n is composite, and let k be a divisor of n, $k \neq 1, n$. Then $[k][n/k] = [0]$ in $\mathbb{Z}_n$, and every zero divisor in $\mathbb{Z}_n$ is of this form. Thus, for n composite, $\mathbb{Z}_n$ is not an integral domain.
(e) If R is an integral domain, then $R[x]$, the ring of polynomials in x with coefficients in R, and $R[[x]]$, the ring of formal power series in x with coefficients in R, are both integral domains.
(f) Let $R = \mathbb{Q}, \mathbb{R}$, or $\mathbb{C}$ and consider the ring $M_n(R)$ of n-by-n matrices with coefficients in R. Then $A \in M_n(R)$ is a zero divisor if A is a nonzero matrix that is not invertible. Similarly, in the rings

$$\{\text{diagonal matrices in } M_n(R)\}$$
$$\{\text{upper triangular matrices in } M_n(R)\}$$

every nonzero matrix in that ring that is not invertible is a zero divisor, and in the ring

$$\{\text{strictly upper triangular matrices in } M_n(R)\}$$

every nonzero matrix in that ring is a zero divisor.
(g) Let R and S be nontrivial rings and let $T = R \times S$. Then every element of T of the form $(r, 0)$ with $r \neq 0$, or of the form $(0, s)$ with $s \neq 0$, is a zero divisor in T. ◊

Here is a second kind of element, that in a way is antithetical to zero divisors.

Definition 3.1.8. A *unit* in a ring R with 1 is an element r of R such that there exists an element r' of R with $rr' = r'r = 1$. We set

$$R^* = \{\text{units of } R\}. \qquad \diamond$$

Remark 3.1.9.

(a) In the noncommutative case, if there is an element r' of R with $r'r = 1$ and an element r'' of R with $rr'' = 1$, then $r'' = 1r'' = (r'r)r'' = r'(rr'') = r'1 = r'$, so r is a unit.

(b) Also in the noncommutative case, it is perfectly possible that there exists an element r' of R with $r'r = 1$, but there does not exist an element r'' of R with $rr'' = 1$, and vice versa. $\qquad \diamond$

Definition 3.1.10. A *field* is a commutative ring with 1 such that $R^* = R - \{0\}$ (i.e., such that every nonzero element of R is a unit).

A *division ring* (or *skew field*) is a ring with 1 such that $R^* = R = \{0\}$. $\qquad \diamond$

Lemma 3.1.11. *Let R be a ring with 1. If r is a unit in R, then r is not a zero divisor in R. In particular, if R is a field then R is an integral domain.*

Proof.   Let r be a unit of R, so that there is an element r' of r with $rr' = r'r = 1$. If $rs = 0$, then

$$s = 1s = (r'r)s = r'(rs) = r'(0) = 0$$

and similarly if $sr = 0$, then $s = 0$, so r is not a zero divisor.

In particular, in a field, every nonzero element of R is a unit, so no nonzero element of R can be a zero divisor. $\qquad \square$

Lemma 3.1.12. *Let R be a ring with 1. Then R^* is a group under multiplication.*

Proof.   We check the group axioms.

(Closure) Let r_1 be a unit in R, so that there is an element r_1' of R with $r_1 r_1' = r_1' r_1 = 1$, and let r_2 be a unit in R, so that there is an element r_2' of r with $r_2 r_2' = r_2' r_2 = 1$. Then

$$(r_1 r_2)(r_2' r_1') = r_1(r_2 r_2')r_1' = r_1(1)r_1' = r_1 r_1' = 1$$

and

$$(r_2' r_1')(r_1 r_2) = r_2'(r_1' r_1)r_2 = r_2'(1)r_2 = r_2' r_2 = 1$$

so $r_1 r_2$ is a unit in R.

(Associativity) Multiplication in R is associative.

(Identity) $1 \in R$ is a unit.

(Inverse) If r is a unit, and $r'r = rr' = 1$, then $r^{-1} = r'$ (also a unit). $\square$

Example 3.1.13.

(a) As we have observed, $\mathbb{Z}$ is an integral domain. $\mathbb{Z} = \{\pm 1\}$.

(b) For the ring with 1, $\mathbb{Z}_n$, the units in $\mathbb{Z}_n$ are $\{[k] \in \mathbb{Z}_n \mid k$ is relatively prime to $n\}$. We denoted this group by $\mathbb{Z}_n^*$ in Chapter 1, anticipating Definition 3.1.8.

(c) We have observed that $\mathbb{Z}_p$, for p a prime, $\mathbb{Q}, \mathbb{R}$, and $\mathbb{C}$ are integral domains. In fact, they are fields.

(d) Let R be a commutative ring with 1. We know $R[x] = \{\Sigma_{n=0}^N a_n x^n \mid a_n \in R\}$.

Then $R[x]^* = \{a_0 \mid a_0 \in R^*\}$.
In particular, if R is a field, $R[x]^* = \{a_0 \mid a_0 \neq 0\}$.
We also know $R[[x]] = \{\Sigma_{n=0}^\infty a_n x^x \mid a_n \in R\}$.
Then $R[[x]]^* = \{\Sigma_{n=0}^\infty a_n x^n \mid a_0 \in R^*\}$.
In particular, if R is a field, $R[[x]]^* = \{\Sigma_{n=0}^\infty a_n x^n \mid a_0 \neq 0\}$.
(Note that difference between these two cases!) $\Diamond$

We don't always have "cancellation" in rings, but it is very useful to note that often we do.

Lemma 3.1.14. *Let R be a ring, and let $a \in R$ be any element that is not a zero divisor. (In particular, if R is an integral domain this is the case for any $a \in R$, $a \neq 0$.) If $ab = ac$ for some elements b, c of R, then $b = c$, and if $ba = ca$ for some elements of R, then $b = c$.*

Proof. Suppose $ab = ac$. Then $0 = ab - ac = a(b - c)$ and since a is not a zero divisor, we must have $b - c = 0$, i.e., $b = c$. Similarly if $ba = ca$. $\square$

In order to obtain new examples of fields, and of integral domains, let us carefully examine the familiar construction of the complex numbers.

Example 3.1.15. Let $i = \sqrt{-1}$. Let

$$\mathbb{R}(i) = \{a + bi \mid a, b \in \mathbb{R}\},$$

with addition and multiplication defined "as usual", where $i^2 = -1$. We recognize that $\mathbb{C} = \mathbb{R}(i)$. It is easy to see that $\mathbb{C}$ is a commutative ring with 1. How do we know that $\mathbb{C}$ is a field? In order to show this, suppose $z \in \mathbb{C}$, $z \neq 0$. Then $z = a + bi$ with not both of a and b equal to 0. Then if $\bar{z} = a - bi$,

$$z\bar{z} = (a + bi)(a - bi) = a^2 + b^2 \neq 0,$$

so

$$z \left(\frac{\bar{z}}{a^2 + b^2} \right) = 1$$

and z is a unit, with $z^{-1} = \bar{z}/(a^2+b^2) = a/(a^2+b^2)+(-b/(a^2+b^2))i$.
   Observe that, by the same logic,

$$\mathbb{Q}(i) = \{a + bi \mid a, b \in \mathbb{Q}\}$$

is a field.
   Note that we may also form

$$\mathbb{Z}(i) = \{a + bi \mid a, b \in \mathbb{Z}\}$$

and then $\mathbb{Z}(i)$ is an integral domain. $\mathbb{Z}(i)$ is called the ring of *Gaussian integers*.   ◊

Example 3.1.16. Let $D \neq 0, 1$ be squarefree integer, i.e., an integer not divisible by any perfect square other than $(-1)^2 = (1)^2 = 1$. Let

$$\mathbb{Q}(\sqrt{D}) = \{a + b\sqrt{D} \mid a, b \in \mathbb{Q}\}$$

with addition and multiplication defined "as usual", where $(\sqrt{D})^2 = D$. It is easy to see that $\mathbb{Q}(\sqrt{D})$ is a commutative ring with 1. In order to show that $\mathbb{Q}(\sqrt{D})$ is a field, let $z = \mathbb{Q}(\sqrt{D})$, $z \neq 0$. Then $z = a+b\sqrt{D}$ with not both of a and b equal to 0. Then if $\bar{z} = a-b\sqrt{D}$,

$$z\bar{z} = (a + b\sqrt{D})(a - b\sqrt{D}) = a^2 - b^2 D.$$

We claim that $a^2 - b^2 D \neq 0$. To see this, first observe that if $D \neq 0$, 1 is squarefree, then D cannot be a perfect square, i.e., cannot be the square of an integer. But then $a^2 - b^2 D = 0$ is equivalent to $D = b^2/a^2 = (b/a)^2$, and it is easy to check that if D is not the square of an integer, it cannot be the square of a rational number either. Thus,

$$z \left(\frac{\bar{z}}{a^2 - b^2 D} \right) = 1$$

and z is a unit, with $z^{-1} = z/(a^2 - b^2 D) = a/(a^2 - b^2 D) + (-b/(a^2 - b^2 D))\sqrt{D}$.

(Note how this exactly generalizes the construction of $\mathbb{Q}(i)$ in Example 3.1.15. We obtain $\mathbb{Q}(i)$ as $\mathbb{Q}(\sqrt{D})$ for $D = -1$.)

Now we form

$$\mathcal{O}(\sqrt{D}) = \begin{cases} \{a + b\sqrt{D} \mid a, b \text{ both integers}\} \text{ if } D \equiv 2, 3 \pmod 4 \\ \{a + b\sqrt{D} \mid a, b \text{ both integers or both half-integers}\} \\ \qquad \text{if } D \equiv 1 \pmod 4 \end{cases}$$

where a half-integer is a rational number $c/2$ with c an odd integer. (Thus, a half-integer is not an integer, but twice it is.) We leave it to the reader to show that $\mathcal{O}(\sqrt{D})$ is an integral domain. (The hard part of this is to show that $\mathcal{O}(\sqrt{D})$ is closed under multiplication in the case $D \equiv 1 \pmod 4$.) $\mathcal{O}(\sqrt{D})$ is called the *ring of algebraic integers* in $\mathbb{Q}(\sqrt{D})$.

(Again, this is a generalization of the construction of $\mathbb{Z}(i)$ in Example 3.1.15. We obtain $\mathbb{Z}(i)$ as $\mathcal{O}(\sqrt{D})$ for $D = -1$.) ◇

Before we go on, there is a subtle difference between rings and rings with 1 that we need to address. It occurs in two forms.

Definition 3.1.17.

(a) Let R be a ring. A subset S of R is a *subring* of R if S is a ring, with the same operations as R.

(b) Let R be a ring with 1. A subset S of R is a *sub(ring with 1)* of R if S is a ring with 1, with the same operations as R, and $1_S = 1_R$. ◇

Definition 3.1.18.

(a) Let R and S be rings. A *homomorphism of rings (or ring homomorphism)* $\varphi: R \to S$ is a function satisfying

$$\varphi(r_1 + r_2) = \varphi(r_1) + \varphi(r_2), \qquad \varphi(r_1 r_2) = \varphi(r_1)\varphi(r_2)$$

for all $r_1, r_2 \in R$.

(b) Let R and S be rings with 1. A *homomorphism of rings with 1* (or *ring with 1 homomorphism*) is a homomorphism of rings $\varphi\colon R \to S$ with $\varphi(1_R) = 1_S$.

(c) As a special case of (b), if R and S are fields a *homomorphism of fields* (or *field homomorphism*) is a homomorphism of rings $\varphi\colon R \to S$ with $\varphi(1_R) = 1_S$.

(d) In either case, an invertible homomorphism is an *isomorphism*. If there is an isomorphism of rings $\varphi\colon R \to S$, or an isomorphism of rings with 1, or of fields, $\varphi\colon R \to S$, then R and S are *isomorphic rings*, or *isomorphic* rings with 1, or *isomorphic* fields.   ◊

Example 3.1.19.

(a) The trivial ring $R = \{0\}$ is a subring of every ring, but is not a sub (ring with 1) of any ring with 1.

(b) For $n > 1$, $n\mathbb{Z}$ is a subgroup of the ring $\mathbb{Z}$, but is not a sub (ring with 1) of the ring with 1 $\mathbb{Z}$.

(c) $\mathbb{Q}$ is a subfield of $\mathbb{R}$ and $\mathbb{R}$ is a subfield of $\mathbb{C}$. $\mathcal{O}(\sqrt{D})$ is a sub (ring with 1) of the ring with 1 $\mathbb{Q}(\sqrt{D})$.

(d) Let R and S be rings with 1 and let $T = R \times S$. Then T has a subring $R_0 = \{(r,0) \mid r \in R\}$ and there is an isomorphism of rings with 1 $\varphi_1\colon R \to R_0$ by $\varphi_1(r) = (r,0)$. Similarly T has a subring $S_0 = \{(0,s) \mid s \in S\}$ and there is an isomorphism of rings with 1 $\varphi_2\colon S \to S_0$ by $\varphi_2(s) = (0,s)$. But, although T is a ring with 1 and R_0 and S_0 are each rings with 1, R_0 and S_0 are *not* sub(rings with 1) of T, as $1_T = (1_R, 1_S)$ while $1_{R_0} = (1_R, 0) \neq 1_T$ and $1_{S_0} = (0, 1_S) \neq 1_T$.   ◊

Lemma 3.1.20. *Let S be a subring of the ring R (resp. let S be a sub (ring with 1) of the ring with 1 R. Then S is the image of a homomorphism of rings (resp. a homomorphism of rings with 1).*

Proof.   In either case S is the image of the inclusion $i\colon S \to R$. ☐

Here are some examples of ring homomorphisms.

Example 3.1.21.

(a) For any positive integer n, we saw in Chapter 2 that we have a homomorphism of groups $\varphi\colon \mathbb{Z} \to \mathbb{Z}_n$ given by $\varphi(k) = [k]_n$ ($= k \pmod n$). As we have seen here, both $\mathbb{Z}$ and $\mathbb{Z}_n$ are rings, and it is easy to check that φ is a homomorphism of rings (and a homomorphism of rings with 1, except when $n = 1$, in which case $\mathbb{Z}_n$ is a trivial ring).

(b) Similarly, for any m we have a ring homomorphism $\varphi\colon \mathbb{Z}_{mn} \to \mathbb{Z}_n$ given by $\varphi([k]_{mn}) = [k]_n$.

(c) Now let m and n be any two relatively prime positive integers. Then we have a homomorphism of rings $\varphi\colon \mathbb{Z}_{mn} \to \mathbb{Z} \times \mathbb{Z}_n$ given by $\varphi([k]_{mn}) = ([k]_m, [k]_n)$, and the proof of Theorem 2.2.15 shows that φ is an isomorphism of rings, and an isomorphism of rings with 1 providing $m, n > 1$.

(d) Consider the polynomial ring $R[x]$. We have been regarding polynomials as expressions, but we can also regard them as functions. If $f(x) = a_0 + a_1 x + \cdots + a_n x^n$, then for any $r \in R$. $f(r) = a_0 + a_1 r + \cdots + a_n r^n$. Then for any fixed element r_0 of R, we have the homomorphism $\varphi\colon R[x] \to R$ given by $\varphi(f(x)) = f(r_0)$.

(e) Let $R = \mathbb{Q}(\sqrt{D})$ as in Example 3.1.16. Then we have a field isomorphism $\varphi\colon R \to R$ given by $\varphi(z) = \bar{z}$, where, if $z = a + b\sqrt{D}$, $\bar{z} = a - b\sqrt{D}$. $\diamondsuit$

We use the same language for ring homomorphisms as we do for group homomorphisms.

Definition 3.1.22. Let $\varphi\colon R \to S$ be a homomorphism of rings, or a homomorphism of rings with 1. Then

$$\mathrm{Ker}(\varphi) = \{r \in R \mid \varphi(r) = 0\},$$

$$\mathrm{Im}(\varphi) = \{s \in S \mid s = \varphi(r) \text{ for some } r \in R\}. \qquad \diamondsuit$$

For general rings, or rings with 1, a homomorphism may be injective, surjective, both, or neither. But for fields the situation is more restrictive.

Lemma 3.1.23. *Let R and S be fields and let $\varphi\colon R \to S$ be a field homomorphism. Then φ is injective. Consequently, φ is an isomorphism if and only if it is surjective.*

Proof. Suppose that $r \in R$, $r \neq 0$. Then r has an inverse r^{-1}, with $rr^{-1} = 1$. But then $1 = \varphi(1) = \varphi(rr^{-1}) = \varphi(r)\varphi(r^{-1})$ so $\varphi(r) \neq 0$. Consequently, $\mathrm{Ker}(\varphi) = \{0\}$ and φ is injective. $\qquad \square$

Example 3.1.24. Our restriction that D be a squarefree integer, $D \neq 0, 1$, in constructing $\mathbb{Q}(\sqrt{D})$ in Example 3.1.16 was just to eliminate duplication. For suppose $D' = e^2 D$ for some nonzero rational number e. Then we have an isomorphism, $\varphi \colon \mathbb{Q}(\sqrt{D'}) \to \mathbb{Q}(\sqrt{D})$ given by $\varphi(a + b\sqrt{D'}) = a + be\sqrt{D}$. However, this would have changed the ring of algebraic integers. The choice we made there turns out to be the right choice. $\Diamond$

3.2   Ideals in rings

While subgroups are very important in studying groups, it turns out that it is not subrings, but rather ideals, that play a key role in studying rings. We introduce ideals in this section, and they will reappear throughout this chapter.

Definition 3.2.1. Let I be a subset of the ring R. If $(I, +)$ is a subgroup of $(R, +)$ and

(1) $ri \in I$ for every $r \in R$, $i \in I$ then I is a *left ideal* of R,
(2) $ir \in I$ for every $r \in R$, $i \in I$ then I is a *right ideal* of R,
(3) $ri \in I$ and $ir \in I$ for every $r \in R$, $i \in I$, then I is a *two-sided ideal*, or simply an *ideal*, of R. $\Diamond$

We will usually just use the term ideal, but we will occasionally use the term two-sided ideal for emphasis. Of course, if R is commutative, the notions of left ideal, right ideal, and two-sided ideal coincide, and in any case a two-sided ideal is both a left ideal and a right ideal.

Note that the condition of being an ideal is *stronger* than the condition of being a subring. A subring must be closed under multiplication of any two of its elements, while an ideal must be closed under multiplication of any of its elements by *any* element of the ring.

Example 3.2.2.

(a) $I = \{0\}$ is an ideal, the *trivial ideal*, of any ring R. Any other ideal of R is *nontrivial*.
(b) $I = R$ is an ideal, the *improper ideal*, of any ring R. Any other ideal of R is *proper*. $\Diamond$

Here are some more interesting examples.

Example 3.2.3.

(a) For any integer n, $n\mathbb{Z}$ is an ideal in $\mathbb{Z}$.
(b) Let $R = \mathbb{Z}[x]$, the ring of polynomials in x with coefficients in $\mathbb{Z}$. We have the following ideals in R:

$$I_1 = \{f(x) = a_0 + \cdots + a_n x^n \mid a_0 = 0\},$$

$$I_2 = \{f(x) = a_0 + \cdots + a_n x^n \mid a_i \text{ is even for every } i\},$$

$$I_3 = \{f(x) = a_0 + \cdots + a_n x^n \mid a_0 \text{ is even}\}.$$

(c) Let $R = \mathbb{Z}_n$ and let k be any integer dividing n. Then $I = \{[0], [k], \ldots, [(n/k - 1)k]\}$ is an ideal of R.
(d) Let R be a ring with 1 and let X be any set. Let $S = \{$functions $f \colon X \to R\}$ as in Example 3.1.4 (j).
For any $x_0 \in X$,

$$I - \{f. \; X \to R \mid f(x_0) = 0\}.$$

is an ideal of S. More generally, for any subset X_0 of X,

$$I = \{f \colon X \to R \mid f(x_0) = 0 \quad \text{for every } x \in X_0\}$$

is an ideal of S.

(e) Let $T = R \times S$ as in Example 3.1.4 (k). Then $R \times \{0_S\}$ and $\{0_R\} \times S$ are ideals in T.
(f) Let R be a commutative ring with 1. Then

$\{A \in M_n(R) \mid$ the first column of A is $0\}$ is a left ideal of $M_n(R)$, and

$\{A \in M_n(R) \mid$ the first row of A is $0\}$ is a right ideal of $M_n(R)$.

(g) Let $R = \{$upper triangular matrices in $M_n(R)\}$. Then

$$I = \{\text{strictly upper triangular matrices in } M_n(R)\}$$

is a two-sided ideal of R. $\diamond$

In fact, we have a sharper result than Example 3.2.3(a).

Lemma 3.2.4. *Let I be an ideal of $\mathbb{Z}$. Then $I = n\mathbb{Z}$ for some integer n.*

Proof. Note that $(I, +)$ is a subgroup of $(\mathbb{Z}, +)$ and we found all subgroups of $(\mathbb{Z}, +)$ in Lemma 2.3.8: They are all given by $n\mathbb{Z}$, for some integer n. Now, while it is certainly not the case that every additive subgroup of $(R, +)$ is an ideal of R, that is the case here: If m is any integer and n' is any element of $n\mathbb{Z}$, then $n' = nk$ for some integer k, and so $mn' = n(mk) \in n\mathbb{Z}$, and hence $n\mathbb{Z}$ is an ideal of $\mathbb{Z}$. $\square$

Here is a result that makes it slightly easier to check that a subset of a ring is an ideal.

Lemma 3.2.5. *Let I be a subset of the ring R with 1. Then the condition that $(I, +)$ is a subgroup of $(R, +)$ in Definition 3.2.1 may be replaced with the condition that I is closed under addition.*

Proof. Suppose that I is closed under addition. We have to verify that $(I, +)$ is a subgroup of $(R, +)$. Now for any $i \in I$, $0 = 0i \in I$ or $0 = i0 \in I$, or $0 = 0i = i0 \in I$, by Lemma 3.1.2(b), so I has the identity 0 for addition. Furthermore, if R is a ring with 1, then if $i \in I$, $-i = (-1)i$ or $-i = i(-1)$ or $-i = (-1)i = i(-1) \in I$ by Lemma 3.1.2(d), so I has the additive inverse of any element of I, so $(I, +)$ is a group. $\square$

Here is an easy, but useful, observation.

Lemma 3.2.6. *Let R be a ring with 1, and let I be a left/right/two-sided ideal of R. Then $I = R$ if and only if $1 \in I$.*

Proof. Suppose I is a left ideal. Certainly if $I = R$ then $1 \in I$. On the other hand, suppose $1 \in I$. Then for every $r \in R$, $r = r1 \in I$, so $I = R$. Similarly for a right ideal. And a two-sided ideal is both. $\square$

Remark 3.2.7. Let R be a ring with 1, and let I be a proper ideal of R. Then we observe from Remark 3.2.2 and Lemma 3.2.6 that I is a subring of R, but not a sub(ring with 1) of R. $\Diamond$

For the sake of simplicity, we will state the next result only in the commutative case. A similar, though definitely *not* identical, result holds in the noncommutative case.

Lemma 3.2.8. *Let R be a commutative ring with 1. Then R is a field if and only if the only ideals of R are $\{0\}$ and R.*

Proof. Suppose that R is a field. Let I be an ideal of R. If $I = \{0\}$, there is nothing to prove. If $I \neq \{0\}$, let i be an element of I, $i \neq 0$. Then $1 = i^{-1}i \in I$, and so, by Lemma 3.2.6, $I = R$.

On the other hand, suppose that R is not a field. Let $r_0 \in R$ with $r_0 \neq 0$ and r_0 not a unit. Consider $I = \{rr_0 \mid r \in R\}$. Then I is an ideal of R. Now $I \neq \{0\}$ as $r_0 = 1r_0 \in I$. Also, $I \neq R$ as if $I = R$ then $1 \in I$ so $1 = r_1 r_0 = r_0 r_1$ for some $r_1 \in R$, in which case r_0 is a unit, contrary to our hypothesis. $\qquad\square$

For the sake of simplicity we will state the next few results just for ideals, although they are also valid for left ideals and right ideals.

Lemma 3.2.9. *Let R be a ring and let I and J be ideals of R. Then*

(a) $I \cap J$ *is an ideal of R,*
(b) $I + J = \{i + j \mid i \in I, j \in J\}$ *is an ideal of R,*
(c) $IJ = \{\Sigma_{k=1}^{n} i_k j_k \mid i_k \in I, j_k \in J \text{ for } k = 1, \ldots, n\}$ *is an ideal of R.*

Proof. We leave this as an exercise for the reader. $\qquad\square$

Remark 3.2.10. Note that, unless $I \subseteq J$ or $J \subset I$, $I \cup J$ is not an ideal of R. $\qquad\Diamond$

Lemma 3.2.11. *Let R be a ring and let $I_1, I_2, \ldots$ be ideals of R with $I_1 \subseteq I_2 \subseteq \ldots$. Then $I = \cup I_j$ is an ideal of R.*

Proof. Let $r_1, r_2 \in I$. Then $r_1 \in I_{j_1}$ for some j_1 and $r_2 \in I$ for some j_2. Let $j_3 = \max(j_1, j_2)$. Then $r_1, r_2 \in I_{j_3}$ (as $I_{j_1} \subseteq I_{j_3}$ and $I_{j_2} \subseteq I_{j_3}$) so $r_1 + r_2 \in I_{j_3} \subseteq I$. Also, let $r \in I$. Then $r \in I_{j_1}$ for some j_1. For any $s \in R$, $sr \in I_{j_1} \subseteq I$, and $rs \in I_{j_1} \subseteq I$. Then, by Lemma 3.2.5, I is an ideal of R. $\qquad\square$

We have an important (rather abstract-looking) condition on a ring.

Definition 3.2.12. Let R be a ring. Then R satisfies the *ascending chain condition* (ACC), or is a *Noetherian ring* if every sequence of ideals $I_1 \subseteq I_2 \subseteq I_3 \subseteq \ldots$ is eventually constant, i.e., if there is some j_0 such that $I_{j_0} = I_{j_0+1} = I_{j_0+2} = \ldots$, or, otherwise said, if every sequence of ideals $I_1 \subset I_2 \subset I_3 \subset \ldots$ is finite. $\qquad\Diamond$

Here is a more concrete-looking equivalent condition. We restrict our selves to rings with 1 to simplify the next definition.

Definition 3.2.13. Let R be a ring with 1 and let I be an ideal of R. A set $\{i_k\}$ of elements, of I *generates* I if every element i of I can be written as

$$i = \sum_k r_k i_k s_k \qquad \text{for some } r_k, s_k \in R.$$

In this case we write $I = < \{i_k\} >$.

(By convention, the empty sum is equal to 0, so $I = \{0\}$ is generated by the empty set.)

If I has a finite generating set then I is called *finitely generated*. ◇

Note that, even if $\{i_k\}$ is infinite, any sum in this definition is (implicitly) a finite sum — infinite sums do not make sense. If $\{i\}$ consists of a single element, we will often say that i (rather than $\{i\}$) generates I.

Theorem 3.2.14. *Let R be a ring with 1. Then R is Noetherian if and only if every ideal of R is finitely generated.*

Proof.   Suppose that every ideal of R is finitely generated. Consider a chain $I_1 \subseteq I_2 \subseteq \ldots$ of ideals of R, and let $I = \cup_j I_j$. I is an ideal of R, by Lemma 3.2.11. Let $\{i_k\}_{k=1,\ldots,n}$ be a finite set of generators of I. Then each i_k is in (at least) one of the ideals in the union, say $i_k \in I_{j_k}$. Let $j_{\max} = \max(j_1, \ldots, j_n)$. Then $i_k \in I_{j_{\max}}$ for every $k = 1, \ldots, n$, so in particular the ideal generated by $\{i_k\}_{k=1,\ldots,n}$ is contained in $I_{j_{\max}}$. But this ideal is just I. Thus we see $I_{j_{\max}} \subseteq I_{j_{\max}+1} \subseteq \cdots \subseteq I \subseteq I_{j_{\max}}$, so we must have $I_{j_{\max}} = I_{j_{\max}+1} = \cdots = I$.

Conversely, suppose that R has an ideal I that is not finitely generated. We construct an infinite chain of ideals $I_1 \subset I_2 \subset \ldots$ as follows:

Let i_1 be any nonzero element of I and let I_1 be the ideal generated by $\{i_1\}$.

Let i_2 be any element of I that is not in I_1 and let I_2 be the ideal generated by $\{i_1, i_2\}$.

Let i_3 be any element of I that is not in I_2 and let I_3 be the ideal generated by $\{i_1, i_2, i_3\}$.

Continue, and note that we can go on forever, as if we had to stop at some stage n, I would be generated by the finite set $\{i_1, \ldots, i_n\}$. □

Example 3.2.15.

(a) Any field R is a Noetherian ring, as it only has two ideals $\{0\}$ (generated by $\{\ \}$) and R (generated by $\{1\}$).
(b) We saw in Lemma 3.2.5 that every ideal of $\mathbb{Z}$ is of the form $I = n\mathbb{Z}$ for some integer n. If $n = 0$, $I = \{0\}$, generated by $\{\ \}$, and if $n \neq 0$, I is generated by $\{n\}$, so $\mathbb{Z}$ is Noetherian.
(c) We shall see that if R is a field, then any nonzero ideal I of $R[x]$ is generated by $\{p(x)\}$ for some nonzero polynomial $p(x)$, so $R[x]$ is Noetherian.
(d) We shall see in Corollary 3.3.14 that, for any positive integer n, the ring $\mathbb{Z}[x_1, \ldots, x_n]$ of polynomials in the (commuting, as usual) variables $x_1, \ldots, x_n$ with coefficients in $\mathbb{Z}$ is Noetherian, and also that for any field R, the ring $R[x_1, \ldots, x_n]$ is Noetherian. ◊

Remark 3.2.16. Observe in Example 3.2.15 (a), (b), and (c) every nonzero ideal was generated by a *single* element. This turns out to be a crucially important property which we will have a lot to say about below. ◊

Here, by contrast, are some non-Noetherian rings.

Example 3.2.17.

(a) Let R be a field and consider the ring $R[x_1, x_2, \ldots]$ where we have a variable x_i for every positive integer i. Thus $R[x_1, x_2, \ldots]$ is a polynomial ring in infinitely many variables. Let I be the ideal of $R[x_1, x_2, \ldots]$ consisting of polynomials whose constant term in zero. (Note that I is generated by the infinite set $\{x_1, x_2, \ldots\}$.) We claim that I is not finitely generated. Suppose it was, that I had a finite generating set $\{p_1, \ldots, p_k\}$. Now each p_j is a polynomial, so is an expression in only finitely many variables. Thus only finitely many variables appear in expressions in the finite set $\{p_1, \ldots, p_k\}$. Choose x_i to be a variable that does not appear. Then $x_i \in I$, but x_i is not in the ideal generated by $\{p_1, \ldots, p_k\}$.

(b) Let $R = \mathbb{Q}$ and $R_0 = \mathbb{Z}$. Let S be the ring

$$S = \{p(x) \in R[x] \mid \text{the constant term of } p(x) \text{ is in } R_0\}.$$

Choose any nonzero element $r_0 \in R_0$ that is not a unit in R_0. (For example, we could choose $r_0 = 2$.) Let I_i be the ideal of S generated by $(r_0)^{-i}x$ for each $i = 0, 1, 2, \ldots$. Then we have an infinite chain $I_0 \subset I_1 \subset I_2 \subset \ldots$, and so S is not Noetherian.

(c) Let R be a field and let S be the ring of "polynomials in positive rational exponents of x with coefficients in R". That is, an element of S is an expression $f(x) = a_0 + \Sigma_{i=1}^{n} a_i x^{q_i}$ where $a_0, a_1, \ldots, a_n \in R$ and q_i is a positive rational number. We define addition and multiplication in S "as usual" (where, "as usual"; $x^{q_1} x^{q_2} = x^{q_1 + q_2}$).

Let I_i be the ideal of S generated by $x^{1/2^i}$. Then we have an infinite chain $I_0 \subset I_1 \subset I_2 \subset \ldots$, and so S is not Noetherian. (Observe that $I_0 \cup I_1 \cup I_2 \cup \ldots = I$, where I is the ideal of S consisting of those $f(x)$ with constant term $a_0 = 0$.)   ◊

Ideals are important for many reasons (as we shall see), but here is one important one.

Lemma 3.2.18. *Let R and S be rings and let $\varphi\colon R \to S$ be a ring homomorphism. Then $\mathrm{Ker}(\varphi)$ is an ideal of R.*

Proof.   First observe that if φ is a ring homomorphism, it is a homomorphism of the additive groups $Q = (R, +) \to (S, +)$. Then we know that $\mathrm{Ker}(\varphi)$ is a subgroup of $(R, +)$. Indeed, we know that it is a normal subgroup, but since $(R, +)$ is an abelian group, subgroups and normal subgroups coincide.

Also, let $i \in \mathrm{Ker}(\varphi)$ and let $r \in R$. Then

$$\varphi(ri) = \varphi(r)\varphi(i) = \varphi(r)0 = 0 \quad \text{so} \quad ri \in \mathrm{Ker}(\varphi)$$

and

$$\varphi(ir) = \varphi(i)\varphi(r) = 0\varphi(r) = 0 \quad \text{so} \quad ir \in \mathrm{Ker}(\varphi)$$

and hence we conclude that $\mathrm{Ker}(\varphi)$ is an ideal   □

Now, forgetting about multiplication for the moment, R is an abelian group under addition and an ideal I of R is a subgroup, so we may form the quotient group R/I. But of course we are working with rings, and we would like R/I to be a ring. Indeed, it is, as we now see.

Let us recall that the elements of R/I are cosets of I, that we can write any coset of I as $r + I$ for some $r \in R$, and that two cosets $r + I$ and $r' + I$ are the same, or equivalently, that r and r' are representatives of the same coset, if and only if $r' = r + i$ for some element i of I. (Here I have deliberately written cosets rather than left cosets to remind you that we are in the abelian case.)

Theorem 3.2.19. *Let R be a ring and let I be an ideal of R. Then R/I is a ring with multiplication given by*

$$(r_1 + I)(r_2 + I) = r_1 r_2 + I.$$

Furthermore, the map $\pi \colon R \to R/I$ given by $\pi(r) = r + I$ is a ring homomorphism.

If R is a ring with 1 and I is a proper ideal of R, then R/I is a ring with 1 and π is a homomorphism of rings with 1.

Proof. The key thing to check is that multiplication is well-defined, i.e., independent of our choice of coset representatives. So suppose $r_1' + I = r_1 + I$ and $r_2' + I = r_2 + I$. We have to show that $(r_1' + I)(r_2' + I) = (r_1 + I)(r_2 + I)$.

Now we know that $r_1' = r_1 + i_1$ and $r_2' = r_2 + i_2$ for some $i_1, i_2 \in I$. We compute

$$r_1' r_2' = (r_1 + i_1)(r_2 + i_2) = r_1 r_2 + r_1 i_2 + i_1 r_2 + i_1 i_2$$

$$= r_1 r_2 + i_3 \quad \text{where } i_3 = r_1 i_2 + i_1 r_2 + i_1 i_2$$

and we observe that $i_3 \in I$ precisely because I is an ideal. Then we see that

$$r_1' r_2' + I = r_1 r_2 + I$$

i.e.,

$$(r_1' + I)(r_2' + I) = (r_1 + I)(r_2 + I)$$

as required.

We leave it to the reader to check the remaining ring axioms, and to check that π is a ring homomorphism.

In the case of rings with 1 there is a little more work to do. Recall that the trivial ring is *not* a ring with 1, so we cannot have $R/I = \{0\}$, i.e., we cannot have $I = R$, i.e., I must be a proper ideal of R. Conversely, if I is a proper ideal of R, we can form R/I, and then check that $1 + I$ is the multiplicative identity in R/I. But, remembering that $I = 0 + I$ is the zero element of R/I, we need $1 + I \neq I$, i.e., $1 \notin I$. That must be the case as if we had $1 \in I$, then by Lemma 3.2.7, we would have $I = R$, which it is not. Given this, we also see in this case that $\pi(1) = 1 + I$, which we just observed was the multiplicative identity in R/I, so π is a homomorphism of rings with 1. □

Definition 3.2.20. The ring (or ring with 1) R/I of Theorem 3.2.19 is the *quotient* of R by I, and the homomorphism $\pi \colon R \to R/I$ is the *quotient map* (or *canonical projection*).   ◇

Remark 3.2.21. If $\pi \colon R \to R/I$ is the quotient map, then π is an epimorphism and $\mathrm{Ker}(\pi) = I$.   ◇

There is one thing that may have bothered you a bit. When we formed R/I, we formed the quotient of additive groups, i.e., we formed the quotient of R under the equivalence relation that two elements r and r' are equivalent if $r' = r + i$ for some $i \in I$. Shouldn't we have taken the multiplication on R into account as well? The answer is that we did. We did not try to take R/I for I simply a subgroup of the additive group of R, but rather required I to be an *ideal*, and the requirements for I to be an ideal involve the multiplication on R. You can see from the proof of Theorem 3.2.19 that these requirements were what we needed for multiplication in R/I to be defined (i.e., these requirements were precisely what we needed to ensure that, in the notation of that proof, $i_3 \in I$).

Example 3.2.22.

(a) For $n > 0$, the ring $\mathbb{Z}_n$ is the quotient of the ring $\mathbb{Z}$ by the ideal $n\mathbb{Z}$. Note that the ideal $n\mathbb{Z}$ is generated by $\{n\}$.

(b) Let $R = \mathbb{Z}[x]$ as in Example 3.2.4 (b). Then we have an isomorphism $\varphi_1 = R/I_1 \to \mathbb{Z}$ given by $\varphi(f(x)) = a_0$. Note that the ideal I_1 is generated by $\{x\}$.

We have an isomorphism $\varphi_2 \colon R/I_2 \to \mathbb{Z}_2[x]$ given by $\varphi_2(f(x)) = [a_0]_2 + \cdots + [a_n]_2 x^n$. Note that the ideal I_2 is generated by $\{2\}$.

We have an isomorphism $\varphi_3 \colon R/I_3 \to \mathbb{Z}_2$ given by $\varphi(f(x)) = [a_0]_2$. Note that the ideal I_3 is generated by $\{2, x\}$. $\Diamond$

When we derived the Chinese remainder theorem (Corollary 2.2.17) in Chapter 2, we mentioned that it is best understood in terms of ring theory. We now derive that here (in a more generalized form).

Definition 3.2.23. Let R be a commutative ring with 1. Two ideals I and J of R are *coprime* if $I + J = R$. $\Diamond$

Example 3.2.24. Let $R = \mathbb{Z}$ and let m and n be relatively prime integers. Then there are integers x and y with $mx + ny = 1$. Now if $I = m\mathbb{Z}$ and $J = n\mathbb{Z}$, then $mx \in I$ and $ny \in J$, so $1 \in I + J$ and hence $I + J = R$. Thus, we see that if m and n are relatively prime integers, the ideals $m\mathbb{Z}$ and $n\mathbb{Z}$ are coprime ideals of $\mathbb{Z}$.

Also note that if m and n are relatively prime,

$$m\mathbb{Z} \cap n\mathbb{Z} = \{\text{integers divisible by both } m \text{ and } n\}$$

$$= \{\text{integers divisible by } mn\} = mn\mathbb{Z}. \qquad \Diamond$$

Lemma 3.2.25. *Let R be a commutative ring with 1 and let I and $J_1, \ldots, J_m$ be ideals of R with I and J_k coprime for each $k = 1, \ldots, m$. Then I and $J = J_1 \cap \cdots \cap J_m$ are coprime.*

Proof. For each value of k, since I and J_k are coprime, there is an element i_k of I and an element j_k of J_k with $i_k + j_k = 1$. Then

$$1 = (i_1 + j_1)(i_2 + j_2) \cdots (i_m + j_m) = i + j$$

where $j = j_1, \ldots, j_m$ and i is the sum of the other terms in the product. Now every term in i has at least one i_k factor, so $i \in I$, and j has a j_k factor for every k, so $j \in J = J_1 \cap \ldots \cap J_m$. Thus, $1 \in I + J$ and hence $I + J = R$. $\square$

Theorem 3.2.26 (Chinese remainder theorem). *Let R be a commutative ring with 1 and let $I_1, \ldots, I_n$ be pairwise coprime* *ideals of R (i.e., I_{k_1} and I_{k_2} are coprime whenever $k_1 \neq k_2$). Let $r_1, \ldots, r_n$ be arbitrary elements of R. Then the system of simultaneous congruences*

$$x \equiv r_1 \pmod{I_1}$$

$$x \equiv r_2 \pmod{I_2}$$

$$\vdots$$

$$x \equiv r_n \pmod{I_n}$$

has a solution $x = b$ in R, and $x = b'$ is a solution if and only if $b' \equiv b \pmod{I_1 \cap \cdots \cap I_n}$.

Proof. In Lemma 3.2.25, let $I = I_1$ and $J_1 = I_2, J_2 = I_3, \ldots, J_{n-1} = I_n$. Then I_1 and $J = J_1 \cap \ldots \cap J_{n-1}$ are coprime, so write $1 = i + j$ with i and j as in that lemma. Since $j \in J, j \in J_k$ for each $k = 1, \ldots, n-1$, i.e., $j \equiv 0 \pmod{J_k}$ for each such k. Also, $i \in I$, i.e., $i \equiv 0 \pmod{I_1}$, so the equation $1 = i + j$ gives

$$1 \equiv i + j \equiv 0 + j \equiv j \pmod{I_1}.$$

Set $h_1 = i$. Then $h_1 \equiv 1 \pmod{I_1}$, $h_1 \equiv 0 \pmod{I_k}$ for $k \neq 1$.

Now repeat the same process with $I = I_2$ and $J_1 = I_1, J_2 = I_3, \ldots, J_{n-1} = I_n$ to obtain an element h_2 of R with $h_2 \equiv 1 \pmod{I_2}$, $h_2 \equiv 0 \pmod{I_k}$ for $k \neq 2$. Keep repeating the process for $I = I_3, \ldots, I = I_n$ to obtain elements $h_1, \ldots, h_n$ of R with $h_k \equiv 1 \pmod{I_k}$ and $h_k \equiv 0 \pmod{I_{k'}}$ whenever $k' \neq k$. Now set

$$b = \sum_{k=1}^{n} h_k r_k.$$

Then b is a solution of this system of congruences.

Now if $b' \equiv b \pmod{I_1 \cap \cdots \cap I_n}$, then, as $I_1 \cap \cdots \cap I_n \subseteq I_k$ for each k, $b' \equiv b \pmod{I_k}$ for each k, so b' is also a solution. On the other hand, if b' is any solution, then $b' \equiv b \pmod{I_k}$ for each k, i.e., $b' - b \in I_k$ for each k, in which case $b' - b \in I_1 \cap \cdots \cap I_n$, i.e., $b' \equiv b \pmod{I_1 \cap \cdots \cap I_n}$. $\square$

Corollary 3.2.27. *Let $m_1, \ldots, m_n$ be pairwise relatively prime integers, and let $r_1, \ldots, r_n$ be arbitrary integers. Then there is an integer b satisfying the system of simultaneous congruences*

$$b \equiv r_1 \ (\mathrm{mod} \ m_1)$$

$$\vdots$$

$$b \equiv r_2 \ (\mathrm{mod} \ m_2)$$

$$b \equiv r_n \ (\mathrm{mod} \ m_n),$$

and an integer b' is a solution of this system if and only if $b' \equiv b$ (mod $m_1, \ldots, m_n$).

Proof. From Example 3.2.24, we see that this is just the special case of the Chinese remainder theorem when $R = \mathbb{Z}$. □

Remark 3.2.28. We will see later (Example 3.7.5) that there is an effective method for obtaining the ring elements $h_1, \ldots, h_n$ in the proof of the Chinese remainder theorem for many rings, including in particular $R = \mathbb{Z}$, and so there is an effective method for solving simultaneous congruences in these cases. ◊

We have stated Theorem 3.2.26 in terms of ideals, but we can restate it in terms of quotients.

Theorem 3.2.29 (Chinese remainder theorem). *Let R be a commutative ring and let $I_1, \ldots, I_n$ be pairwise coprime ideals of R. Then we have a ring isomorphism*

$$\bar{\varphi} \colon R/(I_1 \cap \cdots \cap I_n) \to (R/I_1) \times (R/I_2) \times \ldots (R/I_n).$$

Proof. Let $(r_1+I_1, r_2+I_2, \ldots, r_n+I_n)$ be any element of $(R/I_1) \times \cdots \times (R/I_n)$. Then by Theorem 3.2.26 there is a element b of R with $(b + I_1, b + I_2, \ldots, b + I_n) = (r_1 + I_1, r_2 + I_2, \ldots, r_n + I_n)$. In other words, if we define $\varphi \colon R \to (R/I_1) \times (R/I_2) \times \cdots \times (R/I_n)$ by

$$\varphi(x) = (x + I_1, x + I_2, \ldots, x + I_n)$$

Then φ is onto. What is $\mathrm{Ker}(\varphi)$? Certainly $0 \in \mathrm{Ker}(\varphi)$. But then by Theorem 3.2.26 again $\mathrm{Ker}(\varphi) = \{b' \in R \mid b' \equiv 0 \ (\mathrm{mod} \ I_1 \cap \cdots \cap I_n)\} = I_1 \cap \cdots \cap I_n$, so $(R/J_1) \times \cdots (R/I_n)$ is isomorphic to

$R/\mathrm{Ker}(\varphi) = R/(I_1 \cap \cdots \cap I_n)$. Indeed, tracing through the definition of the quotient we see that the isomorphism $\bar{\varphi}$ is given by

$$\bar{\varphi}(x + I_1 \cap \cdots \cap I_n) = (x + I_1, \ldots, x + I_n).$$ □

We have previously proven that the ring $\mathbb{Z}_{mn}$ is isomorphic to the ring $\mathbb{Z}_m \times \mathbb{Z}_n$ in case m and n are relatively prime. That proof was indeed correct, but here is the "right" proof of that result, which we state here for an arbitrary number of factors, not just two.

Corollary 3.2.30. *Let $m_1, \ldots, m_n$ be pairwise relatively prime integers. Then we have a ring isomorphism*

$$\bar{\varphi} \colon \mathbb{Z}_{m_1 \ldots m_n} \quad \to \quad \mathbb{Z}_{m_1} \times \cdots \times \mathbb{Z}_{m_n}.$$

Proof.   This corollary follows from Theorem 3.2.29 in the same way that Corollary 3.2.27 followed from Theorem 3.2.26.   □

3.3   The integers, and rings of polynomials

In this section, we want to look at some of the basic properties of $\mathbb{Z}$, as well as basic properties of polynomial rings.

Theorem 3.3.1 (The division algorithm in $\mathbb{Z}$). *Let $a \in \mathbb{Z}$ and let $b \in \mathbb{Z}$ with $b \neq 0$. Then there are integers q and r such that*

$$a = bq + r \qquad with \qquad 0 \leq r < |b|.$$

Proof.   We shall prove this in case $a \geq 0$ and $b > 0$. The other cases can be reduced to this case, and we leave them as an exercise.

We prove this by complete induction on a.

If $0 \leq a < b$ then we have $a = b0 + a$ so $a = bq + r$ with $q = 0$ and $r = a$, and by our assumption on a, $0 \leq r < |b|$.

Suppose that $a \geq b$ and the theorem is true for all nonnegative integers $< a$. Since $a \geq b$, $a - b \geq 0$, so by the inductive hypothesis $a - b = bq' + r$ for some integer q' and some r with $0 \leq r < |b|$. But then $a = bq + r$ with $q = q' + 1$ and r unchanged, so we still have $0 \leq r < |b|$. Thus, the theorem is true for a as well.

Then by induction we are done.   □

Now we turn our attention to polynomial rings.

Definition 3.3.2. Let R be a commutative ring with 1 and let $f(x) \in R[x]$ be a nonzero polynomial. If $f(x) = a_0 + \cdots + a_n x^n$, the *degree* of $f(x)$ is n. (In particular, a polynomial of degree 0 is a nonzero constant polynomial.) $\Diamond$

Remark 3.3.3. Note that the degree of the 0 polynomial is undefined. $\Diamond$

Lemma 3.3.4. *Let R be an integral domain. If $f(x)$ and $g(x)$ are polynomials in $R[x]$ of degrees m and n respectively, then $f(x)g(x)$ is a polynomial in $R[x]$ of degree $m + n$.*

Proof. If $f(x) = a_0 + \cdots + a_m x^m$ with $a_m \neq 0$ and $g(x) = b_0 + \cdots + b_n x^n$, with $b_n \neq 0$, then $f(x) = a_0 b_0 + \cdots + a_m b_n x^{m+n}$ and $a_m b_n \neq 0$ as R is an integral domain. $\square$

Definition 3.3.5. A polynomial is *monic* if the coefficient of its high-order term is equal to 1, i.e., a polynomial $f(x) = a_0 + \cdots + a_n x^n$ of degree n is monic if $a_n = 1$. $\Diamond$

Theorem 3.3.6 (The division algorithm for polynomials). *Let R be a commutative ring with 1. Let $f(x) \in R[x]$ be a polynomial and let $g(x) \in R[x]$ be a nonzero polynomial. Suppose that*

(1) $g(x)$ is monic; or
(2) R is a field.

Then there are unique polynomials $q(x)$ and $r(x) \in R[x]$ such that

$$f(x) = g(x)q(x) + r(x) \text{ with } r(x) = 0 \text{ or } \deg r(x) < \deg g(x).$$

Proof. We shall prove existence first and afterwards prove uniqueness.

First of all, if $f(x) = 0$, then $f(x) = g(x)q(x) + r(x)$ with $q(x) = 0$ and $r(x) = 0$.

Suppose that $f(x)$ is nonzero. Let $f(x) = a_0 + \cdots + a_m x^m$ have degree m and $g(x) = b_0 + \cdots + b_n x^n$ have degree n. In case (1) $b_n = 1$.

First we handle the case $n = 0$, In case (1) we have $f(x) = 1f(x) + 0$ and in case (2) we have $f(x) = b_0(b_0^{-1}f(x)) + 0$.

Now suppose $n > 0$. We proceed by complete induction on m. If $m < n$, then $f(x) = g(x)0 + f(x)$ and we let $r(x) = f(x)$.

Suppose the theorem is true for all polynomials of degree $< m$, and let $f(x)$ have degree m. In case (1), let $q_1(x) = a_m x^{m-n}$ and in case (2) let $q_1(x) = (a_m b_n^{-1})x^{m-n}$. Then $f(x)$ and $g(x)q_1(x)$ are both polynomials of degree n with leading coefficient (the coefficient of the high-order term x^n) equal to a_m.

If $f(x) = g(x)q_1(x)$, let $r(x) = 0$, and we are done.

Otherwise let $f'(x) = f(x) - g(x)q_1(x)$. Then $f'(x)$ has degree less than m, so by the inductive hypothesis there are polynomials $q'(x)$ and $r(x)$, with $r(x) = 0$ or $\deg r(x) < \deg g(x)$, such that $f' = g(x)q' + r(x)$. But then, if $q(x) = q_1(x) + q'(x)$,

$$f(x) = g(x)q(x) + r(x) \quad \text{with} \quad r(x) = 0 \quad \text{or} \quad \deg r(x) < \deg g(x)$$

Thus, the theorem is true if $f(x)$ has degree m as well.

Thus, by induction we are done.

Now we must prove uniqueness.

Suppose $f(x) = g(x)q_1(x) + r_1(x)$ and $f(x) = g(x)q_2(x) + r_2(x)$ with $r_1(x) = 0$ or $\deg r_1(x) < \deg g(x)$ and with $r_2(x) = 0$ or $\deg r_2(x) < \deg g(x)$. Then $g(x)q_1(x) + r_1(x) = g(x)q_2(x) + r(x)$ which yields the equation $g(x)(q_1(x) - q_2(x)) = r_2(x) - r_1(x)$.

Now the right hand side is either 0 or a polynomial of degree $< \deg g(x)$. If $q_1(x) \neq q_2(x)$, then $q_1(x) - q_2(x) \neq 0$ and, since $g(x)$ is monic or R is a field, the product of the leading coefficients of $g(x)$ and $q_1(x) - q_2(x)$ is nonzero, so the left hand side is a polynomial of degree $\geq \deg g(x)$. But if so it is impossible for the left hand and right hand side to be equal, so this cannot be the case. Thus, we must have $q_1(x) = q_2(x)$, in which case $r_1(x) = r_2(x)$ as well, and the expression for $f(x)$ is unique. $\square$

(You have surely noticed that the idea of the proof of the existence part of Theorem 3.3.6 was exactly the same as the idea of the proof of Theorem 3.3.1.)

Corollary 3.3.7. *Let R be a commutative ring with 1. Let $f(x) \in R[x]$ be a polynomial. Then for any $a \in R$, $f(x) = (x-a)q(x) + f(a)$ for some polynomial $q(x) \in R[x]$. In particular, $x - a$ divides $f(x)$ in $R[x]$ if and only if a is a root of $f(x)$, i.e., if $f(a) = 0$.*

Proof. By Theorem 3.3.6, $f(x) = (x-a)q(x)+c$ for some constant polynomial c, and then $f(a) = (a-a)q(a) + c = c$. □

Lemma 3.3.8. *Let R be an integral domain. Let $x - a$ divide the product $h(x) = f(x)g(x)$ of the polynomials $f(x)$, $g(x) \in R[x]$. Then $x - a$ divides $f(x)$ or $x - a$ divides $g(x)$.*

Proof. If $x - a$ divides $h(x)$, then $h(a) = 0$. But $h(a) = f(a)g(a)$. Thus $f(a)g(a) = 0$. Since R is an integral domain, (at least) one of the factors must be 0. So $f(a) = 0$, in which case $(x - a)$ divides $f(x)$, or $g(a) = 0$, in which case $x - a$ divides $g(x)$. □

Corollary 3.3.9. *Let R be an integral domain. Let $f(x) \in R[x]$ be a nonzero polynomial. If $f(x)$ has degree n, then $f(x)$ has at most n roots in R.*

Proof. By induction on n. If $n = 0$, $f(x)$ is a nonzero constant polynomial, so has no roots.

Assume the corollary is true for all polynomials of degree $n - 1$, and let $f(x)$ have degree n. If $f(x)$ does not have a root, then we are done. Suppose $f(x)$ has a root a. Then $f(x) = (x - a)q(x)$ for some polynomial $q(x)$ of degree $n - 1$. By the inductive hypothesis, $q(x)$ has at most $n - 1$ roots, so $f(x)$ has at most $n - 1 + 1 = n$ roots.

Then by induction we are done. □

These easy arguments, combined with the group theory we have done, have a powerful consequence.

Theorem 3.3.10. *Let R be an integral domain. Then any finite subgroup of R^*, the group of units of R, is cyclic.*

Proof. Let G be a finite subgroup of R^* and suppose that G has order n. Then $g^n - 1$ for every $g \in G^*$, i.e., every g in G^* is a root of the polynomial $f(x) = x^n - 1 \in R[x]$.

Now let d be any integer dividing n. If $g \in G$ has order dividing d, then by the same logic $g^d = 1$ so g is a root of the polynomial $f(x) = x^d - 1$ in $R[x]$. By Corollary 3.3.9, $f(x)$ has at most d roots in R. Thus, there are at most d elements of R, and hence at most d elements of G, of order dividing d.

Hence, by Corollary 2.2.6, G is cyclic. □

Here is the most important special case of this theorem.

Corollary 3.3.11. *Let R be a finite field. Then $R^* = R - \{0\}$ is cyclic. In particular, for any prime p, $\mathbb{Z}_p^*$ is cyclic.*

Proof.   If R is finite, R^* is finite, so we may directly apply Theorem 3.3.10. $\qquad\qquad\qquad\qquad\qquad\qquad\qquad\qquad\qquad\qquad\qquad\qquad\qquad\qquad$ □

Remark 3.3.12. Note that in Lemma 3.3.8, Corollary 3.3.9, and Theorem 3.3.10 we had the hypothesis that R *is an integral domain.* This hypothesis is necessary and in general these results are false without it.

For example, let $R = \mathbb{Z}_8$. Note that $9 \equiv 1 \pmod 8$. Thus $x^2 - 9 = x^2 - 1$ in $R[x]$. Of course, $x - 3$ divides $x^2 - 9 = (x - 3)(x + 3)$. Thus $x - 3$ divides $x^2 - 1 = (x - 1)(x + 1)$ but clearly $x - 3$ does not divide either factor. Also, note that $1^2 = 3^2 = 5^2 = 7^2 = 1$ in $\mathbb{Z}_8$, so the quadratic polynomial $x^2 - 1$ has the four roots 1, 3, 5 and 7 in $\mathbb{Z}_8$.

Finally, note that $\mathbb{Z}_8^* = 1, 3, 5, 7$ and this group is isomorphic to $\mathbb{Z}_2 \times \mathbb{Z}_2$, not to the cyclic group $\mathbb{Z}_4$. $\qquad\qquad\qquad\qquad\qquad\qquad\qquad$ ◊

We now prove one of the standard results of ring theory.

Theorem 3.3.13 (Hilbert basis theorem). *Let R be a commutative Noetherian ring with 1. Then $R[x]$ is a commutative Noetherian ring with 1.*

Proof.   Let I be an ideal of $R[x]$. If $I = \{0\}$ then I is certainly finitely generated, so suppose not.

For a nonzero polynomial $f(x) \in R[x]$, let $h(f(x))$ be the coefficient of its high-order term, i.e., if $f(x) = a_0 + \cdots + a_n x^n$, then $h(f(x)) = a_n$. Define ideals $I_0 \subseteq I_1 \subseteq I_2 \subseteq \ldots$ of R by

$$I_n = \{0\} \cup \{h(f(x)) \mid f(x) \in R[x]$$

is a polynomial of degree at most $n\}$.

(Note we have to include 0 separately as it cannot be a high-order coefficient.) It is easy to check that each I_n is an ideal of R.

Since R is Noetherian, this sequence is eventually constant.

Choose a value N for which $I_N = I_{N+1} = I_{N+2} = \ldots$.

For any value of n, let

$$I_n' = 0 \cup \{h(f(x)) \mid f(x) \in R[x] \text{ is a polynomial of degree exactly } n\}.$$

Certainly $I'_n \subseteq I_n$. We claim that in fact $I'_n = I_n$. To see this, suppose we have $r \in I_n$, so that there is some polynomial $f(x) \in R[x]$ of degree $k \leq n$ with $h(f(x)) = r$. If $k = n$, Then $r \in I'_n$, If $k < n$, note that $x^{n-k}f(x)$ is a polynomial of degree n with $h(x^{n-k}f(x)) = h(f(x)) = r$, so in this case, too, $r \in I'_n$. Thus $I'_n \subseteq I_n$, and so they are equal.

Now R is Noetherian, so each ideal I_n is finitely generated. For each $n = 0, 1, \ldots, N$ let $\{f_{n1}(x), \ldots, f_{nk_n}\}$ be a set of polynomials of degree exactly n that generate I_n. (We can choose the degree to be exactly n precisely because $I'_n = I_n$.) We claim the set

$$S = \{f_{01}(x), \ldots, f_{0k_0}(x), f_{11}(x), \ldots, f_{1k_1}(x),$$

$$\ldots, f_{N1}(x), \ldots, f_{Nk_N}(x)\}$$

generates I. Since this set is finite, I is finitely generated, and since I was an arbitrary ideal of $R[x]$, this shows $R[x]$ is Noetherian.

Thus we must prove this claim. Let J be the ideal of $R[x]$ generated by S. Since every element of S is in I, we certainly have $J \subseteq I$. We need to show $I \subseteq J$. In other words, what we need to show is:

Let $f(x) \in R[x]$ be a polynomial in I. Then $f(x) \in J$.

This is certainly true if $f(x) = 0$. Suppose $f(x)$ is nonzero. We prove the claim by induction on $n = \deg(f(x))$. Write $f(x) = a_0 + \cdots + a_n x^n$.

If $n = 0$, then $f(x) = a_0$, a constant polynomial, and $h(f(x)) = a_0$. Then $\{f_{01}(x) = a_{01}, \ldots, f_{0k_0}(x) = a_{0k_0}\}$ generates $I'_0 = I_0$, so $a_0 = r_{01}a_{01} + \cdots + r_{0k_0}a_{0k_0}$ for some elements $r_{01}, \ldots, r_{0k_0}$ of R, i.e., $f(x) = r_{01}f_{01}(x) + \cdots + r_{0k_0}f_{0k_0}(x)$, and $f(x) \in J$.

Now suppose the claim is true for all polynomials of degree $< n$, and suppose $f(x)$ has degree n. There are two cases:

Case 1: $n \leq N$. Then by the definition of $\{f_{n1}(x), \ldots, f_{nk_n}(x)\}$, these polynomials all have degree n, and if these polynomials have leading coefficients $a_{n1}, \ldots, a_{nk_n}$ we have $a_n = r_{n1}a_{n1} + \cdots + r_{nk_n}a_{nk_n}$, for some $r_{n1}, \ldots, r_{nk_n}$ in R. In other words, the polynomials $f(x)$ and $g(x) = r_{n1}f_{n1}(x) + \cdots + r_{nk_n}f_{nk_n}(x)$ are polynomials with the same degrees and with the same leading coefficient, so either $f(x) = g(x) \in J$, or $h(x) = f(x) - g(x)$ is a polynomial of degree $< n$, with $h(x) \in I$. But then by the inductive hypothesis $h(x) \in J$, and so $f(x) = g(x) + h(x) \in J$ as well.

Case 2: $n > N$. In this case $I_n = I_N$. Then by the definition of $\{f_{N1}(x), \ldots, f_{Nk_N}(x)\}$ these polynomials all have degree N, and if these polynomials have leading coefficients $a_{N1}, \ldots, a_{Nk_N}$, we have $a_n = r_{N1}a_{N1} + \cdots + r_{Nk_N}a_{Nk_N}$ for some $r_{N1}, \ldots, r_{Nk_N}$ in R. Now the polynomials $f(x)$ and $g(x) = r_{N1}x^{n-N}f_{N1}(x) + \cdots + r_{Nk_N}x^{n-N}f_{Nk_N}(x)$ are polynomials with the same degree and with the same leading coefficient, so, applying the some inductive argument as in Case 1, we conclude $f(x) \in J$ in this case as well.

Then by induction we are done. □

Corollary 3.3.14.

(a) *Let R be field. Then for any positive integer n, $R[x_1, \ldots, x_n]$ is a Noetherian ring.*

(b) *Let $R = \mathbb{Z}$. Then for any positive integer n, $R[x_1, \ldots, x_n]$ is a Noetherian ring.*

Proof. We saw in Example 3.2.15 that any field is a Noetherian ring, as is $\mathbb{Z}$. Then this corollary follows from Theorem 3.3.13 by induction on the number of variables. □

3.4 Euclidean domains and principal ideal domains

We have been assuming, and using, basic properties of the integers $\mathbb{Z}$ all along. Now we would like to prove them. Why have we not proved them before now? The reason is that we want to prove them in a more general context. The context is that of principal ideal domains (PIDs).

That is, we want to show:

(1) $\mathbb{Z}$ is a PID; and
(2) PIDs have important properties.

We can then conclude that $\mathbb{Z}$ has these properties.

Of course, if $\mathbb{Z}$ were the only PID we would not have gained anything by this approach. But it is not. There are a variety of rings that are PIDs. So the advantage of our strategy is that we can prove these properties once and for all, and then they hold for all PIDs, rather than having to go back and reprove them everytime we encounter a new PID. Not only would this be inefficient, it would be repetitious, as the basic ideas of the different proofs would be all the same. Even worse, we could never get a complete proof, as there might always be a new PID we hadn't thought of. These considerations aside, when we examine proofs, we would like to distill out their essential ideas, and then see how to use them most effectively, and that is what we will be doing here. Historically speaking, these results were first proved for $\mathbb{Z}$ (actually, first proved in the context of the positive integers) and then, millennia later, generalized. With the benefit of hindsight, we will be proving these results in general and then obtaining them for $\mathbb{Z}$, and certain other rings, as special cases.

You'll notice I have been talking about step (2) here, and indeed talking very vaguely, as I have not even told you what these properties are. So I will have the ask you to believe me when I tell you that they are important while we carry out step (1). (First things first!) But I have said we do not just want to do this for $\mathbb{Z}$, but for other rings R as well. Thus our step (1) will be

(1) Show that certain rings are PIDs.

But actually, we will be doing step (1) in two sub-steps

(1a) Show that certain rings are Euclidean domains.
(1a) Show that every Euclidean domain is a PID.

The point of doing things this way is that we have a concrete numerical criterion for showing that a ring is a Euclidean domain, and hence a PID, and we will see some concrete applications of this criterion.

Also, the reason some rings are called Euclidean domains is that we can perform Euclid's algorithm in them. Euclid's algorithm is an algorithm, which means we can use it to perform effective (and, it turns out, efficient) computations, which we will illustrate.

I should point out that step (1b) says that every Euclidean domain is a PID, but does *not* say that every PID is a Euclidean domain. In fact, that is false — there are PIDs that are not Euclidean domains. But these do not play as important a role as Euclidean domains, so we will not concern ourselves with any of them.

We now go to work.

Recall that $\mathbb{N}$ denotes the natural numbers, i.e., the positive integers.

We will let $\bar{\mathbb{N}}$ denote the nonnegative integers, $\bar{\mathbb{N}} = \mathbb{N} \cup \{0\}$. (This is not standard notation.)

Definition 3.4.1. Let R be an integral domain. A *norm* δ on R is a function $\delta\colon R - \{0\} \to \bar{\mathbb{N}}$ such that $\delta(a) \leq \delta(ab)$ for all $a, b \in R - \{0\}$. The norm δ on R is a *Euclidean norm* if for any $a \in R$ with $b \neq 0$, there are elements q and r of R with

$$a = bq + r \quad \text{where } r = 0 \text{ or } \delta(r) < \delta(b).$$

If R has a Euclidean norm δ, then R is a *Euclidean domain* (with respect to δ). $\Diamond$

There is a kind of norm that is, as we shall see, particularly easy to work with.

Definition 3.4.2. Let R be an integral domain. A norm, or a Euclidean norm, on R is *multiplicative* if $\delta(0) = 0$, $\delta(1) = 1$, and $\delta(ab) = \delta(a)\delta(b)$ for all $a, b \in R$. $\Diamond$

Theorem 3.4.3. *The function* $\delta(a) = |a|$ *is a multiplicative Euclidean norm on* $\mathbb{Z}$, *and hence* $\mathbb{Z}$ *is a Euclidean domain.*

Proof.   This is Theorem 3.3.1. $\square$

Theorem 3.4.4. *Let* R *be a field. The function* $\delta(p(x)) = \deg p(x)$ *is a Euclidean norm on* $R[x]$, *and hence, if* R *is a field,* $R[x]$ *is a Euclidean domain.*

Proof.   This is Theorem 3.3.6. $\square$

We have the following general properties of norms.

Lemma 3.4.5. *Let* R *be an integral domain with norm* δ.

(a) *For any* $r \in R$, $r \neq 0$, $\delta(r) \geq \delta(1)$.
(b) *If* u *is a unit in* R, $\delta(u) = \delta(1)$.
(c) *If* R *is a Euclidean domain with norm* δ, *and* $r \in R$ *with* $\delta(r) = \delta(1)$, *then* r *is a unit in* R.

Proof.

(a) $\delta(r) = \delta(1r) \geq \delta(1)$.

(b) If u is a unit, let $v \in R$ with $uv = 1$. Then $\delta(1) = \delta(uv) \geq \delta(u)$ and $\delta(u) \geq \delta(1)$ by (a), so $\delta(u) = \delta(1)$.

(c) Since R is a Euclidean domain, we have that $1 = rs + t$ for some $s \in R$ and some $t \in R$ with $t = 0$ or $\delta(t) < \delta(r)$. But if $\delta(r) = \delta(1)$ there are no elements t of R with $\delta(t) < \delta(r)$, so we must have $t = 0$, $1 = rs$, and so r is a unit in R.

$\square$

Now we specialize our attention.

Lemma 3.4.6.

(a) *Let D be a squarefree integer and let δ be the function on $\mathbb{Q}\sqrt{D}$ defined by $\delta(z) = |z\bar{z}|$. (Recall that if $z = a + b\sqrt{D}$, with $a, b \in \mathbb{Q}$, then $\bar{z} = a - b\sqrt{D}$.) Then $\delta(z_1 z_2) = \delta(z_1)\delta(z_2)$ for all z_1, z_2 in $\mathbb{Q}(\sqrt{D})$.*

(b) *The function δ restricts to a multiplicative norm on $\mathcal{O}(\sqrt{D})$*

Proof. Part (a) is direct calculation, which works just like calculations in the complex numbers. (Note that $\delta(a + b\sqrt{D}) = |a^2 - b^2 D|$.) Then to show part (b), we simply need to show that $\delta(z) \in \bar{\mathbb{N}}$ whenever $z \in \mathcal{O}(\sqrt{D})$. This is clear when $D \equiv 1$ or $3 \pmod 4$, as then a and b are integers. We leave the more interesting case $D \equiv 1 \pmod 4$ to the reader.

$\square$

It is certainly not the case that δ is always, or even usually, a Euclidean norm on $\mathcal{O}(\sqrt{D})$. In fact, this is the exception rather than the rule. But it is true in the following particularly important case.

Theorem 3.4.7. *The function δ is a multiplicative Euclidean norm on $\mathcal{O}(\sqrt{-1}) = \mathbb{Z}[i]$. Consequently the Gaussian integers $\mathbb{Z}[i]$ are a Euclidean domain.*

Proof. Let $a = w + xi \in \mathbb{Z}[i]$ and $b = y + zi \in \mathbb{Z}[i]$, $b \neq 0$. Then w, x, y, and z are integers with not both y and z equal to 0.

We have shown that $\mathbb{Q}(i)$ is a field (Example 3.1.16), so we divide a by b in $\mathbb{Q}(i)$. Then

$$\frac{a}{b} = \frac{w + xi}{y + xi} = \frac{w + xi}{y + zi} \cdot \frac{y - zi}{y - zi} = \frac{wy + xz + (-wz + xy)i}{y^2 + z^2} = u + vi$$

where

$$u = \frac{wy + xz}{y^2 + z^2} \in \mathbb{Q} \quad \text{and } v = \frac{-wz + xy}{y^2 + z^2} \in \mathbb{Q}.$$

If (we are very lucky and) $u \in \mathbb{Z}, v \in \mathbb{Z}$, set $q = u + vi$. Then $q \in \mathbb{Z}[i]$ and $a/b = q$, i.e., $a = bq = bq + 0$ so set $r = 0$ and we are done.

Suppose not. Choose integers m and n with $|u - m| \leq 1/2, |v - n| \leq 1/2$, and let $q = m + ni$. Then

$$
\begin{aligned}
a = b(a/b) = b(u + vi) &= b((m + (u - m)) + (n + (v - n))i) \\
&= b(m + ni) + b((u - m) + (v - u)i) \\
&= bq + r
\end{aligned}
$$

where

$$r = b((u - m) + (v - n)i).$$

First let us observe that, since $\mathbb{Z}[i]$ is a ring, $r = a - bq \in \mathbb{Z}[i]$.

The key thing we have to show is that $\delta(r) < \delta(b)$. To this end, let

$$s = (u - m) + (v - n)i \in \mathbb{Q}(i).$$

Then $\delta(s) = (u - m)^2 + (v - n)^2 \leq (1/2)^2 + (1/2)^2 = 1/2$ and so, since δ is multiplicative on $\mathbb{Q}(i)$,

$$\delta(r) = \delta(bs) = \delta(b)\delta(s) \leq (1/2)\delta(b) < \delta(b)$$

as required.   □

Remark 3.4.8. Observe that for any D, if $z \in \mathcal{O}(\sqrt{D})$ with $\delta(z) = 1$, then z is a unit: The equation $\delta(z) = 1$ is the equation $|z\bar{z}| = 1$, i.e., $z\bar{z} = \pm 1$, so $z(\pm\bar{z}) = 1$ and z is a unit.   ◇

Now we come to principal ideal domains. We begin with the definition.

Definition 3.4.9. Let R be a commutative ring with 1. An ideal I of R is *principal* if it is generated by a single element r_0 of R, or, equivalently, if $I = \{rr_0 \mid r \in R\}$.

An integral domain R is a *principal ideal domain* (PID) if every ideal in R is principal.   ◇

Remark 3.4.10. If R is a field, then R is a PID, as R only has two ideals, $\{0\}$, generated by $0 \in R$, and R, generated by $1 \in R$. As we will see, we are mostly interested in PIDs because of questions of divisibility, and these questions are completely uninteresting for fields, as every nonzero element of a field is a unit. $\diamond$

Here is our next sub-step.

Theorem 3.4.11. *Let R be a Euclidean domain. Then R is a PID.*

Proof. We have to show that every ideal I of R is principal.
If $I = \{0\}$, then I is generated by the single element 0.
Suppose that $\{I \neq \{0\}\}$. Consider the following set S:

$$S = \{\delta(r) \mid r \in I, r \neq 0\}.$$

Note that S is a nonempty subset of $\bar{\mathbb{N}}$. Thus, S has a smallest element s_0. Let $r_0 \in I$ with $\delta(r_0) = s_0$. We claim that the single element r_0 generates I, or, in other words, that $I = I_{r_0} = \{rr_0 \mid r \in R\}$. Certainly, $I_{r_0} \subseteq I_0$, as I is an ideal, so we need to show $I \subseteq I_{r_0}$. To this end, let r be an arbitrary element of I. Since R is a Euclidean domain, we know that

$r = r_0 q + t$ for some $q \in R$ and some $t \in R$ with $t = 0$ or $\delta(t) < \delta(r_0)$.

Now I is an ideal, $r \in I$ and $r_0 \in I$, so $t = r - r_0 q \in I$.
If t were not zero, it would be an element of I with $\delta(t) < \delta(r_0) = s_0$, which is impossible, as s_0 is the *smallest* norm of any nonzero element of I. Hence, $t = 0$, $r = r_0 q$, so $r \in I_0$, as required. $\square$

Corollary 3.4.12. *The following rings are PIDs:*

(a) $\mathbb{Z}$
(b) $R[x]$ *for any field R*
(c) $\mathbb{Z}[i]$

Proof.

(a) Immediate from Theorem 3.4.3 and Theorem 3.4.11.
(b) Immediate from Theorem 3.4.4 and Theorem 3.4.11.
(c) Immediate from Theorem 3.4.7 and Theorem 3.4.11. $\square$

3.5   Integral domains and divisibility

One of the main reasons we are interested in PIDs is in connection with questions of divisibility and factorization. But we will start by discussing questions of divisibility in integral domains in general.

Definition 3.5.1. Let R be an integral domain. Let a and b be elements of R with $b \neq 0$. Then a is *divisible by* b, or is a *multiple of* b, or b *divides* a, or is a *factor* of a, if there is an element q of R with $a = bq$. We write this as $b|a$.   ◊

Definition 3.5.2. Let R be an integral domain, and let a be a nonzero element of R.

(a) a is a *unit* of R if a divides 1.
(b) a is *irreducible* in R if a is not a unit of R and whenever $a = bc$, $b, c \in R$, then b is a unit or c is a unit.
(c) a is *prime* in R if a is not a unit and whenever a divides a product bc, $b, c \in R$, then a divides b or a divides c.   ◊

Remark 3.5.3. You may be a bit surprised by Definition 3.5.2. You are undoubtedly familiar with (and, indeed, we have used many times in this book so far) the notion of a prime in $\mathbb{Z}$, and that looks very much like the definition of an irreducible element in a general integral domain rather than that of a prime. But it turns out that the correct generalization of the notion of a prime in $\mathbb{Z}$ is a prime in R as we have stated it.   ◊

The notions of prime and irreducible in a general integral domain are distinct, but there is a close relationship between them.

Lemma 3.5.4. *Let R be an integral domain, and let a be prime in R. Then a is irreducible in R.*

Proof.   Suppose that a is prime in R. Let $a = bc$. We have to show that b is a unit or c is a unit. Now if $a = bc$, then certainly a divides bc ($bc = a = a1$). Since a is a prime, a divides b or a divides c. Suppose that a divides b, and write $b = ab'$. Then

$$a1 = a = bc = (ab')c = a(b'c)$$

so by cancellation in an integral domain (Lemma 3.1.14) $1 = b'c$ and c is a unit. Similarly, if a divides c then b is a unit.   □

Remark 3.5.5. As we shall see, it is *not* always true that an irreducible element of an integral domain R is prime, although in many important cases, including the case of R a PID, it *is*. ◊

Definition 3.5.6. Let R be an integral domain and let a and b be nonzero elements of R. Then a and b are *associates* if a divides b and b divides a. We write this as $a \cong b$. (This is not standard notation.) ◊

Lemma 3.5.7. *Let R be an integral domain.*

(a) $a \cong b$ *is an equivalence relation on* $R - \{0\}$
(b) $a \cong b$ *if and only if* $a = bu$ *for some unit u, or $b = av$ for some unit v*

Proof. We leave this as an exercise for the reader. □

Remark 3.5.8. Note that if b divides a and b' is an associate of b, then b' divides a as well. If $b' = bu$ for a unit u of R, and $uv = 1$, then if $a = bq$, $a = b1q = b(uv)q = (bu)(vq) = b'q'$ for $q' = vq$. ◊

Definition 3.5.9. Let $\{a_i\}$ be a set of elements in an integral domain R. Then g is a *greatest common divisor (gcd)* of $\{a_i\}$ if

(1) g divides a_i for each i; and
(2) if d is any element of R that divides each a_i, then d divides g. ◊

Remark 3.5.10. How do we know that $\{a_i\}$ *has* a gcd? That is, how do we know that there *is* an element g of R satisfying these two conditions? The answer is, we don't. As we shall see, there are integral domains R in which gcd's do not in general exist. But as we shall also see, there are many cases in which they do. In particular, we shall see that if R is a PID, gcd's always exist. Not only that, we will see that they have a stronger property. ◊

Remark 3.5.11. Again you may be puzzled by the term gcd. What does "greatest" mean? In fact, looking at the definition, a gcd should really be called a most divisible common divisor, rather than a greatest one. We use this term for historical reasons. The notion of a gcd goes back to Euclid, who considered the positive integers, and there the gcd of a and b was indeed the greatest (i.e., largest) common divisor of a and b. But when mathematicians generalized this notion to integral domains, we kept the term. ◊

Although we don't know that a gcd exists, let us proceed for the moment and assume that it does, and see what we can say about it.

Lemma 3.5.12. *Suppose that $\{a_i\}$ has a gcd g. Then g' is a gcd of $\{a_i\}$ if and only if g' and g are associates. Thus, if g is a gcd of $\{a_i\}$, all gcd's of $\{a_i\}$ are given by $g' = gu$ for some unit u of R.*

Proof. By property (2) of a gcd, g divides g' and g' divides g, so they are associates. Then the conclusion follows from Remark 3.5.8. □

Remark 3.5.13. From Lemma 3.5.12 we see that it is improper to speak of *the* gcd of $\{a_i\}$; rather we need to speak of *a* gcd of $\{a_i\}$. We will use the (nonstandard) notation $g \cong \gcd(\{a_i\})$ in this case. This is related to, but not the same as, the notation $a \cong b$ in Definition 3.5.5, but we use the same notation because these are closely related. Similarly, we will use the notation $\gcd(\{a_i\}) \cong \gcd(\{b_i\})$ to mean that any gcd of $\{a_i\}$ is a gcd of $\{b_i\}$; again this is another use of this notation, but again we continue to use it because of the close relation of these ideas. (Strictly speaking, we should write $\gcd(\{a_i\})$ for the set of all gcd's of $\{a_i\}$ and then write $g \in \gcd(\{a_i\})$ to mean that g is a gcd of $\{a_i\}$, But this notation is clumsy and unintuitive. Nobody uses it, and we won't either.) ◊

Remark 3.5.14. Observe that, in our language, the principal ideal $\{rr_0 \mid r \in R\}$ in a commutative ring with 1 generated by the element r_0 consists precisely of the multiples of r_0 in R. ◊

Theorem 3.5.15. *Let R be a principal ideal domain (PID). Let $\{a_i\}$ be a set of elements of R, not all zero. Then $\{a_i\}$ has a gcd g. Furthermore, g is a generator of the ideal I generated by $\{a_i\}$, and so g can be written as*

$$g = \sum a_i b_i \quad \text{for some} \quad \{b_i\} \subseteq R,$$

a finite sum (i.e., all but finitely many $b_i = 0$).

Proof. Let I be the ideal of R generated by $\{a_i\}$. Since R is a PID, I is principal, so is generated by a single element g. We claim that g is a gcd of $\{a_i\}$.

First of all, property (1) is true as $a_i \in I$ for each i, and I consists precisely of the multiples of g (Remark 3.5.14).

Now note that, since $g \in I$, we can write g as

$$g = \sum a_i b_i \qquad \text{for some} \quad \{b_i\} \in R,$$

a finite sum (Definition 3.2.13).

Suppose that d divides each a_i, so that $a_i = da_i'$ for each i. Then

$$g - \sum a_i b_i = \sum (da_i') b_i = d \left(\sum a_i' b_i \right)$$

so d divides g, and property (2) is true as well.   □

Corollary 3.5.16. *In any one of the following rings:*

(a) $\mathbb{Z}$
(b) $R[x]$ *for any field R*
(c) $\mathbb{Z}[i]$

the conclusion of Theorem 3.5.15 holds.

Proof.   Immediate from Corollary 3.4.12.   □

In order to aid us in our theoretical development, let us define a new object.

Definition 3.5.17. An integral domain R is a *GCD domain* (resp. an *f-GCD domain* if every set (resp. every finite set) $\{a_i'\}$ of elements of R, not all zero, has a gcd.   ◇

Remark 3.5.18. We observe that every GCD domain is an f-GCD domain, and that every Noetherian f-GCD domain is a GCD domain.   ◇

Remark 3.5.19. We have been focusing on, and will continue for a while to focus on, theoretical properties of gcd's. But you may well ask, how do we in practice go about finding them. We will later see a very effective method, Euclid's algorithm, for computing $\gcd(\{a_1, a_2\})$ for a pair of nonzero elements a_1 and a_2 of a Euclidean domain R.   ◇

We record a couple of results now that will be very useful in our computations later.

Lemma 3.5.20. *Let R be an f-GCD domain.*

(a) *Let $a \in R$, $a \neq 0$. Then $a \cong gcd(\{a\})$.*
(b) *Let $b \in R$, $b \neq 0$. If a is any element of R such that b divides a, then $b \cong gcd(\{a, b\})$. In particular, $b \cong gcd(\{b, 0\})$.*
(c) *Let $b \in R$, $b \neq 0$. Let $a \in R$ and let q and r be any elements of R with $a = bq + r$. Then $gcd(\{a, b\}) \cong gcd(\{b, r\})$.*

Proof.   (a) is immediate and (b) is almost immediate as every element of R divides 0. To show (c), we will show that a and b, and b and r, have exactly the same common divisors. Then they will certainly have the same gcd. To that end, suppose that d divides both a and b. Then d will divide b (by assumption) and $r = a - bq$. On the other hand, suppose d divides both b and r. Then d will divide b (by assumption) and $a = bq + r$.   □

Lemma 3.5.21. *Let R be a GCD domain (resp. an f-GCD domain) and let B and C be sets (resp. finite sets) of elements of R, not all zero. Let $A = B \cup C$. Then $gcd(A) \cong gcd(\{gcd(B), gcd(C)\})$.*

Proof.   Let $g_A = gcd(A)$, $g_B = gcd(B)$, $g_C = gcd(C)$, and $g = gcd(\{g_B, g_C\})$. We want to show that $g_A \cong g$.

By definition, g_A divides every element of A, so it divides every element of B, in which case it divides g_B, and it divides every element of C, in which case it divides g_C. Thus g_A is a common divisor of g_B and g_C, so g_A divides $g = gcd(\{g_B, g_C\})$.

On the other hand, by definition g divides both g_B and g_C. Since g divides g_B, it divides every element of B, and g divides g_C, so g divides every element of C. Thus g divides every element of $A = B \cup C$, so g divides g_A.   □

Henceforth we will simplify our notation (as is standard) and write $g \cong gcd(a_1, a_2)$ rather than $g \cong gcd(\{a_1, a_2\})$, $g \cong gcd(a_1, a_2, a_3)$ rather than $g \cong gcd(\{a_1, a_2, a_3\})$, etc.

Remark 3.5.22. Lemma 3.5.21 has a very practical application. Suppose we can compute the gcd of any two elements of R (see Remark 3.5.19). Then we can inductively compute the gcd of any finite number of elements of R:

$$gcd(a_1, a_2, a_3) = gcd(a_1, gcd(a_2, a_3)),$$

$$gcd(a_1, a_2, a_3, a_4) = gcd(a_1, gcd(a_2, a_3, a_4)), \text{ etc.}   \lozenge$$

We now define a property that will, as we will see, play a key role.

Definition 3.5.23.

(a) Let R be a GCD domain (resp. an f-GCD domain). A set (resp. a finite set) $\{a_i\}$ of elements of R is *relatively prime* if $1 \cong \gcd(\{a_i\})$.

(b) Let R be an f-GCD domain. A set $\{a_i\}$ of elements of R is *pairwise relatively prime* if $1 \cong \gcd(a_i, a_j)$ whenever $i \neq j$. ◇

To see the difference between these two notions, observe that $\{6, 10, 15\}$ is a relatively prime subset of $\mathbb{Z}$ but is *not* pairwise relatively prime.

Lemma 3.5.24. *Let R be a GCD-domain (resp. an f-GCD domain) and let $\{a_i\}$ be a set (resp. a finite set) of elements of R, not all zero. Let $g \cong \gcd(\{a_i\})$.*

(a) *Let d be a common divisor of $\{a_i\}$, (i.e., d is a divisor of each a_i). Then $g/d \cong \gcd(\{a_i/d\})$. In particular, $\{a_i/d\}$ is relatively prime if and only if $d \cong g$.*

(b) *Let m be any nonzero element of R. Then $gm \cong \gcd(\{ma_i\})$.*

Proof. For simplicity, we will assume that our set just consists of two elements of R, which we write as $\{a, b\}$.

(a) Let $a = da'$ and $b = db'$. Since d is a common divisor of a and b, d divides g. Write $g = dg'$. We claim that $g' \cong \gcd(a', b')$. To show this we must show g' satisfies both properties of a gcd:

   (1) We have that g divides a, i.e., that dg' divides da', and hence by cancellation that g' divides a'; similarly g' divides b'.

   (2) Suppose that h' is a common divisor of a' and b'. Then h' divides a' so dh' divides $da' = a$; similarly dh' divides b. Thus dh' is a common divisor of a and b, so dh' divides $g = dg'$. Then, by cancellation, h' divides d'.

(b) This follows from part (a) by changing our point of view. Let $\tilde{a} = ma$, $\tilde{b} = mb$, and $\tilde{g} \cong \gcd(\tilde{a}, \tilde{b})$. Then $a = \tilde{a}/m$, $b = \tilde{b}/m$ so by part (a), if $g \cong \gcd(a, b)$, then $\tilde{g}/m \cong g$, i.e., $\tilde{g} \cong mg$. □

Now we come to a result that plays a key role.

Lemma 3.5.25 (Euclid's lemma). *Let R be an f-GCD domain. Let a be any nonzero element of R. Let b and c be elements of R and suppose that a divides bc. If a and b are relatively prime, then a divides c.*

Proof.   Let $d \cong \gcd(ac, bc)$. Since c divides both ac and bc, c divides d. Write $d = ce$. Now d divides bc, i.e., ce divides bc, so e divides b. Also, d divides ac, i.e., ce divides ac, so e divides a. Thus e divides $\gcd(a, b)$. But a and b are assumed to be relatively prime, i.e., $1 \cong \gcd(a, b)$, so e divides 1, i.e., e is a unit, and so $e \cong 1$, and $d \cong c$.

Now a certainly divides ac, and a divides bc by hypothesis, so a divides d, and hence, since $d \cong c$, a divides c.   □

Remark 3.5.26. This proof was short, though a bit tricky. We will see that in the case of a PID, Euclid's lemma has an even shorter and more straightforward proof.   ◊

Remark 3.5.27. Euclid's lemma is false in general without the assumption that a and b are relatively prime. For example, in $\mathbb{Z}$, 6 divides $210 = 10 \cdot 21$ without dividing either factor.   ◊

Here are two important consequences of Euclid's lemma.

Corollary 3.5.28. *Let R be an f-GCD domain and let a and b be nonzero elements of R. Let c be an element of R and suppose that a divides c and b divides c. If a and b are relatively prime, then ab divides c.*

Proof.   Since a divides c, we may write $c = ad$ for some element d of R. Then b divides ad, and b and a are relatively prime, so, by Euclid's lemma, b divides d. Write $d = be$ for some element e of R. Then

$$c = ad = a(be) = (ab)e$$

and so ab divides c.   □

Remark 3.5.29. This corollary is false in general without the assumption that a and b are relatively prime. For example, 6 divides 30 and 10 divides 30, but $6 \cdot 10 = 60$ does not divide 30.   ◊

Corollary 3.5.30. *Let R be an f-GCD domain and let a, b, and c be elements of R. If a and b are relatively prime, and a and c are relatively prime, then a and bc are relatively prime.*

Proof. Let $d \cong \gcd(a, bc)$. Let $e \cong \gcd(d, b)$. Since d divides a, e divides $\gcd(a, b) \cong 1$. Thus $e \cong 1$, i.e., d and b are relatively prime.

Now d divides bc, and d and b are relatively prime, so, by Euclid's lemma, d divides c. Thus d is a common divisor of a and c. But a and c are relatively prime, so $d \cong 1$, i.e., a and bc are relatively prime. $\square$

Here is a third, particularly important, consequence of Euclid's lemma.

Corollary 3.5.31. *Let R be an f-GCD domain and let a be an element of R. Then a is prime if and only if a is irreducible.*

Proof. We already know, by Lemma 3.5.4, that in any integral domain, every prime is irreducible. So we must show that if R is an f-GCD domain, every irreducible is prime.

Let $a \in R$ be irreducible, and let d be a divisor of a. By the definition of an irreducible element, there are only two (mutually exclusive) possibilities: $d \cong 1$ or $d \cong a$. Now suppose that a divides bc. We must show that a divides b or a divides c. If a divides b, we are done. Suppose not. Then, since $d \cong \gcd(\{a, b\})$ is a divisor of a, we must have $d \cong 1$, i.e., a and b are relatively prime. Then, by Euclid's lemma, a divides c. $\square$

Remark 3.5.32. We want to point out an important detail about the conclusion of Lemma 3.5.24. Let R be an f-GCD domain and let $a, b \in R$, not both zero. Let $g \cong \gcd(a, b)$. Write $a = ga'$, $b = gb'$. Then Lemma 3.5.24 (a) tells us that a' and b' are relatively prime. But it is *not* necessarily the case that a' and g are relatively prime, or that b' and g are relatively prime. Here is an example to illustrate this. Let $R = \mathbb{Z}$, let $a = 12$, and let $b = 18$. Then $g = 6$, and $a = 6 \cdot 2$, $b = 6 \cdot 3$. Sure enough, 2 and 3 are relatively prime, but neither 2 and 6, nor 3 and 6, are relatively prime. $\Diamond$

3.6 Principal ideal domains and unique factorization domains

In this section we focus on principal ideal domains (PIDs) and complete step (2) of our program, to show that

(2) PIDs have important properties.

We can break step (2) into sub-steps, to show that

(2a)   PIDs have a variety of important properties, and

(2b)   PIDs have a particularly important property, that of "unique factorization" (which we have yet to define).

Actually, we have essentially already accomplished the first sub-step. But as these properties were derived in various different points in our development, for both convenience and clarity we will collect them here.

First we recall the basic definition, Definition 3.4.9.

Definition 3.6.1. Let R be a commutative ring with 1. An ideal I of R is *principal* if it is generated by a single element r_0 of R, or, equivalently, if $I = \{rr_0 \mid r \in R\}$.

An integral domain R is a *principal ideal domain* (PID) if every ideal in R is principal.   ◊

Theorem 3.6.2. *Every Euclidean domain is a PID.*

Proof.   This is Theorem 3.4.11.   □

Corollary 3.6.3. *The following rings are PIDs: $\mathbb{Z}$, $R[x]$ for R a field, and $\mathbb{Z}[i]$.*

Proof.   This is Corollary 3.4.12.   □

Theorem 3.6.4. *Let R be a PID. Let $\{a_i\}$ be a set of elements of R, not all zero. Then $\{a_i\}$ has a gcd g. Furthermore, g is a generator of the ideal I generated by $\{a_i\}$, and so g can be written as*

$$g = \sum a_i b_i \quad \text{for some} \quad \{b_i\} \subseteq R,$$

a finite sum (i.e., all but finitely many $b_i = 0$).

Proof.   This is Theorem 3.5.15.   □

In the language of Definition 3.5.17, the first sentence of the conclusion says that every PID is a GCD domain. The second sentence of the conclusion (beginning "Furthermore") is a stronger property of PIDs, not shared by all GCD domains.

We have Euclid's lemma, Lemma 3.5.25, but we remarked there when we proved it that it has an easier and more straightforward proof in the case of a PID. We restate it and give that proof now.

Lemma 3.6.5 (Euclid's lemma). *Let R be a PID. Let a be any nonzero element of R. Let b and c be elements of R and suppose that a divides bc. If a and b are relatively prime, then a divides c.*

Proof. We are given that a and b are relatively prime, i.e., that $1 \cong \gcd(a, b)$ (Definition 3.5.23). Then, by Theorem 3.6.4,

$$1 = ar + bs \quad \text{for some} \quad r, s \in R$$

and then

$$c = c(ar + bs) = a(br) + (bc)s.$$

Now a visibly divides the first term on the right-hand side, and by hypothesis a divides bc, so a divides the second term as well. Hence, a divides their sum, which is c, as claimed. $\square$

Then we have several consequences of Euclid's lemma.

Corollary 3.6.6. *Let R be a PID. Let a and b be nonzero elements of R. Let c be an element of R and suppose that a divides c and b divides c. If a and b are relatively prime, then ab divides c.*

Proof. This is Corollary 3.5.28. $\square$

Corollary 3.6.7. *Let R be a PID. Let $a, b,$ and c be elements of R. If a and b are relatively prime, and a and c are relatively prime, then a and bc are relatively prime.*

Proof. This is Corollary 3.5.30. $\square$

Corollary 3.6.8. *Let R be a PID. Let a be an element of R. Then a is prime if and only if a is irreducible.*

Proof. This is Corollary 3.5.31. $\square$

Now we have a look at the Chinese remainder theorem for PIDs.

Lemma 3.6.9. *Let R be a PID. Let I and J be ideals in R, generated by elements a and b respectively.*

(a) *The ideals I and J of R are coprime if and only if the elements a and b are relatively prime.*

(b) *If a and b are relatively prime then $IJ = I \cap J$ is the ideal generated by ab.*

Proof.

(a) By definition (Definition 3.2.23), I and J are coprime if $I + J = R$. Then this conclusion follows immediately from Theorem 3.6.4 and Lemma 3.2.6.

(b) This follows directly from Corollary 3.6.6. $\square$

Let I be an ideal of the PID R, generated by an element r_0. We adopt the standard notation $a \equiv b \pmod{r_0}$ for $a \equiv b \pmod{I}$.

Theorem 3.6.10 (Chinese remainder theorem). *Let R be a PID. Let $\{a_m, \ldots, a_n\}$ be a set of pairwise relatively prime elements of R. Let $r_1, \ldots, r_n$ be arbitrary elements of R. Then the system of simultaneous congruences*

$$x \equiv r_1 \pmod{a_1}$$

$$x \equiv r_2 \pmod{a_2}$$

$$\vdots$$

$$x \equiv r_n \pmod{a_n}$$

has a solution $x = b$ in R, and $x = b'$ is a solution if and only if $b' \equiv b \pmod{a_1 \cdots a_n}$.

Proof.   This is Theorem 3.2.26, stated in our language, and using Lemma 3.6.9. $\square$

The general results we have stated so far have evidently been closely related to questions of divisibility. Here is one more general result that we have already proved. On the face of it, it doesn't seem to have much to do with divisibility. But appearances are deceiving, and we will soon see the essential role it plays.

Theorem 3.6.11. *Let R be a PID. Then R is a Noetherian ring.*

Proof.   Immediate from Theorem 3.2.14. $\square$

Now we came to our main result, "unique factorization". But let us see what we should mean by that. Thinking about the positive integers, we know we can factor 6 are $6 = 2 \cdot 3$, a product of primes. Now this factorization is, strictly speaking, not unique, as we also have $6 = 3 \cdot 2$. Thus the first thing we see is that we wish to consider two factorizations to be "essentially" the same if they only differ in the order of the factors.

But now let us think about factorization in $\mathbb{Z}$. Here we have

$$6 = 1 \cdot 2 \cdot 3 = 1 \cdot (-2) \cdot (-3) \equiv (-1) \cdot 2 \cdot (-3) = (-1) \cdot (-2) \cdot 3.$$

We see that what we have done here is simply spread unit factors (recall the units in $\mathbb{Z}$ are $\{\pm 1\}$) around in such a way that they cancel, and we want to consider all of these factorizations to be "essentially" the same as well. Recall that two nonzero elements that differ by a unit factor are associates (so here 2 and -2 are associates, as are 3 and -3).

With these considerations in mind we can formulate what it means for factorizations to be "essentially" unique.

But we have one more consideration before we do so. We can always group terms together, so that we can write, for example, $4 = 2 \cdot 2 = 2^2$ if we wish, and it is often convenient to do so. Again, we also have $4 = (-2)^2$. Now of course $2 \neq -2$, but once again they are associates. Thus what we will mean in our second formulation by primes being "distinct" is not just that they are not the same, but also not associates of each other.

It is illuminating to state the property of "unique factorization" separately, and then to formulate our main result as saying that PIDs have this property.

Definition 3.6.12. An integral domain R is a *unique factorization domain (UFD)* if every nonzero element a of R can be written as

$$a = u p_1 \ldots p_r$$

for some unit u and primes $p_1, \ldots, p_r$, and if also

$$a = v q_1 \ldots q_s$$

for some unit v and primes $q_1, \ldots, q_s$, then $r = s$ and after possible reordering, p_i and q_i are associates for $i = 1, \ldots, r$. Equivalently, R

is a UFD if every nonzero element a of R can be written as

$$a = up_1^{e_1} \cdots p_m^{e_m}$$

where u is a unit, $p_1, \ldots, p_m$ are distinct primes, and $e_1, \ldots, e_m$ are positive integers, and if also

$$a = vq_1^{f_1} \cdots q_n^{f_n}$$

where v is a unit, $q_1, \ldots, q_n$ are distinct primes, and $f_1, \ldots, f_n$ are positive integers, then $n = m$ and after possible reordering, p_i and q_i are associates and $f_i = e_i$ for $i = 1, \ldots, m$. $\Diamond$

Theorem 3.6.13. *Let R be a PID. Then R is a UFD.*

Proof.   First we show that every nonzero element a and R *has* a factorization as in Definition 3.6.12, and then we show that the factorization of a is essentially unique.

The first step has two substeps. Then first substep is to prove the following claim:

Claim. Every nonzero element a of R that is not a unit is divisible by some irreducible element p of R.

Proof of claim. If a is irreducible, set $p = a$ and we are done. If a is not irreducible, write $a = a_1 b_1$ with neither a_1 nor b_1 units. If a_1 is irreducible, set $p = a_1$ and we are done. If a_1 is not irreducible, write $a_1 = a_2 b_2$ with neither a_2 nor b_2 units (so that $a = a_1 b_1 = a_2(b_2 b_1)$). If a_2 is irreducible, set $p = a_2$ and we are done. If a_2 is not irreducible, write $a_2 = a_3 b_3$ with neither a_3 nor b_3 units. If a_3 is irreducible, set $p = a_3$ and we are done. If not, continue.

We must show this process stops at some stage, say stage n, in which case $p = a_n$ is irreducible. Suppose not. Let I_1 be the principal ideal generated by a_1, I_2 the principal ideal generated by a_2, I_3 the principal ideal generated by a_3, $\ldots$. Then, since $b_1, b_2, b_3, \ldots$ are not units, we have an infinite chain

$$I_1 \subset I_2 \subset I_3 \subset \cdots$$

which is impossible, as a PID is a Noetherian ring (Theorem 3.6.11).

The second substep is to prove the following claim:

Claim. Every nonzero element a of R can be written as

$$a = up_1 \ldots p_r$$

with u a unit and $p_1, \ldots, p_r$ irreducible.

Proof of claim. If a is a unit, set $u = a$ and we are done.

If a is not a unit, then a is divisible by some irreducible element p_1 of R. Write $a = p_1 q_1$. If q_1 is a unit, set $u = q_1$ and we are done. If not, q_1 is divisible by some irreducible element p_2 of R. Write $p_1 = p_2 q_2$ (so that $a = (p_1 p_2) q_2$). If q_2 is a unit, set $u = q_2$ and we are done. If not, once again, continue.

Again we must show that this process steps at some stage, say stage r, in which case, setting $u = q_r$, $a = up_1 \ldots p_r$. Again, suppose not.

Let I_1 be the principal ideal generated by p_1, I_2 the principal ideal generated by p_2, ... Then, since $q_1, q_2, \ldots$ are not units, we have an infinite chain

$$I_1 \subset I_2 \subset \ldots$$

which is again impossible as R is a Noetherian ring.

Thus, we have finished step 1.

Now for step 2, essential uniqueness. Suppose we have two factorizations as in Definition 3.6.12. Since R is a PID, every irreducible is a prime (Corollary 3.6.8) so we have two factorizations of a into primes

$$a = up_1 \ldots pr = vq_1 \ldots q_s.$$

Now p_1 visibly divides a, so it divides the right-hand product. By the definition of a prime, it must divide one of the factors. It certainly doesn't divide the unit factor, so it must divide one of the other factors. By reordering, if necessary, we may assume it divides q_1. But q_1 is irreducible, so we must have $q_1 = p_1 v'$ for some unit v', in which case p_1 and q_1 are associates. Thus,

$$a = up_1 \ldots p_r = v(v'p_1)q_2 \ldots, q_3$$

so by cancellation

$$a' = up_2 \ldots p_r = (vv')q_2 \ldots q_3.$$

Again we apply the same argument: p_2 divides a', so must divide some term on the right-hand side, which, after possible reordering, we may assume to be q_2, in which case $q_2 = p_2 v''$ for some unit v'', and p_2 and q_2 are associates. Thus,

$$a' = up_2 \ldots p_r = (vv'v'')p_2 q_3 \ldots q_s$$

so by cancellation

$$a'' = up_3 \ldots p_r = (vv'v'')q_3 \ldots q_s$$

and continue, matching up the factors 1 by 1, until we are done. $\square$

We conclude this section by investigating properties of UFDs. As we will see later (Remark 3.11.10) there are important examples of rings that are UFDs but not PIDs.

Lemma 3.6.15. *Let R be a UFD. Let p be an element of R. Then p is prime if and only if p is irreducible.*

Proof.   Recall from Lemma 3.5.4 that if p is prime then p is irreducible. So we must show that if p is irreducible then p is prime. So suppose that p is irreducible and that p divides $a = bc$. We must show that p divides b or p divides c. Since p divides a, a certainly has a factorization into irreducibles

$$a = upp_2 \ldots p_i.$$

Suppose p does not divide b. Then b has a factorization into irreducibles

$$b = vq_1 \ldots q_j,$$

with none of the q's an associate of p. Similarly, if p does not divide c then c has a factorization into irreducibles

$$c = wr_1 \ldots r_k,$$

with none of the r's an associate of p.

Now $a = bc$ so we see that a has the factorizations into irreducibles

$$a = upp_2 \ldots p_i = (vw)q_1 \ldots q_j r_1 \ldots r_k$$

But then a has two distinct factorizations into irreducibles (these two factorizations being distinct as p appears in the first but not in the second); contradiction. $\square$

Lemma 3.6.16. *Let R be a UFD and let a and b be nonzero elements of R. Let a and b have factorizations into powers of distinct primes*

$$a = up_1^{e_1} \cdots p_i^{e_i} q_1^{f_1} \cdots q_j^{f_j}$$
$$b = vq_1^{g_1} \cdots q_j^{g_j} r_1^{h_1} \cdots r_k^{h_k}.$$

Then, if $d_1 = min(f_j, g_j), \ldots, d_j = min(f_j, g_j)$,

$$q_1^{d_1} \cdots q_j^{d_j} \cong gcd(a, b).$$

Proof. We leave this as an exercise for the reader. ◻

Lemma 3.6.17. *Let R be a UFD. Then R is a GCD domain.*

Proof. We also leave this as an exercise for the reader. ◻

Remark 3.6.18. As we have observed, the gcd is only defined up to multiplication by a unit. In the case of $\mathbb{Z}$, the units are ± 1, so the gcd is only defined up to sign. Here we make the *convention* that we always choose the plus sign, so that the gcd is a positive integer. In the case of $R[x]$, R a field, the units are the nonzero elements of R. Here we make the *convention* that the gcd is a monic polynomial. For a general GCD domain, there is no preferred choice, with one exception: If elements are relatively prime, we make the *convention* that their gcd is 1. But we stress that these are matters of convention, *not* of necessity. ◇

Example 3.6.19. We set $R = \mathbb{Z}$ and note that R is a Euclidean domain, and hence a PID.

(a) We see immediately that 2 and 3 are relatively prime, i.e., $1 \cong gcd(2,3)$, and almost as quickly that $1 = 2(-1) + 3(1)$.

(b) We see immediately that $10 = 2 \cdot 5$ and $33 = 3 \cdot 11$ are relatively prime, i.e., $1 \cong gcd(10, 33)$, but it takes a little work to see that $1 = 10(10) + 33(-3)$.

(c) We see that $360 = 2^3 \cdot 3^2 \cdot 5$ and $700 = 2^2 \cdot 5^2 \cdot 7$, and so, by Lemma 3.6.16, $20 = 2^2 \cdot 5 \cong gcd(360, 700)$. Then we can see that $20 = 360(2) + 700(-1)$.

(d) Suppose we want to find $gcd(161, 1001)$. It takes some work to factor these. When we do, we find $161 = 7 \cdot 23$, $1001 = 7 \cdot 11 \cdot 13$, so $7 \cong gcd(161, 1001)$. Then also $7 = 161(56) + 1001(-9)$, but it does not seem at all easy to find this expression.

(e) To find $\gcd(2501, 4551)$ we must factor these numbers, which is not at all easy. But it turns out that $2501 = 41 \cdot 61$ and $4551 = 41 \cdot 111$, so $41 \cong \gcd(2501, 4551)$. Then also $41 = 2501(-20) + 4551(11)$ and again it does not seem at all easy to find this expression.

(f) How about $\gcd(12345, 54321)$? It turns out that $3 \cong \gcd(12345, 54321)$ and $3 \cong 12345(3617) + 54321(-822)$.

(g) How about $\gcd(124816, 618421)$? It turns out that these two numbers are relatively prime, i.e., $1 \cong \gcd(124816, 618421)$, and $1 = 124816(-266427) + 618421(53773)$. $\Diamond$

Remark 3.6.20. Just proceeding as we have so far, to find $g \cong \gcd(a, b)$ we would have to factor a and b, and this quickly becomes impractical as a and b get large. Even if we accomplish this first step, this doesn't at all help us in the more difficult second step of finding x and y with $g = ax + by$. But in the next section we will not only see how to easily find g *without* having to factor a and b, and almost as easily how to find such integers x and y. $\Diamond$

We can now see the essential role that the gcd plays in unique factorization.

Definition 3.6.21. Let R be a ring. Then R is *principally Noetherian* if every sequence of principal ideals $I_1 \subseteq I_2 \subseteq I_3 \subseteq \ldots$ is eventually constant, or, otherwise said, if every sequence of principal ideals $I_1 \subset I_2 \subset I_3 \subset \ldots$ is finite. $\Diamond$

Theorem 3.6.22. Let R be an integral domain. Then R is a UFD if and only if R is a principally Noetherian f-GCD domain.

Proof. Suppose that R is a principally Noetherian f-GCD domain. Consider the proof of Theorem 3.6.13, that every PID is a UFD. Step 1 of the proof goes through unchanged as we are assuming R is principally Noetherian. Step 2 of the proof goes through unchanged as we are assuming that R is an f-GCD domain, and hence, by Corollary 3.5.31, every irreducible element of R is prime.

Conversely, if R is a UFD it is easy to check that R is principally Noetherian, and we have already observed that R is a GCD domain (and hence an f-GCD domain) in Lemma 3.6.17. $\square$

Remark 3.6.23. We close this section by observing that at this point we can breathe a huge sign of relief. We started out by *assuming* properties of the integers that we stated in Appendix A, and used these properties in our earlier development. But now that we have shown that $\mathbb{Z}$ is a PID, and hence a UFD, we have *proved* that these properties hold. So we are now standing on firm logical ground. ◊

3.7 Euclid's algorithm

Let R be a Euclidean domain. In this section we present Euclid's algorithm, a very effective method of first, finding $g \cong \gcd(a, b)$ for any two elements a and b of R, not both zero, and second, writing $g = ax + by$ with x and y elements of R. (Historically speaking, Euclid's algorithm came first, and then we defined Euclidean domains to be the integral domains in which Euclid's algorithm works.) We will develop this algorithm, give a variety of examples in various situations, and show how to apply it in the Chinese remainder theorem.

We fix a Euclidean domain R and a Euclidean norm δ on R. Let $a, b \in R$, not both zero. If $b = 0$ then $a \cong \gcd(\{a, b\})$ and if $a = 0$ then $b \cong \gcd(a, b)$, by Lemma 3.5.20(b). Thus we have found the gcd in these simple (and not very interesting) cases. The interesting case is when a and b are both nonzero.

Algorithm 3.7.1 (Euclid's algorithm). Let $a, b \in R$ be nonzero. Set $a_0 = a$, $a_1 = b$. Then, by the definition of a Euclidean domain (Definition 3.4.1), there are elements q_1 and a_2 of R with

$$a_0 = a_1 q_1 + a_2 \quad \text{with } a_2 = 0 \text{ or } \delta(a_2) < \delta(a_1).$$

Suppose $a_2 \neq 0$. Then, similarly, there are elements q_2 and a_3 of R with

$$a_1 = a_2 q_2 + a_3 \quad \text{with } a_3 = 0 \text{ or } \delta(a_3) < \delta(a_2).$$

Suppose $a_3 \neq 0$. Then, there are elements q_3 and a_4 of R with

$$a_2 = a_3 q_3 + a_4 \quad \text{with } a_4 = 0 \text{ or } \delta(a_4) < \delta(a_3).$$

Continue ...
Claim. This process cannot go on forever.

Proof of claim. If it did, we would have an infinite sequence $a_1, a_2, a_3, \ldots$ with

$$\delta(a_1) > \delta(a_2) > \delta(a_3) > \ldots.$$

But each $\delta(a_i)$ is a nonnegative integer, and it is impossible to have an infinite strictly decreasing sequence of nonnegative integers.   □

Hence, this sequence stops at some stage n, i.e., $a_{n-1} = a_n q_n + a_{n+1}$ with $a_{n+1} = 0$.

Let us write down what we have obtained:

$$a_0 = a_1 q_1 + a_2$$
$$a_1 = a_2 q_2 + a_3$$
$$a_2 = a_3 q_3 + a_4$$
$$\vdots$$
$$a_{n-3} = a_{n-2} q_{n-2} + a_{n-1}$$
$$a_{n-2} = a_{n-1} q_{n-1} + a_n$$
$$a_{n-1} = a_n q_n + 0$$

Now we know (from Lemma 3.5.20(b)) that $a_n \cong \gcd(a_n, 0)$. But we also know

$$\gcd(a_{n-1}, a_n) \cong \gcd(a_n, 0) \text{ by Lemma 3.5.20(c),}$$
$$\gcd(a_{n-2}, a_{n-1}) \cong \gcd(a_{n-1}, a_n) \text{ by Lemma 3.5.20(c),}$$
$$\gcd(a_{n-3}, a_{n-2}) \cong \gcd(a_{n-2}, a_{n-1}) \text{ by Lemma 3.5.20(c),}$$
$$\vdots$$
$$\gcd(a_2, a_3) \cong \gcd(a_3, a_4) \text{ by Lemma 3.5.20(c),}$$
$$\gcd(a_1, a_2) \cong \gcd(a_2, a_3) \text{ by Lemma 3.5.20(c),}$$
$$\gcd(a_0, a_1) \cong \gcd(a_1, a_2) \text{ by Lemma 3.5.20(c),}$$

Following this chain from the bottom up, we see

$$a_n \cong \gcd(a_n, 0) \cong \gcd(a_{n-1}, a_n) \cong \gcd(a_{n-3}, a_{n-1}) \cong \gcd(a_{n-3}, a_{n-2})$$
$$\ldots \cong \gcd(a_2, a_3) \cong \gcd(a_1, a_2) \cong \gcd(a_0, a_1).$$

Thus, $a_n \cong \gcd(a_0, a_1)$ and we have found a gcd of our original two elements of R.

Furthermore, from the next-to-the last equation we see that

$$a_n = a_{n-2}(1) + a_{n-1}(-q_{n-1})$$

and we have expressed a_n in terms of a_{n-2} and a_{n-1}. But we can solve the equation above that one for a_{n-1} and substitute:

$$a_n = a_{n-2}(1) + [a_{n-3}(1) + a_{n-2}(-q_{n-2})](-q_{n-1})$$
$$= a_{n-3}(-q_{n-1}) + a_{n-2}(1 + q_{n-2}q_{n-1}).$$

We don't need to keep track of these exact coefficients for the proof (though we will certainly need to do so in our computations). We just observe that we have now expressed a_n in terms of a_{n-3} and a_{n-2}. We use the equation above that to express a_n in terms of a_{n-4} and a_{n-3}. We continue to "roll up". By the time we get to the first equation we have expressed a_n in terms of a_1 and a_2, and when we have used that equation we have expressed a_n in terms of a_0 and a_1,

$$\gcd(a, b) \cong a_n = ax + by \text{ for some } x, y \in R,$$

as desired. ◊

Example 3.7.1. We let $R = \mathbb{Z}$ and do several examples of increasing complexity.

(a) $\gcd(33, 10)$:

$$33 = 10 \cdot 3 + 3$$
$$10 = 3 \cdot 3 + 1$$
$$3 = 1 \cdot 3$$

so $1 \cong \gcd(33, 10)$, and then

$$1 = 10 + 3(-3) = 10 + [33 + 10(-3)](-3) = 33(-3) + 10(10).$$

(b) gcd(1001,161):

$$1001 = 161 \cdot 6 + 35$$
$$161 = 35 \cdot 4 + 21$$
$$35 = 21 \cdot 1 + 14$$
$$21 = 14 \cdot 1 + 7$$
$$21 = 14 \cdot 1$$
$$14 = 7 \cdot 2$$

so $7 \cong$ gcd(1001,161), and then

$$7 = 21 + 14(-1)$$
$$= 21 + [35 + 21(-1)](-1) = 35(-1) + 21(2)$$
$$= 35(-1) + [161 + 35(-4)](2) = 161(2) + 35(-9)$$
$$= 161(2) + [1001 + 161(-6)](-9) = 1001(-9) + 161(56).$$

Note that if we only wanted to find $\gcd(1001, 161)$, we could have stopped with the second equation $161 = 35 \cdot 4 + 21$ as we can see right away that $7 \cong \gcd(35, 21)$.

(c) gcd(4551,2501):

$$4551 = 2501 \cdot 1 + 2050$$
$$2501 = 2050 \cdot 1 + 451$$
$$2050 = 451 \cdot 4 + 246$$
$$451 = 246 \cdot 1 + 205$$
$$246 = 205 \cdot 1 + 41$$
$$205 = 41 \cdot 5$$

so $41 \cong$ gcd(4551,2501), and then

$$41 = 246 + 205(-1)$$
$$= 246 + [451 + 246(-1)](-1) = 451(-1) + 246(2)$$
$$= 451(-1) + [2050 + 451(-4)](2) = 2050(2) + 451(-9)$$
$$= 2050(2) + [2501 + 2050(-1)](-9) = 2501(-9) + 2050(11)$$
$$= 2501(-9) + [4551 + 2501(-1)](11) = 4551(11) + 2501(-20).$$

(d) gcd(54321,12345):

$$54321 = 12345 \cdot 4 + 4941$$

$$12345 = 4941 \cdot 2 + 2463$$

$$4941 = 2463 \cdot 2 + 15$$

$$2463 = 15 \cdot 164 + 3$$

$$15 = 3 \cdot 5$$

so $3 \cong \gcd(54321,12345)$, and then

$$3 = 2463 + 15(-164)$$

$$= 2463 + [4941 + 2463(-2)](-164) = 4941(-164) + 2463(329)$$

$$= 4941(-164) + [12345 + 4941(-2)](329)$$

$$= 12345(329) + 4941(-822)$$

$$= 12345(329) + [54321 + 12345(-4)](-822)$$

$$= 54321(-822) + 12345(3617).$$

(e) gcd(618421,124816):

Recall that the Euclidean norm on $\mathbb{Z} - \{0\}$ is given by $\delta(n) = |n|$, so given a and b we want to express a as $a = bq + r$ with $r = 0$ or $|r| < |b|$. We have previously always chosen $r > 0$. But we don't have to. In particular, we may always choose r so as to make $|r|$ as small as possible. This can speed up computations. We do that here:

$$618421 = 124816 \cdot 5 + (-5659)$$

$$124816 = (-5659)(-22) + 318$$

$$-5659 = 318(-18) + 65$$

$$318 = 65(5) + (-7)$$

$$65 = (-7)(-9) + 2$$

$$-7 = 2(-3) + (-1)$$

$$2 = (-1)(-2)$$

so we see $-1 \cong \gcd(618421,124816)$, or, equivalently, $1 \cong \gcd(618421,124816)$, i.e., 618421 and 124816 are relatively prime.

(Once again, we could have concluded this from the fourth equation $318 = 65(5) + (-7)$ as we can see right away that 65 and -7 are relatively prime.)

Now, for the second step:

$$-1 = (-7) + 2(3)$$
$$= (-7) + [65 + (-7)9](3) = 65(3) + (-7)(28)$$
$$= 65(3) + [318 + 65(-5)](28) = 318(28) + 65(-137)$$
$$= 318(28) + [-5659 + 318(18)](-137)$$
$$= (-5659)(-137) + 318(-2438)$$
$$= (-5659)(-137) + [124816 + (-5659)(22)](-2438)$$
$$= 124816(-2438) + (-5659)(-53773)$$
$$= 124816(-2438) + [618421 + 124816(-5)](-53773)$$
$$= 618421(-53773) + 124816(266427)$$

and also

$$1 = 618421(53773) + 124816(-266427).$$

(f) gcd(871,455,273):

Here we use Lemma 3.5.21 (or Remark 3.5.22):

$$\gcd(871,455,273) = \gcd(871, \gcd(455,273)).$$

We compute the inner gcd first. We use positive remainders, and leave it to the reader to check that we obtain $91 \cong \gcd(455,273)$ and $91 = 455(-1) + 273(2)$. We then compute $\gcd(871,91)$. Again we use positive remainders and we leave it to the reader to check that we obtain $13 \cong \gcd(871,91)$ and $13 = 871(2) + 91(-19)$. Finally,

$$13 = 871(2) + 91(-19)$$
$$= 871(2) + [455(-1) + 273(2)](-19)$$
$$= 871(2) + 455(19) + 273(-38). \qquad \Diamond$$

Example 3.7.3. We now let $R = \mathbb{Z}[i]$ and we recall that R has the multiplicative Euclidean norm $\delta(a + bi) = a^2 + b^2$ (Theorem 3.4.7). We will follow the strategy of the proof of that theorem in performing Euclid's algorithm.

(a) gcd$(33 + 29i, 9 + 12i)$:

Let $a_0 = 33+29i$, $a_1 = 9+12i$. We observe that $\delta(a_1) = 9^2+12^2 = 225$.

We compute

$$\frac{33 + 29i}{9 + 12i} = \frac{33 + 29i}{9 + 12i} \cdot \frac{9 - 12i}{9 - 12i} = \frac{645 - 135i}{225}.$$

Now the nearest integer to $\frac{645}{225}$ is 3 and the nearest integer to $\frac{-135}{225}$ is -1, so we choose $q_1 = 3 + (-1)i = 3 - i$. Then

$$\begin{aligned}
a_0 = a_1q_1 + a_2 \text{ so } a_2 = a_0 - a_1q_1 &= (33 + 29i) - [(9 + 12i)(3 - i)] \\
&= (33 + 29i) - (39 + 27i) \\
&= -6 + 2i.
\end{aligned}$$

Thus the first step in Euclid's algorithm is

$$33 + 29i = (9 + 12i)(3 - i) + (-6 + 2i).$$

Note $\delta(a_2) = (-6)^2 + 2^2 = 40 < \delta(a_1)$ as we expect.
Next we compute

$$\frac{9 + 12i}{-6 + 2i} = \frac{9 + 12i}{-6 + 12i} \cdot \frac{-6 - 2i}{-6 - 2i} = \frac{-30 - 90i}{40}.$$

Now the nearest integer to $\frac{-30}{40}$ is -1 and the nearest integer to $\frac{-90}{40}$ is -2, so we choose $q_2 = -1 + (-2)i = -1 - 2i$. Then

$$\begin{aligned}
a_1 = a_2q_2 + a_3 \text{ so } a_3 = a_1 - a_2q_2 &= (9 + 12i) \\
&\quad -[(-6 + 2i)(-1 - 2i)] \\
&= (9 + 12i) - (10 + 10i) \\
&= -1 + 2i.
\end{aligned}$$

Thus the next step in Euclid's algorithm is

$$9 + 12i = (-6 + 2i)(-1 - 2i) + (-1 + 2i).$$

Note $\delta(a_3) = (-1)^2 = 5 < \delta(a_2)$ as we expect.

Next we compute

$$\frac{-6+2i}{-1+2i} = \frac{-6+2i}{-1+2i} \cdot \frac{-1-2i}{-1-2i} = \frac{10+10i}{5} = 2+2i.$$

Thus, we see that $-1+2i$ divides $-6+2i$, and we are done. We summarize:

$$33+9i = (9+12i)(3-i) + (-6+2i)$$
$$9+12i = (-6+2i)(-1-2i) + (-1+2i)$$
$$-6+12i = (-1+2i)(2+2i)$$

and we see $-1+2i \cong \gcd(33+29i, 9+2i)$.
Furthermore,

$$-1+2i = (9+12i) + (-6+2i)(1+2i)$$
$$= (9+12i) + [(33+29i) + (9+12i)(-3+i)](1+2i)$$
$$= (33+29i)(1+2i) + (9+12i)(-4-5i).$$

(b) $\gcd(9-2i, 4+7i)$:
   Let $a_0 = 9-2i$, $a_1 = 4+7i$. We observe that $\delta(a_0) = 9^2 + (-2)^2 = 85$ and $\delta(a_1) = 4^2 + 7^2 = 65$.

We compute

$$\frac{9-2i}{4+7i} = \frac{9-2i}{4+7i} \cdot \frac{4-7i}{4-7i} = \frac{22-71i}{65}.$$

Now the nearest integer to $\frac{22}{65}$ is 0 and the nearest integer to $\frac{-71}{65}$ is -1, so we choose $q_1 = -i$. Then

$$a_0 = a_1 q_1 + a_2 \text{ so } a_2 = a_0 - a_1 q_1 = (9-2i) - [(4+7i)(-i)]$$
$$= (9-2i) - (7-4i) = 2+2i.$$

Thus, the first step in Euclid's algorithm is

$$9-2i = (4+7i)(-i) + (2+2i).$$

Note $\delta(a_2) = 2^2 + 2^2 = 8 < \delta(a_1)$ as we expect.

Next we compute

$$\frac{4+7i}{2+2i} = \frac{4+7i}{2+2i} \cdot \frac{2-2i}{2-2i} = \frac{22+6i}{8}.$$

Now the nearest integer to $\frac{22}{8}$ is 3 and the nearest integer to $\frac{6}{8}$ is 1, so we choose $q_2 = 3 + i$. Then

$$a_1 = a_2 q_2 + a_3 \text{ so } a_3 = a_1 - a_2 q_2 = (4+7i) - [(2+2i)(3+i)]$$
$$= (4+7i) - (4+8i) = -i.$$

Now we observe that $-i$ is a unit in $\mathbb{Z}[i]$. So $-i$ certainly divides $2 + 2i$; indeed $2 + 2i = -i(-2 + 2i)$ and we have:

$$9 - 2i = (4+7i)(-i) + (2+2i)$$
$$4 + 7i = (2+2i)(3+i) + (-i)$$
$$2 + 2i = (-i)(-2+2i).$$

More to the point, since $-i$ is a unit, we conclude that $9 - 2i$ and $4 + 7i$ are relatively prime, $1 \cong \gcd(9 - 2i, 4 + 7i)$.
Furthermore,

$$-i = (4+7i) + (2+2i)(-3-i)$$
$$= (4+7i) + [(9-2i) + (4+7i)i](-3-i)$$
$$= (9-2i)(-3-i) + (4+7i)(2-3i)$$

or

$$1 = (9-2i)(1-3i) + (4+7i)(3+2i).$$

Note that if had just wanted to see whether $9 - 2i$ and $7 + 4i$ were relatively prime, we could have stopped this process earlier. Since we have a multiplicative norm, if $g \cong \gcd(a, b)$, then g divides a, so $\delta(g)$ divides $\delta(a)$, and g divides b, so $\delta(g)$ divides $\delta(b)$; hence $\delta(g)$ divides $\gcd(\delta(a), \delta(b))$. In our case here, if $g \cong \gcd(9 - 2i, 7 + 4i)$ then also, from the next step, $g \cong \gcd(4 + 7i, 2 + 2i)$. But $\delta(4 + 7i) = 65$ and $\delta(2 + 2i) = 8$, and these integers are relatively prime, so we must have $\delta(g) = 1$ and hence g is a unit, and we may conclude without further ado that $9 - 2i$ and $4 + 7i$ are relatively prime. $\Diamond$

Example 3.7.4. We now consider the polynomial ring $R[x]$, where R is a field, and recall that this ring has the Euclidean norm $\delta(p(x)) = \deg p(x)$.

(a) Let $R = \mathbb{Q}$, the field of rational numbers. We wish to find $\gcd(x^3 + x + 2, x^2 + 2x + 3)$, and for this we use the usual division algorithm for polynomials. We have:

$$x^3 + x + 2 = (x^2 + 2x + 3)(x - 2) + (2x + 8)$$

$$x^2 + 2x + 2 = (2x + 8)(x/2 - 1) + 11$$

$$x/2 - 1 = 11(x/22 - 1/11).$$

We note that 11 is a unit in $\mathbb{Q}[x]$, so these two polynomials are relatively prime, $1 \cong \gcd(x^3 + x + 2, x^2 + 2x + 3)$. Furthermore,

$$
\begin{aligned}
11 &= (x^2 + 2x + 3) + (2x + 3)(-x/2 + 1) \\
&= (x^2 + 2x + 3) + [(x^3 + x + 2) \\
&\quad + (x^2 + 2x + 3)(-x + 2)](-x/2 + 1) \\
&= (x^3 + x + 2)(-x/2 + 1) + (x^2 + 2x + 3)(x^2/2 - 2x + 3),
\end{aligned}
$$

or

$$1 = (x^3 + x + 2)(-x/22 + 1/11) + (x^2 + 2x + 3)(x^2/22 - 2x/11 + 3/11).$$

(b) Now let $R = \mathbb{Z}_{11}$, the integers modulo 11. Since 11 is a prime, $\mathbb{Z}_{11}$ is a field. Again we wish to find $\gcd(x^3 + x + 2, x^2 + 2x + 3)$. To do so, we can reuse our work from the last computation.

The first step is almost unchanged:

$$x^3 + x + 2 - (x^2 + 2x + 3)(x + 9) + (2x + 8).$$

The only difference is that $-2 \equiv 9 \pmod{11}$, so we have replaced $x - 2$ by $x + 9$.

The second step is more interesting. In our previous computation, when we divided $x^2 + 2x + 3$ by $2x + 8$ we obtained a quotient of $x/2 - 1$ and a remainder of 11. Now the quotient $x/2 - 1 = \frac{1}{2}x - 1 = 2^{-1}(x) - 1$. But in $\mathbb{Z}_{11}$, $2^{-1} = 6$ (as $2 \cdot 6 \equiv 1 \pmod{11}$), and $-1 \equiv 10 \pmod{11}$. Also, in $\mathbb{Z}_{11}$, $0 \equiv 11 \pmod{11}$, so the remainder is 0, i.e., $2x + 8$ divides $x^2 + 2x + 3$. Thus, Euclid's algorithm gives here

$$x^3 + x + 2 = (x^2 + 2x + 3)(x + 9) + (2x + 8)$$

$$x^2 + 2x + 3 = (2x + 8)(6x + 10)$$

Thus, $x + 4 \cong \gcd(x^3 + x + 2, x^2 + 2x + 3)$.

Furthermore, again using congruences (mod 11),

$$2x + 8 = (x^3 + x + 2) + (x^2 + 2x + 3)(10x + 2)$$

or

$$x + 4 = (x^3 + x + 2)(6) + (x^2 + 2x + 3)(5x + 1). \qquad \Diamond$$

Now we will see how to apply Euclid's algorithm in conjunction with the Chinese remainder theorem. To keep things relatively simple, we will restrict our attention to $R = \mathbb{Z}$.

Example 3.7.5.

(a) Let $R = \mathbb{Z}$. We wish to solve the simultaneous congruences:

$$x \equiv 25 \ (\text{mod } 64)$$

$$x \equiv 49 \ (\text{mod } 121)$$

Since 64 and 121 are relatively prime, the Chinese remainder theorem (Corollary 3.2.27) guarantees us that this system has a solution. We wish to find it, and to do so we adopt the strategy of the proof of Theorem 3.2.26. That is, we want to find an element h_1 of $\mathbb{Z}$ with $h_1 \equiv 1 \ (\text{mod } 64)$ and $h_1 \equiv 0 \ (\text{mod } 121)$, and an element h_2 of $\mathbb{Z}$ with $h_2 \equiv 0 \ (\text{mod } 64)$ and $h_2 \equiv 1 \ (\text{mod } 121)$. Then we obtain a solution $b = 25h_1 + 49h_2$. How can we find h_1 and h_2? By Euclid's algorithm!

We apply Euclid's algorithm to 121 and 64, and we obtain

$$1 = 121(9) + 64(-17).$$

Now certainly $121(9) \equiv 0 \ (\text{mod } 121)$, as $121(9)$ is visibly divisible by 121. But the above equation gives the congruence

$$1 \equiv 121(9) + 64(-17) \ (\text{mod } 64).$$

We see that the second summand $64(-17) \equiv 0 \ (\text{mod } 64)$, as $64(-17)$ is visibly divisible by 64. Thus, $121(9) \equiv 1 \ (\text{mod } 64)$. Hence we may choose $h_1 = 121(9) = 1089$.

By exactly the same logic we see that $64(-17) \equiv 0 \pmod{64}$ and $64(-17) \equiv 1 \pmod{121}$, so we may choose $h_2 = 64(-17) = -1088$. Thus we have a solution $b = 25(1089) + 49(-1088) = -26087$. This is a perfectly valid solution, but we wish to be neat about this and get a solution between 0 and $7743 = 64 \cdot 121 - 1$. (Remember that the Chinese remainder theorem tells us that our solution is not unique; it is only unique $\pmod{64 \cdot 121}$.) A little arithmetic shows that $-26087 \equiv 4889 \pmod{64 \cdot 121}$, so our solution is

$$x \equiv 4889 \pmod{7744}.$$

(b) Let $\mathbb{R} = \mathbb{Z}$. We wish to solve the simultaneous congruences:

$$x \equiv 6 \pmod{21}$$
$$x \equiv 5 \pmod{23}$$
$$x \equiv 4 \pmod{25}$$

Again, since $\{21, 23, 25\}$ is a pairwise relatively prime set of integers, the Chinese remainder theorem guarantees us that this system has a solution. We use the same method to find it. First, we want to find an integer h_1 with $h_1 \equiv 1 \pmod{21}$, $h_1 \equiv 0 \pmod{23}$, $h_1 \equiv 0 \pmod{25}$. Now $h_1 \equiv 0 \pmod{23}$ means h_1 is divisible by 23 and $h_1 \equiv 0 \pmod{25}$ means h_1 is divisible by 25. Since 23 and 25 are relatively prime, this means that h_1 is divisible by their product $23 \cdot 25 = 575$. Thus we apply Euclid's algorithm to 21 and 575 to obtain

$$1 = 575(8) + 21(-219)$$

and by the same logic as before we see that if $h_1 = 575(8) = 4600$, $h_1 \equiv 1 \pmod{21}$, $h_1 \equiv 0 \pmod{23}$, $h_1 \equiv 0 \pmod{25}$.

Next we wish to find h_2 with $h_2 \equiv 1 \pmod{23}$, $h_2 \equiv 0 \pmod{21}$, $h_2 \equiv 0 \pmod{25}$. We apply the exact same logic to 23 and $21 \cdot 25 = 425$ to obtain

$$1 = 525(-6) + 23(137)$$

and take $h_2 = 525(-6) = -3150$.

Finally, we wish to find h_3 with $h_3 \equiv 1 \pmod{25}$, $h_3 \equiv 0 \pmod{21}$, $h_3 \equiv 0 \pmod{23}$, and we apply the exact same logic a third time to 25 and $21 \cdot 23 = 483$ to obtain

$$1 = 483(-3) + 25(58)$$

and take $h_3 = 483(-3) = -1449$.

Then we obtain a solution $b = 6(4600) + 5(-3150) + 4(-1449) = 6054$ and note that this solution is unique mod $21 \cdot 23 \cdot 25 = 12075$, so the solution to our system of congruences is

$$x \equiv 6054 \pmod{12075}. \qquad \Diamond$$

Although it is implicit in our previous work, it is worth explicitly writing out the method we have used in finding solutions to simultaneous congruences. Again, for simplicity, we restrict ourselves to the case $R = \mathbb{Z}$.

Theorem 3.7.6. *Let $\{m_1, \ldots, m_n\}$ be a set of pairwise relatively prime nonzero integers. Let $M = m_1, \ldots, m_n$ and let $M_k = M/m_k$, $k = 1, \ldots, n$. Let x_k and y_k be integers with $m_k x_k + M_k y_k = 1$ and set $h_k = 1 = m_k x_k = M_k y_k$ for $k = 1, \ldots, n$. Let $r_1, \ldots, r_n$ be arbitrary integers. Then the system of simultaneous congruences*

$$x_1 \equiv r_1 \pmod{m_1}$$

$$\vdots$$

$$x_n \equiv r_n \pmod{m_n}$$

has the unique solution (mod M) given by

$$x \equiv \sum_{k=1}^{n} h_k r_k \pmod{M}.$$

Proof. Note that m_k and M_k are relatively prime for each k, and so such integers x_k and y_k exist (and we can find them by using Euclid's lemma). Then $h_k \equiv 1 \pmod{m_k}$ and $h_k \equiv 0 \pmod{M_k}$, and this latter condition implies $h_k \equiv 0 \pmod{m_j}$ for all $j \neq k$. Then it is easy to verify that $x \equiv \sum_{k=1}^{n} h_k r_k \pmod{M}$ is indeed a solution of this system, and this solution is unique (mod M). $\qquad \square$

3.8   Applications to number theory

This section consists of two independent parts. They are unified in that each is an application of our results in group and ring theory to prove a beautiful and important theorem in number theory. These theorems are Fermat's theorem on the sum of two integer squares, and the Law of Quadratic Reciprocity. In each case the proofs we present are not the original proofs, but rather proofs that use ideas that were historically developed much later.

Fermat's theorem states that every prime $p \equiv 1 \pmod 4$ is a sum of two squares, $p = a^2 + b^2$ for some integers a and b. Actually, his theorem is more precise, and we will state and prove the precise version below. But here are some examples:

$$5 = 2^2 + 1^2$$
$$13 = 3^2 + 2^2$$
$$17 = 4^2 + 1^2$$
$$29 = 5^2 + 2^2$$
$$37 = 6^2 + 1^2$$
$$41 = 5^2 + 4^2$$
$$99989 = 230^2 + 217^2$$
$$618421 = 786^2 + 25^2$$

Indeed, the representation of p as a sum of squares is "essentially" unique. What do we mean by essentially here? Note, for example, we can write

$$5 = 2^2 + 1^2 = 2^2 + (-1)^2 = (-2)^2 + 1^2 = (-2)^2 + (-1)^2$$
$$= 1^2 + 2^2 = 1^2 + (-2)^2 = (-1)^2 + 2^2 = (-1)^2 + (-2)^2$$

and we want to consider these eight possibilities to be essentially the same, so essentially unique should mean up to the signs and order of a and b.

What about primes $p \equiv 3 \pmod 4$? We can rule these out immediately, as it is easy to check that for any integers a and b, $a^2 + b^2 = 0, 1$, or $2 \pmod 4$.

In fact, Fermat determined exactly which integers can be expressed as a sum of two squares. This follow easily from the above result, and we shall derive this, too.

Fermat lived in the 17th century. He claimed this theorem, although he did not pass along this proof, but we believe him, and credit the theorem to him. He did describe his proof as being by his "method of descent" (a method that involves the contrapositive of mathematical induction) and Euler wrote down a proof along these lines in the 18th century. The proof we will give is that given by Dedekind in 1894. The key to this proof is a fact that on the surface of it appears completely unrelated to this theorem, the fact that $\mathbb{Z}[i]$ is a UFD! Given this fact, we will be able to prove this deep theorem in just a few paragraphs.

Theorem 3.8.1 (Fermat). *Let p be a prime with $p \equiv 1 \pmod 4$. Then $p = a^2 + b^2$ for some integers a and b, unique up to sign and order.*

Proof (Dedekind). Let $R = \mathbb{Z}[i]$ and recall we have the norm δ on R defined as follows: If $z \in R$, $z = r + si$, then $\bar{z} = r - si$ and $\delta(z) = z\bar{z} = r^2 + s^2$. Recall that δ is a multiplicative Euclidean norm on R and hence that R is a Euclidean domain, hence a PID, hence a UFD.

As we have shown, since $p \equiv 1 \pmod 4$, -1 is a quadratic residue $\pmod p$ (Corollary 2.7.13), i.e., there is an integer d with $d^2 \equiv -1 \pmod p$. Then $d^2 + 1 \equiv 0 \pmod p$, i.e., $d^2 + 1$ is divisible by p. But $d^2 + 1 = (1 + di)(1 - di)$, a factorization in R. Thus, p divides this product, but does not divide either factor (as neither $(1 + di)/p$ nor $(1 - di)/p$ is in R), so p is not a prime in R. Hence, p is not irreducible in R (as in any UFD, primes and irreducibles are the same), so $p = rs$ for some $r, s \in R$, neither of which is a unit. But then $\delta(p) = \delta(rs) = \delta(r)\delta(s)$, i.e., $p^2 = \delta(r)\delta(s)$. Since r and s are not units, $\delta(r) \neq 1$ and $\delta(s) \neq 1$ (Remark 3.4.8). Thus we must have $\delta(r) = \delta(s) = p$. Setting $r = a + bi$, $\delta(r) = a^2 + b^2$, so $p = a^2 + b^2$ for some integers a and b (and you can check then that we must have $s = \bar{r} = a - bi$), so, as far as existence goes, we are done!

Now for essential uniqueness. Suppose $p = a^2 + b^2 = c^2 + d^2$. Observe that all of a, b, c, and d are nonzero (as otherwise p would be a perfect square). Let $r = a + bi$ and $s = c + di$ so $p^2 = \delta(p) = \delta(r)\delta(s)$ and $\delta(r) = \delta(s) = p$. We must have r and s irreducible (as if $r = tu$,

$p = \delta(r) = \delta(t)\delta(u)$, so either $\delta(t) = 1$ and t is a unit, or $\delta(u) = 1$ and u is a unit, and similarly for s). Once again, R is a UFD so the irreducibles in R are the same as the prime in R. Thus, r is a prime and r divides $r\bar{r} = p = s\bar{s}$, so r must divide s or r must divide $\bar{s}$, i.e., $s = rv$ or $\bar{s} = rv$ for some $v \in R$. Since $\delta(r) = \delta(s) = \delta(\bar{s})$, $\delta(v) = 1$ and v is a unit. Now the units in R are $\{\pm 1, \pm i\}$. Thus, we have two possibilities (r divides s or r divides $\bar{s}$) and for each possibility we have four choices for v, giving a total of eight possibilities, and direct calculation shows these eight possibilities just give all the possible signs and order of a and b, so we have essential uniqueness, and we are done! □

We now give a proof, from our viewpoint, of a result that was known to Diophantus (3rd century C.E.).

Lemma 3.8.2. *Let m and n be positive integers, each of which is a sum of two squares. Then their product mn is a sum of two squares.*

Proof.　Suppose $m = a^2 + b^2$ and $n = c^2 + d^2$. Set $r = a + bi$ and $s = c + di$. Then $m = r\bar{r}$ and $n = s\bar{s}$, so $mn = (r\bar{r})(s\bar{s}) = (rs)(\bar{r}\bar{s}) = t\bar{t}$ with $t = rs$. Thus if $t = e + fi$, $mn = e^2 + f^2$. □

Remark 3.8.3. This proof gives an explicit formula. We have $t = e + fi = (a + bi)(c + di) = (ac - bd) + (ad + bc)i$ so $e = ac - bd$ and $f = ad + bc$. Thus, we obtain the algebraic identity

$$(a^2 + b^2)(c^2 + d^2) = (ac - bd)^2 + (ad + bc)^2. \qquad \lozenge$$

Corollary 3.8.4 (Fermat). *Let N be a positive integer. Then N is a sum of two integer squares if and only if the highest power of any prime $q \equiv 3 \pmod{4}$ dividing N is even.*

Proof.　If $N = 1$, $N = 1^2 + 0^2$, so the corollary is true for $N = 1$.
If $N = 2$, $N = 1^2 + 1^2$, so the corollary is true for $N = 2$.
Suppose N is as stated. Then N factors as a product of distinct prime powers

$$N = 2^e p_1^{f_1} \cdots p_j^{f_j} q_1^{2g_1} \cdots q_k^{2g_k}$$

where $p_1, \ldots, p_j \equiv 1 \pmod{4}$ and $q_1, \ldots, q_k \equiv 3 \pmod{4}$. Let

$$M = 2^e p_1^{f_1} \cdots p_j^{f_j}$$

and

$$Q = q_1^{g_1} \cdots q_k^{g_k}$$

so that $N = MQ^2$.

Now, as we have just observed, 2 is a sum of two squares, and by Fermat's theorem, each of $p_1, \ldots, p_j$ is a sum of two squares. Thus, applying Lemma 3.8.2 repeatedly, M is a sum of two squares, $M = a^2 + b^2$. But then

$$N = MQ^2 = (a^2 + b^2)Q^2 = (aQ)^2 + (bQ)^2$$

is a sum of two squares as well.

Now suppose N is a sum of two squares.

We make a preliminary observation:

Let q be a prime with $q \equiv 3 \pmod 4$. If u and v are integers not divisible by q, then $u^2 + v^2$ is not divisible by q. For if $u^2 + v^2 = 0 \pmod q$, then $u^2 = (-1)v^2 \pmod q$. Now, since $q \equiv 3 \pmod 4$, -1 is *not* a quadratic residue $\pmod q$ (Corollary 2.7.13). Then, on the one hand, from the left-hand side of this congruence we evidently see that u^2 *is* a quadratic residue $\pmod q$, while from the right-hand side of the congruence we see, with the help of Lemma 2.7.10, that it is *not*; contradiction.

Now suppose $N = x^2 + y^2$ for some integers x and y. Let q be a prime with $q \equiv 3 \pmod 4$ that divides N. If $x = 0$ or $y = 0$, then N is a perfect square, so certainly the highest power of q dividing N is even. Suppose not. Write $x = q^e u$ and $y = q^f v$, where u and v are not divisible by q. We may assume that $e \leq f$ (otherwise interchange x and y). Then

$$N = x^2 + y^2 = q^{2e}u^2 + q^{2f}v^2 = q^{2e}(u^2 + q^{2f-2e}v^2)$$

If $e < f$ then q^{2f-2e} is divisible by q, so the parenthesized expression is congruent to $u^2 \pmod q$; since $u \not\equiv 0 \pmod q$, $u^2 \not\equiv 0 \pmod q$, so this expression is not divisible by q and the highest power of q dividing N is q^{2e}, an even power.

If $e = f$ then the parenthesized expression is $u^2 + v^2$, and by our earlier observation $u^2 + v^2$ is not divisible by q, so the highest power of q dividing N is q^{2e}, again an even power.

Thus, in any case, if N is a sum of two squares, the highest power of q dividing N is even, as claimed. $\qquad\square$

Next we come to the Law of Quadratic Reciprocity. Although we will not be using it again in this book, this is one of the great theorems of number theory, and every student of mathematics should see it (at least) once.

The Law of Quadratic Reciprocity was first proved by Gauss in his *Disquisitiones Arithmeticae*, published in 1801. Gauss returned to this theorem many times in his life, and produced a total of seven different proofs. It has been reproved many times, by many different people, since. We will present a proof due to Zolotarev from 1872. The basic idea of this proof is to look at signs of permutations. The lemma that starts us off, Zolotarev's lemma, has a very easy proof, but, as you will see, the proof of the Law of Quadratic Reciprocity that we give is rather tricky- this deep theorem has no really easy proof.

Let us fix an odd prime p. We recall from Definition 2.7.8 that we have the quadratic residue character on the multiplicative group $\mathbb{Z}_p^*$ defined by $\chi_p(a) = 1$ if a is a quadratic residue (mod p) and $\chi_p(a) = -1$ if a is a quadratic nonresidue (mod p). (In comparison to that definition, we are simplifying our notation here by writing a instead of $[a]$. We will continue to use this simplification throughout this section.) We also recall from that definition that we have the Legendre symbol $\left(\frac{a}{p}\right)$ defined by $\left(\frac{a}{p}\right) = \chi_p(a)$.

Recall from Corollary 3.3.11 that $\mathbb{Z}_p^*$ is cyclic. A generator of $\mathbb{Z}_p^*$ is called a *primitive root* (mod p). For example, 2 is a primitive root (mod 5) as the powers of 2 (mod 5) are $\{1,2,4,3\}$. It is not a primitive root (mod 7) as the powers of 2 (mod 7) are $\{1,2,4\}$, but 3 is a primitive root (mod 7) as the powers of 3 (mod 7) are $\{1,3,2,6,4,5\}$.

Here is the lemma that starts the ball rolling.

Lemma 3.8.5 (Zolotarev's lemma). *Let p be an odd prime. Let a be relatively prime to p, and let $\mu_a \colon \mathbb{Z}_p^* \to \mathbb{Z}_p^*$ be the permutation $\mu_a(x) = ax$ (mod p). Then $\chi_p(a) = \mathrm{sign}(\mu_p)$.*

Proof.   Let r be a primitive root (mod p). Then $a \equiv r^k$ (mod p) for some k, so $\mu_a = \mu_{r^k}$. But $\mu_{r^k} = (\mu_r)^k$ (where by power we mean composition), so $\mathrm{sign}(\mu_a) = \mathrm{sign}(\mu_r)^k$. But also $\chi_p(a) = \chi_p(r^k) = \chi_p(r)^k$. Thus we need only show $\chi_p(r) = \mathrm{sign}(\mu_r)$.

On the one hand, $\chi_p(r) = -1$ as r is a quadratic nonresidue, being an odd power (the power 1) of the generator r, and hence a nonsquare in the group of even order $\mathbb{Z}_p^*$.

On the other hand, let us write the elements of $\mathbb{Z}_p^*$ in the order $(1\, r\, r^2 \ldots r^{p-2})$. Then we see $\mu_r(1) = r, \mu_r(r) = r^2, \ldots$ $\mu_r(r^{p-2}) = r^{p-1} = 1$. In other words, μ_r is a single $(p-1)$-cycle. But $p-1$ is even so this is an odd permutation, i.e., $\mathrm{sign}(\mu_r) = -1$. $\qquad\square$

Corollary 3.8.6. *Let p be an odd prime.*

(a) *If a is relatively prime to p, and $\mu_a \colon \mathbb{Z}_p \to \mathbb{Z}_p$ is the permutation $\mu_a(x) = ax \ (mod\ p)$, then $\chi_p(a) = \mathrm{sign}(\mu_p)$.*
(b) *For any a, if $\alpha_a \colon \mathbb{Z}_p \to \mathbb{Z}_p$ is the permutation $\alpha_a(x) = a + x \ (mod\ p)$, then $\mathrm{sign}(\alpha_a) = 1$.*

Proof.

(a) μ_a on $\mathbb{Z}_p$ only differs from μ_a on $\mathbb{Z}_p^*$ by the fact that $\mu_a(0) = 0$, i.e., we have an additional 1-cycle, and that does not affect the sign of a permutation, so this follows immediately from Lemma 3.8.5.
(b) This is certainly true for $a = 0$, as then α_a is the identity. For $a \neq 0$, note that $\alpha_a = (\alpha_1)^a$ (where by power we again mean composition). Writing the elements of $\mathbb{Z}_p$ in the order $(0\, 1\, 2 \ldots p-1)$, we see that $\alpha_1(0) = 1, \alpha_1(1) = 2, \ldots, \alpha_1(p-1) = 0$. In other words, α_1 is a single p-cycle. But p is odd so this is an even permutation, i.e., $\mathrm{sign}(\alpha_1) = 1$.
$\qquad\square$

Theorem 3.8.7 (Law of Quadratic Reciprocity). *Let p and q be odd primes. Then*

$$\left(\frac{p}{q}\right)\left(\frac{q}{p}\right) = (-1)^{\frac{p-1}{2} \cdot \frac{q-1}{2}}.$$

Proof (Zolotarev). We stated this in terms of Legendre symbols, as this most commonly done, but we will use quadratic residue characters in our proof.

We begin by considering the ring $\mathbb{Z}_p \times \mathbb{Z}_q = \{(a, b) \mid 0 \leq a \leq p-1,\ 0 \leq b \leq q-1\}$. We write the elements of this ring in a p-row, q-column rectangle, rows numbered $0, \ldots, p-1$ and columns numbered $0, \ldots, q-1$, so that the entry in row a, column b, is (a, b).

	0	1	2	$\ldots$	$q-1$
0	$(0,0)$	$(0,1)$	$(0,2)$	$\ldots$	$(0,q-1)$
1	$(1,0)$	$(1,1)$	$(1,2)$	$\ldots$	$(1,q-1)$
2	$\vdots$	$\vdots$	$\vdots$		$\vdots$
$\vdots$					
$p-1$	$(p-1,0)$	$(p-1,1)$	$(p-1,2)$	$\ldots$	$(p-1,q-1)$

We recall that we have an isomorphism of rings (and hence a bijection) $\varphi\colon \mathbb{Z}_{pq} \to \mathbb{Z}_p \times \mathbb{Z}_q$ given by

$$\varphi(c) = (c \ (\mathrm{mod}\ p),\ c \ (\mathrm{mod}\ q)).$$

We first consider the permutations on this array defined by

$$\zeta(a,b) = (qa+b, b) \quad \text{and} \quad \theta(a,b) = (a, a+pb).$$

We note that

$$\varphi(qa+b) = (qa+b, b) \quad \text{and} \quad \varphi(a+pb) = (a, a+pb)$$

and so

$$\varphi^{-1}\zeta(a,b) = qa+b \quad \text{and} \quad \varphi^{-1}\theta(a,b) = a+pb.$$

These are both bijections, so we may consider the bijection

$$\sigma = (\varphi^{-1}\zeta)(\varphi^{-1}\theta)^{-1}.$$

Then $\sigma(\varphi^{-1}\theta) = \varphi^{-1}\zeta$ so we see that $\sigma\colon \mathbb{Z}_{pq} \to \mathbb{Z}_{pq}$ is the bijection (i.e., permutation)

$$\sigma(a+pb) = qa+b.$$

Now from the equation $\sigma(\varphi^{-1}\theta) = \varphi^{-1}\zeta$ we see

$$\mathrm{sign}(\sigma)\,\mathrm{sign}(\varphi^{-1})\,\mathrm{sign}(\theta) = \mathrm{sign}(\varphi^{-1})\,\mathrm{sign}(\zeta)$$

$$\mathrm{sign}(\sigma)\,\mathrm{sign}(\theta) = \mathrm{sign}(\zeta)$$

and since the sign of a permutation is ± 1,

$$\mathrm{sign}(\theta)\,\mathrm{sign}(\zeta) = \mathrm{sign}(\sigma).$$

We now evaluate each of these three signs. We begin with θ. Note that θ leaves each of the rows in this array fixed, and permutes the entries within the rows. Let θ_a be the permutation on row a of the array, so $\theta = \theta_0\theta_1 \ldots \theta_{p-1}$ and hence $\text{sign}(\theta) = \text{sign}(\theta_0)\text{sign}(\theta_1) \ldots \text{sign}(\theta_{p-1})$. Now for each a, θ_a is the permutation on $\mathbb{Z}_q$ given by $\theta_a = \alpha_a\mu_p$ where α_a and μ_p are as in Lemma 3.8.6. Thus, $\text{sign}(\theta_a) = \text{sign}(\alpha_a)\text{sign}(\mu_p) = \chi_q(p)$ by Lemma 3.8.6, and then $\text{sign}(\theta) = \chi_q(p)^p = \chi_q(p)$ as p is odd. By exactly the same logic, $\text{sign}(\zeta) = \chi_p(q)$. Thus, we see that the left-hand side of this equation is $\chi_q(p)\chi_p(q)$. To complete the proof we now show that $\text{sign}(\sigma) = (-1)^{\frac{p-1}{2} \cdot \frac{q-1}{2}}$.

We determine $\text{sign}(\sigma)$ by counting inversions. Let us order the elements of the above array *by columns*. This gives an ordering of $\mathbb{Z}_{pq}$ by counting down the columns. i.e., $(0,0) \leftrightarrow 0, (1,0) \leftrightarrow 1, \ldots, (p-1,0) \leftrightarrow p-1, (0,1) \leftrightarrow p, (1,1) \leftrightarrow p+1, \ldots, (p-1,q-1) \leftrightarrow pq-1$.

Note that under this ordering

$$(a, b) \leftrightarrow a + pb.$$

In other words, if $c \in \mathbb{Z}_{pq}$ is $c = a+pb, 0 \leq a \leq p-1, 0 \leq b \leq q-1$, then c is the cth element in this ordering.

Suppose instead we decide to order the elements of this array *by rows*. Then $(0,0) \leftrightarrow 0$, $(0,1) \leftrightarrow 1$, $\ldots$ are under this ordering

$$(a, b) \leftrightarrow qa + b.$$

In other words, if $d \in \mathbb{Z}_{pq}$ is $d = qa+b, 0 \leq a \leq p-1, 0 \leq b \leq q-1$, then d is the dth element in this ordering.

Now σ is given by $\sigma(a + pb) = qa + b$. So we can describe σ as follows: Let $c \in \mathbb{Z}_{pq}$. Count down the array by columns $0, 1, \ldots$, until we reach position c. Then count $0, 1, \ldots$ across the array by rows until we reach the c-th element in our count. Suppose that the element in that position is d. Then $\sigma(c) = d$.

Now to count inversions. An inversion occurs when c' precedes c in our ordering, but when $\sigma(c) = d$ precedes $\sigma(c') = d'$ in our ordering. In our ordering, c' precedes c when we get to c' in counting by *columns before* we get to c, or, equivalently, when c' is anywhere in a column to the left of where c is, or in the same column as c but above c. In other words, if c is in position (a, b) and c' is in position (a', b'), c' precedes c if $a' < a$, or if $a' = a$ and $b' < b$.

Consider these possibilities for c'. We now make the key observation that if c' is in row b or above, i.e., if $b' \leq b$, then we will reach x' in counting by *rows before* we get to c, i.e., in these cases d' will *precede* d in our ordering — no inversions. But if c' is in a row below row b, then we will reach c' in counting by *rows after* we get to c, i.e., in these cases d' will *follow* d in our order-all inversions.

Thus, we see that for any (a, b), we get an inversion from (a', b') when (a', b') lies in a corner of our array strictly to the left of and below (a, b):

	0	1	....	$b-1$	b		$\cdots q-1$
0							
$\vdots$							
a					$\bullet\,(a,b)$		
$a+1$	$\bullet$	$\bullet$	$\cdots$	$\bullet$			
$\vdots$	$\vdots$	$\vdots$		$\vdots$			
$p-1$	$\bullet$	$\bullet$	$\cdots$	$\bullet$			

Thus, we simply need to count the number of these pairs $((a', b'), (a, b))$.

We will do so by moving along rows.

Suppose we are in row 0 (i.e., $a = 0$). If we are in column 0, there are no entries below and to the left. If we are in column 1, there is a single column to the left, which has entries in rows, $1, \ldots, p-1$ below our entry, i.e., a total of $p-1$ entries. If we are in column 2, there are two columns to the left, each with (again) $p-1$ entries below our entry. Proceeding in this way, we see we obtain a total of $(p-1) + 2(p-1) + \cdots + (q-1)(p-1) = (p-1)(1+2+\cdots+q-1) = (p-1)(q-1)(q)/2$ inversions. Now for row 1. Again we get nothing from column 0. Now we get 1 fewer entry, i.e., $p-2$ entries, below and to the left of the entry in column 1, $2(p-2)$ entries below and to the left of the entry in column 2, etc., for a total of $(p-2)+2(p-2)+\cdots+(q-1)(p-2) = (p-2)(1+2+\cdots q-1) = (p-2)(q-1)(q)/2$ inversions. Proceeding in this way we see we obtain a total of $(p-1)(q-1)(q)/2 + (p-2)(q-1)(q)/2 + \cdots + (q-1)(q)/2 = \frac{p(p-1)}{2} \cdot \frac{q(q-1)}{2}$ inversions. Hence,

$$\text{sign}(\sigma) = (-1)^{\frac{p(p-1)}{2} \cdot \frac{q(q-1)}{2}} = \left((-1)^{\frac{p-1}{2} \cdot \frac{q-1}{2}}\right)^{pq} = (-1)^{\frac{p-1}{2} \cdot \frac{q-1}{2}}$$

as p and q are both odd. $\qquad\square$

Remark 3.8.8. As the determination of the permutation σ, and the computation of its sign, in this proof were rather tricky, we will illustrate these by an example. We take $p = 5$ and $q = 3$ and write the column ordering in the respective positions. We have:

$$
\begin{array}{ccc}
0 & 5 & 10 \\
1 & 6 & 11 \\
2 & 7 & 12 \\
3 & 8 & 13 \\
4 & 9 & 14
\end{array}
$$

Then we see $\sigma(0) = 0, \sigma(1) = 5, \sigma(2) = 10, \sigma(3) = 1, \sigma(4) = 6, \ldots$ Writing the elements in this array (rather than linearly) for clarity, we see:

$$
\sigma\left(\begin{bmatrix} 0 & 5 & 10 \\ 1 & 6 & 11 \\ 2 & 7 & 12 \\ 3 & 8 & 13 \\ 4 & 9 & 14 \end{bmatrix}\right) = \begin{bmatrix} 0 & 11 & 8 \\ 5 & 2 & 13 \\ 10 & 7 & 4 \\ 1 & 12 & 9 \\ 6 & 3 & 14 \end{bmatrix}
$$

and we see we have inversions $(5,1)$, $(5,2)$, $(5,3)$, $(5,4)$, $(10,1)$, $(10,2)$, $(10,3)$, $(10,4)$, $(10,6)$, $(10,7)$, $(10,8)$, $(10,9)$, $(6,2)$, $(6,3)$, $(6,4)$, $\ldots$, $(13,4)$, $(13,9)$, as in the proof. $\diamond$

Corollary 3.8.9. *Let p and q be odd primes.*

(a) *If at least one of p and q is congruent to 1 (mod 4), then either both p is a quadratic residue (mod q) and q is a quadratic residue (mod p) or both p is a quadratic nonresidue (mod q) and q is a quadratic nonresidue (mod p).*

(b) *If both p and q are congruent to 3 (mod 4), then either p is a quadratic residue (mod q) and q is a quadratic nonresidue (mod p), or p is a quadratic nonresidue (mod q) and q is a quadratic residue (mod p).*

Proof. This is simply a restatement of the Law of Quadratic Reciprocity. $\square$

We close this section with a Zolotarev-style proof of Gauss's lemma (Lemma 2.7.14).

Lemma 3.8.10 (Gauss's lemma). *Let p be an odd prime and let a be an integer that is relatively prime to p. Let*

$$T = \{i \mid 1 \leq i \leq (p-1)/2 \quad \text{and}$$

$$ai \equiv k \pmod{p} \text{ for some } k \text{ with } (p+1)/2 \leq k \leq p-1\}.$$

Let $t = \#(T)$. Then $\left(\frac{a}{p}\right) = (-1)^t$.

Proof.   Write $\mathbb{Z}_p^* = \{1, 2, \ldots, p-1\}$ in two rows in the following order

$$\begin{array}{ccc} 1 & 2 & \cdots & (p-1)/2 \\ (p-1) & (p-2) & & (p+1)/2 \end{array}$$

Note that i and j are in the same column if and only if $i + j = 0$ (mod p). Let $\mu_a \colon \mathbb{Z}_p^* \to \mathbb{Z}_p^*$ be as in Zolotarev's lemma. Then $\left(\frac{a}{p}\right) = \chi_p(a) = \text{sign}(\mu_a)$. We will calculate $\text{sign}(\mu_a)$ from its action on this array.

Note that if $i + j \equiv 0 \pmod{p}$, then $\mu_a(i) + \mu_a(j) \equiv 0 \pmod{p}$, so we see that μ_a permutes the columns of this array, and may or may not interchange the entries in a column. Thus we may write $\mu_a = \tau\sigma$, where σ is the permutation on the columns, keeping the entries in every column in their same positions, while τ fixes each column, but interchanges the entries in a column exactly when σ does. Note this happens precisely for column i when i is in the set T. Thus, τ is a product of t transpositions.

Now we may write σ as $\sigma_1\sigma_2$ where σ_1 is the action of σ on the first row and σ_2 is the action of σ on the second row. But, since σ preserves columns, these actions are *exactly the same*. Thus we see $\mu_a = \tau\sigma = \tau\sigma_1\sigma_2$ so

$$\text{sign}(\mu_a) = \text{sign}(\tau)\text{sign}(\sigma_1)\text{sign}(\sigma_2) = \text{sign}(\tau)\text{sign}(\sigma_1)^2 = \text{sign}(\tau),$$

i.e.,

$$\chi_p(a) = (-1)^t$$

as claimed. $\square$

3.9 Some examples of integral domains

In this section we first want to look a little deeper at $\mathbb{Z}[i]$, where we do have unique factorization, and then look at some examples where we don't.

We begin with a lemma that we will use in several cases.

Lemma 3.9.1. *Let D be a squarefree integer. Let $R = \mathcal{O}(\sqrt{D})$ and let δ be the multiplicative norm on R given by $\delta(a + b\sqrt{D}) = |a^2 - b^2 D|$.*

(a) *If a is an element of with $\delta(a) = p$, where p is a prime, then a is irreducible in R.*

(b) *Suppose that p and q are prime (perhaps $q = p$). If R does not have an element of norm p, and a is an element of R with $\delta(a) = pq$, then a is irreducible in R.*

Proof. Let $a = bc$. To show that a is irreducible, we must show that b or c is a unit.

(a) If $\delta(a) = p$, then $\delta(bc) = \delta(b)\delta(c) = p$, so $\delta(b) = 1$, in which case b is a unit, or $\delta(c) = 1$, in which case c is a unit (Lemma 3.4.8).

(b) If $\delta(a) = pq$, then $\delta(bc) = \delta(b)\delta(c) = pq$. Since R does not have an element of norm p, we cannot have $\delta(b) = p$ or $\delta(c) = p$. Thus, we must have $\delta(b) = 1$ or $\delta(c) = 1$, in which case b or c is a unit, as in part (a). □

Now we determine the distinct primes in $\mathbb{Z}[i]$. Recall that two primes are said to be distinct if they are not associates of each other.

Theorem 3.9.2. *Let $R = \mathcal{O}(\sqrt{-1}) = \mathbb{Z}[i]$. The distinct primes in R are:*

(i) *$1 + i$ and its associates. (Note that $1 - i = -i(1 + i)$ is an associate of $1 + i$, and $-i(1 + i)^2 = 2$.)*

(ii) *Let p be a prime congruent to 1 (mod 4) and let a and b be integers with $a^2 + b^2 = p$.*

(ii(a)) *$a + bi$ and its associates.*

(ii(b)) *$a - bi$ and its associates.*

(iii) *For p a prime congruent to 3 (mod 4), p and its associates.*

Proof. First we recall that, since $\mathbb{Z}[i]$ is a UFD, primes and irreducibles in $\mathbb{Z}[i]$ are the same.

If $a \in \mathbb{Z}[i]$ is as in (i), (ii(a)), (ii(b)), or (iii), it follows immediately from Lemma 3.9.1 that a is irreducible, and hence prime.

Now let $a \in \mathbb{Z}[i]$ with $\delta(a) = q$. If $q = 2$ or p for p a prime congruent to 1 (mod 4), or p^2 for p a prime congruent to 3 (mod 4), then a must be one of (i), (ii(a)), (ii(b)), and (iii). Suppose not. Write $\delta(a) = q_0 r$ where q_0 is of this form. Then $a\bar{a} = q_0 r$, $r > 1$. But q_0 is divisible by some prime element z of $\mathbb{Z}[i]$, so z divides a or $\bar{a}$. If z divides $\bar{a}$, switch z and $\bar{z}$. Thus, we may assume z divides a, so $a = zw$ for some $w \in \mathbb{Z}[i]$. But then $\delta(a) = \delta(z)\delta(w)$, i.e., $q_0 r = \delta(z)\delta(w)$, with $\delta(z)$ dividing q_0, so $\delta(w) > 1$ and w is not a unit. Hence, a is not irreducible, so is not prime. $\qquad\square$

Now we turn to some examples where things go wrong.

Example 3.9.3. Let R be a field and let $S = R[\{x^{1/n}\}_{n \in \mathbb{N}}]$ be the ring of Example 3.2.17(c). As we observed there, S is not Noetherian. We claim that the element x of S is not divisible by any irreducible. It is easy to check that if $p(x)$ is an element of S with at least two terms, and $q(x)$ is any nonzero element of S, then $p(x)q(x)$ has at least two terms. Thus, the only divisors of x have a single term, so must be rx^q for some $r \neq 0$ in R (a unit in S) and some positive rational number q. But $x^q = (x^{q/2})^2$ so x^q is not irreducible. We also see that S is not a GCD domain: The set $\{x^q \mid q \in \mathbb{Q}, q > \sqrt{2}\}$ does not have a GCD. On the other hand, S *is* an f-GCD domain. We can see this as follows: Let $\{p_1(x), \ldots, p_k(x)\}$ be a finite set of elements of S, not all zero. Then there is some integer N such that every term in each of these polynomials has exponent of x that is an integer multiple of $x^{1/N}$. Substitute $y = x^{1/N}$ to obtain a set of polynomials $\{p_1(y), \ldots, p_k(y)\}$ in the polynomial ring $R[y]$. Now $R[y]$ is a GCD ring, so this set has a gcd $g(y)$. Now substitute back $x^{1/N} = y$ to obtain that $g(x^{1/N})$ is a gcd of $\{p_1(x), \ldots, p_k(x)\}$. $\qquad\Diamond$

Example 3.9.4. Let R be a field and let $S = R[x^2, x^3]$ be the ring of polynomials in the variables x^2 and x^3 with coefficients in R. Note that every power x^k with $k \geq 2$ is in S ($x^4 = x^2 \cdot x^2, x^5 = x^2 \cdot x^3$, etc.). Thus, we may alternatively write S as the subring of $R[x]$ given by

$$S = \{a_0 + a_2 x^2 + a_3^{x^3} + \cdots + a_n x^n \mid a_0, a_2, a_3, \ldots, a_n \in R\},$$

i.e., the subring of $R[x]$ consisting of all polynomials that do not have an "x" term.

First we observe that x^2 and x^3 are both irreducibles in S. (Any factorization of x^2 or x^3 would have to have an "x" term.) Then $x^6 = x^2 \cdot x^2 \cdot x^2 = x^3 \cdot x^3$ are two distinct factorizations of x^6 into irreducibles, so S is *not* a UFD. Also, observe that the divisors of x^5 (ignoring unit factors) are $\{1, x^2, x^3\}$ and the divisors of x^6 (again ignoring unit factors) are $\{1, x^2, x^3, x^4\}$, so the set of common divisors is $\{1, x^2, x^3\}$. Hence the elements x^5 and x^6 do not have a gcd (as neither of x^2 nor x^3 divides the other in S), so S is *not* an f-GCD domain.

But S *is* Noetherian, as we see from the following argument: Let I be a nonzero ideal in S. Let d be the smallest degree of a nonzero polynomial in I and let $p_d(x)$ be a monic polynomial of degree d in I. Note that $p_d(x)$ is unique as if not, and $p'_d(x)$ were some other, then $p_d(x) - p'_d(x)$ would be a polynomial of lower degree in I; impossible. Now I may or may not contain a polynomial of degree $d + 1$. If not, do nothing. If so, let $p_{d+1}(x)$ be a monic polynomial in I with the coefficient of x^d in $p_{d+1}(x)$ equal to zero (which we can always arrange by subtracting an appropriate multiple of $p_d(x)$), and by the same argument $p_{d+1}(x)$ is unique. We claim that I is generated by the single polynomial $p_d(x)$ in the first case, or by the pair of polynomials $p_d(x)$ and $p_{d+1}(x)$ in the second case. We prove this by induction on the degree n of any nonzero polynomial $q(x)$ in I. If $n = d$, then $q(x)$ must be a multiple of $p_d(x)$. If $n = d+1$, then there are no such polynomials in the first case, and any such polynomial is easily checked to be a sum of multiples of $p_d(x)$ and $p_{d+1}(x)$ in the second case. Now assume the result is true for any polynomial of degree $< n$, and let $q(x)$ be a polynomial of degree $n \geq d + 2$. Then $q(x) = x^2 r(x)$ for some polynomial $r(x)$ of degree $n - 2$. Then $r(x) \in I$ by the inductive hypothesis, and I is an ideal of S, so $q(x) = x^2 r(x) \in I$, and by induction we are done.   ◊

Example 3.9.5.

(a) Let $R = \mathcal{O}(\sqrt{-5}) = \mathbb{Z}[\sqrt{-5}]$. Recall that R has the multiplicative norm $\delta(a + b\sqrt{-5}) = a^2 + 5b^2$. We can easily make a table of elements of R with small norm:

Norm   Elements of R with this norm
1   ± 1
2   none
3   none
4   ± 2
5        $\pm\sqrt{-5}$
6        $\pm(1 + \sqrt{-5}), \pm(1 - \sqrt{-5})$
7   none
8   none
9        $\pm 3, \pm(2 + \sqrt{-5}), \pm(2 - \sqrt{-5})$
10   none

Then we have factorizations:

$$6 = 2 \cdot 3 = (1 + \sqrt{-5})(1 - \sqrt{-5})$$
$$9 = 3 \cdot 3 = (2 + \sqrt{-5})(2 - \sqrt{-5})$$

We observe that these are factorizations into irreducibles by Lemma 3.9.1. Thus both of these two elements have two distinct factorizations into irreducibles, and hence R is not a UFD.

(b) Let $R = \mathcal{O}(\sqrt{-6}) = \mathbb{Z}[\sqrt{-6}]$, with multiplicative norm $\delta(a + b\sqrt{-6}) = a^2 + 6b^2$. We again make a table of elements of R with small norm:

Norm   Elements of R with this norm
1   ± 1
2   none
3   none
4   ± 2
5   none
6        $\pm\sqrt{-6}$
7        $\pm(1 + \sqrt{-6}), \pm(1 - \sqrt{-6})$
8   none
9   ± 3
10      $\pm(2 + \sqrt{-6}), \pm(2 - \sqrt{-6})$

Then we have factorizations:

$$6 = 2 \cdot 3 = -1(\sqrt{-6})(\sqrt{-6})$$
$$10 = 2 \cdot 5 = (2 + \sqrt{-6})(2 - \sqrt{-6})$$

Again, these are factorizations into irreducibles by Lemma 3.9.1. Thus, both of these elements have two distinct factorizations into irreducibles, and hence R is not a UFD.

(c) Let $R = \mathcal{O}(\sqrt{10}) = \mathbb{Z}[\sqrt{10}]$, with multiplicative norm $\delta(a + b\sqrt{10}) = |a^2 - 10b^2|$.

We have the factorization

$$10 = 2 \cdot 5 = (\sqrt{10})(\sqrt{10}).$$

We observe that 2 is not prime, as it divides the product $10 = (\sqrt{10})(\sqrt{10})$ without dividing either of the factors. We observe that 2 has norm 4. We claim that R does not have an element of norm 2. For suppose $x + y\sqrt{10} \in R$ with $\delta(x + y\sqrt{10}) = 2$. Then

$$x^2 - 10y^2 = \pm 2$$
$$x^2 \equiv \pm 2 \ (\mathrm{mod}\ 5)$$

which has no solution.

Then by Lemma 3.9.1(b) 2 is irreducible.

We also observe that 5 has norm 25. We claim that R does not have an element of norm 5. For suppose $x + y\sqrt{10} \in R$ with $\delta(x + y\sqrt{10}) = 5$. Then

$$x^2 - 10y^2 = \pm 5$$
$$x^2 = \pm 5 + 10y^2 = 5(2y^2 \pm 1)$$

Thus, x must be divisible by 5, and then x^2 is divisible by 25, so

$$2y^2 \pm 1 \equiv 0 \ (\mathrm{mod}\ 5)$$

which has no solution.

Then by Lemma 3.9.1(b) 5 is irreducible.

Also, since $\sqrt{10}$ has norm 10, again, by Lemma 3.9.1(b), $\sqrt{10}$ is irreducible.

Thus, 10 has two distinct factorizations into irreducibles and hence R is not a UFD. $\diamond$

These three rings are Noetherian, however, as we now see.

Lemma 3.9.6. *Let D be a squarefree integer and let $R = \mathcal{O}(\sqrt{D})$.*
Then every nonzero ideal I of R is of one of the following two forms:

(i) *I is a principal ideal generated by a single element a of R.*
(ii) *I is an ideal generated by an element a of R and an integer g with g dividing $\delta(a)$.*

Proof. Let I be a nonzero ideal of R. If I is principal, we are done. Suppose not.

Let z be any element of I, $z \neq 0$. Then I contains the integer $z\bar{z}$, and it contains $z\bar{z}\sqrt{D}$ as well. Let

$$S_1 = \{|k| \neq 0 \mid k \text{ is an integer in } I\}$$

$$S_2 = \{|n| \neq 0 \mid m + n\sqrt{D} \text{ is in } I \text{ for some } m\}.$$

Now S_1 is a nonempty set of positive integers, so has a smallest element k_0. Also, S_2 is a nonempty set of positive integers or half-integers, so has a smallest element n_0. Let $a = m_0 + n_0\sqrt{D} \in I$.

Now let $b = m + n\sqrt{D}$ be any element of I. We claim that n is an integer multiple of n_0. We can write $n = n_0 q + r$ with $0 \leq r < n_0$. (If n_0 is an integer, this is just the division algorithm. If n_0 is a half-integer, this is true as well-just apply the division algorithm to $2n$ and $2n_0$). Then $b - aq = m' + r\sqrt{D}$ for some m', and with $0 \leq r < n_0$. But $b - aq \in I$, so by the minimality of n_0 we must have $r = 0$. Thus we see that n is always divisible by n_0. But then $b - aq = m'$. Thus $m' \in S_1$, so again by the division algorithm we must have that m' is an integer multiple of k_0, $m' = jk_0$ for some integer j. Thus, $b = aq + jk_0$ so I is generated by the two elements a and k_0.

Now to finish the proof we "neaten up". Since $a \in I$, $\delta(a) = a\bar{a} \in I$. Since $\delta(a)$ and k_0 are integers in I, $g = \gcd(\delta(a), k_0)$ is in I (as $g = \delta(a)s + k_0 t$ for some $s, t \in \mathbb{Z}$, and $\mathbb{Z} \subseteq R$). Now k_0 is a multiple of g, so if I is generated by a and k_0 it is also generated by a and g, and, finally g certainly divides $\delta(a)$. $\qquad\square$

3.10　Quotient fields and localization

To motivate our constructions in this section, let us look at a few examples of rings.

First we have the ring (indeed, the field) of rational numbers

$$\mathbb{Q} = \{a/b \mid a, b \in \mathbb{Z}, b \neq 0\},$$

which we are familiar with. Here are a couple of less familiar examples. Fix a prime $p \in \mathbb{Z}$. We have the rings

$$R = \{a/b \mid a, b \in \mathbb{Z}, b \neq 0, b \text{ a power of } p\},$$

$$S = \{a/b \mid a, b \in \mathbb{Z}, b \neq 0, b \text{ relatively prime to } p\}.$$

In each case, what have we done? We have chosen a "suitable" subset of $\mathbb{Z}$ that we will allow as denominators in our fractions. Suitable means that we want the result to be a ring, so that it must be closed under the operations of addition and multiplication. How do we do addition and multiplication in $\mathbb{Q}$, R, or S? By the "usual" rules for adding and multiplying fractions.

But actually we have jumped the gun, because we have already written the elements of $\mathbb{Q}$, R and S as a/b, which presumes that we already know how to do some arithmetic. If we think more basically about this, we should really write them as ordered pairs (a, b), and then mimic the "usual" operations of arithmetic on these ordered pairs. But if we are careful to do so, the first thing we run up against is the fact that different ordered pairs can represent the same fraction (e.g., $1/2 = 2/4 = 3/6 = \ldots$). So what we should do is to put a relation (in fact an equivalence relation) on ordered pairs, saying that two ordered pairs (a, b) and (c, d) are equivalent if $a/b = c/d$. But of course we can't phrase it that way, as that would be going around in circles, since we are trying to define a/b and c/d. How can we do this in a noncircular fashion? Again we can think about when "usual" fractions a/b and c/d are equal, and remember that this is true when $ad = bc$. With all this in mind, we can now proceed in a very straightforward way.

Definition 3.10.1. Let R be an integral domain and let A be a nonempty subset of R that is closed under multiplication (i.e., if $a_1 \in A$ and $a_2 \in A$ then $a_1 a_2 \in A$), and assume that $0 \notin A$, $1 \in A$.

Define a relation $\sim$ on $R \times A$ by

$$(r_1, a_1) \sim (r_2, a_2) \text{ if } r_1 a_2 = r_2 a_1. \qquad \Diamond$$

Lemma 3.10.2. *The relation $\sim$ is an equivalence relation on $R \times A$.*

Proof. We leave this as an exercise for the reader. $\qquad \square$

Definition 3.10.3. Let R_A be the set of equivalence classes of $R \times A$ under the relation $\sim$. R_A is called the *localization of R away from A*. We write the equivalence class of (r, a) as r/a. $\Diamond$

Lemma 3.10.4. *The operations of addition and multiplication on R_A given by*

$$\frac{a_1}{r_1} + \frac{a_2}{r_2} = \frac{a_1 r_2 + a_2 r_1}{r_1 r_2}$$

$$\frac{a_1}{r_1} + \frac{a_2}{r_2} = \frac{a_1 a_2}{r_1 r_2}$$

are well defined.

Proof. These are operations on equivalence classes, so we must check they are independent of the choice of representatives. Again we leave this for the reader. $\square$

Theorem 3.10.5. *In the above situation, R_A is an integral domain. The zero element of R_A is $0/1$ and the identity element of R_A is $1/1$.*

Proof. We must verify that the operations on R_A satisfy all the properties of addition and multiplication in an integral domain. Again we leave this for the reader. $\square$

Remark 3.10.6. We regard R as a subset of R_A by identifying $r \in R$ with $r/1 \in R_A$. $\Diamond$

Lemma 3.10.7. R_A^*, *the units in R_A, is $R_A^* = \{a_1/a_2 \mid a_1 \in A\}$.*
In particular, if $A = R - \{0\}$, $R_A^* = R_A - \{0\}$, so that in this case R_A is a field.

Proof. Again we leave this for the reader. $\square$

Definition 3.10.8. In the above situation, if $A = R - \{0\}$, R_A is called the *quotient field* of R. $\Diamond$

Example 3.10.9.

(a) If $R = Z$, the quotient field of R is $\mathbb{Q}$.
(b) If $R = \mathcal{O}(\sqrt{D})$, the quotient field of R is $\mathbb{Q}(\sqrt{D})$.
(c) Let R be a field. Then the quotient field of the polynomial ring $R[x]$ is $\{p(x)/q(x) \mid p(x), q(x) \in R[x], \ q(x) \neq 0\}$. This field is the field of *rational functions* in x with coefficients in R.

(d) More generally, if R is an integral domain with quotient field S, the quotient field of $R[x]$ is the field of rational functions in x with coefficients in S.

(e) Let R be a field. Recall that $R[[x]]$ is the ring of formal power series in x with coefficients in R,

$$R[[x]] = \left\{ \sum_{n=0}^{\infty} a_n x^n \mid a_n \in R \right\}.$$

As we have observed, $R[[x]]^* = \{\Sigma_{n=0}^{\infty} \mid a_n \in R, a_0 \neq 0\}$. Then the quotient field of $R[[x]]$ is

$$\left\{ \sum_{n=-N}^{\infty} a_n x^n \mid a_n \in R \right\}. \qquad \qquad \lozenge$$

Remark 3.10.10. We have restricted our attention here to integral domains for simplicity (and because it is the only case we will need). But the construction of R_A goes through more generally. We can begin with R any commutative ring with 1. We just need to require that the subset A does not contain any zero divisors of R. (Of course, if R has zero divisors, R_A will also have zero divisors.) $\qquad \lozenge$

3.11 Polynomial rings: Unique factorization and related matters

In this section, we fix a UFD (unique factorization domain) R that is not a field, and we let F be its quotient field. The most important special case of this is when $R = \mathbb{Z}$, in which case $F = \mathbb{Q}$. But the argument in general is *exactly* the same-word for word and symbol for symbol-in the general case as it is in this special case, so we may as well give it in general.

We recall that $F[x]$ is a PID, and hence a UFD. Done!

Now $R[x]$ is not a PID: If $a \neq 0$ is any nonunit in R,

$$I = \{p(x) = a_0 + a_1 x + \cdots + a_n x^n \mid a_0 \text{ is divisible by } a\}$$

is an ideal of $R[x]$ that is generated by $\{a, x\}$, but is not generated by any single element of $R[x]$, so $R[x]$ is *not* a PID.

234   *An Introduction to Abstract Algebra: Sets, Groups, Rings, and Fields*

Our goal is to show that $R[x]$ *is* a UFD. We will show this by first looking at polynomials in $R[x]$, and then, at the crucial stage, by relating the situation in $R[x]$ to the situation in $F[x]$.

Definition 3.11.1. Let $f(x) = a_n x^n + \cdots + a_0$ be a nonzero polynomial in $R[x]$. Then $f(x)$ is *primitive* if its coefficients $\{a_0, \ldots, a_n\}$ are relatively prime. $\Diamond$

Lemma 3.11.2 (Gauss's lemma). *If $g(x)$ and $h(x)$ are primitive polynomials in $R[x]$, then their product $f(x) = g(x)h(x)$ is a primitive polynomial in $R[x]$.*

Proof.   We prove this by contradiction. Let

$$g(x) = b_m x^m + \cdots + b_0,$$

$$h(x) = c_k x^k + \cdots + c_0,$$

$$f(x) = a_n x^n + \cdots + a_0.$$

Suppose that $f(x)$ is not primitive. Choose a prime p that divides $\gcd(a_0, \ldots, a_n)$.

Now $g(x)$ is assumed to be primitive, so not all of its coefficients are divisible by p. Let i be the smallest value such that b_i is not divisible by p.

Similarly, $h(x)$ is assumed to be primitive, so not all of its coefficients are divisible by p. Let j be the smallest value such that c_j is not divisible by p.

Consider the coefficient a_{i+j} of $f(x)$. This coefficient is given by

$$a_{i+j} = (b_{i+j}c_0 + b_{i+j-1}c_1 + \cdots + b_{i+1}c_{j-1})$$
$$+ b_i c_j + (b_{i-1}c_{j+1} + \cdots + b_1 c_{i+j-1} + b_0 ci + j).$$

Now a_{i+j} is assumed to be divisible by p. The "c" coefficients in every term in the first parenthesized expression are assumed to be divisible by p. The "b" coefficients in every term in the second parenthesized expression are assumed to be divisible by p. Hence the remaining term, $b_i c_j$, must be divisible by p. But p is a prime, so that implies that b_i is divisible by p or c_j is divisible by p; contradiction. $\square$

Definition 3.11.3. Let $f(x) = a_n x^n + \cdots + a_0$ be a nonzero polynomial in $R[x]$. The *content* $c(f(x))$ is $c(f(x)) \cong \gcd(a_0, \ldots, a_n)$. $\diamond$

Lemma 3.11.4. *Let* $f(x) \in R[x]$, $f(x) \neq 0$ *and let* $d \cong c(f(x))$. *Then* $\tilde{f}(x) = (1/d)f(x)$ *is a primitive polynomial in* $R[x]$.

Proof. If $d \cong \gcd(a_0, \ldots, a_n)$, let $a_0 = da_0', \ldots, a_n = da_n'$. Then $\{a_0', \ldots, a_n'\}$ is a relatively prime set of elements of R, so $\tilde{f}(x) = (1/d)f(x) = a_n' x^n + \cdots + a_0'$ is a primitive polynomial in $R[x]$. $\square$

Lemma 3.11.5. *Let* $f(x), g(x) \in R[x]$, $f(x) \neq 0$, $g(x) \neq 0$. *Then* $c(f(x)g(x)) \cong c(f(x))c(g(x))$.

Proof. Let $c = c(f(x))$ and $d = c(g(x))$ and write $f(x) = c\tilde{f}(x)$, $g(x) = d\tilde{g}(x)$ as in Lemma 3.11.4. Then $h(x) = f(x)g(x) = (cd)\tilde{f}(x)\tilde{g}(x)$. Let $e \cong c(h(x))$ and write $h(x) = e\tilde{h}(x)$ as in that lemma. Then

$$e\tilde{h}(x) = (cd)(\tilde{f}(x)\tilde{g}(x)).$$

Now by Gauss's lemma (Lemma 3.11.2), $\tilde{f}(x)\tilde{g}(x)$ is primitive. Thus the right-hand side has content cd, while the left-hand side has content e, so we must have $e \cong cd$. $\square$

Now suppose $h(x) \in R[x]$ and $h(x)$ is a product $h(x) = f(x)g(x)$ with $f(x), g(x) \in F[x]$. We would like to conclude $f(x), g(x) \in R[x]$. For example, in case $R = \mathbb{Z}$, $x^2 + x = x(x+1)$. But that can't always be right as, again in case $R = \mathbb{Z}$, we also have $x^2 + x = (2x)(1/2x + 1/2)$. However, this is the only sort of thing that can go wrong.

Corollary 3.11.6. *Let* $h(x) \in R[x]$ *and suppose that* $h(x) = f(x)g(x)$ *with* $f(x), g(x) \in F[x]$. *Then there are polynomials* $f_1(x), g_1(x) \in R[x]$ *with* $h(x) = f_1(x)g_1(x)$, *where* $f_1(x)$ *is a constant multiple of* $f(x)$ *and* $g_1(x)$ *is a constant multiple of* $g(x)$.

Proof. We may write $h(x) = e\tilde{h}(x)$ with $e \in R$ and $\tilde{h}(x) \in R[x]$ a primitive polynomial. We may write $f(x) = c\tilde{f}(x)$ with $c \in F$ and $\tilde{f}(x) \in R[x]$ a primitive polynomial, and $g(x) = d\tilde{g}(x)$ with $d \in F$ and $\tilde{g}(x) \in R[x]$ a primitive polynomial. Then $e\tilde{h}(x) = cd\tilde{f}(x)\tilde{g}(x)$.

Again, by Gauss's lemma, $\tilde{f}(x)\tilde{g}(x)$ is a primitive polynomial in $R[x]$, so we must have $cd \in R$. Then, as in the proof of Lemma 3.11.5, we have $e \cong cd$, i.e., $e = ucd$ for some unit $u \in R$.

Let $f_1(x) = \tilde{f}(x) = (1/c)f(x)$ and $g_1(x) = (e/du)g(x) = (e/u)\tilde{g}(x)$. Then $f_1(x) \in R[x]$, $g_1(x) \in R[x]$ and

$$f_1(x)g_1(x) = \left(\frac{1}{c}\right)f(x)\left(\frac{e}{du}\right)g(x) = \left(\frac{e}{cdu}\right)f(x)g(x)$$

$$= f(x)g(x) = h(x)$$

as claimed. $\qquad\square$

Corollary 3.11.7. *Let $h(x)$ be a primitive polynomial in $R[x]$. Then $h(x)$ is irreducible in $R[x]$ if and only if $h(x)$ is irreducible in $F[x]$.*

Proof. First of all, note that, since $h(x)$ is primitive, $h(x)$ has no nonunit constant factors in $R[x]$, so any nonunit factor of $h(x)$ in $R[x]$ must be a nonconstant polynomial.

Now if $h(x)$ is irreducible in $F[x]$, i.e., if $h(x)$ cannot be written as a product $h(x) = f(x)g(x)$ with $f(x), g(x) \in F[x]$ nonunits, i.e., nonconstant polynomials, then $h(x)$ certainly cannot be written as such a product $h(x) = f(x)g(x)$ with $f(x), g(x) \in R[x]$ (as every polynomial in $R[x]$ is a polynomial in $F[x]$).

On the other hand, if $h(x)$ is not irreducible in $F[x]$, so that $h(x) = f(x)g(x)$ with $f(x), g(x) \in F[x]$ nonunits, i.e., nonconstant polynomials, then, by Corollary 3.11.6, $h(x) = f_1(x)g_1(x)$ with $f_1(x)$, $g_1(x) \in R[x]$ nonconstant polynomials, and hence nonunits, and so $h(x)$ is not irreducible in $R[x]$. $\qquad\square$

Now we arrive at the result to which we have been heading.

Theorem 3.11.8. *Let R be a UFD. Then the ring $R[x]$ of polynomials in x with coefficients in R is a UFD.*

Proof. Once again we must first show that any nonzero polynomial $h(x)$ in $R[x]$ has a factorization into irreducibles, and then show that such a factorization is essentially unique.

For the first step: Write $h(x) = e\tilde{h}(x)$ where $e \cong c(h(x))$. Now $e \in R$ so e has a factorization into irreducibles (i.e., primes) $e = u_1 p_1 \ldots p_k$ with u_1 a unit and $p_1, \ldots, p_k$ primes. Then $\tilde{h}(x) \in F[x]$ and we know that $F[x]$ is a UFD, i.e., $\tilde{h}(x)$ has a factorization into irreducible in $F[x]$. But putting Corollary 3.11.6 and Corollary 3.11.7 together, we see that $\tilde{h}(x)$ has a factorization into irreducibles in $R[x]$, i.e.,

$$\tilde{h}(x) = u_2 f_1(x) \ldots f_m(x) \text{ with each } f_i(x) \in R[x] \text{ irreducible}$$

and u_2 a unit in $R[x]$, i.e., a unit in R

and then, if $u = u_1 u_2$, a unit in R,

$$h(x) = u p_1 \ldots p_k f_1(x) \ldots f_m(x)$$

is a factorization of $h(x)$ into irreducibles in $R[x]$.

For the second step: Suppose we have two factorizations

$$h(x) = u p_1 \ldots p_k f_1(x) \ldots f_m(x).$$

and

$$h(x) = v q_1 \ldots q_l q_1(x) \ldots g_n(x).$$

First notice that, since $\tilde{h}(x)$ is primitive, each $f_i(x)$ is primitive, and so $p_1 \ldots p_k \cong c(h(x))$. Similarly, each $g_j(x)$ is primitive, so $q_1 \ldots q_l \cong c(h(x))$. Hence $p_1 \ldots p_k = w q_1 \ldots q_l$ for some unit $w \in R$. But by unique factorization in R, we must have $l = k$, and, after possible reordering, $p_i \cong q_i$ for $i = 1, \ldots, k$.

But these are two factorizations of $h(x)$ in $R[x]$, hence in $F[x]$. Since each factor $f_i(x)$ is irreducible in $R[x]$, it is irreducible in $F[x]$ (Corollary 3.11.7) and similarly, since each factor $g_j(x)$ is irreducible in $R[x]$, it is irreducible in $F[x]$. But we have unique factorization in $F[x]$, so $m = n$, and, after possible reordering, $f_i(x) \cong g_i(x)$ in $F[x]$ for $i = 1, \ldots, m$, i.e., $f_i(x) = u_i g_i(x)$ for some unit $u_i \in F$. But $f_i(x)$ and $g_i(x)$ are both primitive polynomials in $R[x]$, so u_i is a unit in R, in which case $f_i(x) \cong g_i(x)$ in $R[x]$, and we are done. $\qquad \square$

Corollary 3.11.9. *Let R be a UFD.*

(a) *For any n, $R[x_1, \ldots, x_n]$ is a UFD.*
(b) *For any set of variables $\{x_i\}$, $R[\{x_i\}]$ is a UFD.*

Proof.

(a) Since $R[x_1, \ldots, x_n] = (R[x_1, \ldots, x_{n-1}])[x_n]$, this follows directly from Theorem 3.11.18 by induction.

(b) Consider any polynomial in $R[\{x_i\}]$. This can only involve finitely many variables, so must be in $R[x_1, \ldots, x_n]$ for some n. Also, any factor of it must be in $R[x_1, \ldots, x_n]$. So, by part (a), this polynomial has an essentially unique factorization in $R[\{x_i\}]$. $\square$

Remark 3.11.10. If R is a UFD that is not a field, and $\{x_i\}$ has at least one (and possibly infinitely many) elements, then $R[\{x_i\}]$ is a UFD that is not a PID.

If R is a field, and $\{x_i\}$ has at least two (and possibly infinitely many) elements, then $R[\{x_i\}]$ is a UFD that is not a PID. $\Diamond$

Remark 3.11.11. If R is Noetherian, and $\{x_i\}$ is finite, then $R[\{x_i\}]$ is Noetherian. This is the Hilbert basis theorem, Theorem 3.3.13.

If R is not Noetherian, the $R[\{x_i\}]$ is certainly not Noetherian.

If R is Noetherian, and $\{x_i\}$ is infinite, then $R[\{x_i\}]$ is not Noetherian. This is Example 3.2.17. $\Diamond$

Now we turn to practical questions about factorization of polynomials in $R[x]$. First we can ask when such a polynomial has a root in R, or in F. This question is easy to answer.

Lemma 3.11.12. *Let R be a UFD.*

(a) *Let $f(x) \in R[x]$ be a monic polynomial,*

$$f(x) = x^n + a_{n-1}x^{n-1} + \cdots + a_0.$$

Then any root of $f(x)$ in F must be an element s of R. Furthermore, s must divide a_0 in R.

(b) *Let $f(x) \in R[x]$ be arbitrary,*

$$f(x) = a_n x^n + a_{n-1}x^{n-1} + \cdots + a_0.$$

Then any root s/t of $f(x)$ in F with s/t in lowest terms (i.e., $s \in R$, $t \in R$ and s and t relatively prime) must have s dividing a_0 in R and t dividing a_n in R.

Proof.   Note that (a) is a special case of (b). If (b) is true, and $f(x)$ is monic, then t must divide 1 in R, i.e., t is a unit in R, and so s/t is an element of R, and s/t divides a_0 if and only if s divides a_0. Thus, we need only prove (b).

Let $f(x)$ have root s/t with s and t relatively prime. Substituting, $0 = a_n(s/t)^n + a_{n-1}(s/t)^{n-1} + \cdots + a_1(s/t) + a_0$. Multiplying through by t^n,

$$0 = a_n s^n + a_{n-1}s^{n-1}t + \cdots + a_1 st^{n-1} + a_0 t^n.$$

Now the left-hand side, and every term on the right-hand side, except possibly the last, is divisible by s. So the last term, $a_0 t^n$, must be divisible by s as well. We are assuming that s and t are relatively prime, so s and t^n are relatively prime as well, and so s must divide a_0.

Similarly, by looking at the first term on the right-hand side, we see that t must divide a_n. □

Lemma 3.11.12 was first thought of in the case $R = \mathbb{Z}$, $F = \mathbb{Q}$ (long before people thought about UFD's in general) so it is often called the rational root test.

The second question we can ask is when a polynomial in $R[x]$ is irreducible. In general, this is a difficult question to answer. But we do have the following very useful criterion.

Lemma 3.11.13 (Eisenstein's criterion). *Let R be a UFD. Let $h(x) = a_n x^n + \cdots + a_0 \in R[x]$ be an arbitrary polynomial. Suppose there is some prime p in R such that:*

(i) *p does not divide a_n,*
(ii) *p divides $a_{n-1}, \ldots, a_0$,*
(iii) *p^2 does not divide a_0.*

Write $h(x) = c(h(x))\tilde{h}(x)$, so that $\tilde{h}(x)$ is a primitive polynomial in $R[x]$. Then $\tilde{h}(x)$ is irreducible in $F[x]$ (and hence in $R[x]$).

In particular, if $h(x)$ is a primitive polynomial in $R[x]$ then $h(x)$ is irreducible in $F[x]$ (and hence in $R[x]$).

Proof. Note, by Corollary 3.11.7, that $\tilde{h}(x)$ is irreducible in $F[x]$ if and only if it is irreducible in $R[x]$.

We prove this theorem by contradiction. Let

$$f(x) = b_m x^m + \cdots + b_0$$

$$g(x) = c_k x^k + \cdots + c_0$$

with $m, k \geq 1$ and suppose $\tilde{h}(x) = f(x)g(x)$. We may assume $f(x), g(x) \in R[x]$.

Observe that $a_0 = b_0 c_0$. Then, from conditions (ii) and (iii), we see that p divides exactly one of b_0 and c_0. Suppose that p divides b_0 but not c_0.

We claim that p divides $b_0, \ldots, b_m$. We prove this by induction. In case $i = 0$, we are assuming p divides b_0.

Now suppose p divides $b_0, \ldots, b_{i-1}$ and consider b_i. We see that

$$a_i = b_0 c_i + b_1 c_{i-1} + \cdots + b_{i-1} c_1 + b_i c_0.$$

Now p divides a_i (by condition (ii)) and, by the inductive hypothesis, p divides every term on the right-hand side except for possibly the last one, so p must divide the last term $b_i c_0$ as well. But p does not divide c_0, so p must divide b_i.

Now note that $a_n = b_m c_k$, so, since p divides b_m, p must divide a_n. But this contradicts condition (i).

Thus, it is impossible to factor $\tilde{h}(x)$, i.e., $\tilde{h}(x)$ is irreducible. □

Example 3.11.14. Let p be a prime and let n be any positive integer. Then $h(x) = x^n - p$ is an irreducible polynomial of degree n in $\mathbb{Z}[x]$ (or $\mathbb{Q}[x]$). ◇

Remark 3.11.15. The integers $\mathbb{Z}$ are a UFD, and the polynomial rings $\mathbb{Z}[x]$ and $\mathbb{Q}[x]$ are UFDs. In a UFD, as we know, primes and irreducibles are the same. Nevertheless, it is common to refer to integers as prime and polynomials as irreducible. This is an accident of mathematical history. ◇

3.12   Ideals: Maximal and prime

We now let R be an arbitrary commutative ring with 1. We want to consider two kinds of ideals in R.

Definition 3.12.1. An ideal I of R is *maximal* if $I \neq R$ and there is no ideal J of R with $I \subset J \subset R$. Equivalently, I is maximal if $I \subset R$ and if J is an ideal of R such that $I \subseteq J$, then $J = I$ or $J = R$. ◇

Definition 3.12.2. An ideal I of R is *prime* if whenever $a, b \in R$ with $ab \in I$, then $a \in I$ or $b \in I$. ◇

These two notions are closely related.

Lemma 3.12.3. *Let I be an ideal of R. If I is maximal, then I is prime.*

Proof. Let I be a maximal ideal of R. Suppose $a, b \in R$ with $ab \in I$. We need to show that $a \in I$ or $b \in I$.

If $a \in I$ we are done. Suppose not. Let J be the ideal generated by a and I. Concretely, $J = \{ra + i \mid r \in R, i \in I\}$. Now $I \subset J$, so, since I is maximal, we must have $J = R$. In particular, $I \in J$, so $1 = r_0 a + i_0$ for some $r \in R$, $i_0 \in I$. But then

$$b = b \cdot 1 = b(r_0 a + i_0) = r_0(ab) + b i_0$$

But $ab \in I$ by assumption, and $i_0 \in I$. Thus $b \in I$, as required. $\square$

As we shall see, the converse of this result is sometimes, but not always, true. But first, an easy observation.

Lemma 3.12.4. *The ideal $I = \{0\}$ of R is a prime ideal if and only if R is an integral domain.*

Proof. By definition R is an integral domain if it has no zero divisors, i.e., if a and b are elements of R with $ab = 0$, then $a = 0$ or $b = 0$. $\square$

We have used the word "prime" before, in connection with elements of an integral domain. We are using it again, in connection with ideals. Since we are using the same word in two different contexts, we would expect these to be a close relation between the two, and there is.

Lemma 3.12.5. *Let R be an integral domain and let $r \in R$, $r \neq 0$. Then the principal ideal I generated by r is a prime ideal if and only if r is a prime in R.*

Proof. Note that $I \neq R$ if and only if r is not a unit. But I consists exactly of the multiples of r, so $a \in I$ if and only if r divides a, and $b \in I$ if and only if r divides b, and $ab \in I$ if and only if r divides ab, and so we see the two conditions are equivalent. $\square$

Example 3.12.6. Let $R = \mathbb{Z}[x]$. As we have seen, 2 and x are both primes in R, so if I_1 is the (principal) ideal generated by 2, and

I_2 is the (principal) ideal generated by x, then I_1 and I_2 are both prime ideals in R. But they are not maximal ideals, as they are both contained in the ideal J generated by $\{2, x\}$. We have already seen that $J = \{$polynomials $a_n x^n + \cdots + a_0$ in $\mathbb{Z}[x]$ with a_0 even$\}$, so J is indeed a proper ideal in R. Now J is in fact a maximal ideal in R, as we now see: Let $f(x)$ be any polynomial not in J, i.e., any polynomial with odd constant term. Then it is easy to check that we can write $1 = f(x)g(x) + h(x)$ for some polynomial $g(x) \in R$ and some polynomial $h(x) \in I$. But then if K is the ideal generated by $f(x)$ and J, we have $K = R$. $\Diamond$

Thus, in general not every prime ideal is maximal. But in one important case it is.

Lemma 3.12.7. *Let R be a principal ideal domain and let I be a nonzero prime ideal of R. Then I is maximal.*

Proof.   In this case, by Lemma 3.12.5, I is generated by a prime $p \in R$. Let $a \in R$, $a \notin I$. Let J be the ideal generated by a and I. We need to show that $J = R$. Let $d \cong \gcd(a, p)$. Now p is a prime, and p does not divide a (as $a \notin I$), so a and p are relatively prime, i.e., $d \cong 1$. But R is a PID, so we know we can write

$$1 = ab + pq \text{ for some } b, q \in R.$$

But that implies $1 \in J$ and hence $J = R$. $\square$

We now return to the situation of a general commutative ring with 1.

Theorem 3.12.8. *Let R be a commutative ring with 1 and let I be an ideal of R.*

(a) *I is a prime ideal if and only if the quotient R/I is an integral domain.*

(b) *I is a maximal ideal if and only if the quotient R/I is a field.*

Proof.   Let π be the quotient map $\pi\colon R \to R/I$, so $\pi(a) = a + I$. For simplicity, we will write $\pi(a) = [a]$. Observe that $[a] = 0$ if and only if $a \in I$.

(a) Suppose that I is a prime ideal and let $[a]$, $[b] \in R/I$ with $[a][b] = 0$. We need to show $[a] = 0$ or $[b] = 0$. Now $[a][b] = [ab]$, so

$0 = [a][b] = [ab]$ which is true if and only if $ab \in I$. But I is a prime ideal, so that implies $a \in I$, in which case $[a] = 0$, or $b \in I$, in which case $[b] = 0$.

On the other hand, suppose I is not a prime ideal. Let $a, b \in R$ with $a \notin I$, $b \notin I$, but $ab \in I$. Then $[a] \neq 0$, $[b] \neq 0$, but $[a][b] = [ab] = 0$.

(b) Suppose that I is a maximal ideal and let $[a] \in R/I$, $[a] \neq 0$. We need to show that $[a]$ has an inverse $[b]$ in R/I. Now $a \in R$ with $a \notin I$ (as $[a] \neq 0$) and I is maximal, so, as we have seen in the proof of Lemma 3.12.3, there is an element b of R and an element i of I with $ab + i = 1$. But then $[a][b] = [1]$ in R/I, and $[1] = 1 + I$ is the identity element of R/I.

On the other hand, if I is not a maximal ideal, let J be an ideal of R with $I \subset J \subset R$. Let $a \in J$, $a \notin I$. We claim $[a] \in R/I$ does not have an inverse. Suppose it did, $[a][b] = [1]$ for some $[b] \in R/I$. Then $ab \in 1 + I$, i.e., $ab = 1 + i$ for some element i of I, and then $1 = ab + (-i)$. Now $a \in J$, $-i \in I$ and $I \subseteq J$, so $-i \in J$; hence $1 \in J$ and $J = R$, a contradiction. $\qquad \square$

We will be using part (b) of Theorem 3.12.8 extensively, as it is a very effective method of constructing fields.

Example 3.12.9. Let $R = F[x]$ with F a field. Then R is a PID. Let $p(x) \in R$ be an irreducible polynomial, i.e., a prime in R. Let I be the ideal generated by $p(x)$. Then I is a nonzero prime ideal in the PID R, and so, by Lemma 3.12.7, a maximal ideal in R. Then R/I is a field. $\qquad \Diamond$

We have seen that $\mathbb{Z}[x]$ is an example of a UFD that is not a PID, and also an example of a ring in which not every nonzero prime ideal is maximal. It is no coincidence that we used the same example twice!

Theorem 3.12.10. *Let R be an integral domain. The following are equivalent:*

(1) *R is a PID.*
(2) *R is a UFD and every nonzero prime ideal of R is maximal.*

Proof. We have already shown that if R is a PID, then R is a UFD (Theorem 3.6.13), and also that every nonzero prime ideal of

R is maximal (Lemma 3.12.7), so we know that condition (1) implies condition (2).

We must show that condition (2) implies condition (1). This is a long proof, and we will break it up into several steps.

Let R be a UFD in which every nonzero prime ideal is maximal. Let I be an ideal of R. We must show I is principal. This is certainly true if $I = \{0\}$, so assume I is nonzero.

Step 1: Since R is a UFD, it is a GCD domain (Lemma 3.6.17). Let $g \cong \gcd(\{ \text{ elements of } I \})$. Write $g = up_1^{e_1} \ldots p_k^{e_k}$ with u a unit and $p_1, \ldots, p_k$ distinct primes. From Lemma 3.6.16 we see that there must be an element a_1 of I such that the highest power of p_1 dividing a_1 is $p_1^{e_1}$, an element a_2 of I such that the highest power of p_2 dividing a_2 is $p_2^{e_2}$, ..., an element a_k of I such that the highest power of p_k dividing a_k is $p_k^{e_k}$.

Then we see that $g \cong \gcd(a_1, \ldots, a_k)$. The point here is that even if the ideal I is not finitely generated, g is a gcd of a *finite* number of elements of I. (It may be that some of the a_i's coincide, or that some a_i divide another, in which case this set is redundant. We could be "neat" about it and discard the redundant elements, but we don't have to-with or without redundancies, the point is that this set is finite.)

Let J be the ideal of R generated by g. Since g divides every element of I, $I \subseteq J$. We will show that $g \in I$, in which case $J \subseteq I$. Thus, $I = J$, a principal ideal (as J is generated by the single element g).

Let $A = \{a_1, \ldots, a_k\}$. If A consists only of a single element $\{a_1\}$, then $g \cong a_1$ so in this case certainly $I = J$.

The crucial case is when A has two elements, $A = \{a_1, a_2\}$, which we rename $\{a, b\}$ for clarity.

Step 2: Let $A = \{a, b\}$, and $g \cong \gcd(a, b)$. We claim $g \in I$.

Step 2a: Suppose a is a prime p. If p divides b, then $g \cong a$ and once again $I = J$.

Suppose that a does not divide b. Since we are assuming that a is a prime, that means that a and b are relatively prime, and so $1 \cong g$, and hence $J = R$. Let I_0 be the ideal generated by a and I_1 be the ideal generated by $A = \{a, b\}$. Since a does not divide b, $I_0 \subset I_1$, and since $a, b \in I$, $I_1 \subseteq I$. Now a is a prime, so I_0 is a nonzero prime ideal, and we are assuming that every nonzero prime ideal of R is maximal. Thus $I_1 = R$, and so $I = R$, and so $I = J$. We observe for future use that in this case, $1 = ar + bs$ for some $r, s \in R$.

Step 2b: Suppose that a and b are relatively prime. Again in this case, $1 \cong g$ and hence $J = R$. Write $a \cong q_1^{f_1} \ldots q_j^{f_j}$ where $q_1, \ldots, q_j$ are distinct primes (some subset of $p_1, \ldots p_k$). Since a and b are relatively prime, q_i and b are relatively for each i. Thus, by step 2a, we have

$$1 = q_i r_i + b s_i \text{ for some } r_i, s_i \in R,$$

for each $i = 1, \ldots, j$. But then

$$1 = \prod_{i=1}^{j} (q_i r_i + b s_i)^{f_i}$$

Now observe that in this product, there is one term that has $q_1^{f_i} \ldots q_j^{f_j} \cong a$ as a factor, and every other term has b as a factor. Thus, we see that in this case, $1 = ar + bs$ for some $r, s \in R$, so once again $I = R$, and so $I = J$.

Step 2c: Let a and b be arbitrary. By considering common prime factors of a and b, we see we can write

$$a = q_1^{f_1} \ldots q_t^{f_t} r$$
$$b = q_1^{g_1} \ldots q_t^{g_t} s$$

where r and s are relatively prime, r and $q_1^{g_1} \ldots q_t^{g_t}$ are relatively prime, and s and $q_1^{f_1} \ldots q_t^{f_t}$ are relatively prime. Let $e_1 = \min(f_1, g_1), \ldots, e_t = \min(f_t, g_t)$. Then $g \cong q_1^{e_1} \ldots q_t^{e_t}$. Again let I_0 be the ideal generated by a and b.

Now set

$$a' = q_1^{f_1 - e_1} \ldots q_t^{f_t - e_t} r$$
$$b' = q_1^{g_1 - e_1} \ldots q_t^{g_t - e_t} s$$

and let I_0' be the ideal generated by a' and b'. We observe that $I_0 = \{gi \mid i \in I_0'\}$.

Now $a' = a/g$, $b' = b/g$, so a' and b' are relatively prime. Hence, by step 2b, $1 = a'r + b's$ for some $r_1 s \in R$, and then

$$g = g \cdot 1 = g(a'r + b's) = (ga')r + (gb')s = ar + bs$$

so $g \in I$, and hence $I = J$.

Step 3: Let $A = \{a_1, \ldots, a_k\}$. We argue by induction on k. We saw that in case $k = 1$ there was nothing to prove, and the case $k = 2$ was step 2. Now assume that any ideal I for which the set A consists of $k-1$ elements is principal, and suppose that A consists of k elements. Recall that $\gcd(a_1, \ldots, a_k) = \gcd(\gcd(a_1, a_2), a_3, \ldots, a_k)$. By the $k = 2$ case, if $g_{12} \cong \gcd(a_1, a_2)$, then $g_{12} \in I$, so we may replace A by $A' = \{g_{12}, a_3, \ldots, a_k\}$, a set of $k-1$ elements of I with a same gcd, g, so by the $k-1$ case $g \in I$, $I = J$, and by induction we are finally done.   □

3.13   Exercises

1. Let R be a ring. For a set X, let
$$R^X = \{f{:}X \to R\}$$
Define addition and multiplication in R^X by $(f + g)(x) = f(x) + g(x)$ and $(fg)(x) = f(x)g(x)$. Show that R^X with these operations is a ring.

2. (a) Let X be a set and let $P(X)$ be the set of subsets of X. Define addition and multiplication on $P(X)$ by
$$A + B = (A \cap B^C) \cup (A^C \cup B), \text{the symmetric}$$
$$\text{difference of} A \text{ and } B,$$
$$AB = A \cap B, \text{ the intersection of } A \text{ and } B.$$
Show that $P(X)$ with these operations is a ring.

   (b) Let $R = \mathbb{Z}_2$. Show that $P(X)$ is isomorphic to R^X as defined above.

3. A *Boolean ring* is a ring B such that $b^2 = b$ for every $b \in B$. (For example, $P(X)$ is a Boolean ring.)
   Let B be a Boolean ring.

   (a) Show that B is commutative.

   (b) Show that $2b = 0$ for every $b \in B$.

   (c) Let B be finite. For a nonempty subset A of B, let
$$b_A = \prod_{b \in A} b.$$
Let $e = \Sigma_A b_A$ where the sum is taken over all nonempty subsets A of B. Show that $eb = be = b$ for every $b \in B$.

(Thus, a nontrivial finite Boolean ring is automatically a ring with 1.)

(d) Give an example of an infinite Boolean ring that is a ring with 1, and one that is not.

4. Let R be an arbitrary ring. Let $S = R \times \mathbb{Z}$ with the following operations

$$(a, m) + (b, n) = (a + b, m + n)$$
$$(a, m) \cdot (b, n) = (ab + na + mb, mn).$$

Show that S is a ring with 1. Note that R is isomorphic to the subring $R_0 = \{(r, 0)\}$ of S. (However, even if R is a ring with 1, R_0 is not a subring- with-1 of S.)

5. (a) Show that $\mathcal{O}(\sqrt{D})$, as defined in Example 3.1.16, is a ring.

(b) Show that $\mathcal{O}(\sqrt{D})$ is an integral domain.

6. Prove Lemma 3.2.10.

7. We proved the Noether isomorphism theorems for groups. Prove the analogous theorems for rings:

(a) (First isomorphism theorem) Let $\varphi \colon R \to S$ be a ring homomorphism. Then $\mathrm{Im}(\varphi)$ is isomorphic to $R/\mathrm{Ker}(\varphi)$.

(b) (Second isomorphism theorem) Let I and J be ideals in a ring R. Then $I/I \cap J$ is isomorphic to $(I + J)/J$.

(c) (Third isomorphism theorem) Let I and J be ideals in a ring R with $J \subseteq I$. Then R/I is isomorphic to $(R/J)/(I/J)$.

(d) (Correspondence theorem) Let J be an ideal in a ring R. Then there is a $1-1$ correspondence between $\{$ideals of R containing $J\}$ and $\{$ideals of $R/J\}$ given by $I \to I/J$.

8. Let R be an integral domain that has the structure of a finite dimensional vector space over some field $\mathbb{F}$. Show that R is a field.

9. Let $\mathbb{F}$ be a field and let a and b be fixed elements of $\mathbb{F}$. Let

$$I = \begin{bmatrix} 1 & 0 \\ 0 & 1 \end{bmatrix} \quad \text{and} \quad C = \begin{bmatrix} 0 & -b \\ 1 & -a \end{bmatrix}$$

Let $R = \{sI + tC \mid s, t \in \mathbb{F}\}$.

(Note that R is a 2-dimensional $\mathbb{F}$-vector space. We may consider $\mathbb{F} \subset R$ by identifying $s \in \mathbb{F}$ with $sI \in R$.)

(a) Show that R is a commutative ring with 1.

(b) If the quadratic polynomial $p(x) = ax^2 + bx + c$ does not have a root in $\mathbb{F}$, show that R is a field.

(c) Show that C is a root of $p(x)$ in R, i.e., that $p(C) = 0$.

10. The *center* $C(R)$ of a ring R is

$$C(R) = \{r \in R \mid rs = sr \text{ for all } s \in R\},$$

i.e., the center of R is the subring of R consisting of those elements of R that commute with every element of R.

Let A be an arbitrary commutative ring with 1, and let $R = M_n(A)$ be the ring of n-by-n matrices with entries in A, with the usual matrix operations. Show that $C(R) = \{\text{scalar matrices in } M_n(A)\}$.

11. Let G be an arbitrary group and let R be a commutative ring with 1. The *group ring* of G with coefficients in R is

$$R[G] = \left\{ \sum_{g \in G} r_g g \mid \text{ only finite many } r_g \neq 0 \right\}$$

with operations

$$\sum_{g \in G} r_g g + \sum_{g \in G} s_g g = \sum_{g \in G} (r_g + s_g)g$$

$$\left(\sum_{g \in G} r_g g \right) \left(\sum_{h \in G} s_h h \right) = \sum_{g, h \in G} (r_g s_h)gh.$$

(a) The augmentation ideal I of $R[G]$ is the kernel of the augmentation map $\epsilon\colon R[G] \to R$ given by $\epsilon(\Sigma r_g g) = \Sigma r_g$. Show that I is generated as an abelian group by $\{g - 1 \mid g \in G\}$.

(b) Let $\{C_i\}$ be the sets of conjugacy classes of elements of G that contains only finitely many elements, and for each such conjugacy class C_i, let $c_i = \Sigma_{g \in C_i} g$. Show that the center $C(R[G])$ is generated as an abelian group by $\{c_i\}$.

12. Let R be a ring with 1.

(a) Show that R has a nontrivial proper left ideal if and only if R has a nontrivial proper right ideal.

(b) Show that R is a skew field if and only if the only left ideals of R are $\{0\}$ and R, or, equivalently (by part (a)) if and only if the only right ideals of R are $\{0\}$ and R. (If R is a skew field, the only two-sided ideals of R are $\{0\}$ and R, but the converse of this statement is false, as we see from the next problem.)

13. A ring is called *simple* if it has no nontrivial proper two-sided ideals.

    (a) Let $\mathbb{F}$ be a field and let $R = M_n(\mathbb{F})$. Show that R is a simple ring.

    (b) More generally, let A be a commutative ring with 1 and let $R = M_n(A)$. Show that every two-sided ideal of R is given by $R = M_n(I)$ for some ideal I of A.

14. (a) Let R be a ring with 1 and let r and s be elements of R with $rs = 1$. Of course, if $sr = 1$ then r and s are units, $s = r^{-1}$ and $r = s^{-1}$. Suppose that $sr \neq 1$. Show that neither r, s, nor sr are units.

    (b) Give an example of this situation.

15. Let R be a ring. An element $r \in R$, $r \neq 0$, is *nilpotent* if $r^k = 0$ for some $k > 0$.

    (a) If R is commutative, show that

$$\{\text{nilpotent elements of } R\} \cup \{0\}$$

    is an ideal of R.

    (b) Give an example to show that (a) may be false if R is not commutative.

16. An *idempotent* in a ring R with 1 is an element $e \neq 0$ or 1 with $e^2 = e$.

    (a) If e is an idempotent of R show that $f = 1 - e$ is also an idempotent of R with $ef = fe = 0$.

    (b) A set $\{e_1, \ldots, e_n\}$ of elements of R is a complementary set of idempotents of R if $e_i^2 = e_i$ for each i and $e_i e_j = e_j e_i = 0$ for $i \neq j$. If $\{e_1, \ldots, e_n\}$ is such a set, show that

$$R = Re_1 \oplus \cdots \oplus Re_n \text{ as left ideals,}$$
$$R = e_1 R \oplus \cdots \oplus e_n R \text{ as right ideals.}$$

(c) An idempotent e of R is central if $e \in C(R)$, the center of R. If e is a central idempotent of R, show that $eR = Re = eRe$ is a subring of R, and is a ring with 1. If $\{e_1, \ldots, e_n\}$ is a complementary set of central idempotents of R, show that

$$R = e_1 R e_1 \oplus \ldots \oplus e_n R e_n \text{ as rings.}$$

17. (a) Let G be a cyclic group of order n, which we write multiplicatively as $\{1, g, \ldots, g^{n-1}\}$. Let $R = \mathbb{C}[G]$ be the complex group ring of G. Set $\zeta = \exp(2\pi i/n) \in \mathbb{C}$. Let

$$e_k = \frac{1}{n} \sum_{i=0}^{n-1} \zeta^{ki} g^i \qquad \text{for } k = 0, \ldots, n-1.$$

Show that $\{e_0, \ldots, e_{n-1}\}$ is a complementary set of central idempotents of $\mathbb{C}[G]$.

(b) Let n be odd and let G be the dihedral group $G = D_{2n}$ of order $2n$. Following our previous notation, $G = \{1, \alpha, \ldots, \alpha^{n-1}, \beta, \alpha\beta, \ldots, \alpha^{n-1}\beta\}$. Let H be the subgroup $H = \{1, \alpha, \ldots, \alpha^{n-1}\}$ of G. Set $\zeta = \exp(2\pi i/n) \in \mathbb{C}$. Let

$$e_0 = \frac{1}{2n} \sum_{g \in G} g,$$

$$e_1 = \frac{1}{2n} \left[\sum_{g \in H} g - \sum_{g \notin H} g \right],$$

$$f_k = \frac{1}{n} \sum_{g \in H} (\zeta^{ki} + \zeta^{-ki}) g^i \qquad \text{for } k = 1, \ldots, (n-1)/2.$$

Show that $\{e_0, e_1, f_1, \ldots, f_{(n-1)/2}\}$ is a complementary set of central idempotents in $\mathbb{C}[G]$. (There is a similar, but slightly more complicated, formula in case n is even.)

18. Let $R = \mathbb{Z}$. In each case, find $d = \gcd(a, b)$, and express d in the form $d = ar_0 + bs_0$ for some r_0, s_0 in R.

(a) $a = 25, b = 18$
(b) $a = 1223, b = 541$
(c) $a = 12599, b = 8557$

    (d) $a = 126749, b = 28805$

    (e) $a = 1079909, b = 404689$

19. Let $R = \mathbb{Z}[i]$. In each case, find $d = \gcd(a, b)$, and express d in the form $d = ar_0 + bs_0$ for some r_0, s_0 in R.

    (a) $a = 8 + 13i, b = 10 + 17i$

    (b) $a = 1 + 31i, b = 16 + 41i$

    (c) $a = 14 + 17i, b = 26 - 3i$

20. Let $R = \mathbb{Q}[x]$. In each case, find $d = \gcd(a, b)$, and express d in the form $d = ar_0 + bs_0$ for some r_0, s_0 in R.

    (a) $a = x^4 + x^3 + 3x^2 + 4x + 2,$ $b = x^3 + 2x + 1$

    (b) $a = x^4 + 2x^3 - x^2 + 2x - 2,$ $b = x^3 + 3x^2 + x + 3$

21. Let R be a PID. Let $a, n \in R$ with $n \neq 0$, and let $d = \gcd(a, n)$. Show that the congruence $ax \equiv b \pmod{n}$ has a solution if and only if b is divisible by d. In that case, show that the solution is unique $(\bmod\ n/d)$. (In particular, if a and n are relatively prime, the congruence $ax \equiv b \pmod{n}$ has a unique solution $(\bmod\ n)$ for any b.)

    (b) In case $R = \mathbb{Z}$, suppose that b is divisible by d. Show that the congruence $ax \equiv b \pmod{n}$ has exactly d solutions $(\bmod\ n)$.

22. Let $R = \mathbb{Z}$. Consider each of the following congruences $(\bmod\ n)$. Find all solutions (if any) $(\bmod\ n)$.

    (a) $16x \equiv 9 \pmod{25}$

    (b) $18x \equiv 3 \pmod{47}$

    (c) $95x \equiv 21 \pmod{683}$

    (d) $35x \equiv 77 \pmod{140}$

    (e) $65x \equiv 91 \pmod{120}$

23. Let R be a PID, let $a, b \in R$ and let $d = \gcd(a, b)$. Write $a = da'$ and $b = db'$. Suppose that $d = ar_0 + bs_0$. Show that

$$r = r_0 + tb', \quad s = s_0 - ta', \quad t \in R$$

are solutions to $d = ar + bs$, and furthermore that all solutions to $d = ar + bs$ are of this form.

24. Let $R = \mathbb{Z}$. Find all solutions to $\gcd(693, 819, 1001) = 693x + 819y + 1001z$ with x, y, z in R.

25. Solve each of the following systems of simultaneous congruences in $\mathbb{Z}$.

(a) $x \equiv 19 \pmod{32}$
$\quad\ x \equiv\ \ 9 \pmod{45}$
(b) $x \equiv 87 \pmod{163}$
$\quad\ x \equiv 56 \pmod{257}$
(c) $x \equiv 5 \pmod 9$
$\quad\ x \equiv 7 \pmod{19}$
$\quad\ x \equiv 9 \pmod{29}$
(d) $x \equiv 10 \pmod{47}$
$\quad\ x \equiv 17 \pmod{55}$
$\quad\ x \equiv 25 \pmod{91}$
(e) $x \equiv 3 \pmod 5$
$\quad\ x \equiv 4 \pmod 7$
$\quad\ x \equiv 5 \pmod{11}$
$\quad\ x \equiv 6 \pmod{13}$

26. Solve the following system of simultaneous congruences in $\mathbb{Z}[i]$:

$$x \equiv 3 - i \qquad (\mathrm{mod}\ 3 + 7i)$$
$$x \equiv 2 + i \qquad (\mathrm{mod}\ 4 + 5i).$$

27. Let n be a positive integer. Show that the following are equivalent:

   (a) For every integer a, there is some positive integer t with $a^{t\varphi(n)+1} \equiv a \pmod n$.
   (b) For every integer a, and every positive integer t, $a^{t\varphi(n)+1} \equiv a \pmod n$.
   (c) n is a product of distinct primes.

28. (a) An integer is squarefree if it is not divisible by any perfect square except 1. Show that for any integer k, there is a sequence of k consecutive integers none of which is squarefree.
   (b) Call an integer n exactly divisible by a prime p if p divides n but p^2 does not. Show that for any integer k, there is a sequence of k consecutive positive integers $x_1, \ldots, x_k$ and a sequence of primes $p_1, \ldots, p_k$ such that p_i exactly divides x_i but p_i does not divide x_j for $j \neq i$, for each $i = 1, \ldots, k$.

29. Prove the following generalization of the Chinese remainder theorem:

    Theorem. *Let R be a commutative ring with 1, and let I_1 and I_2 be two ideals of R. Let r_1 and r_2 be two elements of R. Then the system of simultaneous congruences*

    $$x \equiv r_1 \pmod{I_1}$$
    $$x \equiv r_2 \pmod{I_2}$$

    has a solution if and only if $r_1 \equiv r_2 \pmod{I_1 + I_2}$.
    If that is the case, and $x = b$ is any solution, then $x = b'$ is a solution if and only if $b' \equiv b \pmod{I_1 \cap I_2}$.

30. Let $\mathbb{F}$ be a field and let $R = \mathbb{F}[x]$. Let a and b be positive integers. Show that $\gcd(x^a - 1, x^b - 1) = x^{\gcd(a,b)} - 1$.

31. Let R be a PID.

    (a) If $a \equiv b \pmod{n_1}$ and $a \equiv b \pmod{n_2}$, show that $a \equiv b \pmod{\operatorname{lcm}(n_1, n_2)}$. (In particular, if n_1 and n_2 are relatively prime, $a \equiv b \pmod{n_1 n_2}$.)

    (b) If $a \equiv b \pmod{n_1}$ and $a \equiv c \pmod{n_2}$, show that $b \equiv c \pmod{\gcd(n_1, n_2)}$. (In particular, if n_1 and n_2 are relatively prime, this gives no information.)

32. Let $p(x)$ be any polynomial with integer coefficients. Show that

    $$\sum_{i=1}^{n^2} p(i) \equiv 0 \pmod{n}$$

    for every positive integer n.

33. Let R be an f-GCD domain and let $a, b, c, d \in R$ with $\gcd(a, c) = \gcd(a, d) = \gcd(b, c) = \gcd(b, d) = 1$. Show that $\gcd(ab, cd) = 1$.

34. (a) Let R be an f-GCD domain. Let $a, b \in R$ and suppose that a^k divides b^k for some positive integer k. Show that a divides b.

    (b) Give an example of an integral domain R and elements a, b of R with a^k dividing b^k for some $k > 0$ but a not dividing b.

35. (a) Let R be an f-GCD domain and let $\mathbb{F}$ be its quotient field. Deduce from the previous exercise that if $r \in R$, and $r = s^k$ for some $s \in \mathbb{F}$, then $s \in R$. (In other words, r is a k-th power in $\mathbb{F}$ if and only if it is a k-th power in R.)

(b) Give an example of an integral domain R with quotient field $\mathbb{F}$, and an element a of R that is a k-th power in $\mathbb{F}$, but not in R.

36. Let R be an f-GCD domain. Let $a, b \in R$ and suppose that ab is a k-th power in R, i.e., $ab = c^k$ for some $c \in R$. If a and b are relatively prime, show that there are elements e and f of R such that a is an associate of e^k and b is an associate of f^k.

37. Let $\mathbb{F}$ be a field and let $R = F[x]$. Let $p(x) \in R$ with the property that $p(f) = 0$ for every $f \in \mathbb{F}$.

   (a) If $\mathbb{F}$ is infinite, show that $p(x)$ is the 0 polynomial.
   (b) If $\mathbb{F}$ is finite, give an example of a nonzero polynomial with this property.

38. (a) Let R be a UFD that is not field. Suppose that R has only finitely many units. Show that if $p(x) \in R[x]$ is any nonconstant polynomial, then $p(r)$ is composite for infinitely many $r \in R$.
   (b) Give an example of a UFD R, that is not a field, and a nonconstant polynomial $p(x) \in R[x]$ such that for every $r \in R$, $p(r)$ is either 0, a unit, or a prime.

39. Prove Lemma 3.5.7.

40. (a) Let R be a UFD. Prove that two elements a and b of R are relatively prime if and only if they have no common prime factor.
   (b) Prove Lemma 3.6.16 and Lemma 3.6.17.

41. Let R be an integral domain and let a and b be nonzero elements of R. We say that $a|b^\infty$ if $a|b^N$ for some N. Write $a \sim b$ if $a|b^\infty$ and $b|a^\infty$.

   (a) Show that $\sim$ is an equivalence relation on $R - \{0\}$.
   (b) Let R be an f-GCD domain. If $a|b^\infty$ and c is relatively prime to b, show that c is relatively prime to a. Conclude that if $a \sim b$, then c is relatively prime to a if and only if c is relatively prime to b.
   (c) Suppose that R is a UFD. Let a and b be nonzero elements of R. Show that we can write $a = d_a \tilde{a}$, $b = d_b \tilde{b}$ with $d_a \sim d_b$, $\tilde{a}$ and $\tilde{b}$ relatively prime and $\tilde{a}$ and $\tilde{b}$ each relatively prime to both d_a and d_b. In the notation of Lemma 3.6.16, identify $d_a, \tilde{a}, d_b$, and $\tilde{b}$.

42. Let R be a UFD. Let a, b, $c \in R$ with $c \neq 0$, and suppose that $\gcd(a, b, c) \cong 1$. Show that there is some $d \in R$ such that $\gcd(a + bd, c) \cong 1$.

43. Let R be an integral domain. Analogous to the notion of greatest common divisor (gcd) we have the notion of least common multiple (lcm).

    Definition. Let $\{a_i\}$ be a finite set of nonzero elements in an integral domain R. Then m is ar *least common multiple* (lcm) of $\{a_i\}$ if

    (1) a_i divides m for each i_i and
    (2) if n is any element of R such that each a_i divides n, then m divides n.

    Similarly to the situation with the gcd, if $\{a_i\}$ has a least common multiple m, then $m' \in R$ is an lcm of $\{a_i\}$ if and only if m and m' are associates.

    (a) Let R be a UFD. In the situation, and notation, of Lemma 3.6.16, let $m_1 = \max(f_1, g_1), \ldots, m_j = \max(f_j, g_j)$. Show that

    $$\operatorname{lcm}(a, b) \cong p_1^{e_1} \ldots p_i^{e_i} q_1^{m_1} \ldots q_j^{m_j} r_1^{h_1} \ldots r_k^{h_k}.$$

    (b) More generally, let R be an f-GCD domain. Let a and b be nonzero elements of R. Show that a and b have an lcm, and moreover that $\operatorname{lcm}(a, b) \cong ab/\gcd(a, b)$.

44. Let R be an f-GCD domain and let a and b be nonzero elements of R.

    (a) Show that for any nonzero element c of R,

    $$\operatorname{lcm}(ca, cb) \cong c \cdot \operatorname{lcm}(a, b).$$

    (b) Let c be any nonzero element of R that divides both a and b. Show that $\operatorname{lcm}(a/c, b/c) \cong \operatorname{lcm}(a, b)/c$.

    (Compare Lemma 3.5.24).

45. Let R be an f-GCD domain and let B and C be finite sets of nonzero elements of R, each of which has an lcm. Let $A = B \cup C$. Show that A has an lcm, and moreover that $\operatorname{lcm}(A) \cong \operatorname{lcm}(\operatorname{lcm}(B), \operatorname{lcm}(C))$. (Compare Lemma 3.5.21.) Note

that, by induction on the number of elements of A, this shows that any finite set of nonzero elements in an f-GCD domain has an lcm.

46. Let $\{a_1, \ldots, a_n\}$ be a finite set of nonzero elements in an f-GCD domain R. Let A be the product $A = a_1, \ldots, a_n$. Show that $\mathrm{lcm}(a_1, \ldots, a_n) \cong A/\gcd(A/a_1, \ldots, A/a_n)$.

47. Derive the method of "partial fractions" as follows:

    (a) Let R be a PID with quotient field $\mathbb{F}$. Let $s \in \mathbb{F}$, $s \notin R$, and write s in the form $s = a/b$, where b has prime factorization $b = p_1^{e_1} \ldots p_k^{e_k}$, with $p_1, \ldots, p_k$ distinct primes and $e_1, \ldots, e_k$ positive integers, and a and b relatively prime. Show that

    $$s = \frac{a_1}{p_1^{e_1}} + \cdots + \frac{a_k}{p_k^{e_k}}$$

    for some elements $a_1, \ldots, a_k$ of R with a_i and p_i relatively prime, for each $i = 1, \ldots, k$.

    (b) Now suppose that R is a Euclidean domain with Euclidean norm δ. Let p be a prime in R and let a be an element of R that is relatively prime to p. Let e be a positive integer. Show that

    $$\frac{a}{p} = a_0 + \frac{a_1}{p} + \cdots + \frac{a_e}{p^e}$$

    for some elements $a_0, a_1, \ldots a_e$ of R, with $a_j = 0$ or $\delta(a_j) < \delta(p)$ for each $j = 1, \ldots, e$.

    (c) Assembling (a) and (b), let R be a Euclidean domain with Euclidean norm δ, let $\mathbb{F}$ be the quotient field of R, and let $s \in \mathbb{F}$. If $s \notin R$, write $s = a/b$ as in (a). Conclude that there is an element c_0 of R, and for each $i = 1, \ldots, k$ there are elements c_{ij} of R, $j = 1, \ldots, e_i$, with $c_{ij} = 0$ or $\delta(c_{ij}) < \delta(p_i)$, such that

    $$s = c_0 + \sum_{i=1}^{k} \sum_{j=1}^{e_i} \frac{c_{ij}}{p_i^j}.$$

    (If $s \in R$, we just have the expression $s = c_0$ with $c_0 = s \in R$.)

48. (a) Find a partial fraction decomposition of $4321/6000$ in $\mathbb{Q}$.

    (b) Find a partial fraction decomposition of $(x^4 + 1)/((x + 1)^2(x^3 - x + 2))$ in $\mathbb{Q}[x]$.

49. A vector $v \in \mathbb{Z}^n$, $v = \begin{bmatrix} a_1 \\ \vdots \\ a_n \end{bmatrix}$, is primitive if $\gcd(a_1, \ldots, a_n) = 1$.

    Show that the group $GL_n(\mathbb{Z})$ acts transitively on {primitive vectors in $\mathbb{Z}^n$}.

    (Observe that this is equivalent to the claim that for any primitive vector $v \in \mathbb{Z}^n$, there is a matrix $A \in GL_n(\mathbb{Z})$ whose first column is v.)

50. (a) Let p be a prime. Show that $GL_n(\mathbb{Z}_p)$ acts transitively on {nonzero vectors in $\mathbb{Z}_p^n$}.

    (b) Use (a) and an inductive argument to determine $|GL_n(\mathbb{Z}_p)|$.

51. Let p be a prime congruent to 1 (mod 4), and let a and b be integers with $a^2 + b^2 = p$. Find an explicit isomorphism $\varphi \colon \mathbb{Z}_p \to \mathbb{Z}[i]/I$ where $I = <a + bi>$ is the principal ideal of $\mathbb{Z}[i]$ generated by $a + bi$.

52. (a) Show that $\delta(z) = |z\bar{z}|$ is a multiplicative Euclidean norm on $\mathcal{O}(\sqrt{-2})$. Conclude that $\mathcal{O}(\sqrt{-2})$ is a Euclidean domain, and hence a PID.

    (b) Show that $\delta(z) = |z\bar{z}|$ is a multiplicative Euclidean norm on $\mathcal{O}(\sqrt{2})$. Conclude that $\mathcal{O}(\sqrt{2})$ is a Euclidean domain, and hence a PID.

53. (a) Show that any prime p that is congruent to 1 or 3 (mod 8) can be written as $p = a^2 + 2b^2$ for some integers a and b.

    (b) Show that any prime p that is congruent to 1 or 7 (mod 8) can be written as $p = a^2 - 2b^2$ for some integers a and b.

54. Let $U(D)$ be the group of units in $\mathcal{O}(\sqrt{D})$.

    (a) Show that $U(-1) = \{\pm 1, \pm i\}$.

    (b) Show that $U(-3) = \{\pm 1, (\pm 1 \pm \sqrt{-3})/2\}$.

    (c) Show that $U(D) = \{\pm 1\}$ for $D < 0$, $D \neq -1, -3$.

    (d) It is known that $U(D)$ is infinite for $D > 0$. Find a unit other than ± 1 in $U(D)$ for $D = 2, 3, 5, 6, 7$.

    (e) Show that, for $D > 0$, every unit of $U(D)$ other than ± 1 is an element of infinite order in $U(D)$.

55. Prove the following general test, which we can often use to show that $\mathcal{O}(\sqrt{D})$ is not a UFD.

    Theorem. *Let $R = \mathcal{O}(\sqrt{D})$. If there is some prime $p \in \mathbb{Z}$ such that*

    (1) R *does not have an element β with norm $\delta(\beta) = p$, and*

(2) R has an element α that is not divisible by p, but with norm $\delta(\alpha)$ divisible by p,

then p is irreducible but not prime in R. Consequently, R is not a UFD.

56. Use this test to show that $\mathcal{O}(\sqrt{D})$ is not a UFD in the following cases:

(a) $D < 0$, $D \neq -2$, D is even, $p = 2$.

(b) $D < 0$, $D \neq -1$, $D \equiv 3 \pmod 4$, $p = 2$.

(c) $D < 0$, $D \neq -7$, $D \equiv 1 \pmod 8$, $p = 2$.

(d) $D < 0$, $D \equiv 5 \pmod 8$, D is composite and p is the smallest prime factor of D.

(e) $D < 0$, $D \equiv 5 \pmod 8$, $m = (1 - D)/4$ is composite and p is the smallest prime factor of m.

(f) $D > 0$, D is divisible by a prime $q \equiv 5 \pmod 8$, $p = 2$.

(g) $D > 0$, D is divisible by a prime $q_1 \equiv 3 \pmod 8$ and by a prime $q_2 \equiv 7 \pmod 8$, $p = 2$.

$\mathcal{O}(\sqrt{D})$ is a UFD (in fact, a PID) for $D = -1, -2, -3, -7$. Parts (a), (b), and (c) show that, except for these values, $\mathcal{O}(\sqrt{D})$ is not a UFD for $D < 0, D \not\equiv 5 \pmod 8$. Parts (d) and (e) give partial information about the case $D \equiv 5 \pmod 8$. Gauss knew that $\mathcal{O}(\sqrt{D})$ is a UFD (in fact, a PID) for $D = -1, -2, -3, -7, -11, -19, -43, -67, -163$ and conjectured that these 9 values of D are the only negative values of D for which that is the case. That his conjecture is true is a deep and justly famous theorem of 20th century mathematics.

Part (f) and (g) give very partial information about the case $D > 0$. Here Gauss conjectured that there are infinitely many positive values of D such that $\mathcal{O}(\sqrt{D})$ is a UFD (in fact, a PID). This conjecture is still completely open.

(While we have seen examples of UFDs that are not PIDs, we will see in Chapter 5 that rings of algebraic integers, which include $\mathcal{O}(\sqrt{D})$, are UFDs if and only if they are PIDs.)

57. (a) Let $\{f_1(x), \ldots, f_k(x)\}$ be a set of primitive polynomials in $\mathbb{Z}[x]$. If these polynomials have a common integer root r, then they are all divisible by the polynomial $x - r$ and so are not relatively prime in $\mathbb{Z}[x]$. Suppose these polynomials do not have a common integer root. Show that these polynomials

are relatively prime in $\mathbb{Z}[x]$ if and only if there is some positive integer N such that $\gcd(f_1(n), \ldots, f_k(n))$ divides N for every integer n.

    (b) If that is the case, let $N_0(f_1(x), \ldots, f_k(x))$ be the smallest such positive integer. Find $N_0(x(x+1), (x+2)(x+3))$.

58. Let $p(x) \in \mathbb{Q}[x]$ be an arbitrary nonconstant polynomial. Show that there are infinitely many $a \in \mathbb{Q}$ such that the polynomial $q(x) = p(x) + a$ is irreducible in $\mathbb{Q}[x]$.

59. Prove Lemma 3.10.2, Lemma 3.10.4, Theorem 3.10.5, and Lemma 3.10.7.

60. Let R be a commutative ring with 1.

    (a) Let M be a maximal ideal of R. Let $a \in R$, $a \notin M$. Show that for any $b \in R$, the congruence $ax \equiv b \pmod{M}$ has a solution, and that that solution is unique $\pmod{M}$.

    (b) Let P be a prime ideal of R. Let $a \in R$, $a \notin P$. Let $b \in R$. Suppose that the congruence $ax \equiv b \pmod{P}$ has a solution. Show that that solution is unique $\pmod{P}$.

    (c) Let P be a prime ideal of R that is not maximal. Show that there are elements a, b of R with $a \notin P$, such that the congruence $ax \equiv b \pmod{P}$ does not have a solution.

    (d) Give an example of (c).

61. (a) Let $R = \mathbb{Z}[x]$. For every $n \geq 1$, give an example of an ideal in R that is generated by n elements, but not by $n-1$ elements. Prove your example is correct.

    (b) Same for $R = \mathbb{Q}[x, y]$.

62. Show that every maximal ideal M in $\mathbb{Z}[x]$ is $M = <p, f(x)>$ where p is a prime in $\mathbb{Z}$ and $f(x)$ is a polynomial that is irreducible $\pmod{p}$.

63. Let R be a commutative ring with 1.

Definition. R is a *local ring* if it has a unique maximal ideal M.

    (a) Show that if $M = \{\text{nonunits of } R\}$ is an ideal, then R is a local ring with maximal ideal M.

    (b) Suppose that R is an integral domain with quotient field $\mathbb{F}$, and for every $x \in \mathbb{F}$, $x \neq 0$ or $x^{-1} \in R$. (In this situation, R is called a *valuation ring* of $\mathbb{F}$.) Show that $M = \{\text{nonunits of } R\}$ is an ideal of R (and hence, by (a), that R is a local ring).

64. Let $\mathbb{F}$ be a field. A *discrete valuation* v on $\mathbb{F}$ is a function $v\colon \mathbb{F}^* \to$ $\mathbb{Z}$ that is onto $\mathbb{Z}$, with the properties that

   (1) $v(xy) = v(x) + v(y)$ for all $x, y \in \mathbb{F}^*$, i.e., v is a homomorphism from the multiplicative group of $\mathbb{F}$ to the additive group of $\mathbb{Z}$; and

   (2) $v(x + y) \geq \min(v(x), v(y))$ for all $x, y \in \mathbb{F}^*$ with $x + y \neq 0$.

   Let $R = \{x \in \mathbb{F}^* \mid v(x) \geq 0\} \cup \{0\}$.

   (a) Show that R is a ring. R is called the *valuation ring* of v.
   (b) Show that R is a valuation ring of $\mathbb{F}$.
   (c) Show that $x \in R$ is a unit of R if and only if $v(x) = 0$.
   (d) If $x \in \mathbb{F}$ with $v(x) = 0$, show that $x \in R$ (and hence, by (c), that x is a unit of R).
   For each $k \geq 1$, let $M_k = \{x \in R \mid v(x) \geq k\}$.
   (e) Show that M_1 is the unique maximal ideal of R, and hence that R is a local ring.
   (f) Show that every nonzero proper ideal of R is M_k for some k. Also, show that $M_k = (M_1)^k$ for each k.
   (g) Show that M_1 is the only nonzero prime ideal of R.
   (h) Show that v is a Euclidean norm on R. Conclude that R is a PID.
   (i) More precisely, let $x \in R$ with $v(x) = 1$. Show that M_k is the ideal generated by x^k, for each $k \geq 1$.
   (j) Show that $M_1 \supset M_2 \supset M_3 \supset \ldots$, and that $\cap_{i=1}^{\infty} M_i = \{0\}$.

65. Let R_0 be a PID with quotient field $\mathbb{F}$. Let p be a prime in R_0. Define $v_p\colon \mathbb{F}^* \to \mathbb{Z}$ as follows: Let $x \in \mathbb{F}$, $x \neq 0$, and write $x = p^n a/b$ with a and b relatively prime to p. Then $v_p(x) = n$.

   (a) Show that v_p is a discrete valuation on $\mathbb{F}$, with valuation ring $R = \{x \in \mathbb{F} \mid x = a/b \text{ with } b \text{ relatively prime to } p\} \cup \{0\}$.
   (b) Let $\mathbb{F} = \mathbb{Q}$. Show that every discrete valuation v on $\mathbb{F}$ is $v = v_p$ for some prime p in $\mathbb{Z}$.

66. Let R be a valuation ring of a discrete valuation v. We have seen that R is a local ring with maximal ideal M, that M is the only nonzero prime ideal of R, and that $M \supset M^2 \supset M^3 \supset \ldots$ and $\cap_{i=1}^{\infty} M^i = \{0\}$.

   (a) Give an example of an integral domain R, not a field, that is a local ring with maximal ideal M where $M = M^2 = M^3 = \ldots$.

(b) Give an example of an integral domain R that is a local ring with maximal ideal M where $M \supset M^2 \supset M^3 \supset \ldots$ but $\bigcap_{i=1}^{\infty} M^i \neq \{0\}$.

(c) Give an example of an integral domain R that is a local ring with maximal ideal M that has a nonzero prime ideal $P \neq M$.

67. Let $f(x) = a_0 + a_1 x + \ldots$ and $g(x) = b_0 + b_1 x + \ldots$ be formal power series with integer coefficients and suppose that a_0 and b_0 are relatively prime. Let t be a nonzero integer. Let $h(x) = c_0 + c_1 x + \ldots$ be a formal power series with integer coefficients with $c_0 = a_0 b_0$ and $h(x) \equiv f(x)g(x) \pmod{t}$.

(a) Show that for every $n \neq 0$ there are formal power series $f_n(x)$ and $g_n(x)$ with integer coefficients, with $f_0(x) = f(x)$, $g_0(x) = g(x)$, and for $n > 0$:

$$f_n(x) = f_{n-1}(x) + dtx^n \text{ for some integer } d,$$

$$g_n(x) = g_{n-1}(x) + etx^n \text{ for some integer } e,$$

so that $f_n(x) \equiv f(x) \pmod{t}$ and $g_n(x) \equiv g(x) \pmod{t}$, and hence $f_n(x)g_n(x) \equiv h(x) \pmod{t}$, and furthermore that

$$f_n(x)g_n(x) = h(x) + \text{ terms of degree } > n.$$

For example, let $f(x) = 3 + x$, $g(x) = 5 + x$. Let $t = 11$ and let $h(x) = 15 - 3x + x^2$. Then:

$$(3 + x)(5 + x) = (15 - 3x + x^2) + 11x$$

$$(3 - 21x)(5 + 34x) = (15 - 3x + x^2) - 715x^2$$

$$(3 - 21x + 143x^2)(5 + 34x) = (15 - 3x + x^2) + 4862x^3$$

$$(3 - 21x + 143x^2 + 880x^3)(5 + 34x - 154x^3)$$

$$= (15 - 3x + x^2) + 1166x^4 + 220022x^5 - 135520x^6$$

(b) (i) Let $f(x) = x + 2$, $g(x) = x - 7$. Let $t = 5$ and $h(x) = x^2 - 14$. Find $f_n(x)$, $g_n(x)$ for $n = 0, 1, 2, 3$.

(ii) Let $f(x) = 7x + 2$, $g(x) = x^2 + 2x + 3$. Let $t = 7$ and $h(x) = 9x^2 + 11x + 6$. Find $f_n(x)$, $g_n(x)$ for $n = 0, 1, 2, 3$.

Note that in a natural sense we may form

$$f_\infty(x) = \lim {}_{n\to\infty} f(x) \text{ and } g_\infty x = \lim {}_{n\to\infty} g(x)$$

and then we have $f_\infty(x) \equiv f(x)$ (mod t), $g_\infty(x) \equiv g(x)$ (mod t), and $f_\infty(x)g_\infty(x) = h(x)$. (Even if we begin with $f(x)$ and $g(x)$ polynomials, as in the above examples, $f_\infty(x)$ and $g_\infty(x)$ will in general be formal power series but not polynomials.)

68. Fix a prime p. Let $S = \{(a_1, a_2, a_3, \ldots)\}$ be the set of infinite sequences with $a_i \in \mathbb{Z}$ for every i. Define addition and multiplication on S coordinatewise, i.e., $(a_1, a_2, a_3, \ldots) + (b_1, b_2, b_3, \ldots) = (a_1 + b_1, a_2 + b_2, a_3 + b_3, \ldots$ and $(a_1, a_2, a_3, \ldots)(b_1, b_2, b_3 \ldots) = (a_1 b_1, a_2 b_2, a_3 b_3, \ldots)$. Observe that S is a commutative ring with 1.

Let T be the subring of S defined by

$$T = \{(a_1, a_2, a_3, \ldots) \mid a_{i+1} \equiv a_i \text{ (mod } p^i) \text{ for each } i \geq 1\}$$

Let I be the ideal of T defined by

$$I = \{(a_1, a_2, a_3, \ldots) \mid a_i \equiv 0 \text{ (mod } p_i) \text{ for each } i \geq 1\}$$

(a) Show that I is a prime ideal of T that is not maximal. Conclude that $R = T/I$ is an integral domain that is not a field. We denote R by $\hat{\mathbb{Z}}_p$. R is called the ring of *p-adic integers*. While elements of R are usually just written as sequences, we will put brackets around them to make clear that they are really equivalence classes of sequences. Note that any $r \in R$ has a unique representative of the form

$$(a_1, a_2, a_3, \ldots) \quad \text{with} \quad 0 \leq a_i < p^i \text{ for each } i \geq 1.$$

(This may or may not be the best representative to use.)

(b) Show that $\varphi \colon \mathbb{Z} \to R$ by $\varphi(n) = [(n, n, n, \ldots)]$ is a 1-1 homomorphism. (Thus, by identifying n with $\varphi(n)$, we may regard $\mathbb{Z}$ as a subring of R.)

(c) Show that $[(a_1, a_2, a_3, \ldots)]$ is a unit of R if and only if $a_1 \not\equiv 0$ (mod p). (The preceding exercise may be helpful. Otherwise, you can do this "by hand".)

(d) Let $v\colon R - \{0\} \to \mathbb{Z}$ by $v([(a_1, a_2, a_3, \ldots)]) = i - 1$ if i is the smallest value of k such that $a_k \not\equiv 0 \pmod{p^k}$. Show that v is a Euclidean norm on R. Conclude that R is a PID.

(e) Show that R is a local ring with unique maximal ideal

$$M = \{[(a_1, a_2, a_3, \ldots)] \mid a_1 \equiv 0 \pmod p\} = \{\text{nonunits of } R\}.$$

Show that every ideal of R is

$$M_i = \{[(a_1, a_2, a_3, \ldots)] \mid a_k \equiv 0 \pmod{p^k} \text{ for } k \leq i\}$$

and that $M_i = M^i$.

(f) Let $\mathbb{F}$ be the quotient field of R. We denote $\mathbb{F}$ by $\hat{\mathbb{Q}}_p$. $\mathbb{F}$ is called the field of *p-adic numbers*. Show that every element of R (resp. $\mathbb{F}$) can be written uniquely as

$$p^k[(a_1, a_2, a_3, \ldots)] \text{ with } a_1 \not\equiv 0 \pmod p$$

for some nonnegative integer (resp. integer) k.

(g) Show that as a set, R (and hence $\mathbb{F}$) is uncountable.

69. (a) Prove the following lemma.

Lemma (Hensel's Lemma). Fix a prime p. Let $f(x) = \Sigma_i c_i x^i$ be a polynomial with integer coefficients. Let a be an integer.

  (i) Suppose that there is an integer x_1 such that $f(x_1) \equiv a \pmod p$ and $f'(x_1) = q \not\equiv 0 \pmod p$. Then for every $n \geq 1$ there exists an integer x_n, unique $\pmod{p^n}$, such that $f(x_n) \equiv a \pmod{p^n}$ and $x_{n+1} \equiv x_n \pmod{p^n}$.

  (ii) More generally, suppose that there is an integer x_m such that $f(x_m) \equiv a \pmod{p^m}$ and $f'(x_m) = p^k q$, q not divisible by p, with $m \geq 2k+1$. Then for every integer $n \geq m$ there exists an integer x_n, unique $\pmod{p^n}$, such that $f(x_n) \equiv a \pmod{p^n}$ and $x_{n+1} \equiv x_n \pmod{p^{n-k}}$.

(b) Conclude the following corollary.

Corollary. In the situation of Hensel's lemma, there exists a unique $r \in \hat{\mathbb{Z}}_p$, $r = [(a_1, a_2, a_3, \ldots)]$ with $a_m \equiv x_m \pmod{p^m}$ and $f(r) = a$.

70. Fix a prime p.

    (a) Let a be an integer relatively prime to p.

        (i) Let m be a positive integer relatively prime to p. Show that a has an m-th root in $\hat{\mathbb{Z}}_p$ if and only if the congruence $x^m \equiv a \pmod{p}$ has a solution in $\mathbb{Z}$.

        (ii) Show that a has a p-th root in $\hat{\mathbb{Z}}_p$ if and only if the congruence $x^p \equiv a \pmod{p^3}$ has a solution in $\mathbb{Z}$.

    (b) In general, let a be a nonzero integer and let m be a positive integer. Find and prove a necessary and sufficient condition for a to have an m-th root in $\hat{\mathbb{Z}}_p$.

Chapter 4

Field Theory

We now turn our attention to fields. Fields, of course, are particular kinds of rings, but field theory has a completely different flavor than ring theory.

In field theory, we are principally interested in investigating fields $\mathbb{E}$ that are "extensions" of a field $\mathbb{F}$. Simply put, $\mathbb{E}$ is an extension of $\mathbb{F}$ if $\mathbb{F}$ is contained in $\mathbb{E}$. As we will see, we will often begin with $\mathbb{F}$, and a polynomial $p(x)$ with coefficients in $\mathbb{F}$, and obtain $\mathbb{E}$ by "adjoining", i.e., adding in, a root, or roots, of $p(x)$ to $\mathbb{F}$. (Of course, we will have to make this precise). But from this short description you can already see that field theory is intimately related to questions about roots of polynomials — indeed, this was its historical origin.

Our study of field theory will culminate in Galois theory, where, as we will also see, group theory plays a fundamental role in studying field extensions.

4.1 Definition, examples, and basic properties

We begin by recalling the definition of a field, which we restate here for convenience.

Definition 4.1.1. A *field* $\mathbb{F}$ is a commutative ring with 1 such that $\mathbb{F}^* = \mathbb{F} - \{0\}$ (i.e., such that every nonzero element of $\mathbb{F}$ is a unit). $\Diamond$

Let us begin by looking at some familiar (and not so familiar) examples.

Example 4.1.2.

(a) The rational numbers $\mathbb{Q}$ are a field.
(b) The real numbers $\mathbb{R}$ are a field.
(c) The complex numbers $\mathbb{C}$ are a field.
(d) $\mathbb{C}$ can be described as $\mathbb{C} = \mathbb{R}(i)$ as in Example 3.1.15.
(e) For $D \neq 0, 1$ a squarefree integer, $\mathbb{Q}(\sqrt{D})$ as in Example 3.1.16 is a field.
(f) For p a prime, $\mathbb{Z}_p = \{0, 1, \ldots, p-1\}$ with addition being addition (mod p) and multiplication being multiplication (mod p) is a field. Henceforth we will denote this field by $\mathbb{F}_p$.   ◊

Note that if $j \in \mathbb{F}_p$, then $k = j^{-1}$ is the element of $\mathbb{F}_p$ with $jk = 1$. How can we find j^{-1}? Let us see.

Example 4.1.3.

(a) In $\mathbb{F}_{11}$, it is easy to find inverses by trial and error. Certainly $1^{-1} = 1$. Then $2^{-1} = 6$ as $2 \cdot 6 = 12 \equiv 1 \pmod{11}$, and so $6^{-1} = 2$. Also, $3^{-1} = 4$ as $3 \cdot 4 = 12 \equiv 1 \pmod{11}$, and so $4^{-1} = 3$. Next, $5^{-1} = 9$ as $5 \cdot 9 = 45 \equiv 1 \pmod{11}$, and $9^{-1} = 5$. Next, $7^{-1} = 8$, as $7 \cdot 8 = 56 \equiv 1 \pmod{11}$, and so $8^{-1} = 7$. Finally, $10 \equiv -1 \pmod{11}$, and hence $(10)^2 \equiv (-1)^2 \equiv 1 \pmod{11}$, giving $10^{-1} = 10$.
(b) Let $p = 618421$, a prime. We wish to find 124816^{-1} in $\mathbb{F}_p$. Here we resort to Euclid's algorithm. We computed in Example 3.7.2 (e) that

$$1 = 618421(53773) + 124816(-266427)$$

so

$$1 \equiv 124816(-266427) \pmod{618421}$$

so

$$124816^{-1} \equiv -266427 \pmod{618421}$$

and easy arithmetic shows

$$-266427 \equiv 351994 \pmod{618421}$$

so

$$124816^{-1} = 351994$$

in $\mathbb{F}_p$.   ◊

Given a finite field, (i.e., a field with a finite number of elements) we can write out its addition and multiplication tables.

Example 4.1.4. Addition and multiplication in $\mathbb{F}_7$ are given by:

+	0	1	2	3	4	5	6
0	0	1	2	3	4	5	6
1	1	2	3	4	5	6	0
2	2	3	4	5	6	0	1
3	3	4	5	6	0	1	2
4	4	5	6	0	1	2	3
5	5	6	0	1	2	3	4
6	6	0	1	2	3	4	5

·	0	1	2	3	4	5	6
0	0	0	0	0	0	0	0
1	0	1	2	3	4	5	6
2	0	2	4	6	1	3	5
3	0	3	6	2	5	1	4
4	0	4	1	5	2	6	3
5	0	5	3	1	6	4	2
6	0	6	5	4	3	2	1

$\Diamond$

We now define an important invariant of fields.

Definition 4.1.5. Let $\mathbb{F}$ be a field. The *characteristic* char($\mathbb{F}$) is the smallest positive integer n such that $n \cdot 1 = 0 \in \mathbb{F}$, or 0 if no such positive integer exists. $\Diamond$

Lemma 4.1.6. *Let $\mathbb{F}$ be a field. Then char($\mathbb{F}$) = 0 or is a prime.*

Proof. Suppose $n = $ char($\mathbb{F}$) $\neq 0$ and is not a prime. Certainly $n \neq 1$ as $1 \neq 0$ in $\mathbb{F}$. Write $n = ab$, $1 < a, b < n$. Then $a \cdot 1 \neq 0 \in \mathbb{F}$ and $b \cdot 1 \neq 0 \in \mathbb{F}$, but $(a \cdot 1)(b \cdot 1) = ab \cdot 1 = n \cdot 1 = 0$ in $\mathbb{F}$, which is impossible as any field is an integral domain. $\square$

Example 4.1.7.

(a) $\mathbb{Q}, \mathbb{R}, \mathbb{C}$ and $\mathbb{Q}(\sqrt{D})$ are all fields of characteristic 0.
(b) $\mathbb{F}_p$ is a field of characteristic p. $\Diamond$

Now suppose $\mathbb{F}$ and $\mathbb{F}'$ are fields. Then they are both commutative rings with 1, and a homomorphism $\varphi\colon \mathbb{F} \to \mathbb{F}'$ of fields is simply a homomorphism rings with 1. But in the case of fields we have a strong property (Lemma 3.1.23), which we restate (and reprove) here.

Lemma 4.1.8. *Let $\mathbb{F}$ and $\mathbb{F}'$ be fields and let $\varphi\colon \mathbb{F} \to \mathbb{F}'$ be a homomorphism of fields. Then φ is injective. Consequently, φ is an isomorphism if and only if it is surjective.*

Proof.   Suppose that $a \in \mathbb{F}$, $a \neq 0$. Then a has an inverse a^{-1}, with $aa^{-1} = 1$. But then $1 = \varphi(1) = \varphi(aa^{-1}) = \varphi(a)\varphi(a^{-1})$ so $\varphi(a) \neq 0$. Consequently, $\mathrm{Ker}(\varphi) = \{0\}$ and φ is injective. $\qquad\qquad\square$

Definition 4.1.9. Let $\mathbb{F}$ be a field. A homomorphism $\varphi\colon \mathbb{F} \to \mathbb{F}$ is an *endomorphism* of $\mathbb{F}$, and an isomorphism $\varphi\colon \mathbb{F} \to \mathbb{F}$ is an *automorphism* of $\mathbb{F}$. $\Diamond$

Example 4.1.10.

(a) The inclusions $i\colon \mathbb{Q} \to \mathbb{R}$, $i\colon \mathbb{R} \to \mathbb{C}$, and $i\colon \mathbb{Q} \to \mathbb{Q}(\sqrt{D})$ are all field homomorphisms.
(b) Let $\mathbb{F} = \mathbb{C}$. Then $\varphi\colon \mathbb{F} \to \mathbb{F}$ by $\varphi(a+bi) = a-bi$ is an automorphism of $\mathbb{F}$. (This is just complex conjugation.)
(c) Let $\mathbb{F} = \mathbb{Q}(\sqrt{D})$. Then $\varphi\colon \mathbb{F} \to \mathbb{F}$ by $\varphi(a + b\sqrt{D}) = a - b\sqrt{D}$ is an automorphism of $\mathbb{F}$. (This φ is called conjugation in the field $\mathbb{F}$.) $\Diamond$

Remark 4.1.11. We leave it to the reader to check that it is only possible to have a homomorphism $\varphi\colon \mathbb{F} \to \mathbb{F}'$ of fields if $\mathbb{F}$ and $\mathbb{F}'$ have the same characteristic. $\Diamond$

Remark 4.1.12.

(a) Let $\mathbb{F}$ be any field of characteristic 0. Then we have a homomorphism of rings with 1 $i\colon \mathbb{Z} \to \mathbb{F}$ given by $i(1) = 1$, which then forces $\varphi(n) = \varphi(n \cdot 1) = n \cdot \varphi(1) = n \cdot 1 = n$, and then i extends to a unique homomorphism of fields $i\colon \mathbb{Q} \to \mathbb{F}$ given by $\varphi(n/m) = \varphi(nm^{-1}) = \varphi(n)\varphi(m^{-1}) = \varphi(n)\varphi(m)^{-1} = n/m$ for any $n/m \in \mathbb{Q}$ (i.e., with $m \neq 0$).
(b) Let $\mathbb{F}$ be any field of characteristic p. Then we have a homomorphism of rings with 1 $i\colon \mathbb{Z}_p \to \mathbb{F}$, i.e., $i(1) = 1$, which then forces $\varphi(n) = n$ for any $n \in \mathbb{F}_p$. $\Diamond$

Definition 4.1.13. We identify $\mathbb{F}_0 = \mathbb{Q}$ with a subset of any field of characteristic 0 as in Remark 4.1.12(a), and $\mathbb{F}_0 = \mathbb{F}_p$ with a subset of any field of characteristic p as in Remark 4.1.12(b). In either case, $\mathbb{F}_0$ is called the *prime* field. ◊

We now turn our attention to fields of positive characteristic.

Here is a (surprising) fact about arithmetic in fields of characteristic p.

Lemma 4.1.14. *Let $\mathbb{F}$ be a field of characteristic p and let a, $b \in \mathbb{F}$. Then $(a + b)^p = a^p + b^p$.*

Proof. By the binomial theorem,

$$(a + b)^p = \sum_{i=0}^{p} \binom{p}{i} a^{p-i} b^i.$$

Now $\binom{p}{0} = \binom{p}{p} = 1$, and, since p is prime,

$$\binom{p}{i} = \frac{p!}{i! \, (p - i)!}$$

is divisible by p for every i with $1 \leq i \leq p - 1$. □

Lemma 4.1.15. *Let $\mathbb{F}$ be a finite field and let $\varphi \colon \mathbb{F} \to \mathbb{F}$ be an endomorphism. Then φ is an isomorphism.*

Proof. A 1–1 function from a finite set to itself must be onto. □

Lemma 4.1.16. *Let $\mathbb{F}$ be a field of characteristic p. Then $\Phi \colon \mathbb{F} \to \mathbb{F}$ by $\Phi(a) = a^p$ is an endomorphism of $\mathbb{F}$.*

Proof. We clearly have $\Phi(1) = 1$ and $\Phi(ab) = (ab)^p = a^p b^p = \Phi(a)\Phi(b)$. But also $\Phi(a + b) = (a + b)^p = a^p + b^p = \Phi(a) + \Phi(b)$ by Lemma 4.1.14. □

Definition 4.1.17. The map $\Phi(a) = a^p$ of Lemma 4.1.16 is the *Frobenius* endomorphism (or automorphism, as the case may be) of $\mathbb{F}$. ◊

Lemma 4.1.18. *Let $\mathbb{F}$ be a field of characteristic p.*

(a) If $\mathbb{F} = \mathbb{F}_p$, then the Frobenius map $\Phi \colon \mathbb{F} \to \mathbb{F}$ is the identity.
(b) If $\mathbb{F} \neq \mathbb{F}_p$, then the Frobenius map $\Phi \colon \mathbb{F} \to \mathbb{F}$ is not the identity.

Proof.

(a) By Fermat's little theorem (Theorem 2.7.1) $a^{p-1} \equiv 1 \pmod{p}$ for every integer a that is not divisible by p, so $a^p \equiv a \pmod{p}$ for every such integer, and this is certainly also true if a is divisible by p.

(b) Consider $\{a \in \mathbb{F} \mid \Phi(a) = a\}$. This is the set of roots of the polynomial $f(x) = x^p - x$ in $\mathbb{F}$. By this is a polynomial of degree p, so can have at most p roots in $\mathbb{F}$, by Corollary 3.3.9. By (a), every element of $\mathbb{F}_p$ is a root of this polynomial, so if $a \in \mathbb{F}$, $a \notin \mathbb{F}_p$, $\Phi(a) \neq a$. □

Remark 4.1.19. Admittedly, the only field of characteristic p we have seen so far is $\mathbb{F}_p$ itself. But we will be seeing other fields of characteristic p later (even soon), and we are preparing for them by introducing the Frobenius map now. ◇

Remark 4.1.20. We will see that for every prime p and every positive integer n there is a field $\mathbb{F}_{p^n}$ of p^n elements (as well as infinite fields of characteristic p), and indeed that $\mathbb{F}_{p^n}$ is unique up to isomorphism. We take this opportunity to caution the reader that while $\mathbb{F}_p = \mathbb{Z}_p$, $\mathbb{F}_{p^n} \neq \mathbb{Z}_{p^n}$ for $n > 1$. Note that $\mathbb{Z}_{p^n}$ is not an integral domain for $n > 1$ (as for $n > 1$, $p \neq 0$ in $\mathbb{Z}_{p^n}$ and $p^{n-1} \neq 0$ in $\mathbb{Z}_{p^n}$, but $p \cdot p^{n-1} = p^n = 0$ in $\mathbb{Z}_{p^n}$) so cannot possibly be a field. ◇

4.2   Extension fields

Definition 4.2.1. Let $\mathbb{E}$ and $\mathbb{F}$ be fields with $\mathbb{F} \subseteq \mathbb{E}$. Then $\mathbb{F}$ is a *subfield* of $\mathbb{E}$, or $\mathbb{E}$ is an *extension field* (or simply *extension*) of $\mathbb{F}$. ◇

Example 4.2.2. $\mathbb{R}$ is an extension of $\mathbb{Q}$. $\mathbb{C}$ is an extension of $\mathbb{R}$, and of $\mathbb{Q}$. $\mathbb{Q}(\sqrt{D})$ is an extension of $\mathbb{Q}$. ◇

We now present a general method of constructing field extensions. Actually, we already saw this method in Example 3.12.9, but we present it again here.

Theorem 4.2.3. *Let $\mathbb{F}$ be a field and let $p(x) \in \mathbb{F}[x]$ be an irreducible polynomial. Then the quotient $\mathbb{F}[x]/\langle p(x) \rangle$, where $\langle p(x) \rangle$ denotes the ideal of $\mathbb{F}[x]$ generated by $p(x)$, is a field.*

Proof. Let $R = \mathbb{F}[x]$ and let $I = \langle p(x) \rangle$. Recall that R is a Euclidean ring with norm $\delta(p(x)) = \deg p(x)$.

Since R is a Euclidean ring, it is a PID, and in a PID every irreducible element is prime (Lemma 3.6.15). Thus $p(x) \in \mathbb{F}[x]$ is prime, and so I is a prime ideal (Lemma 3.12.5). Again, R is a PID so every nonzero prime ideal of R is maximal (Lemma 3.12.7). But then R/I is a field (Theorem 3.12.8). $\qquad\square$

As an immediate consequence of this theorem, we have the following general result.

Theorem 4.2.4 (Kronecker). *Let $\mathbb{F}$ be a field, and let $p(x) \in \mathbb{F}[x]$ be any nonconstant polynomial. Then there is an extension field $\mathbb{E}$ of $\mathbb{F}$ in which $p(x)$ has a root.*

Proof. First assume $p(x)$ is irreducible. Let $\mathbb{E} = \mathbb{F}[x]/\langle p(x) \rangle$ as in Theorem 4.2.3. We may regard (and have regarded) $\mathbb{F}$ as contained in $\mathbb{F}[x]$ by identifying $a \in \mathbb{F}$ with the constant polynomial a. Since no element of $\mathbb{F}$ is in the ideal generated by $p(x)$ (since $p(x)$, being nonconstant, has positive degree) this identification gives us a $1-1$ map from $\mathbb{F}$ into $\mathbb{E}$, and we regard $\mathbb{F}$ as a subfield of $\mathbb{E}$ via this identification. In other words, $\mathbb{E}$ is an extension of $\mathbb{F}$. Let $\pi \colon \mathbb{F}[x] \to \mathbb{F}[x]/\langle p(x) \rangle = \mathbb{E}$ be the canonical projection, and set $\alpha = \pi(x)$. Then

$$p(\alpha) = p(\pi(x)) = \pi(p(x)) = 0 \text{ in } \mathbb{E}$$

(as $p(x) \in \langle p(x) \rangle$). In other words, $\alpha \in \mathbb{E}$ is a root of the polynomial $p(x)$, and so we see that $\mathbb{E}$ is an extension field of $\mathbb{F}$ in which $p(x)$ has a root.

If $p(x)$ is not irreducible, factor $p(x)$ into a product of irreducibles, say $p(x) = p_1(x) \ldots p_k(x)$.

Then, as above, $\mathbb{E} = \mathbb{F}[x]/\langle p_1(x) \rangle$ is a field in which the polynomial $p_1(x)$ has a root. But if α is a root of $p_1(x)$, i.e., $p_1(\alpha) = 0$, then certainly $p(\alpha) = 0$, i.e., α is a root of $p(x)$. $\qquad\square$

Now let us look at a couple of examples. The first one shows us how to recover an old example from this new viewpoint. The second is a new example, which, moreover, concretely shows us how to do arithmetic in extension fields.

Example 4.2.5. Let $\mathbb{F} = \mathbb{Q}$ and let $D \in \mathbb{Q}$, D not a perfect square. Then the polynomial $p(x) = x^2 - D \in \mathbb{F}[x]$ is irreducible. (If it were not, it would have a root in $\mathbb{Q}$, which would make D a perfect square.) Let $\mathbb{E} = \mathbb{Q}[x]/\langle p(x)\rangle$, and let $\pi\colon \mathbb{Q}[x] \to \mathbb{E}$ be the canonical projection. Let $\pi(x) = \alpha$.

First we observe that from the division algorithm, $\{f(x) = a_0 + a_1 x \mid a_0, a_1 \in \mathbb{Q}\}$ is a complete set of coset representatives of $\langle p(x)\rangle$ in $\mathbb{Q}[x]$. Then $\pi(a_0 + a_1 x) = a_0 + a_1 \alpha$.

Next, and key, we observe that, as in the proof of Theorem 4.2.4, we have $\alpha^2 - D = \pi(x^2 - D) = \pi(p(x)) = 0 \in \mathbb{Q}[x]/\langle p(x)\rangle$. In other words, $\alpha^2 = D$ in $\mathbb{E}$. Thus we have constructed a field, more precisely, an extension field of $\mathbb{Q}$, in which D has a square root. $\Diamond$

Now you might object that we already know a field in which D has a square root, namely $\mathbb{Q}(\sqrt{D})$, and you would be right in that the field we have constructed in "essentially the same as" $\mathbb{Q}(\sqrt{D})$, though we have to be careful about what we mean by "essentially the same as". But the point of this construction is that we may use it to obtain fields that *are* new.

Example 4.2.6. Let $\mathbb{F} = \mathbb{Q}$ and let $p(x) = x^3 + 7x + 1$, an irreducible polynomial in $\mathbb{Q}[x]$. (If $p(x)$ were not irreducible, it would have to have linear factor, i.e., a root in $\mathbb{Q}$, and it does not.)

Let $E = \mathbb{Q}[x]/\langle p(x)\rangle$ and let $\pi\colon \mathbb{Q}[x] \to \mathbb{E}$ be the canonical projection. Set $\pi(x) = \alpha$ and note that $\alpha^3 + 7\alpha + 1 = \pi(x^3 + 7x + 1) = 0$.

Since $\mathbb{E}$ is a field, α must have an inverse in $\mathbb{E}$, and it does (and we can find it):

$$\alpha^3 + 7\alpha + 1 = 0$$

$$-\alpha^3 - 7\alpha = 1$$

$$\alpha\left(-\alpha^2 - 7\right) = 1$$

so $\alpha^{-1} = -\alpha^2 - 7$.

Let's look at some other elements. Let $\beta_1 = \alpha + 3$ and $\beta_2 = \alpha^2 + 1$. Then $\beta_1 + \beta_2 = \alpha^2 + \alpha + 4$. More interestingly, $\beta_1\beta_2 = (\alpha + 3)(\alpha^2 + 1) = \alpha^3 + 3\alpha^2 + \alpha + 3$. Now $(x + 3)(x^2 + 1) = x^3 +$

$3x^2 + x + 3$ and we simply compute

$$x^3 + 3x^2 + x + 3 = \left(x^3 + 7x + 1\right)(1) + \left(3x^2 - 6x + 2\right)$$

so

$$a^3 + 3\alpha^2 + \alpha + 3 = \left(\alpha^3 + 7\alpha + 1\right)(1) + \left(3\alpha^2 - 6\alpha + 2\right)$$
$$= 0\,(1) + 3\alpha^2 - 6\alpha + 2$$
$$= 3\alpha^2 - 6\alpha + 2$$

and we see

$$\beta_1\beta_2 = 3\alpha^2 - 6\alpha + 2.$$

Now β_1 is a nonzero element of $\mathbb{E}$ (as $x + 3$ is not divisible by $x^3 + 7x + 1$) so it must have an inverse. To find it, we perform Euclid's algorithm on these two polynomials. This only takes one step:

$$x^3 + 7x + 1 = (x + 3)\left(x^2 - 3x + 16\right) - 47$$

so, solving, we find

$$1 = \left(x^3 + 7x + 1\right)\left(-\frac{1}{47}\right) + (x + 3)\left(\frac{1}{47}\left(x^2 - 3x + 16\right)\right)$$

and then

$$1 = \left(\alpha^3 + 7\alpha + 1\right)\left(-\frac{1}{47}\right) + (\alpha + 3)\left(\frac{1}{47}\left(\alpha^2 - 3\alpha + 16\right)\right)$$

so

$$\beta_1^{-1} = \frac{\alpha^2 - 3\alpha + 16}{47}.$$

Similarly, β_2 is a nonzero element of $\mathbb{E}$, and we find its inverse from Euclid's algorithm, which now takes two steps:

$$x^3 + 7x + 1 = \left(x^2 + 1\right)x + (6x + 1)$$

$$x^2 + 1 = (6x + 1)\left(\frac{1}{36}(6x - 1)\right) + \frac{37}{36}$$

so

$$1 = \left(x^2 + 1\right)\left(\frac{36}{37}\right) + (6x + 1)\left(\frac{-6x + 1}{37}\right)$$

$$= \left(x^2 + 1\right)\left(\frac{36}{37}\right) + \left[\left(x^3 + 7x + 1\right)(1)\right.$$

$$\left. + \left(x^2 + 1\right)(-x)\right]\left(\frac{-6x + 1}{37}\right)$$

$$= \left(x^3 + 7x + 1\right)\left(\frac{-6x + 1}{37}\right) + \left(x^2 + 1\right)\left(\frac{6x^2 - x + 36}{37}\right)$$

and then

$$\beta_2^{-1} = \frac{6\alpha^2 - \alpha + 36}{37}. \qquad \diamond$$

Guided by these examples, we will see how to do arithmetic in extension fields. But first we need to introduce some general notions.

Remark 4.2.7. Let $\mathbb{E}$ be an extension field of $\mathbb{F}$. We will (usually) denote elements of $\mathbb{F}$ by Roman letters $(a, b, c, \ldots)$ and elements of $\mathbb{E}$ by Greek letters $(\alpha, \beta, \gamma, \ldots)$ $\qquad \diamond$

Here is an easy lemma, but one which plays a crucial role.

Lemma 4.2.8. *Let $\mathbb{E}$ be an extension fields of $\mathbb{F}$. Then $\mathbb{E}$ is an $\mathbb{F}$-vector space.*

Proof. Since $\mathbb{E}$ is a field, we may add any two elements α_1 and α_2 of $\mathbb{E}$ to obtain the element $\alpha_1 + \alpha_2$ of $\mathbb{E}$. Also, since $\mathbb{F} \subseteq \mathbb{E}$, for any element a of $\mathbb{F}$ and any element α of $\mathbb{E}$, we may multiply α by a to obtain the element $a\alpha$ of $\mathbb{E}$. We leave it to the reader to check that with these definitions of vector addition and scalar multiplication, $\mathbb{E}$ becomes a vector space over $\mathbb{F}$. $\qquad \square$

This leads us to the basic invariant of field extensions.

Definition 4.2.9. Let $\mathbb{E}$ be an extension field of $\mathbb{F}$. The *degree* of $\mathbb{E}$ over $\mathbb{F}$, $(\mathbb{E}/\mathbb{F})$, is the dimension of $\mathbb{E}$ as an $\mathbb{F}$-vector space, $(\mathbb{E}/\mathbb{F}) = \dim_{\mathbb{F}} \mathbb{E}$. (The degree $(\mathbb{E}/\mathbb{F})$ is often denoted by $[\mathbb{E}: \mathbb{F}]$.) $\qquad \diamond$

Let us now see how to determine $(\mathbb{E}/\mathbb{F})$.

Lemma 4.2.10. *Let $p(x) \in \mathbb{F}[x]$ be a nonzero polynomial. The set*

$$S = \{r(x) \in \mathbb{F}(x) \mid r(x) = 0 \quad \text{or} \quad \deg r(x) < \deg p(x)\}$$

is a complete set of left coset representatives of the ideal $I = \langle p(x) \rangle$ of $F[x]$.

Proof. Let $f(x) \in F[x]$. Then we have, by the division algorithm, that

$$f(x) = p(x)q(x) + r(x) \quad \text{with } r(x) = 0 \text{ or } \deg r(x) < \deg p(x)$$

so $f(x) \in r(x) + I$.

On the other hand, if $r_1(x) + I = r_2(x) + I$, then $r_2(x) - r_1(x) \in I$, i.e., $r_2(x) - r_1(x)$ is divisible by $p(x)$. Since $r_1(x) = 0$ or $\deg r_1(x) < \deg p(x)$, and $r_2(x) = 0$ or $\deg r_2(x) < \deg p(x)$, we see that $r_2(x) - r_1(x) = 0$ or $\deg r_2(x) - r_1(x) < \deg p(x)$. But the polynomial $p(x)$ cannot divide a polynomial of lower degree, so we must have $r_2(x) - r_1(x) = 0$, i.e., $r_2(x) = r_1(x)$. □

Theorem 4.2.11. *Let $p(x) \in \mathbb{F}[x]$ be an irreducible polynomial of degree n, and let $\mathbb{E} = \mathbb{F}[x]/\langle p(x) \rangle$. Then $\mathbb{E}$ is an extension of $\mathbb{F}$ of degree n.*

Proof. Let $\pi \colon \mathbb{F}[x] \to \mathbb{E} = \mathbb{F}[x]/\langle p(x) \rangle$ be the canonical projection. Then, since the set S of Lemma 4.2.10 is a complete set of coset representatives of $\langle p(x) \rangle$, $\pi \colon S \to \mathbb{E}$ is 1–1 and onto. It is also a linear transformation, as $\pi(r_1(x) + r_2(x)) = \pi(r_1(x)) + \pi(r_2(x))$ and $\pi(ar(x)) = a\pi(r(x))$, since $\langle p(x) \rangle$ is an ideal of $\mathbb{F}[x]$. Thus it is an isomorphism of vector spaces, so $\dim_{\mathbb{F}} \mathbb{E} = \dim_{\mathbb{F}} S$. But S has basis $\{1, x, \ldots, x^{n-1}\}$, so $\dim_{\mathbb{F}} S = n$. □

Corollary 4.2.12. *In the situation of Theorem 4.2.11, let $\alpha = \pi(x)$. Then $\{1, \alpha, \ldots, \alpha^{n-1}\}$ is a basis for $\mathbb{E}$ as an $\mathbb{F}$-vector space.*

Proof. Since π is a vector space isomorphism, the image of a basis under π is a basis. □

Corollary 4.2.13. *Let $\mathbb{F}_p$ be the field with p elements, p a prime, and let $\mathbb{E}$ be an extension of $\mathbb{F}_p$ of degree n. Then $\mathbb{E}$ has p^n elements.*

Proof.   An n-dimensional vector space over $\mathbb{F}_p$ has p^n elements. $\square$

Remark 4.2.14. We showed in the proof of Theorem 4.2.11 that $\pi\colon S \to \mathbb{E}$ is an isomorphism of $\mathbb{F}$-vector spaces. But we should point out that the structure of an $\mathbb{F}$-vector space is all the structure S has. Namely, we may add elements of S and multiply elements of S by elements of $\mathbb{F}$, so we indeed have an $\mathbb{F}$-vector space structure on S. But that is all. To be sure, we can multiply two elements of S (after all, S is a subset of $\mathbb{F}[x]$, and we may multiply any two polynomials in $\mathbb{F}[x]$), but the result will not necessarily be an element of S. Thus, S is *not* closed under multiplication, so is not a field. $\Diamond$

However, we do have the following explicit description of arithmetic operations in $\mathbb{E}$.

Corollary 4.2.15. *In the situation of Theorem 4.2.11, the arithmetic operations on $\mathbb{E}$ are given as follows: Let β_1, $\beta_2 \in \mathbb{E}$.*

If $\beta_1 = 0$, then $\beta_1 + \beta_1 = \beta_2$ and $\beta_1\beta_2 = 0$, and similarly if $\beta_2 = 0$.
Suppose that β_1 and β_2 are both nonzero. Write $\beta_1 = f_1(\alpha)$ with $\deg f_1(\alpha) < n$ and $\beta_2 = f_2(\alpha)$ with $\deg f_2(\alpha) < n$. Then:

(1) Let $g(x) = f_1(x) + f_2(x)$. Then $\beta_1 + \beta_2 = g(\alpha)$.
(2) Let $h(x) = f_1(x)f_2(x)$. Write $h(x) = p(x)q(x)+r(x)$ with $r(x) = 0$ or $\deg r(x) < n$. Then $\beta_1\beta_2 = r(\alpha)$.
(3) Let $s(x)$ and $t(x)$ be polynomials with $p(x)s(x) + f_1(x)t(x) = 1$. Write $t(x) = p(x)u(x) + v(x)$ with $\deg v(x) < n$. Then $\beta_1^{-1} = v(\alpha)$.

Proof.   Let I be the ideal $I = \langle p(x) \rangle$, and recall that $\beta_1 = f_1(\alpha) = f_1(\pi(x)) = \pi(f_1(x)) = f_1(x) + I$ and similarly for $f_2(x)$. Then:

(1) $(f_1(x) + I) + (f_2(x) + I) = (f_1(x) + f_2(x)) + I = g(x) + I$.

(2) $(f_1(x) + I)(f_2(x) + I) = f_1(x)f_2(x) + I$
$$= (p(x)q(x) + r(x)) + I$$
$$= r(x) + I$$

(3) $p(x)s(x) + f_1(x)t(x) = 1 \in 1 + I$

so

$$f_1(x)t(x) \in 1 + I$$
$$f_1(x)(p(x)u(x) + v(x)) \in 1 + I$$

so

$$f_1(x)v(x) \in 1 + I$$

so

$$\pi(f_1(x)v(x)) = 1 \in \mathbb{E} = \mathbb{F}[x]/I$$

i.e.,

$$\beta_1 v(\alpha) = 1 \text{ in } \mathbb{E}. \qquad \square$$

With this concrete description of arithmetic in $\mathbb{E}$ in hand, let us give one more example.

Example 4.2.16. We construct a field $\mathbb{F}_4$ with $4 = 2^2$ elements, or equivalently, by Corollary 4.2.13, an extension of $\mathbb{F}_2$ degree 2.

We look for a monic irreducible quadratic in $\mathbb{F}_2[x]$. Let $p(x) = x^2 + ax + b$ with $a, b \in \mathbb{F}_2$ yet to be determined. Now a quadratic is irreducible if and only if it does not have a linear factor, so we want $p(0) \neq 0$ in $\mathbb{F}_2$ and $p(1) \neq 0$ in $\mathbb{F}_2$. Calculation shows there is exactly one such polynomial, $p(x) = x^2 + x + 1 \in \mathbb{F}_2[x]$. Then $\mathbb{F}_4 = \mathbb{F}_2[x]/\langle p(x) \rangle$.

We give the addition and multiplication tables for $\mathbb{F}_4$, which you may easily verify.

Note that $p(x)$ has the roots α and $1 + \alpha$ in $\mathbb{F}_4$, i.e., $p(x) = (x - \alpha)(x - (1 + \alpha))$ in $\mathbb{F}_4[x]$.

+	0	1	α	$1 + \alpha$
0	0	1	α	$1 + \alpha$
1	1	0	$1 + \alpha$	α
α	α	$1 + \alpha$	0	1
$1 + \alpha$	$1 + \alpha$	α	1	0

$\cdot$	0	1	α	$1 + \alpha$
0	0	0	0	0
1	0	1	α	$1 + \alpha$
α	0	α	$1 + \alpha$	1
$1 + \alpha$	0	$1 + \alpha$	1	α

Also, the Frobenius map $\Phi \colon \mathbb{F}_4 \to \mathbb{F}_4$, $\Phi(\beta) = \beta^2$, (see Lemma 4.1.16) is given by $\Phi(0) = 0$, $\Phi(1) = 1$, $\Phi(\alpha) = 1+\alpha$, $\Phi(1+\alpha) = \alpha$. $\Diamond$

We record the following result here for future use.

Lemma 4.2.17. *Let* $\mathbb{F}$ *be a field and let* $\mathbb{E}$ *be an extension field of* $\mathbb{F}$. *Let* $f(x)$ *and* $g(x)$ *be polynomials in* $\mathbb{F}[x]$ *and let* $h(x) = gcd(f(x), g(x))$ *in* $\mathbb{F}[x]$. *Then* $f(x)$ *and* $g(x)$ *are polynomials in* $\mathbb{E}[x]$, *so have a gcd* $\tilde{h}(x) = \gcd(f(x), g(x))$ *in* $\mathbb{E}[x]$. *Then* $\tilde{h}(x) = h(x)$.

Proof. By the definition of a gcd, $\tilde{h}(x)$ is divisible in $\mathbb{E}[x]$ by every polynomial in $\mathbb{E}[x]$ that is a common divisor of $f(x)$ and $g(x)$. Now $h(x)$ is such a polynomial, so $h(x)$ divides $\tilde{h}(x)$ in $\mathbb{E}[x]$.

On the other hand, by Corollary 3.5.16, we may write

$$h(x) = f(x) a(x) + g(x) b(x)$$

for some polynomials $a(x)$, $b(x)$ in $\mathbb{F}[x]$.

Now also by the definition of a gcd, $\tilde{h}(x)$ divides both $f(x)$ and $g(x)$ in $\mathbb{E}[x]$, so from this equation we see that $\tilde{h}(x)$ divides $h(x)$ in $\mathbb{E}[x]$.

Hence, $\tilde{h}(x) = h(x)$. □

4.3   Finite extensions

We now wish to investigate finite extensions, which we first define.

Definition 4.3.1. The extension field $\mathbb{E}$ of $\mathbb{F}$ is *finite* over $\mathbb{F}$, or $\mathbb{E}$ is a *finite extension* of $\mathbb{F}$, if $(\mathbb{E}/\mathbb{F}) = \dim_{\mathbb{F}} \mathbb{E}$ is finite. $\Diamond$

We have already seen an example of this, which will turn out to be the prototype.

Example 4.3.2. Let $p(x) \in F[x]$ be an irreducible polynomial of degree n. Then, by Theorem 4.2.11, $\mathbb{E} = \mathbb{F}[x]/\langle p(x) \rangle$ is extension of degree n, so in particular $\mathbb{E}$ is a finite extension of $\mathbb{F}$. $\Diamond$

Let us prove a couple of results which will be very useful for us in studying finite extensions. First we have a definition.

Definition 4.3.3. Let $\mathbb{F}$, $\mathbb{B}$, and $\mathbb{E}$ be fields with $\mathbb{F} \subseteq \mathbb{B} \subseteq \mathbb{E}$. Then $\mathbb{B}$ is *intermediate* between $\mathbb{E}$ and $\mathbb{F}$. $\Diamond$

Lemma 4.3.4. *Let $\mathbb{E}$ be an extension of $\mathbb{F}$ and let $\mathbb{B}$ be intermediate between $\mathbb{E}$ and $\mathbb{F}$. Then $\mathbb{E}$ is a finite extension of $\mathbb{F}$ if and only if $\mathbb{E}$ is a finite extension of $\mathbb{B}$ and $\mathbb{B}$ is a finite extension of $\mathbb{F}$. In this situation, $(\mathbb{E}/\mathbb{F}) = (\mathbb{E}/\mathbb{B})(\mathbb{B}/\mathbb{F})$.*

Proof. First suppose that $\mathbb{E}$ is a finite extension of $\mathbb{F}$.

Then $\dim_{\mathbb{F}} \mathbb{E}$ is finite. Now $\mathbb{B} \subseteq \mathbb{E}$ so $\mathbb{B}$ is a subspace of $\mathbb{E}$ as an $\mathbb{F}$-vector space, and so $\dim_{\mathbb{F}} \mathbb{B} \subseteq \dim_{\mathbb{F}} \mathbb{E}$. In particular, $\dim_{\mathbb{F}} \mathbb{B}$ is finite.

Let $\dim_{\mathbb{F}} \mathbb{E} = n$, so that $\mathbb{E}$ has a basis $S = \{\alpha_1, \ldots, \alpha_n\}$ as an $\mathbb{F}$-vector space. Then S spans $\mathbb{E}$, so every element of $\mathbb{E}$ can be written as an $\mathbb{F}$-linear combination of elements of S. That is, if $\varepsilon \in \mathbb{E}$ is any element, then $\varepsilon = \Sigma_{i=1}^{n} f_i \alpha_i$ for some $f_1, \ldots, f_n \in \mathbb{F}$. But $\mathbb{F} \subseteq \mathbb{B}$ so $f_i \in \mathbb{B}$ for each i. Thus, this expression is an expression for $\mathbb{E}$ as a $\mathbb{B}$-linear combination of elements of S, and so we see that S spans $\mathbb{E}$ as a $\mathbb{B}$-vector space. But then S has a subset T that is a basis for $\mathbb{E}$ as a $\mathbb{B}$-vector space, so $\dim_{\mathbb{B}} \mathbb{E}$ is equal to the number of elements in T, which is at most n, i.e., $\dim_{\mathbb{B}} \mathbb{E} \leq n$ and in particular $\dim_{\mathbb{B}} \mathbb{E}$ is finite.

Now suppose that $\mathbb{E}$ is a finite extension of $\mathbb{B}$ and that $\mathbb{B}$ is a finite extension of $\mathbb{F}$. Let $\dim_{\mathbb{B}} \mathbb{E} = m$ and $\dim_{\mathbb{F}} \mathbb{E} = k$. We will show the equality in the lemma, $\dim_{\mathbb{F}} \mathbb{E} = mk$, and then in particular $\dim_{\mathbb{F}} \mathbb{E}$ is finite.

Choose a basis $\{\varepsilon_1, \ldots, \varepsilon_m\}$ for $\mathbb{E}$ as a $\mathbb{B}$-vector space and a basis $\{\beta_1, \ldots, \beta_k\}$ for $\mathbb{B}$ as an $\mathbb{F}$-vector space. Let

$$A = \{\beta_j \varepsilon_i \mid i = 1, \ldots, m, \; j = 1, \ldots, k\}.$$

We will show that A is a basis for $\mathbb{E}$ as an $\mathbb{F}$-vector space. In order to show this, we must show that A spans $\mathbb{E}$ and that A is linearly independent.

First we show that A spans $\mathbb{E}$. Let $\alpha \in \mathbb{E}$. Then, since $\{\varepsilon_1, \ldots, \varepsilon_m\}$ spans $\mathbb{E}$ as a $\mathbb{B}$-vector space, we can write

$$\alpha = \sum_{i=1}^{m} \gamma_i \varepsilon_i \quad \text{with} \quad \gamma_i \in \mathbb{B}.$$

Now $\{\beta_1, \ldots, \beta_k\}$ spans $\mathbb{B}$ as an $\mathbb{F}$-vector space, so we can write each γ_i as $\gamma_i = \Sigma_{j=1}^{k} f_{ij} \beta_j$ with $f_{ij} \in \mathbb{F}$.

Substituting, we see that

$$\alpha = \sum_{i=1}^{m} \left(\sum_{j=1}^{k} f_{ij} \beta_j \right) \varepsilon_i = \sum_{i,j} f_{ij} \beta_j \varepsilon_i$$

and so A spans $\mathbb{E}$ as an $\mathbb{F}$-vector space.

Next we show that A is linearly independent over $\mathbb{F}$. So suppose

$$\sum_{i,j} f_{ij} \beta_j \varepsilon_i = 0.$$

We wish to show each $f_{ij} = 0$. To do so, we regroup terms.

$$0 = \sum_{i,j} f_{ij} \beta_j \varepsilon_i = \sum_{i=1}^{m} \left(\sum_{j=1}^{k} f_{ij} \beta_j \right) \varepsilon_i = \sum_{i=1}^{m} \delta_i \varepsilon_i$$

$$\text{with } \delta_i = \sum_{j=1}^{k} f_{ij} \beta_j.$$

Now each δ_i is in $\mathbb{B}$, and $\{\varepsilon_1, \ldots, \varepsilon_m\}$ is a set of linearly independent elements of $\mathbb{E}$ over $\mathbb{B}$, so $\delta_i = 0$ for each i.

Thus for each i,

$$0 = \sum_{j=1}^{k} f_{ij} \beta_j.$$

But $\{\beta_1, \ldots, \beta_k\}$ is a set of linearly independent elements of $\mathbb{B}$ over $\mathbb{F}$, so for each i, every $f_{ij} = 0$. Thus $f_{ij} = 0$ for all i, j, and A is linearly independent over $\mathbb{F}$.

Finally, A has mk elements, so $\dim_{\mathbb{F}} \mathbb{E} = mk$, as claimed. $\quad\square$

Corollary 4.3.5. *Let $\mathbb{E}$ be a finite extension of $\mathbb{F}$ and let $\mathbb{B}$ be intermediate between $\mathbb{E}$ and $\mathbb{F}$. Then:*

(1) $\mathbb{B} = \mathbb{E}$ *if and only if* $(\mathbb{B}/\mathbb{F}) = (\mathbb{E}/\mathbb{F})$.
(2) $\mathbb{B} = \mathbb{F}$ *if and only if* $(\mathbb{B}/\mathbb{F}) = 1$, *or, equivalently, if and only if* $(\mathbb{E}/\mathbb{B}) = (\mathbb{E}/\mathbb{F})$.

Proof. Recall that if V is a finite dimensional vector space (over any field) and W is a subspace of V, then $W = V$ if and only if dim W = dim V.

Then (1) follows, setting $W = \mathbb{B}$ and $V = \mathbb{E}$.

Also, the first claim in (2) follows, setting $W = \mathbb{F}$ and $V = \mathbb{B}$.

As for the second claim, since $(\mathbb{E}/\mathbb{F}) = (\mathbb{E}/\mathbb{B})(\mathbb{B}/\mathbb{F})$ by Lemma 4.3.4, we see that $(\mathbb{E}/\mathbb{B}) = (\mathbb{E}/\mathbb{F})$ if and only if $(\mathbb{B}/\mathbb{F}) = 1$. □

Corollary 4.3.6. *Let* $\mathbb{E}$ *be a finite extension of* $\mathbb{F}$ *and let* $\mathbb{B}$ *be intermediate between* $\mathbb{E}$ *and* $\mathbb{F}$. *Then* $(\mathbb{B}/\mathbb{F})$ *divides* $(\mathbb{E}/\mathbb{F})$.

Proof. This is immediate from the equation $(\mathbb{E}/\mathbb{F}) = (\mathbb{E}/\mathbb{B})(\mathbb{B}/\mathbb{F})$. □

Remark 4.3.7. With the convention that $n \cdot \infty = \infty \cdot n = \infty$ for any positive integer n, and $\infty \cdot \infty = \infty$, then $(\mathbb{E}/\mathbb{F}) = (\mathbb{E}/\mathbb{B})(\mathbb{B}/\mathbb{F})$ in general, i.e., Lemma 4.3.4 holds in general. But we need $(\mathbb{E}/\mathbb{F})$ finite in order for Corollary 4.3.5 to be valid and in order for Corollary 4.3.6 to give any information. ◇

We now want to investigate the notion of composition of field extensions. To prepare for doing so, we prove the next lemma, which is interesting and useful in its own right.

Lemma 4.3.8. *Let* R *be an integral domain that is also a finite dimensional vector space over a field* $\mathbb{F}$. *Then* R *is a field.*

Proof. We need to show that every nonzero element of R is invertible. So let $r \in R$, $r \neq 0$. Consider the set $\{1, r, r^2, \dots\}$. This is an infinite set of elements is a finite dimensional vector space, so is linearly dependent. Consider any nontrivial linear dependence relation $a_k r^k + a_{k+1} r^{k+1} + \cdots + a_m r^m = 0$, with $a_k \neq 0$. Then

$$0 = a_k r^k + a_{k+1} r^{k+1} + \cdots + a_m r^m = a_k r^k (1 + b_1 r + \cdots + b_n r^n)$$

where $b_i = a_{k+i}$ and $n = m - k$. Since R is an integral domain. The first factor on the right hand side is nonzero, and then the second factor must be 0. This gives

$$1 = -b_1 r - \cdots - b_n r^n = r\left(-b_1 \cdots - b_n r^{n-1}\right)$$

so

$$r^{-1} = -b_1 - \cdots - b_n r^{n-1} \in R.$$ □

Definition 4.3.9. Let $\mathbb{E}$ be an extension of $\mathbb{F}$ and let $\mathbb{B}_1$ and $\mathbb{B}_2$ be intermediate fields between $\mathbb{E}$ and $\mathbb{F}$. Then $\mathbb{B}_1$ and $\mathbb{B}_2$ are *disjoint* extensions of $\mathbb{F}$ if $\mathbb{B}_1 \cap \mathbb{B}_2 = \mathbb{F}$.   ◊

Definition 4.3.10. Let $\mathbb{E}$ be an extension of $\mathbb{F}$ and let $\mathbb{B}_1$ and $\mathbb{B}_2$ be intermediate fields between $\mathbb{E}$ and $\mathbb{F}$. Their *composite* $\mathbb{D} = \mathbb{B}_1\mathbb{B}_2$ is the smallest subfield of $\mathbb{E}$ containing $\mathbb{B}_1$ and $\mathbb{B}_2$.   ◊

Observe that there is some subfield of $\mathbb{E}$, namely $\mathbb{E}$ itself, that contains both $\mathbb{B}_1$ and $\mathbb{B}_2$, and hence there is a smallest such subfield, namely the intersection of all such subfields.

This is an abstract definition. Let us see how to make it concrete.

Lemma 4.3.11. *Let $\mathbb{E}$ be an extension of $\mathbb{F}$ and let $\mathbb{B}_1$ and $\mathbb{B}_2$ be intermediate fields between $\mathbb{E}$ and $\mathbb{F}$, both of which are finite extensions of $\mathbb{F}$. Then their composite $\mathbb{D} = \mathbb{B}_1\mathbb{B}_2$ is give by*

$$\mathbb{D} = \left\{ \sum b_1^i b_2^i \mid b_1^i \in \mathbb{B}_1, b_2^i \in \mathbb{B}_2 \right\}.$$

Proof.   Call the right hand side R. Clearly any field that contains both $\mathbb{B}_1$ and $\mathbb{B}_2$ must contain R. So we need only show that R is a field.

Now R contains 1 and is clearly closed under addition and multiplication. Also, R has no zero divisors as $R \subseteq \mathbb{E}$ and $\mathbb{E}$ has no zero divisors (as $\mathbb{E}$ is a field). Thus R is an integral domain.

R is also clearly an $\mathbb{F}$-vector space, and if $\mathbb{B}_1$ has basis $\{\beta_{1i} \mid i = 1, \ldots, j\}$ and $\mathbb{B}_2$ has basis $\{\beta_{2k} \mid k = 1, \ldots, l\}$ (as $\mathbb{F}$-vector spaces) then R is spanned by the finite set $\{\beta_{1i}\beta_{2k} \mid i = 1, \ldots, j, \; k = 1, \ldots, l\}$ so is a finite dimensional $\mathbb{F}$-vector space. Thus, by Lemma 4.3.8, R is a field.   □

Corollary 4.3.12. *Let $\mathbb{E}$ be an extension of $\mathbb{F}$ and let $\mathbb{B}_1$ and $\mathbb{B}_2$ be intermediate fields between $\mathbb{E}$ and $\mathbb{F}$, both of which are finite extensions of $\mathbb{F}$. Then:*

(a)   (i) $(\mathbb{B}_1\mathbb{B}_2/\mathbb{F}) \leq (\mathbb{B}_1/\mathbb{F})(\mathbb{B}_2/\mathbb{F})$
      (ii) $(\mathbb{B}_1\mathbb{B}_2/\mathbb{B}_1) \leq (\mathbb{B}_2/\mathbb{F})$
      (iii) $(\mathbb{B}_1\mathbb{B}_2/\mathbb{B}_2) \leq (\mathbb{B}_1/\mathbb{F})$

(b) *If $\mathbb{B}_1$ and $\mathbb{B}_2$ are disjoint extensions of $\mathbb{F}$, the following are equivalent:*

   (i) $(\mathbb{B}_1\mathbb{B}_2/\mathbb{F}) = (\mathbb{B}_1/\mathbb{F})(\mathbb{B}_2/\mathbb{F})$
  (ii) $(\mathbb{B}_1\mathbb{B}_2/\mathbb{B}_1) = (\mathbb{B}_2/\mathbb{F})$
 (iii) $(\mathbb{B}_1\mathbb{B}_2/\mathbb{B}_2) = (\mathbb{B}_1/\mathbb{F})$

(c) *If $\mathbb{B}_1$ and $\mathbb{B}_2$ are not disjoint extensions of $\mathbb{F}$, none of the equalities in (b) are true.*

Proof. (a) (i) follows immediately from the proof of Lemma 4.3.11. We know from Lemma 4.3.4 that

$$(\mathbb{B}_1\mathbb{B}_2/\mathbb{F}) = (\mathbb{B}_1\mathbb{B}_2/\mathbb{B}_1)\,(\mathbb{B}_1/\mathbb{F}) = (\mathbb{B}_1\mathbb{B}_2/\mathbb{B}_2)\,(\mathbb{B}_2/\mathbb{F})$$

from which the rest of (a) and (b) follows.

    (c) We have that

$$
\begin{aligned}
(\mathbb{B}_1\mathbb{B}_2/\mathbb{F}) &= (\mathbb{B}_1\mathbb{B}_2/\mathbb{B}_1 \cap \mathbb{B}_2)\,(\mathbb{B}_1 \cap \mathbb{B}_2/\mathbb{F}) \\
&\leq (\mathbb{B}_1/\mathbb{B}_1 \cap \mathbb{B}_2)\,(\mathbb{B}_2/\mathbb{B}_1 \cap \mathbb{B}_2)\,(\mathbb{B}_1 \cap \mathbb{B}_2/\mathbb{F}) \\
&= (\mathbb{B}_1/\mathbb{B}_1 \cap \mathbb{B}_2)\,(\mathbb{B}_2/\mathbb{B}_1 \cap \mathbb{B}_2) \\
&\qquad (\mathbb{B}_1 \cap \mathbb{B}_2/\mathbb{F})^2 \Big/ (\mathbb{B}_1 \cap \mathbb{B}_2/\mathbb{F}) \\
&= (\mathbb{B}_1/\mathbb{B}_1 \cap \mathbb{B}_2)\,(\mathbb{B}_1 \cap \mathbb{B}_2/\mathbb{F})\,(\mathbb{B}_2/\mathbb{B}_1 \cap \mathbb{B}_2) \\
&\qquad (\mathbb{B}_1 \cap \mathbb{B}_2/\mathbb{F}) \Big/ (\mathbb{B}_1 \cap \mathbb{B}_2/\mathbb{F}) \\
&= (\mathbb{B}_1/\mathbb{F})\,(\mathbb{B}_2/\mathbb{F}) \Big/ (\mathbb{B}_1 \cap \mathbb{B}_2/\mathbb{F}).
\end{aligned}
$$

    If $\mathbb{B}_1$ and $\mathbb{B}_2$ are not disjoint extensions of $\mathbb{F}$, i.e., if $\mathbb{F} \subset \mathbb{B}_1 \cap \mathbb{B}_2$, then $(\mathbb{B}_1 \cap \mathbb{B}_2/\mathbb{F}) > 1$ and we see that $(\mathbb{B}_1\mathbb{B}_2/\mathbb{F}) < (\mathbb{B}_1/\mathbb{F})(\mathbb{B}_2/\mathbb{F})$. $\square$

Corollary 4.3.13. *Let $\mathbb{E}$ be an extension of $\mathbb{F}$ and let $\mathbb{B}_1$ and $\mathbb{B}_2$ be intermediate fields between $\mathbb{E}$ and $\mathbb{F}$, each of which is a finite extension of $\mathbb{F}$. If the degrees $(\mathbb{B}_1/\mathbb{F})$ and $(\mathbb{B}_2/\mathbb{F})$ are relatively prime, then*

(a) $\mathbb{B}_1$ *and* $\mathbb{B}_2$ *are disjoint extension of* $\mathbb{F}$; *and*
(b) $(\mathbb{B}_1\mathbb{B}_2/\mathbb{F}) = (\mathbb{B}_1/\mathbb{F})(\mathbb{B}_2/\mathbb{F})$.

Proof. By Corollary 4.3.6, $(\mathbb{B}_1 \cap \mathbb{B}_2/\mathbb{F})$ must divide both $(\mathbb{B}_1/\mathbb{F})$ and $(\mathbb{B}_2/\mathbb{F})$, so $(\mathbb{B}_1 \cap \mathbb{B}_2/\mathbb{F}) = 1$ and then $\mathbb{B}_1 \cap \mathbb{B}_2 = \mathbb{F}$ by Corollary 4.3.5.

Also, $(\mathbb{B}_1\mathbb{B}_2/\mathbb{F}) \le (\mathbb{B}_1/\mathbb{F})(\mathbb{B}_2/\mathbb{F})$ by Lemma 4.3.12 and is divisible by both $(\mathbb{B}_1/\mathbb{F})$ and $(\mathbb{B}_2/\mathbb{F})$ by Corollary 4.3.6, so $(\mathbb{B}_1\mathbb{B}_2/\mathbb{F}) = (\mathbb{B}_1/\mathbb{F})(\mathbb{B}_2/\mathbb{F})$. $\qquad\square$

4.4   Algebraic elements

Definition 4.4.1. Let $\mathbb{E}$ be an extension of $\mathbb{F}$ and let $\alpha \in \mathbb{E}$. Then α is *algebraic* over $\mathbb{F}$ if $f(\alpha) = 0$ for some nonzero polynomial $f(x) \in \mathbb{F}[x]$.

An extension $\mathbb{E}$ of $\mathbb{F}$ is *algebraic* if every element of $\mathbb{E}$ is algebraic over $\mathbb{F}$. $\qquad\Diamond$

In other words, α is algebraic over $\mathbb{F}$ (which we will abbreviate to algebraic, if $\mathbb{F}$ is understood) if it is a root of some nonzero polynomial $f(x)$ with coefficients in $\mathbb{F}$.

Example 4.4.2.

(a) Every element a of $\mathbb{F}$ is algebraic over $\mathbb{F}$ as it is a root of the polynomial $x - a$ in $\mathbb{F}[x]$.

(b) $\sqrt{2}$ and $-\sqrt{2}$ are both algebraic over $\mathbb{Q}$ as they are both roots of the polynomial $x^2 - 2$ in $\mathbb{Q}[x]$. Indeed, every element of $\mathbb{Q}[\sqrt{2}]$ is algebraic over $\mathbb{Q}$. To see this, let $\alpha \in \mathbb{Q}\left[\sqrt{2}\right]$, so $\alpha = a + b\sqrt{2}$ with $a, b \in \mathbb{Q}$. If $b = 0$, then $\alpha = a \in \mathbb{Q}$ is algebraic over $\mathbb{Q}$ by part (a). Suppose not. Then $\alpha - a = b\sqrt{2}$, so $(\alpha - a)^2 = 2b^2$, $\alpha^2 - 2a\alpha + a^2 = 2b^2$, and then $\alpha^2 - 2a\alpha + (a^2 - 2b^2) = 0$, so α is a root of the polynomial $f(x) = x^2 - 2ax + (a^2 - 2b^2)$ in $\mathbb{Q}[x]$. (This polynomial is irreducible and its other root is $\bar{\alpha} = a - b\sqrt{2}$.)

(c) Let $p(x) \in \mathbb{F}[x]$ be an irreducible polynomial and let $\mathbb{E} = \mathbb{F}[x]/\langle p(x)\rangle$. Let $\pi: \mathbb{F}[x] \to \mathbb{E}$ be the projection and let $\alpha = \pi(x)$. Then, as we have seen, $p(\alpha) = 0$, and so we conclude that α is algebraic over $\mathbb{F}$. $\qquad\Diamond$

Comparing (c) with (b), we might hope that not only α, but in fact, every element of $\mathbb{E}$, is algebraic over $\mathbb{F}$. Indeed, not only is this true, but a more general result is true.

Theorem 4.4.3. *Let $\mathbb{E}$ be a finite extension of $\mathbb{F}$. Then $\mathbb{E}$ is an algebraic extension of $\mathbb{F}$.*

Proof.   Let $(\mathbb{E}/\mathbb{F}) = n$, so that $\mathbb{E}$ is an n-dimensional vector space over $\mathbb{F}$. Let α be any element of $\mathbb{E}$, and consider the set $\{1, \alpha, \ldots, \alpha^n\}$.

This is a set of $n+1$ elements in $\mathbb{E}$, so must be linearly dependent. Thus there are scalars (i.e., elements of $\mathbb{F}$) $a_0, a_1, \ldots, a_n$, not all zero, with $a_0 \cdot 1 + a_1 \cdot \alpha + \cdots + a_n \alpha^n = 0$. Then α is a root of the nonzero polynomial $f(x) = a_n x^n + \cdots + a_1 x + a_0$ in $\mathbb{F}[x]$. □

Lemma 4.4.4. *Let $\mathbb{E}$ be an extension of $\mathbb{F}$ and let $\alpha \in \mathbb{E}$ be algebraic over $\mathbb{F}$. Let $f(x) \in \mathbb{F}[x]$. The following are equivalent:*

(1) $f(x)$ *is the unique monic generator of the ideal* $I = \{g(x) \in \mathbb{F}[x] \mid g(\alpha) = 0\}$ *of* $\mathbb{F}[x]$.
(2) $f(x)$ *is the unique monic polynomial in* $\mathbb{F}[x]$ *of lowest degree with* $f(\alpha) = 0$.
(3) $f(x)$ *is the unique monic irreducible polynomial in* $\mathbb{F}[x]$ *with* $f(\alpha) = 0$.

Proof. First recall that $\mathbb{F}[x]$ is a PID so every ideal is principal. The ideal I is nonzero precisely because α is algebraic over $\mathbb{F}$. Thus I has a unique monic generator $f(x)$.

If $g(x)$ is any polynomial with $g(\alpha) = 0$, then $g(x) \in I$, so $f(x)$ divides $g(x)$. Assume $g(x)$ is monic. Then either $g(x) = f(x)$ or $\deg g(x) > \deg f(x)$, so (1)⇔(2).

Let $g(x)$ be any monic polynomial with $g(\alpha) = 0$. Then $g(x)$ factors uniquely into a product of monic irreducible polynomials $g(x) = f_1(x) \ldots f_k(x)$. Then $0 = g(\alpha) = f_1(\alpha), \ldots, f_k(\alpha)$, so $f_i(\alpha) = 0$ for at least one value of i, i.e., $g(x)$ has a monic irreducible factor $f(x) = f_i(x)$ with $f(\alpha) = 0$, so (1)⇔(3). □

Definition 4.4.5. The unique monic polynomial satisfying the equivalent conditions of the above lemma is called the *minimal polynomial* of α, denoted $m_\alpha(x)$. ◊

(This language comes from property (2), which says that it is of minimal degree.)

Example 4.4.6.

(a) $m_\alpha(x) = x - \alpha$ is linear if and only if $\alpha \in \mathbb{F}$.
(b) If $\mathbb{F} = \mathbb{Q}$, $m_{\sqrt{2}}(x) = x^2 - 2 \in \mathbb{Q}[x]$.
(c) In the situation of Example 4.4.2(c), $m_\alpha(x) = p(x)$. ◊

Remark 4.4.7. The polynomial $m_\alpha(x)$ depends on $\mathbb{F}$ (e.g., if $\mathbb{F} = \mathbb{Q}$ then $m_{\sqrt{2}}(x) = x^2 - 2$ while if $\mathbb{F} = \mathbb{Q}\sqrt{2}$ then $m_{\sqrt{2}}(x) = x - \sqrt{2}$) but we suppress that dependence from the notation. ◊

Definition 4.4.8. Let $\mathbb{E}$ be an extension of $\mathbb{F}$ and let $\alpha \in \mathbb{E}$ be algebraic over $\mathbb{F}$. The *degree* of α (over $\mathbb{F}$) is the degree of its minimal polynomial $m_\alpha(x) \in \mathbb{F}[x]$.   ◇

Lemma 4.4.9. *Let $\mathbb{E}$ be an extension of $\mathbb{F}$ and let be algebraic over $\mathbb{F}$. Let $\mathbb{F}(\alpha)$ be the subfield of $\mathbb{E}$*

$$\mathbb{F}[\alpha] = \{f(\alpha) \mid f(x) \in \mathbb{F}[x]\}.$$

Then $\mathbb{F}[\alpha]$ is an extension of $\mathbb{F}$ of degree equal to the degree of α. Also, if $\mathbb{E}$ is a finite extension of $\mathbb{F}$, then the degree of α divides $(\mathbb{E}/\mathbb{F})$.

Proof.   We have a homomorphism $\varphi \colon \mathbb{F}[x] \to \mathbb{F}[\alpha]$ defined by $\varphi(f(x)) = f(\alpha)$. This homomorphism is onto and its kernel is $I = \langle m_\alpha(x) \rangle$. Thus $\mathbb{F}[x]/\langle m_\alpha(x) \rangle$ is isomorphic to $\mathbb{F}(\alpha)$. Now $\mathbb{F}[x]/\langle m_\alpha(x) \rangle$ is an extension of $\mathbb{F}$ of degree $m_\alpha(x)$, which by definition is the degree of α, by Theorem 4.2.11. Since $\mathbb{F} \subseteq \mathbb{F}(\alpha) \subseteq \mathbb{E}$, the second conclusion follows directly from Corollary 4.3.6.   □

Now let us turn to a concrete question. Let $p(x) \in \mathbb{F}[x]$ be an irreducible polynomial and let $\mathbb{E} = \mathbb{F}[x]/\langle p(x) \rangle$. Given an arbitrary element β of $\mathbb{E}$, how can we find $m_\beta(x)$? The answer is simply trial and error, guided by our proof of the theorem that β must be algebraic over $\mathbb{F}$ (Theorem 4.4.3).

As we have seen, $(\mathbb{E}/\mathbb{F}) = n$ where $\deg p(x) = n$, and $\mathbb{E}$ has basis $\{1, \alpha, \ldots, \alpha^{n-1}\}$, where $\alpha = \pi(x)$. Let $\beta \in \mathbb{E}$, so that we can express $\beta = q_1(x)$ for $q_1(x)$ a polynomial in $\mathbb{F}[x]$ of degree at most $n-1$. Test if $\{1, q_1(\alpha)\}$ is linearly dependent. (Of course, we know the answer here: This will be the case if and only if $\alpha \in \mathbb{F}$.) If so, a nontrivial linear dependence relation will give us the coefficients of $m_\beta(x)$.

If not, express $\beta^2 = q_2(x)$ with $q_2(x)$ a polynomial in $\mathbb{F}[x]$ of degree at most $n - 1$. If n is not divisible by 2, β cannot have degree 2, by Lemma 4.4.9. If n is divisible by 2, test if $\{1, q_1(\alpha), q_2(\alpha)\}$ is linearly dependent. If so, a nontrivial linear dependence relation will give us the coefficient of $m_\beta(x)$.

If not, keep going. We must stop no later than $\{1, q_1(\alpha), \ldots, q_n(\alpha)\}$ so this procedure will eventually come to an end.

Example 4.4.10. Let $\mathbb{F} = \mathbb{Q}$ and let $p(x) = x^3 + 7x + 1$, an irreducible polynomial in $\mathbb{Q}[x]$. Let $\mathbb{E} = \mathbb{Q}[x]/\langle p(x) \rangle$ and let $\pi \colon \mathbb{Q}[x] \to$

E. Set $\pi(x) = \alpha$. We note that $m_\alpha(x) = x^3 + 7x + 1$. Thus α has degree 3 (over $\mathbb{Q}$).

(a) Let $\beta_1 = \alpha + 3$. We wish to find $m_{\beta_1}(x)$. In this case we begin with an observation that will save us some effort. Of course, $\mathbb{F}[\beta_1] \subseteq \mathbb{E} = \mathbb{F}[\alpha]$. But notice that $\alpha = \beta_1 - 3$ so $\mathbb{E} = \mathbb{F}[\alpha] \subseteq \mathbb{F}[\beta_1]$. Hence, $\mathbb{F}[\beta_1] = \mathbb{F}[\alpha]$ and hence β_1 has degree 3 (over $\mathbb{Q}$) as well. Now α is a root of $m_\alpha(x)$, i.e., $\beta_1 - 3$ is a root of $m_\alpha(x)$, so $(\beta_1 - 3)^3 + 7(\beta_1 - 3) + 1 = 0$. In other words, β_1 is a root of the polynomial $(x - 3)^3 + 7(x - 3) + 1 = x^3 - 9x^2 + 34x - 47$ and since this is a monic polynomial of degree 3, it must be $m_{\beta_1}(x)$. Thus, we conclude that $m_{\beta_1}(x) = x^3 - 9x^2 + 34x - 47$.

(b) Let $\beta_2 = \alpha^2 + 1$. Since $\beta_2 \notin \mathbb{Q}$, we cannot have $\deg(\beta_2) = 1$. Then, since $\deg(\beta_2)$ must divide 3, we must have $\deg(\beta_2) = 3$. We compute $\beta_2^2 = (\alpha^2 + 1)^2 = \alpha^4 + 2\alpha^2 + 1 = (\alpha^3 + 7\alpha + 1)\alpha + (-5\alpha^2 - \alpha + 1)$ so $\beta_2^2 = -5\alpha^2 - \alpha + 1$. We compute $\beta_2^3 = \beta_2^2 \beta_2 = (-5\alpha^2 - \alpha + 1)(\alpha^2 + 1) = -5\alpha^4 - \alpha^3 - 4\alpha^2 - \alpha + 1 = (\alpha^3 + 7\alpha + 1)(-5\alpha - 1) + (31\alpha^2 + 11\alpha + 2)$ so $\beta_2^3 = 31\alpha^2 + 11\alpha + 2$. Thus,

$$\{1, \beta_2, \beta_2^2, \beta_2^3\} = \{1, \alpha^2 + 1, -5\alpha^2 - \alpha + 1, 31\alpha^2 + 11\alpha + 2\}.$$

Now if $m_{\beta_2}(x) = x^3 + Ax^2 + Bx + C$, $m_{\beta_2}(\beta_1) = 0$ is the equation

$$(31\alpha^2 + 11\alpha + 2) + A(-5\alpha^2 - \alpha + 1) + B(\alpha^2 + 1) + C = 0.$$

Equating the coefficients of the powers of α, this gives the linear system

$$31 - 5A + B = 0$$
$$11 - A = 0$$
$$2 + A + B + C = 0$$

with solution $A = 11$, $B = 24$, $C = -37$. Thus, $m_{\beta_2}(x) = x^3 + 11x^2 + 24x - 37$. $\diamond$

Example 4.4.11. Let $\mathbb{F} = \mathbb{Q}$ and let $p(x) = x^4 - 10x^2 + 1$. We first see that this is an irreducible polynomial in $\mathbb{Q}[x]$, as follows: By the rational root rest, Lemma 3.11.12, we see that the only possible rational roots of $p(x)$ are $x = \pm 1$. Since $p(1) \neq 0$ and

$p(-1) \neq 0$, neither of these is a root of $p(x)$. Thus the only possible nontrivial factorization of $p(x)$ in $\mathbb{Q}[x]$ is as a product of two quadratics. Now $p(x)$ is a monic polynomial with integer coefficients, so it follows from Gauss's Lemma (specifically, an application of Corollary 3.11.6) that in this case $p(x)$ must have a factorization as a product of two monic quadratics with integer coefficients, $p(x) = \left(x^2 + a_1 x + a_0\right)\left(x^2 + b_1 x + b_0\right)$ with $a_0, a_1, b_0, b_1 \in \mathbb{Z}$.

Now $a_0 b_0 = 1$ so we must have either $a_0 = b_0 = 1$ or $a_0 = b_0 = -1$. In either case, the fact that the "x" coefficient of $p(x)$ is 0 forces $b_1 = -a_1$. But then examining the "x^2" coefficient of $p(x)$ gives the equation $2 - a_1^2 = -10$ in case $a_0 = b_0 = 1$ and $-2 - a^2 = -10$ in case $a_0 = b_0 = -1$, but in neither case does this have solution with $a_0 \in \mathbb{Z}$. Thus, $p(x)$ is irreducible.

Let $\mathbb{E} = \mathbb{Q}[x] / \langle p(x) \rangle$, $\pi \colon \mathbb{Q}[x] \to \mathbb{E}$, and $\pi(x) = \alpha$. Then α has degree 4 with $m_\alpha(x) = x^4 - 10x^2 + 1$.

(a) Let $\beta_1 = \alpha^3 - 9\alpha + 2$. Since $\beta_1 \notin \mathbb{Q}$, we cannot have $\deg(\beta_1) = 1$. We compute $\beta_1^2 = \alpha^6 - 18\alpha^4 + 4\alpha^3 + 81\alpha^2 - 36\alpha + 4 = \left(\alpha^4 - 10\alpha^2 + 1\right)\left(\alpha^2 - 8\right) + \left(4\alpha^2 - 36\alpha + 12\right)$ so $\beta_1^2 = 4\alpha^2 - 36\alpha + 12$. Thus,

$$\left\{1, \beta_1, \beta_1^2\right\} = \left\{1, \alpha^3 - 9\alpha + 2, 4\alpha^3 - 36\alpha + 12\right\}.$$

Now if $m_{\beta_1}(x) = x^2 + Ax + B$, $m_{\beta_1}(\beta_1) = 0$ is the equation

$$\left(4\alpha^3 - 36\alpha + 12\right) + A\left(\alpha^3 - 9\alpha + 2\right) + B = 0.$$

Equating the coefficients of the powers of α, this gives the linear system

$$4 + A = 0$$
$$0 = 0$$
$$-36 - 9A = 0$$
$$12 + 2A + B = 0$$

with solution $A = -4$, $B = -4$. Thus $m_{\beta_1}(x) = x^2 - 4x - 4$.

(b) Let $\beta_2 = \alpha^2 + 2\alpha - 3$. Since $\beta_2 \notin \mathbb{Q}$, we cannot have $\deg(\beta_2) = 1$. We compute $\beta_2^2 = \alpha^4 + 4\alpha^3 - 2\alpha^2 - 12\alpha + 9 = \left(\alpha^4 - 10\alpha^2 + 1\right)(1) + \left(4\alpha^3 + 8\alpha^2 - 12\alpha + 8\right)$ so $\beta_2^2 = 4\alpha^3 + 8\alpha^2 - 12\alpha + 8$. Then

$$\{1, \beta_2, \beta_2^2\} = \{1, \alpha^2 + 2\alpha - 3, 4\alpha^3 + 8\alpha^2 - 12\alpha + 8\}.$$

Now if $m_{\beta_2}(x) = x^2 + Ax + B$, $m_{\beta_2}(\beta_2) = 0$ is the equation

$$(4\alpha^3 + 8\alpha^2 - 12\alpha + 8) + A(\alpha^2 + 2\alpha - 3) + B(1) = 0.$$

Equating the coefficients of the powers of α, this gives the system

$$4 = 0$$

$$8 + A = 0$$

$$-12 + 2A = 0$$

$$8 - 3A + B = 0$$

which has no solution. Thus we conclude β_2 does not have degree 2.

We proceed further. We next compute $\beta_2^3 = \beta_2^2 \beta_2 = (4\alpha^3 + 8\alpha^2 - 12\alpha + 8)(\alpha^2 + 2\alpha - 3) = 4\alpha^5 + 16\alpha^4 - 8\alpha^3 - 40\alpha^2 + 52\alpha - 24 = (\alpha^4 - 10\alpha^2 + 1)(4\alpha + 16) + (32\alpha^3 + 120\alpha^2 + 48\alpha - 40)$ so $\beta_2^3 = 32\alpha^3 + 120\alpha^2 + 48\alpha - 40$. We then compute $\beta_2^4 = \beta_2^3 \beta_2 = (32\alpha^2 + 120\alpha^2 + 48\alpha - 40)(\alpha^2 + 2\alpha - 3) = 32\alpha^5 + 184\alpha^4 + 192\alpha^3 - 304\alpha^2 - 224\alpha + 120 = (\alpha^4 - 10\alpha^2 + 1)(32\alpha + 184) + (-528\alpha^3 - 1536\alpha^2 + 256\alpha + 64)$ so $\beta_2^4 = -528\alpha^3 - 1536\alpha^2 + 256\alpha + 64$. Thus

$$\{1, \beta, \beta^2, \beta^3\} = \{1, \alpha^2 + 2\alpha - 3, \ 4\alpha^3 + 8\alpha^2 - 12\alpha + 8,$$

$$32\alpha^3 + 120\alpha^2 + 48\alpha - 40, \ -528\alpha^3$$

$$-1536\alpha^2 + 256\alpha + 64\}.$$

Now if $m_{\beta_2}(x) = x^4 + Ax^3 + Bx^2 + Cx + D$, $m_{\beta_2}(\beta_2) = 0$ is the equation

$$(-528\alpha^3 - 1536\alpha^2 + 256\alpha + 64) + A(32\alpha^3 + 120\alpha^2 + 48\alpha - 40)$$
$$+ B(4\alpha^3 + 8\alpha^2 - 12\alpha + 8) + C(\alpha^2 + 2\alpha - 3) + D = 0 .$$

Equating the coefficient of the powers of α, this gives the system

$$-528 + 32A + 4B = 0$$

$$-1536 + 120A + 8B + C = 0$$

$$256 + 48A - 12B + 2C = 0$$

$$64 - 40A + 8B - 3C + D = 0$$

with solution $A = -8$, $B = -96$, $C = 448$, $D = 624$. Thus $m_{\beta_2}(x) = x^4 - 8x^3 - 96x^2 + 448x + 624$. $\Diamond$

Example 4.4.12. Let $\mathbb{F} = \mathbb{F}_7$ and let $p(x) = x^3 + x + 6$, an irreducible polynomial in $\mathbb{F}_7[x]$. (We easily see that $p(x)$ is irreducible, as it is a cubic which does not have a root in $\mathbb{F}_7$.)

Let $\mathbb{E} = \mathbb{F}_7[x]/\langle p(x) \rangle$, $p\colon \mathbb{F}_7[x] \to \mathbb{E}$ and $\alpha = \pi(x)$.

Let $\beta = \alpha^2 + 1$. Since $\beta \notin \mathbb{F}_7$, it must have degree 3.

We compute $\beta^2 = \alpha^4 + 2\alpha^2 + 1 = \left(\alpha^3 + \alpha + 6\right)\alpha + \left(\alpha^2 + \alpha + 1\right)$ so $\beta^2 = \alpha^2 + \alpha + 1$. We compute $\beta^3 = \beta^2\beta = \left(\alpha^2 + \alpha + 1\right)(\alpha + 1) = \alpha^4 + \alpha^3 + 2\alpha^2 + \alpha + 1 = \left(\alpha^3 + \alpha + 6\right)(\alpha + 1) + \left(\alpha^2 + \alpha + 5\right)$ so $\beta^3 = \alpha^2 + \alpha + 5$. Then

$$\{1, \beta, \beta^2, \beta^3\} = \{1, \alpha^2 + 1, \alpha^2 + \alpha + 1, \alpha^2 + \alpha + 5\}.$$

Now if $m_\beta(x) = x^3 + Ax^2 + Bx + C$, $m_\beta(\beta) = 0$ is the equation

$$\left(\alpha^2 + \alpha + 5\right) + A\left(\alpha^2 + \alpha + 1\right) + B\left(\alpha^2 + 1\right) + C = 0$$

which gives the system

$$1 + A + B = 0$$
$$1 + A = 0$$
$$5 + A + B + C = 0$$

with solution $A = 6$, $B = 0$, $C = 3$. Thus $m_\beta(x) = x^3 + 6x^2 + 3$. $\Diamond$

Remark 4.4.13. Let us step back from these particular computations and think about what we have been doing. We have a field $\mathbb{E}$ that is an extension of $\mathbb{F}$ of some finite degree n. In our case $\mathbb{E} = \mathbb{F}[x]/\langle p(x) \rangle$ for some irreducible polynomial $p(x)$ in $\mathbb{F}[x]$, so we had an explicit basis $\{1, \alpha, \ldots, \alpha^{n-1}\}$ of $\mathbb{E}$. We considered an element $\beta = b_{n-1}\alpha^{n-1} + \cdots + b_1\alpha + b_0$ of $\mathbb{E}$, and asked about its degree. Suppose that β has degree k, or equivalently that its minimum polynomial $m_\beta(x)$ has degree k, $m_\beta(x) = x^k + A_{k-1}x^{k-1} + \cdots + A_0$. Then

$$B = \{1, \beta, \ldots, \beta^k\}$$

is a linearly dependent set of $(k + 1)$ elements in an n-dimensional vector space, and when try to find the coefficients of $m_\beta(x)$, equating the coefficients of the n powers of α (from 1 to α^{n-1}) give us a non-homogenous system of n equations in the k unknowns $A_0, \ldots, A_{k-1}$.

Now let us think about this situation from a linear algebra perspective. If $k < n$, then B is a set of at most k elements in an n-dimensional vector space, and we expect such a set to be linearly independent. Alternatively, if $k < n$ then we expect that a non-homogenous system of n equations in k unknowns will not have a solution. Thus, either way we would expect that if we pick an element "at random", we will not have $k < n$, and so $k = n$ and $m_\beta(x)$ has degree n. But also in this case we have $\mathbb{F} \subseteq \mathbb{F}[\beta] \subseteq \mathbb{E}$ with $(\mathbb{F}[\beta]/\mathbb{F}) = (\mathbb{E}/\mathbb{F}) = n$, and then $\mathbb{E} = \mathbb{F}[\beta]$ by Corollary 4.3.5.

Now of course in our situation B is not an arbitrary set of elements of $\mathbb{E}$, but rather consists of powers of a single element β, and in our situation the coefficients of the powers of α in our linear system are not arbitrary but rather are expressions (as it happens, complicated polynomial expressions) in the coefficients $b_{n-1}, \ldots, b_0$. So it is possible that there is something mysterious going on that always confounds our expectations, but that seems highly unlikely.

In fact, our linear algebra institution is just about right. We will see later on that in great generality (thought not quite always) that if $\mathbb{E}$ is a finite extension of $\mathbb{F}$ and we pick an element of $\mathbb{E}$ "at random", then $\mathbb{E} = \mathbb{F}[\beta]$. (The fact that at least one such β (almost always) exists is called the theorem of the primitive element, Theorem 4.12.2.) Indeed, our proof of this theorem will be a linear algebra proof (one involving theoretical linear algebra, not writing down systems of linear equations), showing how perspicacious this view point is. ◇

We conclude this section by recording a simple but very useful general result.

Lemma 4.4.14. *Let $f(x) \in \mathbb{F}[x]$ be an irreducible polynomial of degree n. Let $\mathbb{B}$ be an extension of $\mathbb{F}[x]$ of degree m. If m and n are relatively prime, then $f(x)$ is irreducible in $\mathbb{B}[x]$.*

Proof. Since $f(x)$ is irreducible in $\mathbb{F}[x]$, we know that $(\mathbb{F}(\alpha)/\mathbb{F}) = n$, where $\mathbb{F}(\alpha)$ is the field obtained from $\mathbb{F}$ by adjoining a root α of $\mathbb{F}(x)$. Now $(\mathbb{B}/\mathbb{F}) = m$, and m and n are relatively prime, so by Corollary 4.3.13 $(\mathbb{B}/\mathbb{F}(\alpha)/\mathbb{F}) = mn$, i.e., $(\mathbb{B}(\alpha)/\mathbb{F}) = mn$. But $(\mathbb{B}(\alpha)/\mathbb{F}) = (\mathbb{B}(\alpha)/\mathbb{B})(\mathbb{B}/\mathbb{F})$ so $(\mathbb{B}(\alpha)/\mathbb{B}) = n$. But that means that α has degree n over $\mathbb{B}$, and so $f(x)$ is irreducible in $\mathbb{B}[x]$. □

4.5   Adjoining elements

One of the most common, most important, and most useful ways of obtaining field extensions is by adjoining elements. Actually, we have already used this procedure, without giving it a name. But now we will study it more intensively and more systematically.

Before stating the basic definition, we recall that a rational function $r(x)$ is a quotient of polynomials, $r(x) = p(x)/q(x)$ with $p(x)$ and $q(x)$ polynomials.

Definition 4.5.1. Let $\mathbb{F}$ be a subfield of $\mathbb{E}$, and let α be an element of $\mathbb{F}$. The field

$$\mathbb{F}(\alpha) = \{r(x) \mid r(x) = p(x)/q(x) \text{ is a rational function}$$
$$\text{with coefficients in } \mathbb{F}, \text{ and } q(\alpha) \neq 0\}$$

is the field obtained from $\mathbb{F}$ by *adjoining* α. $\Diamond$

It is easy to check that $\mathbb{F}(\alpha)$ is indeed a field. Also, it the smallest field extension of $\mathbb{F}$ that contains α in the sense that any extension $\mathbb{F}$ containing α must contain $\mathbb{F}(\alpha)$.

Observe that if $\alpha \in \mathbb{F}$ then $\mathbb{F}(\alpha) = \mathbb{F}$. Thus, adjoining an element of $\mathbb{F}$ to $\mathbb{F}$ gives us nothing new. But if $\alpha \notin \mathbb{F}$, then $\mathbb{F}(\alpha) \neq \mathbb{F}$, so if $\alpha \notin \mathbb{F}$ we do indeed get a new field.

We remarked that we have already seen and used this construction–compare Lemma 4.4.9. But in that lemma we only needed to use polynomials, while in Definition 4.5.1 we needed to use rational functions. What's the difference? It is because in Lemma 4.4.9 we were requiring that the element α was algebraic. We now recall why that is true (from a slightly different viewpoint, though one which, if you trace it back far enough, turns out to be the same) and show that the converse is true as well.

Lemma 4.5.2. *Let $\mathbb{F}$ be a subfield of $\mathbb{E}$ and let α be an element of $\mathbb{E}$. Then $\mathbb{F}(\alpha) = \{p(x) \mid p(x)$ is a polynomial with coefficients in $\mathbb{F}\}$ if and only if α is algebraic over $\mathbb{F}$.*

Proof.   Both conditions are certainly true if $\alpha = 0$, so we may restrict our attention to the case $\alpha \neq 0$.

First suppose that α is algebraic over $\mathbb{F}$, and consider $\beta = r(\alpha) = p(\alpha)/q(\alpha)$. (Note $q(\alpha) \neq 0$.) Since α is algebraic, it has a minimal polynomial $m_\alpha(x)$. Note that $m_\alpha(x)$ is irreducible (by Lemma 4.4.4) and hence prime in $\mathbb{F}[x]$. Since $q(\alpha) \neq 0$, $m_\alpha(x)$ does not divide $q(x)$, and hence $m_\alpha(x)$ and $q(x)$ are relatively prime. But $\mathbb{F}[x]$ is a Euclidean domain, so there are polynomials $s(x)$ and $t(x)$ in $\mathbb{F}[x]$ with

$$1 = m_\alpha(x)s(x) + q(x)t(x).$$

But then $1 = m_\alpha(\alpha)s(\alpha) + q(\alpha)t(\alpha)$. Of course, $m_\alpha(\alpha) = 0$, so $1 = q(\alpha)t(\alpha)$ and $t(\alpha) = 1/q(\alpha)$. Thus $\beta = r(\alpha) = p(\alpha)/q(\alpha) = p(\alpha)t(\alpha)$ can be expressed as a polynomial in α.

Conversely, suppose that every rational function in α can be expressed as a polynomial in α. In particular, then, $1/\alpha = g(\alpha)$ for some polynomial $g(x)$ with coefficients in $\mathbb{F}$. But then $1 = \alpha g(\alpha)$ so $\alpha g(\alpha) - 1 = 0$ and α is a root of the polynomial $f(x) = xg(x) - 1 \in \mathbb{F}[x]$ and so α is algebraic over $\mathbb{F}$. $\quad\square$

Now let us adjoin more than one element. The basic definition is essentially the same.

Definition 4.5.3. Let $\mathbb{F}$ be a subfield of $\mathbb{E}$, and let $\{\alpha_1, \alpha_2, \ldots\}$ be a subset of $\mathbb{E}$. The field

$$\mathbb{F}(\{\alpha_1, \alpha_2, \ldots\}) = \{r(\alpha_1, \alpha_2, \ldots) \mid r(x_1, x_2, \ldots)$$
$$= p(x_1, x_2, \ldots)/q(x_1, x_2, \ldots)$$

is a rational function of $x_1, x_2, \ldots$ with

coefficients in $\mathbb{F}$, and $q(\alpha_1, \alpha_2, \ldots) \neq 0\}$

is the field obtained from $\mathbb{F}$ by *adjoining* $(\alpha_1, \alpha_2, \ldots)$. $\quad\Diamond$

For convenience, we state the following lemma for two elements, though it holds for any finite number of elements.

Lemma 4.5.4. *Let $\mathbb{F}$ be a subfield of $\mathbb{E}$ and let $\alpha_1, \alpha_2 \in \mathbb{E}$. Then*

$$\mathbb{F}(\alpha_1, \alpha_2) = (\mathbb{F}(\alpha_1))(\alpha_2) = (\mathbb{F}(\alpha_2))(\alpha_1).$$

Furthermore, $\mathbb{F}(\alpha_1, \alpha_2)$ is the composite $\mathbb{F}(\alpha_1)\mathbb{F}(\alpha_2)$.

Proof. $\mathbb{F}(\alpha_1, \alpha_2)$ is the set of rational functions in α_1 and α_2 with coefficients in $\mathbb{F}$. By gathering terms together, we may regard any rational function in α_1 and α_2 with coefficients in $\mathbb{F}$ as a rational function in α_2 with coefficients in $\mathbb{F}(\alpha_1)$, or as a rational function in α_1 with coefficients in $\mathbb{F}(\alpha_2)$, and vice-versa.

Furthermore, it is clear that $\mathbb{F}(\alpha_1, \alpha_2)$ is the smallest subfield of $\mathbb{E}$ containing both $\mathbb{F}(\alpha_1)$ and $\mathbb{F}(\alpha_2)$, so $\mathbb{F}(\alpha_1, \alpha_2) = \mathbb{F}(\alpha_1)\mathbb{F}(\alpha_2)$.   □

Remark 4.5.5. The way we view this lemma is it says that if $\{\alpha_1, \ldots, \alpha_n\}$ is a finite set of elements of $\mathbb{E}$, we may obtain the field $\mathbb{F}(\alpha_1, \ldots, \alpha_n)$ by adjoining all of these elements at once, or by adjoining them one at a time, and in the latter case, the order in which we adjoin them does not matter.   ◇

Theorem 4.5.6. *Let $\{\alpha_1, \ldots, \alpha_n\}$ be a finite set of elements of $\mathbb{E}$, with each α_i algebraic over $\mathbb{F}$. Then $\mathbb{F}(\alpha_1, \ldots, \alpha_n)$ is a finite, and hence algebraic, extension of $\mathbb{F}$. More precisely, if α_i has degree d_i, $i = 1, \ldots, n$, then $(\mathbb{F}(\alpha_1, \ldots, \alpha_n)/\mathbb{F}) \leq d_1 \cdots d_n$.*

Conversely, every finite extension $\mathbb{E}$ of $\mathbb{F}$ is obtained by adjoining finitely many algebraic elements, i.e., $\mathbb{E} = \mathbb{F}(\alpha_1, \ldots, \alpha_n)$ for some finite set $\{\alpha_1, \ldots, \alpha_n\}$ of algebraic elements of $\mathbb{E}$.

Proof.   We proceed by induction on n. We have already seen the case $n = 1$: $\mathbb{F}(\alpha_1)$ is an extension of $\mathbb{F}$ of degree d_1 (Lemma 4.4.9).

Now suppose the theorem is true for any set of $n-1$ elements, and consider a set of n elements $\{\alpha_1, \ldots, \alpha_n\}$. Let $\mathbb{B} = \mathbb{F}(\alpha_1, \ldots, \alpha_{n-1})$ so that $\mathbb{F} \subseteq \mathbb{B} \subseteq \mathbb{F}(\alpha_1, \ldots, \alpha_n)$.

As we observed in Lemma 4.5.4, $\mathbb{F}(\alpha_1, \ldots, \alpha_n) = (\mathbb{F}(\alpha_1, \ldots, \alpha_{n-1}))(\alpha_n) = \mathbb{B}(\alpha_n)$. Then by Lemma 4.3.4 we know that

$$(\mathbb{F}(\alpha_1, \ldots, \alpha_n)/\mathbb{F}) = (\mathbb{B}(\alpha_n)/\mathbb{F}) = (\mathbb{B}(\alpha_n)/\mathbb{B})(\mathbb{B}/\mathbb{F}).$$

By the inductive hypothesis we know that $\mathbb{B}$ is a finite extension of $\mathbb{F}$ of degree $D \leq d_1 \cdots d_{n-1}$, so to prove the theorem we need only show that $\mathbb{B}(\alpha_n)$ is a finite extension of $\mathbb{B}$ of degree $d' \leq d_n$.

But, by Lemma 4.3.4, $\mathbb{B}(\alpha_n)$ is the composite $\mathbb{B}\mathbb{F}(\alpha_n)$, and then, by Corollary 4.3.12,

$$d' = (\mathbb{B}(\alpha_n)/\mathbb{B}) = (\mathbb{B}\mathbb{F}(\alpha_n)/\mathbb{B}) \leq (\mathbb{F}(\alpha_n)/\mathbb{F}) = d_n,$$

and we are done by induction.

For the other direction, suppose that $\mathbb{E}$ is a finite extension of $\mathbb{F}$, i.e., that $\mathbb{E}$ is a finite dimensional vector space over $\mathbb{F}$. Let $(\mathbb{E}/\mathbb{F}) = n$, and choose a basis $\{\alpha_1, \ldots, \alpha_n\}$ of $\mathbb{E}$ as an $\mathbb{F}$-vector space.

Certainly $\mathbb{F}(\alpha_1,\ldots,\alpha_n) \subseteq \mathbb{E}$ as each $\alpha_i \in \mathbb{E}$. On the other hand, since $\alpha_i \in \mathbb{E}$ and $(\mathbb{E}/\mathbb{F})$ is finite, we know that α_i is algebraic over $\mathbb{F}$, for each i. But since $\{\alpha_1,\ldots,\alpha_n\}$ is a basis for $\mathbb{E}$ over $\mathbb{F}$, every element ε of $\mathbb{E}$ can be written as $\varepsilon = f_1 x_1 + \cdots + f_n \alpha_n$ with $f_1,\ldots,f_n \in \mathbb{F}$. Now this expression is a linear function of $\alpha_1,\ldots,\alpha_n$ with coefficients in $\mathbb{F}$, so is certainly a rational function of $\alpha_1,\ldots,\alpha_n$ with coefficients in $\mathbb{F}$, and so $\mathbb{E} \subseteq \mathbb{F}(\alpha_1,\ldots,\alpha_n)$. Hence $\mathbb{E} = \mathbb{F}(\alpha_1,\ldots,\alpha_n)$. $\square$

Corollary 4.5.7. *Let $\{\alpha_1,\ldots,\alpha_n\}$ be a finite set of elements of $\mathbb{E}$, each of which is algebraic over $\mathbb{F}$. Let α_i have degree d_i, $i = 1,\ldots,n$. If $\{d_1,\ldots,d_n\}$ is relatively prime, then $(\mathbb{F}(\alpha_1,\ldots,\alpha_n)/\mathbb{F}) = d_1,\ldots,d_n$.*

Proof. Since $d_i = (\mathbb{F}(\alpha_i)/\mathbb{F})$, this follows immediately from Corollary 4.3.13. $\square$

Now we come to a very important notion.

Definition 4.5.8. Let $f(x) \in \mathbb{F}[x]$ be a polynomial of degree $n \geq 1$, $f(x) = a_n x^n + \cdots + a_0$, and suppose that $\mathbb{E}$ is an extension of $\mathbb{F}$ such that $f(x) = a_n(x - \alpha_1)\cdots(x - \alpha_n)$ in $\mathbb{E}[x]$. Then we say that $f(x)$ *splits* in $\mathbb{E}[x]$, and that the extension $\mathbb{F}(\alpha_1,\ldots,\alpha_n)$ of $\mathbb{F}$ is a *splitting field* for $f(x)$. $\Diamond$

Remark 4.5.9.

(a) We observe that $f(x)$ splits in $\mathbb{E}$ if and only if $f(x)$ has n (not necessarily distinct) roots in $\mathbb{E}$.
(b) An intermediate field $\mathbb{B}$ between $\mathbb{F}$ and $\mathbb{E}$ is a splitting field for $f(x)$ if and only if $\mathbb{B}$ is the smallest such field in which $f(x)$ splits. To see this, note that any field in which $f(x)$ splits must contain $\alpha_1,\ldots,\alpha_n$, and hence must contain $\mathbb{F}(\alpha_1,\ldots,\alpha_n)$, and conversely $\alpha_1,\ldots,\alpha_n$, and hence $\mathbb{F}(\alpha_1,\ldots,\alpha_n)$, must be contained in any such field. $\Diamond$

Theorem 4.5.10. *Let $\mathbb{B}$ be a splitting field for $f(x) \in \mathbb{F}[x]$, a polynomial of degree n. If $f(x)$ is irreducible, then $(\mathbb{B}/\mathbb{F})$ is divisible by n. In any case $(\mathbb{B}/\mathbb{F}) \leq n!$*

Proof. Let $f(x)$ have roots $\alpha_1,\ldots,\alpha_n$. Then $\mathbb{F}(\alpha_1)$ is intermediate between $\mathbb{F}$ and $\mathbb{B}$, so $(\mathbb{B}/\mathbb{F}) = (\mathbb{B}/\mathbb{F}(\alpha_1))(\mathbb{F}(\alpha_1)/\mathbb{F})$.

If $f(x)$ is irreducible, we know that $(\mathbb{F}(\alpha_1)/\mathbb{F}) = n$, and so in this case $(\mathbb{B}/\mathbb{F})$ is divisible by n.

We prove the general inequality by induction on n. It is certainly true when $f(x)$ has degree 1. Now suppose it is true for all polynomials of degree $n - 1$ over any field, and let $f(x)$ have degree n. Consider a root α_1 of $f(x)$. Then α_1 is a root of some irreducible factor $f_1(x)$ of some degree $n' \leq n$ and so $(\mathbb{F}(\alpha_1)/\mathbb{F}) = n' \leq n$. But α_1 is a root of $f(x)$, so $f(x)$ is divisible by $x - \alpha_1$, and so we may write $f(x) = (x - \alpha_1)g(x)$, where $g(x)$ is a polynomial of degree $n - 1$. Note that the coefficients of $g(x)$ are in $\mathbb{F}(\alpha_1)$, i.e., $g(x)$ is a polynomial of degree $n - 1$ in $\mathbb{F}(x_1)[x]$. But $\mathbb{B}$ is also a splitting field for $g(x)$ over $\mathbb{F}(\alpha_1)$, as $\mathbb{B} = \mathbb{F}(\alpha_1, \ldots, \alpha_n) = \mathbb{F}(\alpha_1)(\alpha_2, \ldots, \alpha_n)$. Then by the inductive hypothesis $(\mathbb{B}/\mathbb{F}(\alpha_1)) \leq (n - 1)!$ so $(\mathbb{B}/\mathbb{F}) = (\mathbb{B}/\mathbb{F}(\alpha))(\mathbb{F}(\alpha)/\mathbb{F}) \leq n!$ □

Corollary 4.5.11. *Let $\mathbb{B}$ be a splitting field for $f(x) \in \mathbb{F}[x]$, a polynomial of degree n. Let $f(x)$ have roots $\alpha_1, \ldots, \alpha_n$ in $\mathbb{B}$. Then $(\mathbb{B}/\mathbb{F}) = n!$ if and only if $f(x)$ is irreducible in $\mathbb{F}[x]$, and for every $j \geq 1$, if $g_j(x)$ is defined by $f(x) = (x - \alpha_1) \cdots (x - \alpha_j)g_j(x)$, then $g_j(x)$ is irreducible in $\mathbb{F}(\alpha_1, \ldots, \alpha_j)[x]$.*

Proof.   The proof of Theorem 4.5.10 shows this stronger result in this case.   □

Corollary 4.5.12. *Let $f(x) \in \mathbb{F}[x]$ be a polynomial of degree $n \geq 1$. Suppose that $f(x) = f_1(x) \cdots f_k(x)$ is a factorization of $f(x)$ into irreducible polynomials in $\mathbb{F}[x]$. Let $f_i(x)$ have degree n_i. Let $\mathbb{B}$ be a splitting field of $f(x)$. Then $(\mathbb{B}/\mathbb{F})$ is divisible by n_i for each i, and $(\mathbb{B}/\mathbb{F}) \leq (n_1!) \cdots (n_k!)$.*

Proof.   The case $k = 1$ is Theorem 4.5.10.

Suppose $k = 2$. Then $(\mathbb{B}/\mathbb{F})$ is divisible by both n_1 and n_2, as in the proof of that theorem. Also, if $\alpha_1, \ldots, \alpha_{n_1}$ are the roots of $f_1(x)$, and $\alpha'_1, \ldots, \alpha'_{n_2}$ are the roots of $f_2(x)$, then

$$
\begin{aligned}
(\mathbb{B}/\mathbb{F}) &= \big(\mathbb{F}(\alpha_1, \ldots, \alpha_{n_1}, \alpha'_1, \ldots, \alpha'_{n_2})/\mathbb{F}\big) \\
&= \big(\mathbb{F}(\alpha_1, \ldots, \alpha_{n_1}, \alpha'_1, \ldots, \alpha'_{n_2})/\mathbb{F}(\alpha_1, \ldots, \alpha_{n_1})\big) \\
&\quad \big(\mathbb{F}(\alpha_1, \ldots, \alpha_{n_1}/\mathbb{F})\big) \\
&\leq (n_2!)\,(n_1!).
\end{aligned}
$$

The general case follows by induction on k.   □

The careful reader will note that in this section, we started out not only with a field $\mathbb{F}$ but also with a larger field (i.e., an extension) $\mathbb{E}$ in which some polynomial $f(x) \in \mathbb{F}[x]$ had some (or all) roots. But how do we know some such field $\mathbb{E}$ exists? In particular, how do we know that an arbitrary polynomial $f(x) \in \mathbb{F}[x]$ *has* a splitting field?

If you think about it, there was one situation where we started just with the field $\mathbb{F}$. It was the situation of Theorem 4.2.4, where we began with an irreducible polynomial $f(x) \in \mathbb{F}[x]$ and produced a field where $f(x)$ had a single root.

So we will now apply that result inductively to show that every polynomial has a splitting field.

Theorem 4.5.13. *Let $\mathbb{F}$ be a field and let $f(x) \in \mathbb{F}[x]$ be an arbitrary polynomial. Then $f(x)$ has a splitting field $\mathbb{E}$.*

Proof. Let $f(x)$ have degree n. We prove the theorem by induction on n.

If $n = 1$, then $f(x) = a_1x + a_0$ has root $\alpha = -a_0/a_1$, and $\alpha \in \mathbb{F}$, so $f(x)$ has splitting field $\mathbb{F}$.

Now suppose the theorem is true for all polynomials of degree $n - 1$, and all fields. Let $f(x)$ have degree n.

By Theorem 4.2.4 (Kronecker's theorem) there is a field $\mathbb{E}_1$ in which $f(x)$ has a root α_1, so that $x - \alpha_1$ is a factor of $f(x)$; moreover, $\mathbb{E}_1 = \mathbb{F}(\alpha_1)$. Write $f(x) = (x - \alpha_1)g(x)$, so that $g(x) \in \mathbb{E}_1[x]$ is of degree $n - 1$. Then by the inductive hypothesis $g(x)$ has a splitting field $\mathbb{E}$, i.e., $\mathbb{E} = \mathbb{E}_1(\alpha_2, \ldots, \alpha_n)$ where $\alpha_2, \ldots, \alpha_n$ are the (not necessarily district) roots of $g(x)$ in $\mathbb{E}$. But then

$$\mathbb{E} = \mathbb{E}_1(\alpha_2, \ldots, \alpha_n) = \mathbb{F}(\alpha_1)(\alpha_2, \ldots, \alpha_n) = \mathbb{F}(\alpha_1, \ldots, \alpha_n)$$

is a splitting field for $f(x)$ over $\mathbb{F}$.

Then by induction we are done. $\square$

4.6 Examples of field extensions

In this section, we present a number of examples of field extensions.

Example 4.6.1. The field $\mathbb{Q}(\sqrt{2})$. This is a familiar field, obtained by adjoining the single element $\sqrt{2}$ to $\mathbb{Q}$. This element is algebraic over $\mathbb{Q}$, with minimal polynomial $m_{\sqrt{2}}(x) = x^2 - 2$. $\mathbb{Q}(\sqrt{2})$ is an extension of $\mathbb{Q}$ of degree 2, and as a $\mathbb{Q}$-vector space, $\mathbb{Q}(\sqrt{2})$ has basis $\{1, \sqrt{2}\}$. ◊

Example 4.6.2. The field $\mathbb{Q}(\sqrt{2}, \sqrt{3})$, obtained by adjoining the two elements $\sqrt{2}$ and $\sqrt{3}$ to $\mathbb{Q}$. We claim that this field is an extension of $\mathbb{Q}$ of degree 4. To see this, note that $\sqrt{2}$ is algebraic over $\mathbb{Q}$ with minimal polynomial $m_{\sqrt{2}}(x) = x^2 - 2$, and $\sqrt{3}$ is algebraic over $\mathbb{Q}$ with minimal polynomial $m_{\sqrt{3}}(x) = x^2 - 3$. Thus $\sqrt{2}$ and $\sqrt{3}$ both have degree 2 over $\mathbb{Q}$, so we see from Corollary 4.3.12 that $(\mathbb{Q}(\sqrt{2}, \sqrt{3})/\mathbb{Q}) \le 2 \cdot 2 = 4$.

We also know from Corollary 4.3.6 that $(\mathbb{Q}(\sqrt{2}, \sqrt{3})/\mathbb{Q})$ is divisible by $(\mathbb{Q}(\sqrt{2}/\mathbb{Q}) = 2$. Hence, $(\mathbb{Q}(\sqrt{2}, \sqrt{3})/\mathbb{Q}) = 2$ or 4.

But $(\mathbb{Q}(\sqrt{2}, \sqrt{3})/\mathbb{Q}) = (\mathbb{Q}(\sqrt{2}, \sqrt{3})/\mathbb{Q}(\sqrt{2})) (\mathbb{Q}(\sqrt{2})/\mathbb{Q})$ and $(\mathbb{Q}(\sqrt{2})/\mathbb{Q}) = 2$, so we just need to show that $(\mathbb{Q}(\sqrt{2}, \sqrt{3})/\mathbb{Q}(\sqrt{2})) = 2$. Now $\mathbb{Q}(\sqrt{2}, \sqrt{3}) = (\mathbb{Q}(\sqrt{2})(\sqrt{3}))$ so to show that $(\mathbb{Q}(\sqrt{2}, \sqrt{3})/\mathbb{Q}) \ne 1$, in which case it must be 2, we need only show that $\sqrt{3} \notin \mathbb{Q}(\sqrt{2})$.

We prove this by contradiction. Suppose that $\sqrt{3} \in \mathbb{Q}(\sqrt{2})$, so that $\sqrt{3} = a + b\sqrt{2}$ for some $a, b \in \mathbb{Q}$. Then $3 = (a + b\sqrt{2})^2 = (a^2 + 2b^2) + 2ab\sqrt{2}$, so $2ab = 0$, and hence $a = 0$, $b = 0$. If $a = 0$ we obtain the equation $3 = 2b^2$, which has no solution with $b \in \mathbb{Q}$, and if $b = 0$ we obtain the equation $3 = a^2$, which has no solution with $a \in \mathbb{Q}$, so this is impossible.

Also, $\mathbb{Q}(\sqrt{2})$ has basis $\{1, \sqrt{2}\}$ as a $\mathbb{Q}$-vector space, and $\mathbb{Q}(\sqrt{3})$ has basis $\{1, \sqrt{3}\}$ as a $\mathbb{Q}$-vector space, so it follows from Lemma 4.3.11 that $\mathbb{Q}(\sqrt{2}, \sqrt{3})$ has basis $\{1 \cdot 1, \sqrt{2} \cdot 1, 1 \cdot \sqrt{3}, \sqrt{2} \cdot \sqrt{3}\} = \{1, \sqrt{2}, \sqrt{3}, \sqrt{6}\}$ as a $\mathbb{Q}$-vector space.

Finally, we observe that $\mathbb{Q}(\sqrt{2}), \mathbb{Q}(\sqrt{3})$, and also $\mathbb{Q}(\sqrt{6})$ are intermediate fields between $\mathbb{Q}$ and $\mathbb{Q}(\sqrt{2}, \sqrt{3})$. ◊

We now present a family of examples that generalizes these two.

Example 4.6.3. Let $T = \{a_1, \ldots, a_t\}$ be a set of integers. For a subset S of T, let P_S be the product of the elements of S. (If $S = \phi$, let $P_S = 1$). Suppose that T has the property that for no nonempty subset S of T is P_S a square. For example, this will be the case if the elements of T are pairwise relatively prime nonsquares and at most one of them is the negative of a square.

Let $\mathbb{E} = \mathbb{Q}\left(\sqrt{a_1}, \ldots, \sqrt{a_t}\right)$. We claim that $\mathbb{E}$ is an extension of $\mathbb{Q}$ of degree 2^t. We prove this by induction on t.

For $t = 0$ this is trivial and we know this for $t = 1$. Assume it is true for any set of $t - 1$ elements as above and suppose that T has t elements. Set $\mathbb{D} = \left(\sqrt{a_1}, \ldots, \sqrt{a_{t-1}}\right)$. Then $(\mathbb{E}/\mathbb{Q}) = (\mathbb{E}/\mathbb{D})(\mathbb{D}/\mathbb{Q})$ and by the inductive hypothesis $(\mathbb{D}/\mathbb{Q}) = 2^{t-1}$. Thus, we need to show $(\mathbb{E}/\mathbb{D}) = 2$. Now $\mathbb{E} = \mathbb{D}(\sqrt{a_t})$ so to show this we need only show that $\sqrt{a_t} \notin \mathbb{D}$.

We prove this by contradiction. Suppose $\sqrt{a_t} \in \mathbb{D}$. Then we may write $\sqrt{a_t} = x + y\sqrt{a_{t-1}}$ with $x, y \in \mathbb{B} = \mathbb{Q}\left(\sqrt{a_1}, \ldots, \sqrt{a_{t-2}}\right)$. Squaring, we obtain $a_t = (x^2 + a_{t-1}y^2) + 2xy\sqrt{a_{t-1}}$. Now $a_t \in \mathbb{Q} \subseteq \mathbb{B}$, $x^2 + a_{t-1}y^2 \in \mathbb{B}$, and $2xy \in \mathbb{B}$, so by the inductive hypothesis, which implies that $\sqrt{a_{t-1}} \notin \mathbb{B}$, we must have $2xy = 0$. Then either $x = 0$, in which case $a_t = y^2 a_{t-1}$, $a_t a_{t-1} = y^2 a_{t-1}^2 = (ya_{t-1})^2$ which is impossible by our condition on T, or $y = 0$ in which case $\sqrt{a_t} = x \in \mathbb{B}$, contradicting the inductive hypothesis (applied to the field $\mathbb{Q}\left(\sqrt{a_1}, \ldots, \sqrt{a_{t-2}}, \sqrt{a_t}\right)$).

We may also argue inductively that $\mathbb{D}$ has basis $\{\sqrt{P_S} \mid S$ a subset of $T'\}$, where $T' = \{a_1, \ldots, a_{t-1}\}$, and $\mathbb{Q}(\sqrt{a_t})$ has basis $\{1, \sqrt{a_t}\}$, so we see that $\mathbb{E}$ has basis $\{\sqrt{P_S} \mid S$ a subset of $T\}$.

We note that $\mathbb{E}$ is a splitting field of the polynomial $p(x) = (x^2 - a_1) \cdots (x^2 - a_t) \in \mathbb{Q}[x]$.

Finally, there are many fields intermediate between $\mathbb{Q}$ and $\mathbb{E}$. For example, we have $2^t - 1$ quadratic extensions of $\mathbb{Q}$ contained in $\mathbb{E}$, the fields $\mathbb{Q}(\sqrt{P_S})$ for every nonempty subset S of T. $\Diamond$

We now turn over attention to a different family of examples.

We know that the complex number 1 has n complex nth roots. If we set $\zeta_n = \exp(2\pi i/n)$, they are given by ζ_n^k, $k = 0, \ldots, n - 1$. We thus see that $x^n - 1 = (x - 1)(x - \zeta_n) \cdots (x - \zeta_n^{n-1})$.

Definition 4.6.4. A complex number ζ is a *primitive* nth root of 1 if ζ is an nth root of 1 but ζ is not an mth root of 1 for any $m < n$, or, equivalently, for any m properly dividing n. The nth *cyclotomic polynomial* $\Phi_n(x)$ is the polynomial

$$\prod(x - \zeta)$$

where the product is taken over the primitive nth roots of 1. $\Diamond$

We will be investigating the polynomials $\Phi_n(x)$ in general later on. Here we restrict our attention to the case $n = p$ is a prime.

For p a prime, a pth root ζ of 1 is either $\zeta = 1$ or ζ a primitive pth root of 1. Thus, we see $x^p - 1 = (x-1)\Phi_p(x)$ and hence $\Phi_p(x) = (x^p - 1)/(x-1) = x^{p-1} + x^{p-2} + \cdots + x + 1$.

Lemma 4.6.5. *For p a prime, the pth cyclotomic polynomial $\Phi_p(x)$ is irreducible in $\mathbb{Q}[x]$.*

Proof.   Since $\Phi_p(x)$ is a monic polynomial with integer coefficients, we know that it is irreducible in $\mathbb{Q}[x]$ if and only if it is irreducible in $\mathbb{Z}[x]$.

We show that $\Phi_p(x)$ is irreducible by using a well-known trick. The polynomial $\Phi_p(x)$ is irreducible if and only if the polynomial $\Phi_p(x+1)$ is irreducible. But, by the binomial theorem

$$\Phi_p(x+1) = \frac{(x+1)^p - 1}{(x+1) - 1} = \frac{(x+1)^p - 1}{x} = \sum_{k=1}^{p} \binom{p}{k} x^{k-1}$$

Then $\binom{p}{p} = 1$, $\binom{p}{k}$ is divisible by p for $1 \leq k \leq p$, and $\binom{p}{1} = p$, so we see that $\Phi_p(x+1)$ is irreducible by Eisenstein's Criterion (Lemma 3.11.13).   □

Example 4.6.6. Let $\mathbb{E}$ be the subfield of $\mathbb{C}$ that is the splitting field of $\Phi_p(x)$ over $\mathbb{Q}$. Then $\mathbb{E} = \mathbb{Q}(\zeta_p)$ and $\mathbb{E}$ is an extension of $\mathbb{Q}$ of degree $p-1$. (In fact, $\mathbb{E} = \mathbb{Q}(\zeta_p^k)$ for any k with $1 \leq k \leq p-1$.) As a $\mathbb{Q}$-vector space $\mathbb{E}$ has basis $\{1, \zeta_p, \ldots, \zeta_p^{p-2}\}$. Note that $0 = \Phi_p(\zeta_p) = \zeta_p^{p-1} + \cdots + \zeta_p + 1 = 0$ so we have that $\zeta_p + \cdots + \zeta_p^{p-1} = -1$. Then we see that $\mathbb{E}$ also has basis $\{\zeta_p, \ldots, \zeta_p^{p-1}\}$ as a $\mathbb{Q}$-vector space.   ◊

We now look at a couple of special cases of cyclotomic fields.

Example 4.6.7. Let $p = 3$. The field $\mathbb{Q}(\zeta_3)$ is an extension of $\mathbb{Q}$ of degree 2. Now $\Phi_3(x) = x^2 + x + 1$ is a quadratic, so we may find its roots by the quadratic formula. They are $(-1 \pm \sqrt{-3})/2$. Thus, $\zeta_3 = (-1 + \sqrt{-3})/2$ (where we choose the sign so that $\sqrt{-3}$ has positive imaginary part) and then $\zeta_3^2 = (-1 - \sqrt{-3})/2$ is the other root of this polynomial. Thus, we see $\mathbb{Q}(\zeta_3) = \mathbb{Q}(\sqrt{-3})$.   ◊

Example 4.6.8. Let $p = 5$. The field $\mathbb{Q}(\zeta_5)$ is an extension of $\mathbb{Q}$ of degree 4. We will find ζ_5 explicitly. To this end, consider $\mathbb{B} = \mathbb{Q}(\zeta_5) \cap \mathbb{R}$. Then $4 = (\mathbb{Q}(\zeta_5)/\mathbb{Q}) = (\mathbb{Q}(\zeta_5)/\mathbb{B})(\mathbb{B}/\mathbb{Q})$, and so either $(\mathbb{B}/\mathbb{Q}) = 1$, in which case $\mathbb{B} = \mathbb{Q}$, or $(\mathbb{B}/\mathbb{Q}) = 2$, in which case we also have $(\mathbb{Q}(\zeta_5)/\mathbb{B}) = 2$. We will show the latter case occurs.

To that end, note that $\zeta_5^4 = \bar{\zeta}_5$, so $\theta = \zeta_5 + \zeta_5^4 \in \mathbb{R}$. We find θ as follows: We compute $\theta^2 = (\zeta_5 + \zeta_5^4)^2 = \zeta_5^2 + \zeta_5^3 + 2$. Now $1 + \zeta_5 + \zeta_5^2 + \zeta_5^3 + \zeta_5^4 = 0$, so $\zeta_5^2 + \zeta_5^3 = -1 - \zeta - \zeta_4 = -1 - \theta$.

Hence, $\theta^2 = (-1 - \theta) + 2$, or $\theta^2 + \theta - 1 = 0$. In other words, θ is a root of the quadratic polynomial $x^2 + x - 1$. From the quadratic formula we see that this polynomial has roots $(-1 \pm \sqrt{5})/2$, so $\theta = (-1 + \sqrt{5})/2$. (Since ζ and ζ^4 have positive real part, their sum is a positive real number.) You can easily check that $\theta' = \zeta_5^2 + \zeta_5^3$ is also a root of this polynomial, so $\theta' = (-1 - \sqrt{5})/2$. Thus we see that $\mathbb{Q}(\theta) = \mathbb{Q}(\theta') = \mathbb{Q}(\sqrt{5})$.

Now $\zeta_5^4 = \zeta_5^{-1}$ so $\theta = \zeta_5 + \zeta_5^{-1}$ which gives the equation $\zeta_5^2 - \theta\zeta_5 + 1 = 0$. In other words, ζ_5 is a root of the quadratic $x^2 - \theta x + 1 \in \mathbb{B}[x]$ (and the other root of this quadratic is ζ_5^2). Again we apply the quadratic formula to find that this polynomial has roots $[(-1 + \sqrt{5}) \pm \sqrt{-5 - 2\sqrt{5}}]/2$.

Since ζ_5 has positive imaginary part, we see that

$$\zeta_5 = \frac{(-1 + \sqrt{5}) + \sqrt{-5 - 2\sqrt{5}}}{2}.$$

We observe, as expected, that $(\mathbb{Q}(\zeta_5)/\mathbb{B}) = 2$, as

$$\mathbb{Q}(\zeta_5) = \mathbb{B}\left(\sqrt{-5 - 2\sqrt{5}}\right) \text{ and } -5 - 2\sqrt{5} \in \mathbb{B}. \qquad \Diamond$$

We turn to a third family of examples. To prepare for them, we have the following result, which we state in complete generality.

Lemma 4.6.9 (Abel). *Let $\mathbb{F}$ be a field and let $a \in \mathbb{F}$. If p is a prime and a is not a pth power in $\mathbb{F}$, then the polynomial $x^p - a$ is irreducible in $\mathbb{F}[x]$.*

Proof. We shall show that if $x^p - a$ is not irreducible, then a is a pth power in $\mathbb{F}$. So suppose $x^p - a = f(x)g(x)$ in $\mathbb{F}[x]$, where $f(x)$ has degree d and $g(x)$ has degree $p - d$, for some d with $1 \leq d \leq p - 1$.

Let $\mathbb{E}$ be a splitting field of the polynomial $x^p - a$. Then we see that $\mathbb{E} = \mathbb{F}(\zeta, \alpha)$ where $\zeta^p = 1$ and $\alpha^p = a$, so $x^p - a = (x - \alpha)(x - \zeta\alpha) \cdots (x - \zeta^{p-1}\alpha)$.

(Although this is not strictly necessary for the proof, we should observe that if char $(\mathbb{F}) \neq p$, then $\zeta \neq 1$, as if $\zeta = 1$, then we would have $x^p - 1 = (x-1)^p = x^p - px^{p-1} \cdots$, which is not the case, as $p \neq 0$ in $\mathbb{F}$. On the other hand, if char$(\mathbb{F}) = p$ then $\zeta = 1$, as in this case we do have $x^p - 1 = (x-1)^p$ by Lemma 4.1.14.)

Thus,

$$f(x)g(x) = x^p - a = (x - \alpha)(x - \zeta\alpha) \cdots (x - \zeta^{p-1}\alpha).$$

But we have unique factorization in $\mathbb{E}[x]$, so $f(x)$ must be the product of d of these terms and $g(x)$ must be the product of $p - d$ of these terms. But the constant term of $f(x)$ is $\pm \zeta^k \alpha^d$ for some integer k, so $\beta = \zeta^k \alpha^d \in \mathbb{F}$. Now d and p are relatively prime, so there is a positive integer e with $de \equiv 1 \pmod{p}$, i.e., $de = mp + 1$, so

$$\beta^e = (\zeta^k \alpha^d)^e = \zeta^{ke} \alpha^{de} = \zeta^{ke} \alpha^{mp+1} = \zeta^{ke} a^m \alpha \in \mathbb{F}$$

and then $\gamma = \zeta^{ke} \alpha \in \mathbb{F}$. But

$$\gamma^p = (\zeta^{ke} \alpha)^p = \zeta^{pke} \alpha^p = (1)^{ke} \alpha^p = a \in \mathbb{F}$$

and a is a pth power in $\mathbb{F}$.   □

Example 4.6.10. Let p be a prime and let $n \in \mathbb{Q}$ be a rational number that is not a pth power in $\mathbb{Q}$. Let $f(x) = x^p - n \in \mathbb{Q}[x]$ and let $\mathbb{E} \subseteq \mathbb{C}$ be the splitting field of $f(x)$. Then $\mathbb{E} = \mathbb{Q}(\zeta_p, \sqrt[p]{n})$. (Note that $f(x)$ splits in this field, and that any field in which $f(x)$ splits must contain ζ_p and $\sqrt[p]{n}$, so this field is indeed the splitting field).

Now $\mathbb{E}$ is the composite $\mathbb{E} = \mathbb{Q}(\zeta_p)\mathbb{Q}(\sqrt[p]{n})$. Now ζ_p is a root of the cyclotomic polynomial $\Phi_p(x)$, of degree $p - 1$, and we showed in Lemma 4.6.5 that $\Phi_p(x)$ is irreducible. Hence, $(\mathbb{Q}(\zeta_p)/\mathbb{Q}) = p - 1$. Also, $\sqrt[p]{n}$ is a root of the polynomial $x^p - n$, of degree n, and we showed in Lemma 4.6.9 that this polynomial is irreducible. Hence $(\mathbb{Q}(\sqrt[p]{n})/\mathbb{Q}) = p$. Then $(\mathbb{E}/\mathbb{Q}) = p(p - 1)$ by Corollary 4.3.13. We know that $\mathbb{Q}(\zeta_p)$ has basis $\{1, \zeta_p, \ldots, \zeta_p^{p-2}\}$ (or basis $\{\zeta_p, \ldots, \zeta_p^{p-1}\}$) as a $\mathbb{Q}$-vector space, and that $\mathbb{Q}(\sqrt[p]{n})$ has basis $\{1, \sqrt[p]{n}, (\sqrt[p]{n})^2, \ldots, (\sqrt[p]{n})^{p-1}\}$ as a $\mathbb{Q}$-vector space, so $\mathbb{E}$ has basis $\{\zeta^i(\sqrt[p]{n})^j \mid i = 0, \ldots, p - 2, j = 0, \ldots, p - 1\}$ or $\{\zeta^i(\sqrt[p]{n})^j \mid i = 1, \ldots, p - 1, j = 0, \ldots, p - 1\}$.   ◇

Now we return to reconsider some of our earlier examples.

Example 4.6.11. Let us consider the polynomial $f(x) = x^3 + 7x + 1 \in \mathbb{Q}[x]$, as we did in Example 4.4.10. From elementary calculus we see that $f(x)$ has exactly one real root, which we shall call α_1. Let $\mathbb{B} = \mathbb{Q}(\alpha_1)$. Then $(\mathbb{B}/\mathbb{Q}) = 3$, since $f(x)$ is irreducible in $\mathbb{Q}[x]$, as we have already seen. But we note that $\mathbb{B}$ is *not* a splitting field of $f(x)$ over $\mathbb{Q}$, as $\mathbb{B} \subset \mathbb{R}$ and, as we have just remarked, only one of the roots of $f(x)$ is in $\mathbb{R}$. Let $\mathbb{E} \supseteq \mathbb{B}$ be a splitting field of $f(x)$, so that $\mathbb{E} = \mathbb{Q}(\alpha_1, \alpha_2, \alpha_3)$ where α_1, α_2, and α_3 are the roots of $f(x)$ in $\mathbb{E}$. Then $(\mathbb{E}/\mathbb{Q}) \leq 3! = 6$ by Theorem 4.5.10, and $(\mathbb{E}/\mathbb{Q})$ is divisible by $(\mathbb{B}/\mathbb{Q})$, so $(\mathbb{E}/\mathbb{Q}) = 6$. But then $(\mathbb{E}/\mathbb{B}) = 2$. Now, setting $\alpha = \alpha_1$, we see that $f(x)$ has a factor of $x - \alpha$ in $\mathbb{B}[x]$, i.e., $f(x) = (x - \alpha)g(x) \in \mathbb{B}[x]$ for some polynomial $g(x) \in \mathbb{B}[x]$, and simply dividing polynomials we see that $g(x) = x^2 + \alpha x + (7 + \alpha^2) \subset \mathbb{B}[x]$, and $g(x)$ is an irreducible quadratic in $\mathbb{B}[x]$. Indeed, we would obtain the exact same results, and formula, if we set $\alpha = \alpha_2$, or $\alpha = \alpha_3$. $\Diamond$

Example 4.6.12. Let us consider the polynomial $f(x) = x^4 + nx^2 + 1 \in \mathbb{Q}[x]$, for n an integer, $n \neq \pm 2$, generalizing Example 4.4.11. Let α be a root of this polynomial, and let $\mathbb{B} = \mathbb{Q}(\alpha)$. Then $(\mathbb{B}/\mathbb{Q}) = 4$, since $f(x)$ is irreducible in $\mathbb{Q}[x]$, as we will show below. Note that, since all of the powers of x in $f(x)$ are even, $-\alpha$ is also a root of $f(x)$. Also notice that since this polynomial is palindromic, i.e., the coefficients are the same left-to-right as they are right-to-left $(1, 0, n, 0, 1)$, $1/\alpha$ is a root, and then $-1/\alpha$ is a root as well. Thus, we see that $\mathbb{B}$ *is* a splitting field of $f(x)$, that $f(x) = (x - \alpha)(x - (-\alpha))(x - 1/\alpha)(x - (-1/\alpha)) \in \mathbb{B}[x]$, and that if $\alpha_1, \ldots, \alpha_4$ are the roots of $f(x)$ in a splitting field $\mathbb{E} = \mathbb{Q}(\alpha_1, \alpha_2, \alpha_3, \alpha_4)$, then in fact $\mathbb{E} = \mathbb{Q}(\alpha, -\alpha, 1/\alpha, -1/\alpha) = \mathbb{Q}(\alpha) = \mathbb{B}$.

We now show our claim that $f(x)$ is irreducible in $\mathbb{Q}[x]$. The only possible rational roots of $f(x)$ are ± 1, by Lemma 3.11.12, and neither of these is a root, so $f(x)$ cannot have a linear factor in $\mathbb{Q}[x]$. Thus if $f(x)$ is not irreducible, it must be a product of two quadratics in $\mathbb{Q}[x]$. We cannot have a factor $(x - \alpha)(x - (-\alpha)) = x^2 - \alpha^2$ as then we would have to have $\alpha^2 = \pm 1$, which we have excluded. We cannot have a factor $(x - \alpha)(x - 1/\alpha) = x^2 - (\alpha + 1/\alpha) + 1$, as then $\alpha + 1/\alpha = k$ for some integer k, and then $\alpha^2 - k\alpha + 1 = 0$, with roots $(k \pm \sqrt{k^2 - 4})/2$, and if $k \neq \pm 2$, $k^2 - 4$ is never a perfect square. (If $k = \pm 2$, then $\alpha = \pm 1$ again.) Similarly, we cannot have a factor

$(x - \alpha)(x - (-1/\alpha)) = x^2 - (\alpha - 1/\alpha)x - 1$ as this gives $\alpha - 1/\alpha = k$, $x^2 - k\alpha - 1 = 0$ with roots $(k \pm \sqrt{k^2 + 4})/2$, and if $k \neq 0$, $k^2 + 4$ is never a perfect square. (If $k = 0$, then $\alpha = \pm 1$ again.)

Finally, in the cases we have excluded, $f(x)$ is reducible:

$$x^4 + 2x^2 + 1 = (x^2 + 1)^2 \text{ and } x^4 - 2x^2 + 1 = (x^2 - 1)^2 \qquad \diamond$$

Let us now see how to obtain some infinite algebraic extensions.

Example 4.6.13. Let p_1, p_2, $p_3, \ldots$, be distinct primes, and let $\mathbb{E} = \mathbb{Q}(\sqrt{p_1}, \sqrt{p_2}, \sqrt{p_3}, \ldots)$. Then for any k, $\mathbb{E} \supset \mathbb{B}_k = \mathbb{Q}(\sqrt{p_1}, \ldots, \sqrt{p_k})$ and by Example 4.6.3, $(\mathbb{B}_k/\mathbb{Q}) = 2^k$. Thus $(\mathbb{E}/\mathbb{Q}) \geq 2^k$ for every k, so $(\mathbb{E}/\mathbb{Q}) = \infty$. But note also that, considering the bases we wrote down in that example, every element α of $\mathbb{E}$ must have an expression that involves only finitely many of $\sqrt{p_1}$, $\sqrt{p_2}, \ldots$, so is in the finite extension $\mathbb{B}_k$ for some k, and hence is algebraic. $\qquad \diamond$

Example 4.6.14. Let p_1, p_2, $p_3, \ldots$, be distinct primes, and let $\mathbb{B} = \mathbb{Q}(\sqrt[p_1]{2}, \sqrt[p_2]{2}, \sqrt[p_3]{2}, \ldots)$. Then for any k, $\mathbb{E} \supset \mathbb{B}_k = \mathbb{Q}(\sqrt[p_1]{2}, \ldots, \sqrt[p_k]{2})$. By Lemma 4.6.9, $\mathbb{Q}(\sqrt[p_i]{2}/\mathbb{Q}) = p_i$ for each i, so by Corollary 4.3.13, $(\mathbb{B}_k/\mathbb{Q}) = p_1, \ldots, p_k$. Thus $(\mathbb{E}/\mathbb{Q}) \geq p_1, \ldots, p_k$ for every k, so again $(\mathbb{E}/\mathbb{Q}) = \infty$. Also, by the same argument as in the preceding example, every element α of $\mathbb{E}$ is algebraic. $\qquad \diamond$

Now let us return to an example of finite fields.

Example 4.6.15. Let us consider the polynomial $f(x) = x^3 + x + 6 \in \mathbb{F}_7[x]$, as we did in Example 4.4.12. Let α be a root of this polynomial and consider $\mathbb{F}_7[\alpha]$. Then $(\mathbb{F}_7(\alpha)/\mathbb{F}_7) = 3$, since $f(x)$ is irreducible in, $\mathbb{F}_7[x]$, as we have seen. Once again we may ask whether $\mathbb{F}_7(\alpha)$ is a splitting field of $f(x)$. The answer turns out to be yes, as $f(x)$ has the roots $\alpha_1 = \alpha$, $\alpha_2 = 1 + 5\alpha^2$, and $\alpha_3 = 6 + 6\alpha + 2\alpha^2$, all in $\mathbb{F}_7(\alpha)$. Thus $f(x) = (x - \alpha_1)(x - \alpha_2)(x - \alpha_3)$ in $\mathbb{F}_7(\alpha)$, and $\mathbb{F}_7(\alpha_1, \alpha_2, \alpha_3) = \mathbb{F}_7(\alpha)$. $\qquad \diamond$

Remark 4.6.16. Once you are given α_2 and α_3, you may check by direct substitution that $f(\alpha_2) = 0$ and $f(\alpha_3) = 0$. But of course you may-indeed you should-ask, how we know that $f(x)$ splits in $\mathbb{F}_7(\alpha)$, and, given that, how to find the other roots of $f(x)$ in $\mathbb{F}_7(\alpha)$. We will be answering these questions below. $\qquad \diamond$

4.7   Isomorphisms of fields

Before getting to work in this section, let us reflect on some of our examples.

Let us begin by considering $\mathbb{E} = \mathbb{Q}[x]/\langle p(x) \rangle$ where $p(x) \in \mathbb{Q}[x]$ is the irreducible polynomial $p(x) = x^2 - 2$. Then, as we have seen, $\mathbb{E} = \mathbb{Q}(\alpha)$, i.e., $\mathbb{E}$ is obtained from $\mathbb{Q}$ by adjoining an element α with $\alpha^2 = 2$. We have also looked at the field $\mathbb{Q}(\sqrt{2})$, obtained from $\mathbb{Q}$ by adjoining $\sqrt{2}$, an element with $(\sqrt{2})^2 = 2$. But we could also have obtained $\mathbb{Q}(\sqrt{2})$ by adjoining a different element, $-\sqrt{2}$, which also satisfies $(-\sqrt{2})^2 = 2$. Now somehow these should be algebraically "all the same", that is, somehow, from a purely algebraic point of view, we should not be able to distinguish between them.

Similarly we could consider $\mathbb{E} = \mathbb{Q}[x]/\langle p(x) \rangle$ where $p(x) \in \mathbb{Q}[x]$ is the irreducible polynomial $p(x) = x^3 - 2$. Then $\mathbb{E} = \mathbb{Q}(\alpha)$ is obtained from $\mathbb{Q}$ by adjoining an element α with $\alpha^3 = 2$. But we also have the field $\mathbb{Q}(\sqrt[3]{2})$, obtained by adjoining $\sqrt[3]{2}$, an element with $(\sqrt[3]{2})^3 = 2$, the field $\mathbb{Q}(\zeta_3 \sqrt[3]{2})$, obtained by adjoining $\zeta_3 \sqrt[3]{2}$, where $(\zeta_3 \sqrt[3]{2})^3 = 2$, and the field $\mathbb{Q}(\zeta_3^2 \sqrt[3]{2})$, obtained by adjoining $\zeta_3^2 \sqrt[3]{2}$, where also $(\zeta_3^2 \sqrt[3]{2})^3 = 2$, and again they should somehow be algebraically "all the same".

Moreover, we have the notion of a splitting field of a polynomial $f(x) \in \mathbb{F}[x]$, and we constructed such a splitting field $\mathbb{E}$. But again, somehow all splitting fields of $f(x)$ should somehow be algebraically "the same".

This intuition turns out to absolutely correct, once we make the notion of "the same" precise. And the precise way to say that is that "the same" means isomorphic. Thus we are led to study the notion of isomorphism of fields, which we first introduced in Definition 4.1.9.

It turns out that the notion of field isomorphism is a crucially important one, one that lies at the very heart of Galois theory, which is our ultimate goal in this chapter.

So, with a view toward future developments, we will be stating (and proving) our results in a more general form than is just needed to make these intuitions precise.

Before proceeding, there is a point we need to stress to the reader, one that appears throughout mathematics, and one whose importance cannot be overstated. Suppose we have two objects, of whatever sort, X and Y, and we have an isomorphism, of whatever sort,

$f: X \to Y$. Then X and Y are indeed isomorphic, but that is *not* (repeat, *not*) (repeat again, *not*) to say they are identical. True, the isomorphism f may give a way of identifying X with Y, but that identification *depends on* the isomorphism f, and different isomorphisms will give different identifications.

Before constructing field isomorphisms, let us see an essential restriction on how they must behave.

Lemma 4.7.1. *Let $\mathbb{F}_1$ and $\mathbb{F}_2$ be fields and let $\varphi_0 \colon \mathbb{F}_1 \to \mathbb{F}_2$ be a field isomorphism. Let $\mathbb{E}_1$ be an extension of $\mathbb{F}_1$ and let $\mathbb{E}_2$ be an extension of $\mathbb{F}_2$. Suppose that $\varphi \colon \mathbb{E}_1 \to \mathbb{E}_2$ is a field isomorphism extending φ_0. Let β be an arbitrary element of $\mathbb{E}_1$. If $p_1(x) \in \mathbb{F}_1[x]$ is any polynomial having β as a root, and $p_2(x) = \varphi_0(p_1(x)) \in \mathbb{F}_2[x]$, then $\gamma = \varphi(\beta)$ must be a root of the polynomial $p_2(x)$.*

In particular, if $\mathbb{E}_1$ and $\mathbb{E}_2$ are extensions of $\mathbb{F}$, $\varphi \colon \mathbb{E}_1 \to \mathbb{E}_2$ is an isomorphism that is the identity on $\mathbb{F}$ (perhaps $\mathbb{E}_2 = \mathbb{E}_1$ and φ is an automorphism of $\mathbb{E}_1$ that is the identity on $\mathbb{F}$), β is an element of $\mathbb{E}_1$, and $p(x) \in \mathbb{F}[x]$ is a polynomial having β as a root, then $\gamma = \varphi(\beta)$ must be a root of $p(x)$ in $\mathbb{E}_2$.

Proof.   Let $p_1(x) = a_n x^n + \cdots + a_0 \in \mathbb{F}_1[x]$, in which case $p_2(x) = \varphi_0(a_n)x^n + \cdots + \varphi_0(a_0) \in \mathbb{F}_2[x]$. We then have

$$
\begin{aligned}
p_2(\gamma) = p_2(\varphi(\beta)) &= \varphi_0(a_n)(\varphi(\beta))^n + \cdots + \varphi_0(a_0) \\
&= \varphi(a_n)(\varphi(\beta))^n + \cdots + \varphi(a_0) \\
&= \varphi(a_n \beta^n) + \cdots + \varphi(a_0) \\
&= \varphi(a_n \beta^n + \cdots + a_0) \\
&= \varphi(p_1(\beta)) = \varphi(0) = 0. \qquad \square
\end{aligned}
$$

We would like to prove a converse of this lemma. But, thinking about it, we can see two restrictions. If $p_1(x)$ were not irreducible, but instead, for example, $p_1(x) = q_1(x)r_1(x)$ for distinct irreducible polynomials $q_1(x)$ and $r_1(x)$, we could not have an isomorphism taking a root of $q_1(x)$ to a root of $r_1(x)$. Also, if $\mathbb{F}_1(\beta) \subset \mathbb{E}_1$ and $\mathbb{F}_2(\gamma) \subset \mathbb{E}_2$, we could expect to know $\varphi \colon \mathbb{F}_1(\beta) \to \mathbb{F}_2(\gamma)$, but could not expect to say anything at all about φ (not even whether it exists) on all of $\mathbb{E}_1$ or all of $\mathbb{E}_2$. But these restrictions turn out to be the only ones.

Lemma 4.7.2. *Let* $\mathbb{F}_1$ *and* $\mathbb{F}_2$ *be fields and let* $\varphi_0 \colon \mathbb{F}_1 \to \mathbb{F}_2$ *be a field isomorphism. Let* $p_1(x) \in \mathbb{F}_1[x]$ *be an irreducible polynomial and let* $p_2(x) = \varphi_0(p_1(x)) \in \mathbb{F}_2[x]$. *If* β *is any root of* $p_1(x)$ *in an extension field* $\mathbb{E}_1$ *of* $\mathbb{F}_1$, *and* γ *is any root of* $p_2(x)$ *in an extension field* $\mathbb{E}_2$ *of* $\mathbb{F}_2$, *then there is a unique isomorphism* $\varphi \colon \mathbb{F}_1(\beta) \to \mathbb{F}_2(\gamma)$ *extending* φ_0 *with* $\varphi(\beta) = \gamma$.

In particular, if $\mathbb{E}_1$ *and* $\mathbb{E}_2$ *are extensions of* $\mathbb{F}$, *perhaps* $\mathbb{E}_2 = \mathbb{E}_1$, $p(x) \in \mathbb{F}[x]$ *is an irreducible polynomial,* β *is any root of* $p(x)$ *in* $\mathbb{E}_1$ *and* γ *is any root of* $p(x)$ *in* $\mathbb{E}_2$, *then there is a unique isomorphism* $\varphi \colon \mathbb{F}(\beta) \to \mathbb{F}(\gamma)$ *that is the identity on* $\mathbb{F}$ *with* $\varphi(\beta) = \gamma$.

Proof. We have a ring homomorphism $\epsilon_1 \colon \mathbb{F}_1[x] \to \mathbb{F}_1(\beta)$ given by $\epsilon_1(f(x)) = f(\beta)$. Note that $\epsilon_1(x) = \beta$. The homomorphism ϵ_1 is onto with kernel $I_1 = \{f(x) \mid f(\beta) = 0\} = \langle p_1(x) \rangle$, the principal ideal of $\mathbb{F}_1[x]$ generated by $p_1(x)$, and so we obtain an isomorphism $\bar{\epsilon}_1 \colon \mathbb{F}_1[x]/I_1 \to \mathbb{F}_1(\beta)$ with $\bar{\epsilon}_1(x) = \beta$. Similarly, if $I_2 = \langle p_2(x) \rangle$, the principal ideal of $\mathbb{F}_2[x]$ generated by $p_2(x)$, we obtain an isomorphism $\bar{\epsilon}_2 \colon \mathbb{F}_2[x]/I_2 \to \mathbb{F}_2(\gamma)$ with $\bar{\epsilon}_2(x) = \gamma$.

Now the isomorphism $\varphi_0 \colon \mathbb{F}_1 \to \mathbb{F}_2$ extends to an isomorphism $\varphi_0 \colon \mathbb{F}_1[x] \to \mathbb{F}_2[x]$ with $\varphi_0(x) = x$, and if $p_2(x) = \varphi_0(p_1(x))$, this induces an isomorphism $\bar{\varphi}_0 \colon \mathbb{F}_1[x]/I_1 \to \mathbb{F}_2[x]/I_2$.

Then the isomorphism φ is given by $\varphi = \bar{\epsilon}_2(\bar{\varphi}_0)(\bar{\epsilon}_1)^{-1}$, and $\varphi(\beta) = \gamma$.

Furthermore, φ is unique, as it is determined by its values on elements of $\mathbb{F}_1$ and on the element β. $\qquad\square$

Now we turn our attention to isomorphisms, and automorphisms, of splitting fields.

Lemma 4.7.3. *Let* $\mathbb{F}_1$ *and* $\mathbb{F}_2$ *be fields and let* $\varphi_0 \colon \mathbb{F}_1 \to \mathbb{F}_2$ *be an isomorphism. Let* $p_1(x) \in \mathbb{F}_1[x]$ *be an arbitrary polynomial and let* $p_2(x) = \varphi_0(p_1(x)) \in \mathbb{F}_2[x]$. *If* $\mathbb{E}_1$ *is any splitting field of* $p_1(x)$ *and* $\mathbb{E}_2$ *is any splitting field of* $p_2(x)$, *then there is an isomorphism* $\varphi \colon \mathbb{E}_1 \to \mathbb{E}_2$ *extending* φ_0.

In particular, if $\mathbb{F}$ *is a field and* $\mathbb{E}_1$ *and* $\mathbb{E}_2$ *are any two splitting fields of* $p(x) \in \mathbb{F}[x]$, *then* $\mathbb{E}_1$ *and* $\mathbb{E}_2$ *are isomorphic via an isomorphism* φ *that restricts to the identity on* $\mathbb{F}$.

Proof. By induction on $n = \deg p_1(x) = \deg p_2(x)$.

If $n = 1$, $\mathbb{E}_1 = \mathbb{F}_1$, $\mathbb{E}_2 = \mathbb{F}_2$, and we let $\varphi = \varphi_0$.

Assume now the lemma is true for all polynomials of degree less that n, and all fields, and let $p_1(x)$ have degree n.

Let $\beta_1, \ldots, \beta_n$ be the (not necessarily distinct) roots of $p_1(x)$ in $\mathbb{E}_1$, so that $\mathbb{E}_1 = \mathbb{F}_1(\beta_1, \ldots, \beta_n)$. Let $\gamma_1, \ldots, \gamma_n$ be the (not necessarily distinct) roots of $p_2(x)$ in $\mathbb{E}_2$, so that $\mathbb{E}_2 = \mathbb{F}_2(\gamma_1, \ldots, \gamma_n)$.

Now β_1 is a root of some irreducible factor $q_1(x)$ of $p_1(x)$ in $\mathbb{F}_1[x]$. After possible reordering, we may assume that γ_1 is a root of the irreducible factor $q_2(x) = \varphi(q_1(x))$ of $p_2(x)$ in $\mathbb{F}_2[x]$. Then by Lemma 4.7.2 there is an isomorphism $\psi \colon \mathbb{F}_1(\beta_1) \to \mathbb{F}_2(\gamma_1)$ extending φ_0.

Observe that $\mathbb{E}_1 = \mathbb{F}_1(\beta_1, \ldots, \beta_n) = \mathbb{B}_1(\beta_2, \ldots, \beta_n)$, where $\mathbb{B}_1 = \mathbb{F}_1(\beta_1)$, and similarly $\mathbb{E}_2 = \mathbb{F}_2(\gamma_1, \ldots, \gamma_n) = \mathbb{B}_2(\gamma_2, \ldots, \gamma_n)$, where $\mathbb{B}_2 = \mathbb{F}_2(\gamma_1)$. Note that $p_1(x)$ is divisible by $x - \beta_1$ in $\mathbb{B}_1[x]$ and $p_2(x)$ is divisible by $x - \gamma_1$ in $\mathbb{B}_2[x]$. Write $p_1(x) = (x - \beta_1)r_1(x)$ and $p_2(x) = (x - \gamma_1)r_2(x)$. We observe that $r_2(x) = \psi(r_1(x))$, polynomials of degree $n - 1$.

Now $\mathbb{E}_1$ is a splitting field of $r_1(x)$ (regarded as an extension of $\mathbb{B}_1$) and $\mathbb{E}_2$ is a splitting field of $r_2(x)$ (regarded as an extension of $\mathbb{B}_2$). Thus, by the inductive hypothesis, there is an isomorphism $\varphi \colon \mathbb{E}_1 \to \mathbb{E}_2$ extending ψ, and thus extending φ_0 as well. $\square$

We want to emphasize that the isomorphism φ constructed in Lemma 4.7.3 is almost never unique-quite the opposite! We have the following result, which, for simplicity, we state in the most important special case.

Corollary 4.7.4. *Let $p(x) \in \mathbb{F}[x]$ be an irreducible polynomial, and let $\mathbb{E}$ be a splitting field of $p(x)$. Then for any two roots α_1 and α_2 of $p(x)$ in $\mathbb{E}$, there is an automorphism $\varphi \colon \mathbb{E} \to \mathbb{E}$ extending the identity on $\mathbb{F}$ with $\varphi(\alpha_1) = \alpha_2$.*

Proof. We showed this in the proof of Lemma 4.7.3. In the situation here, $\{\beta_1, \ldots, \beta_n\} = \{\gamma_1, \ldots, \gamma_n\}$ but we may choose $\beta_1 = \alpha_1$ and $\gamma_1 = \alpha_2$. $\square$

Remark 4.7.5. The isomorphism φ in Corollary 4.7.4 may or may not be unique. Suppose that $p(x)$ has roots $\alpha_1, \ldots, \alpha_n$ in $\mathbb{E}$. Then $\mathbb{E} = \mathbb{F}(\alpha_1, \ldots, \alpha_n)$. If it happens that $\mathbb{E} = \mathbb{F}(\alpha_1)$, then φ is unique. But otherwise, φ is not, as we also see from the proof of Lemma 4.7.3. $\Diamond$

4.8 The Galois group: Definition and examples

Our goal is the fundamental theorem of Galois theory. But we have come for enough to be able to define the Galois group, to compute it in some cases of field extensions that we have already seen, and to draw some general conclusions about it.

Here is the basic definition.

Definition 4.8.1. Let $\mathbb{E}$ be an extension of the field $\mathbb{F}$. The *Galois group* $\mathrm{Gal}(\mathbb{E}/\mathbb{F})$ is the group

$$\mathrm{Gal}(\mathbb{E}/\mathbb{F}) = \{\text{automorphisms } \varphi \colon \mathbb{E} \to \mathbb{E} \mid \varphi \text{ is the identity on } \mathbb{F}\},$$

a group under composition. ◇

Example 4.8.2. Let $a \in \mathbb{Q}$ not be a perfect square and let $\mathbb{E} = \mathbb{Q}(\sqrt{a})$, an extension of $\mathbb{Q}$ of degree 2. We observe that $\mathbb{E}$ is a splitting field of the irreducible quadratic $p(x) = x^2 - a \in \mathbb{Q}[x]$. By Lemma 4.7.1 we observe that any automorphism φ of $\mathbb{E}$ must have $\varphi(\sqrt{a}) = \pm\sqrt{a}$, and from Lemma 4.7.2 that either of these is possible. Hence

$$G = \mathrm{Gal}(\mathbb{E}/\mathbb{Q}) = \{\sigma_0, \sigma_1\}$$

with $\sigma_0 = \mathrm{id}$ (the identity) and σ_1 determined by $\sigma_1(\sqrt{a}) = -\sqrt{a}$. We observe that G is a group of order 2, isomorphic to $\mathbb{Z}_2$. ◇

Example 4.8.3. Let $T = \{a_1, \ldots, a_t\}$ be a set of integers as in Example 4.6.3 and let $\mathbb{E} = \mathbb{Q}(\sqrt{a_1}, \ldots, \sqrt{a_t})$ as in that example. We saw there that $\mathbb{E}$ is an extension of $\mathbb{Q}$ of degree 2^t. We observe that $\mathbb{E}$ is a splitting field of the polynomial $p(x) = (x^2 - a_1)(x^2 - a_2) \ldots (x^2 - a_t) \in \mathbb{Q}[x]$. By Lemma 4.7.1 we observe that any automorphism φ of $\mathbb{E}$ must have $\varphi(\sqrt{a_i}) = \epsilon_i\sqrt{a_i}, \epsilon_i = \pm 1$, for each i. Thus, we see that $G = \mathrm{Gal}(\mathbb{E}/\mathbb{Q})$ will have order at most 2^t and will have order exactly 2^t if and only if we can choose all these signs independently. We show we can do so, by induction on t.

In case $t = 1$ this is just the previous example. Now suppose it is true for any set $T' = \{a_1, \ldots, a_{t-1}\}$ and consider $T = \{a_1, \ldots, a_t\}$. Let $\mathbb{D} = \mathbb{Q}(\sqrt{a_1}, \ldots, \sqrt{a_{t-1}})$. Then $(\mathbb{D}/\mathbb{Q}) = 2^{t-1}$ and $(\mathbb{E}/\mathbb{Q}) = 2^t$, so $(\mathbb{E}/\mathbb{D}) = 2$. Clearly $\mathbb{E} = \mathbb{D}(\sqrt{a_t})$ so a_t has degree 2 over $\mathbb{D}$, and so the polynomial $x^2 - a_t$ is irreducible in $\mathbb{D}[x]$ (not just in $\mathbb{Q}[x]$). Now let $\epsilon_1 = \pm 1, \ldots, \epsilon_t = \pm 1$ be any collection of signs. By the induction hypothesis there is an isomorphism $\varphi_0 \colon \mathbb{D} \to \mathbb{D}$ with $\varphi_0(\sqrt{a_i}) = \epsilon_i \sqrt{a_i}$ for $i = 1, \ldots, t-1$. But then by Lemma 4.7.2 there is a (unique) isomorphism $\varphi \colon \mathbb{E} \to \mathbb{E}$ extending φ_0 and with $\varphi(\sqrt{a_t}) = \epsilon_t \sqrt{a_t}$ for each choice $\epsilon_t - \pm 1$ or $\epsilon = -1$. Then by induction we are done. Thus, G has order 2^t.

We can concretely describe G. For each $i = 1, \ldots, t$, let $\sigma_i \colon \mathbb{E} \to \mathbb{E}$ be the automorphism with $\sigma_i(\sqrt{a_i}) = -\sqrt{a_i}$ and $\sigma_i(\sqrt{a_j}) = \sqrt{a_j}$ for $j \neq i$. Then G is the group generated by $\{\sigma_1, \ldots, \sigma_t\}$. We observe that $\sigma_i^2 = \mathrm{id}$ for each i, and that $\sigma_i \sigma_j = \sigma_j \sigma_i$ for each i, j, and so G is isomorphic to the group $(\mathbb{Z}_2)^t$. $\Diamond$

We now return to Example 4.6.6 and investigate it more deeply.

Example 4.8.4. Let $\Phi_p(x)$ be the pth cyclotomic polynomial, p a prime, and let $\mathbb{E} \subset \mathbb{C}$ be a splitting field of $\Phi_p(x)$. Then $\mathbb{E} = \mathbb{Q}(\zeta_p)$ and, since $\Phi_p(x)$ is an irreducible polynomial of degree $p-1$, $\mathbb{E}$ is an extension of $\mathbb{Q}$ of degree $p-1$. Now $\Phi_p(x)$ has roots $\zeta_p, \ldots, \zeta_p^{p-1}$ in $\mathbb{E}$. Then, by Lemma 4.7.2 (or Lemma 4.7.3) there is an automorphism σ_k of $\mathbb{E}$ with $\sigma_k(\zeta_p) = \zeta_p^k$ for each $k = 1, \ldots, p-1$, and, since $\mathbb{E} = \mathbb{Q}(\zeta_p)$, such an automorphism is unique. (Compare Remark 4.7.4.) Thus, $G = \mathrm{Gal}(\mathbb{E}/\mathbb{Q})$ is a group of order $p-1$. To further investigate the structure of G, note that

$$\sigma_j \sigma_k(\zeta_p) = \sigma_j(\zeta_p^k) = (\sigma_j(\zeta_p))^k = (\zeta_p^j)^k = \zeta_p^{jk}$$

and so we see that G is isomorphic to the multiplicative group $\mathbb{Z}_p^*$ (a cyclic group of order $p-1$). $\Diamond$

We shall continue to look at examples of Galois groups of extensions of $\mathbb{Q}$, but will do so in a different order than in Section 4.6, in order to work our way up from easier to harder.

Example 4.8.5. Let us look at a splitting field $\mathbb{E}$ of a polynomial $p(x) = x^4 + nx^2 + 1 \in \mathbb{Q}[x]$, n an integer, $n \neq \pm 2$, as in Example 4.6.12. We observed there that $p(x)$ is irreducible. If we let $\mathbb{Q}(\alpha)$ be a field obtained by adjoining a root α of $p(x)$ to $\mathbb{Q}$, we observed that $p(x)$ splits in $\mathbb{Q}(\alpha)[x]$; to be precise

$$p(x) = (x - \alpha)(x - (-\alpha))(x - 1/\alpha)(x - (-1/\alpha)) \text{ in } \mathbb{Q}(\alpha)[x].$$

Thus, $\mathbb{E} = \mathbb{Q}(\alpha, -\alpha, 1/\alpha, -1/\alpha) = \mathbb{Q}(\alpha)$, and $\mathbb{E}$ is an extension of $\mathbb{Q}$ of degree 4. Once again, by Lemma 4.7.2, there are automorphisms

φ of $\mathbb{E}$ with $\varphi(\alpha) = \alpha, \varphi(\alpha) = -\alpha, \varphi(\alpha) = 1/\alpha$, and $\varphi(\alpha) = -1/\alpha$, and once again, since $\mathbb{E} = \mathbb{Q}(\alpha)$, these automorphisms are unique. Thus, $G = \text{Gal}(\mathbb{E}/\mathbb{Q})$ is a group of order 4, and it is easy to check that it is isomorphic to $\mathbb{Z}_2 \oplus \mathbb{Z}_2$. $\Diamond$

Example 4.8.6. Let us consider a splitting field $\mathbb{E}$ of the irreducible polynomial $p(x) = x^3 + 7x + 1 \in \mathbb{Q}[x]$ of Example 4.6.10. As we saw in that example, $p(x)$ has three distinct roots, which we shall call α, β, and γ, in $\mathbb{E}$, and $\mathbb{E}$ is an extension of $\mathbb{Q}$ of degree 6. By Lemma 4.7.1 any $\sigma \in \text{Gal}(\mathbb{E}/\mathbb{Q})$ must permute α, β, and γ.

Let $\mathbb{B} = \mathbb{Q}(\alpha)$. Then $(\mathbb{B}/\mathbb{Q}) = 3$ as $p(x)$ is irreducible. Then $p(x) = (x - \alpha)g(x)$ in $\mathbb{B}[x]$ and $g(x)$ must be an irreducible quadratic in $\mathbb{B}[x]$, as otherwise $g(x)$ would split in $\mathbb{B}[x]$ and we would have $\mathbb{E} = \mathbb{B}$. Then by Lemma 4.7.2 we have an automorphism σ of $\mathbb{E}$ extending the identity σ_0 on $\mathbb{B}$ (where of course $\sigma_0(\alpha) = \alpha$) with $\sigma(\beta) = \beta$ and $\sigma(\gamma) = \gamma$ (of course, this automorphism of $\mathbb{E}$ is just the identity) and an automorphism σ of $\mathbb{E}$ extending σ_0 with $\sigma(\beta) = \gamma$ and $\sigma(\gamma) = \beta$.

Now α and β are both roots of the irreducible polynomial $p(x)$, so again by Lemma 4.7.2 we have an isomorphism $\sigma_0 \colon \mathbb{Q}(\alpha) \to \mathbb{Q}(\beta)$ with $\sigma_0(\alpha) = \beta$, and once again by Lemma 4.7.2 we have an automorphism σ of $\mathbb{E}$ extending σ_0 with $\sigma(\beta) = \alpha$ and $\sigma(\gamma) = \gamma$, and also an automorphism σ of $\mathbb{E}$ extending σ_0 with $\sigma(\beta) = \gamma$ and $\sigma(\gamma) = \alpha$.

By exactly the same logic we obtain automorphisms σ of $\mathbb{E}$ with $\sigma(\alpha) = \gamma$, $\sigma(\beta) = \beta$, and $\sigma(\gamma) = \alpha$, and with $\sigma(\alpha) = \gamma$, $\sigma(\beta) = \alpha$, and $\sigma(\gamma) = \beta$.

Thus we see that $\text{Gal}(\mathbb{E}/\mathbb{Q})$ is isomorphic to the symmetric group S_3 acting as the full group of permutations of $\{\alpha, \beta, \gamma\}$. We observe that this group has order 6. $\Diamond$

Example 4.8.7. Let p be a prime and let $n \in \mathbb{Q}$ be a rational number that is not a pth power in $\mathbb{Q}$. Let $f(x) = x^p - n$ and let $\mathbb{E} = \mathbb{Q}(\zeta_p, \sqrt[p]{n})$ as in Example 4.6.10. For simplicity, set $\alpha = \sqrt[p]{n}$.

As we observed there, $f(x)$ is irreducible in $\mathbb{Q}[x]$ and $\mathbb{E}$ is a splitting field of $f(x)$. We note that $\mathbb{E}$ has subfields $\mathbb{B} = \mathbb{Q}(\zeta_p)$ and $\mathbb{D} = \mathbb{Q}(\alpha)$, and that $(\mathbb{B}/\mathbb{Q}) = p - 1$ as the cyclotomic polynomial $\Phi_p(x)$, of degree $p - 1$, is irreducible in $\mathbb{Q}[x]$ by Lemma 4.6.5, and also that $(\mathbb{D}/\mathbb{Q}) = p$ as the polynomial $f(x) = x^p - n$ is irreducible in $\mathbb{Q}[x]$ by Lemma 4.6.9. Then, by Lemma 4.4.14, $\Phi_p(x)$ is irreducible in $\mathbb{D}[x]$ and $f(x)$ is irreducible in $\mathbb{B}[x]$.

Thus, by Lemma 4.7.2, there is an element $\sigma \in \mathrm{Gal}(\mathbb{E}/\mathbb{B})$ with $\sigma(\alpha) = \zeta_p \alpha$; regarded as an element of $\mathrm{Gal}(\mathbb{E}/\mathbb{Q})$ σ is defined by $\sigma(\alpha) = \zeta_p \alpha$, $\sigma(\zeta_p) = \zeta_p$. Note that σ is an element of order p.

Also, by Lemma 4.7.2, there is an element $\tau \in \mathrm{Gal}(\mathbb{E}/\mathbb{D})$ with $\tau(\zeta_p) = \zeta_p^r$, where r is a primitive root (mod p); regarded as an element of $\mathrm{Gal}(\mathbb{E}/\mathbb{Q})$ τ is defined by $\tau(\zeta_p) = \zeta_p^r$, $\tau(\alpha) = \alpha$. Note that τ is an element of order $p - 1$.

Now since $\mathbb{E} = \mathbb{Q}(\zeta_p, \alpha)$, an element of $\mathbb{E}$ is determined by its action on ζ_p and α. We compute:

$$\sigma^r \tau(\alpha) = \sigma^r(\tau(\alpha)) = \sigma^r(\alpha) = \zeta_p^r \alpha$$
$$\sigma^r \tau(\zeta_p) = \sigma^r(\tau(\zeta_p)) = \sigma^r(\zeta_p^r) = \zeta_p^r$$

and

$$\tau\sigma(\alpha) = \tau(\sigma(\alpha)) = \tau(\zeta_p \alpha) = \zeta_p^r \alpha$$
$$\tau\sigma(\zeta_p) = \tau(\sigma(\zeta_p)) = \tau(\zeta_p) = \zeta_p^r$$

so we see that $\sigma^r \tau = \tau\sigma \in \mathrm{Gal}(\mathbb{E}/\mathbb{Q})$, i.e., that $\tau\sigma\tau^{-1} = \sigma^r$ in $\mathrm{Gal}(\mathbb{E}/\mathbb{Q})$.

Finally, we note from Lemma 4.7.1 that any automorphism of $\mathbb{E}$ fixing $\mathbb{Q}$ must take a root of $\Phi_p(x)$ to another root of $\Phi_p(x)$, giving $p-1$ possibilities, and also take any automorphism of $\mathbb{E}$ fixing $\mathbb{Q}$ must take a root of $f(x)$ to another root of $f(x)$, giving p possibilities, so the order of $\mathrm{Gal}(\mathbb{E}/\mathbb{Q})$ is at most $p(p-1)$, and exactly $p(p-1)$ if these choices can be made independently. But the group generated by σ and τ has order $p(p-1)$. Thus, we see that

$$\mathrm{Gal}(\mathbb{E}/\mathbb{Q}) = \langle \sigma, \tau \mid \sigma^p = 1, \tau^{p-1} = 1, \tau\sigma\tau^{-1} = \sigma^r \rangle$$

a nonabelian group of order $p(p-1)$, and that for any j with $1 \le j \le p - 1$ and any k with $0 \le j \le p - 1$, there is a unique $\rho \in \mathrm{Gal}(\mathbb{E}/\mathbb{Q})$ with $\rho(x) = \zeta^k \alpha$ and $\rho(\zeta_p) = \zeta_p^j$. (As an abstract group, this group is independent of the choice of primitive root r; changing r amounts to changing generators.) $\Diamond$

We conclude this section with a general construction for finite fields, and apply it to an example.

Recall that if $\mathbb{E}$ is a field of characteristic p, we introduced the Frobenius endomorphism $\Phi \colon \mathbb{E} \to \mathbb{E}$ given by $\Phi(\epsilon) = \epsilon^p$ for every

$\epsilon \in \mathbb{E}$ in Definition 4.1.17, and that if $\mathbb{E}$ is finite then Φ is an automorphism.

Lemma 4.8.8. *Let p be a prime and let $\mathbb{E}$ be a field of p^n elements. Then $\Phi: \mathbb{E} \to \mathbb{E}$ is an automorphism of order n.*

Proof. For any positive integer i, Φ^i is given by $\Phi^i(\epsilon) = \epsilon^{p^i}$. If Φ^i is the identity, then $\Phi^i(\epsilon) = \epsilon$ for every ϵ in $\mathbb{E}$, i.e., $\epsilon^{p^i} - \epsilon = 0$ for every $\epsilon \in \mathbb{E}$, so the polynomial $f(x) = x^{p^i} - x$ has p^n roots. But this is a polynomial of degree p^i, so has at most p^i roots, so if $i < n$ Φ^i cannot be the identity.

On the other hand, if $i = n$, $\Phi^i(\epsilon) = \epsilon$ is the equation $\epsilon^{p^n} - \epsilon = \epsilon(\epsilon^{p^n-1} - 1) = 0$. If $\epsilon = 0$ this is certainly true. If $\epsilon \neq 0$ then ϵ is an element of the multiplicative group $\mathbb{E}^*$ of order $p^n - 1$, so ϵ has order dividing $p^n - 1$, and hence $\epsilon^{p^n-1} = 1$ and then $\epsilon^{p^n-1} - 1 = 0$. Hence $\Phi^n(\epsilon) = \epsilon$ for every $\epsilon \in \mathbb{E}$. ⊔

Corollary 4.8.9. *Let p be a prime and let $f(x) \in \mathbb{F}_p[x]$ be an irreducible polynomial of degree n. Let $\mathbb{E} = \mathbb{F}_p(\alpha)$ be obtained by adjoining a root α of $f(x)$ to $\mathbb{F}_p$. Then $f(x)$ has the distinct roots $\alpha, \Phi(\alpha), \ldots, \Phi^{n-1}(\alpha)$ in $\mathbb{E}$, so that $\mathbb{E}$ is a splitting field of $f(x)$. Furthermore, $Gal(\mathbb{E}/\mathbb{F}_p)$ is a cyclic group of order n generated by Φ.*

Proof. We have observed that $\mathbb{E}$ always has the Frobenius automorphism. Then by Lemma 4.7.1, $\Phi^i(\alpha)$ is a root of $f(x)$ for each i. We cannot have $\Phi^i(\alpha) = \alpha$ for any $0 < i < n$ as any automorphism of $\mathbb{E}$ is determined by its action on α, and that would give $\Phi^i = \text{id}$ for $i < n$, which is impossible by Lemma 4.8.8. Moreover, we cannot have $\Phi^j(\alpha) = \Phi^i(\alpha)$ for any $0 \leq i < j < n$ as that would given $\Phi^{j-i}(\alpha) = \alpha$ which is similarly impossible. Thus, we see that $f(x)$ has the n distinct roots $\alpha, \Phi(\alpha), \ldots, \Phi^{n-1}(\alpha)$ in $\mathbb{E}$, so $f(x)$ splits in $\mathbb{E}$. And again by Lemma 4.7.1, if σ is any automorphism of $\mathbb{E}$, we must have $\sigma(\alpha)$ a root of $f(x)$ in $\mathbb{E}$, i.e., $\sigma(\alpha) = \Phi^i(\alpha)$ for some i, in which case $\sigma = \Phi^i$. □

Example 4.8.10. Let $f(x) = x^3 + x + 6 \in \mathbb{F}_7[x]$ as in Example 4.6.15. As we saw there, this is an irreducible cubic in $\mathbb{F}_7[x]$. Let $\mathbb{E} = \mathbb{F}_7(\alpha)$ be a field obtained by adjoining a root α of $f(x)$ to $\mathbb{F}_7$. Then $(\mathbb{E}/\mathbb{F}_7(\alpha)) = 3$. But as we have just seen from Corollary 4.8.9, in this situation $\mathbb{E}$ is a splitting field of $f(x)$. Furthermore,

$f(x)$ has the roots α, $\Phi(\alpha) = \alpha^7$, and $\Phi^2(\alpha) = (\alpha^7)^7$ in $\mathbb{E}$. We compute (as we did in Example 4.4.12) and we find that $\alpha^7 = 1 + 5\alpha^2$, and $(\alpha^7)^7 = (1 + 5\alpha^2)^7 = 6 + 6\alpha + 2\alpha^2$, as we claimed in Example 4.6.15. ◊

Remark 4.8.11. We call the reader's attention to the fact that in every example in this section, where $\mathbb{E}$ was the splitting field of a polynomial $f(x) \in \mathbb{F}[x]$, and $f(x)$ had distinct roots in $\mathbb{E}$ (which is often, but not always, automatic) we had the equality $|\text{Gal}(\mathbb{E}/\mathbb{F})| = (\mathbb{E}/\mathbb{F})$. As we shall see, this was no accident! ◊

4.9   Normal, separable, and Galois extensions

In this section, we introduce the key notion of a Galois extension $\mathbb{E}$ of a field $\mathbb{F}$. Then we are immediately faced with the question of how to decide when an extension is Galois, and we answer that question as well.

Definition 4.9.1. Let $\mathbb{E}$ be a field and let G be a group of automorphisms of $\mathbb{E}$. Then $\mathbb{F} = \text{Fix}(G)$ is the subfield of $\mathbb{E}$ given by

$$\text{Fix}(G) = \{\alpha \in \mathbb{E} \mid \sigma(\alpha) = \alpha \text{ for every } \sigma \in G\}. \qquad ◊$$

Remark 4.9.2. It is easy to check that $\text{Fix}(G)$ is indeed a field, as for any $\sigma \in G$ we have $\sigma(0) = 0, \sigma(1) = 1, \sigma(\alpha^{-1}) = \sigma(\alpha)^{-1}, \sigma(\alpha + \beta) = \sigma(\alpha) + \sigma(\beta)$, and $\sigma(\alpha\beta) = \sigma(\alpha)\sigma(\beta)$ for $\alpha, \beta \in \mathbb{E}$. ◊

Definition 4.9.3. Let $\mathbb{E}$ be an extension of $\mathbb{F}$. Then $\mathbb{E}$ is a *Galois extension* of $\mathbb{F}$ if

$$\text{Fix}(\text{Gal}(\mathbb{E}/\mathbb{F})) = \mathbb{F}. \qquad ◊$$

Remark 4.9.4. By the definition of the Galois group, every element σ of the Galois group fixes $\mathbb{F}$. Thus we automatically have $\text{Fix}(\text{Gal}(\mathbb{E}/\mathbb{F})) \supseteq \mathbb{F}$, and so $\mathbb{E}$ is Galois exactly when this fixed field is as small as possible, i.e., equal to $\mathbb{F}$. ◊

Our criterion for $\mathbb{E}$ to be Galois is in terms of two other properties of field extensions.

Definition 4.9.5. An extension $\mathbb{E}$ of $\mathbb{F}$ is *normal* if every irreducible polynomial $p(x) \in \mathbb{F}[x]$ that has a root in $\mathbb{E}$ splits in $\mathbb{E}[x]$. ◊

Remark 4.9.6. We see that this definition is equivalent to: An extension $\mathbb{E}$ of $\mathbb{F}$ is normal if for every $\alpha \in \mathbb{E}$, its minimal polynomial $m_\alpha(x) \in \mathbb{F}[x]$ splits in $\mathbb{E}$. ◇

Remark 4.9.7. There are plenty of extensions that are normal, but there are plenty of extensions that are not. For example, let p be an odd prime, $n \in \mathbb{Q}$ not a pth power, and consider the polynomial $f(x) = x^p - n \in \mathbb{Q}[x]$. As we have seen in Lemma 4.6.9, $f(x)$ is irreducible in $\mathbb{Q}[x]$. Let $\mathbb{E} = \mathbb{Q}(\sqrt[p]{n})$. Then $\mathbb{E}$ contains one pth root of n, but no others, so $f(x)$ does not split in $\mathbb{E}[x]$ and $\mathbb{E}$ is not a normal extension of $\mathbb{Q}$. ◇

Definition 4.9.8. An irreducible polynomial $p(x) \in \mathbb{F}[x]$ is *separable* if all of the roots of $p(x)$ in some, or equivalently every, splitting field of $p(x)$ are distinct.

An arbitrary polynomial $p(x) \in \mathbb{F}[x]$ is *separable* if all of its irreducible factors are separable.

An extension $\mathbb{E}$ of $\mathbb{F}$ is *separable* if every element of $\mathbb{E}$ is a root of a separable polynomial. ◇

Remark 4.9.9. We see that the last part of this definition is equivalent to: An extension $\mathbb{E}$ of $\mathbb{F}$ is separable if for every $\alpha \in \mathbb{E}$, its minimal polynomial $m_\alpha(x)$ has distinct roots in some, or every, splitting field. ◇

Remark 4.9.10. There are plenty of extensions that are separable, but extensions that are not are much "rarer". In fact, for many fields $\mathbb{F}$ *every* extension of $\mathbb{F}$ is separable. We do not want to interrupt our main line of argument now, but we will return to this point at the end of this section. ◇

Remark 4.9.11. We see that an extension $\mathbb{E}$ of $\mathbb{F}$ is normal and separable if for every $\alpha \in \mathbb{E}$, its minimal polynomial splits into a product of distinct linear factors in $\mathbb{E}[x]$. ◇

Here is our criterion for an extension $\mathbb{E}$ of $\mathbb{F}$ to be Galois.

Theorem 4.9.12. *Let $\mathbb{E}$ be a finite extension of $\mathbb{F}$. The following are equivalent:*

(1) $\mathbb{E}$ *is a Galois extension of* $\mathbb{F}$.
(2) $\mathbb{E}$ *is a normal and separable extension of* $\mathbb{F}$.
(3) $\mathbb{E}$ *is a splitting field of a separable polynomial* $p(x) \in \mathbb{F}[x]$.

Proof.   (1) $\Rightarrow$ (2): Let $\mathbb{E}$ be a Galois extension of $\mathbb{F}$. Let $\alpha \in \mathbb{E}$ be arbitrary, and let $\alpha_1 = \alpha, \alpha_2, \ldots, \alpha_n$ be the *distinct* elements of $\{\sigma(\alpha) \mid \sigma \in \mathrm{Gal}(\mathbb{E}/\mathbb{F})\}$. Let $m(x) = (x - \alpha_1) \ldots (x - \alpha_n)$. Then, since $\mathrm{Gal}(\mathbb{E}/\mathbb{F})$ permutes $\{\alpha_1, \ldots, \alpha_n\}$, $\sigma(m(x)) = m(x)$ for every $\sigma \in \mathrm{Gal}(\mathbb{E}/\mathbb{F})$. Writing $m(x) = x^n + a_{n-1}x^{n-1} + \cdots + a_0$, we have that $\sigma(a_i) = a_i$ for each $i = 0, \ldots, n-1$. Since $\mathbb{E}$ is a Galois extension of $\mathbb{F}$, this implies that $a_i \in \mathbb{F}$ for each i, and so $m(x) \in \mathbb{F}[x]$. Thus $m(x)$ is a separable polynomial (as its roots are distinct) that splits in $\mathbb{E}[x]$.

(2) $\Rightarrow$ (3): Let $\{\epsilon_1, \ldots, \epsilon_e\}$, be a vector space basis for $\mathbb{E}$ over $\mathbb{F}$. For each $i = 1, \ldots, e$, let $m_i(x) \in \mathbb{F}[x]$ be an irreducible polynomial having ϵ_i as a root. Then each $m_i(x)$ is separable (as $\mathbb{E}$ is a separable extension of $\mathbb{F}$) and splits in $\mathbb{E}$ (as it has the root $\epsilon_i \in \mathbb{E}$ and $\mathbb{E}$ is a normal extension of $\mathbb{F}$). Let $p(x) = m_1(x) \ldots m_e(x)$. Then $p(x)$ is a separable polynomial that splits in $\mathbb{E}$. If $\mathbb{B} \subseteq \mathbb{E}$ is a splitting field for $p(x)$, then $\epsilon_i \in \mathbb{B}$ for each i, so $\mathbb{B} \supseteq \mathbb{F}(\epsilon_1, \ldots, \epsilon_e) = \mathbb{E}$ and thus $\mathbb{B} = \mathbb{E}$.

(3) $\Rightarrow$ (1). Let $n = (\mathbb{E}/\mathbb{F})$. We prove this by complete induction on n, and all fields.

If $n = 1$ the claim is trivial. ($\mathbb{F}$ is certainly a Galois extension of $\mathbb{F}$.) Suppose it is true whenever $(\mathbb{E}/\mathbb{F}) = n' < n$ and let $(\mathbb{E}/\mathbb{F}) = n$.

Let $p(x)$ have roots $\alpha_1, \ldots, \alpha_r$ in $\mathbb{E}$. Some $\alpha_i \notin \mathbb{F}$ as otherwise $\mathbb{E} = \mathbb{F}$. Renumbering if necessary, we may assume $\alpha_1 \notin \mathbb{F}$. For simplicity, let $\alpha = \alpha_1$. Let $\mathbb{B} = \mathbb{F}(\alpha)$.

Then $\mathbb{E}$ is a splitting field of $p(x) \in \mathbb{B}[x]$ and $(\mathbb{E}/\mathbb{B}) < n$, so by the inductive hypothesis $\mathbb{E}$ is a Galois extension of $\mathbb{B}$, i.e., $\mathbb{B} = \mathrm{Fix}(\mathrm{Gal}(\mathbb{E}/\mathbb{B}))$. Since $\mathrm{Gal}(\mathbb{E}/\mathbb{B})$ is a subgroup of $\mathrm{Gal}(\mathbb{E}/\mathbb{F})$ we certainly have that $\mathrm{Fix}(\mathrm{Gal}(\mathbb{E}/\mathbb{F})) = \mathbb{D} \subseteq \mathbb{B}$. We wish to show that $\mathbb{D} = \mathbb{F}$.

Let $m_\alpha(x) \in \mathbb{F}[x]$ be the minimum polynomial of α over $\mathbb{F}$, and let $\tilde{m}_\alpha(x) \in \mathbb{D}[x]$ be the minimum polynomial of α over $\mathbb{D}$. Since $\mathbb{D} \subseteq \mathbb{B} = \mathbb{F}(\alpha)$, we see that $\mathbb{D}(\alpha) = \mathbb{F}(\alpha)$, and so

$$\deg \tilde{m}_\alpha(x) = (\mathbb{D}(\alpha)/\mathbb{D}) = (\mathbb{F}(\alpha)/\mathbb{D}) \le (\mathbb{F}(\alpha)/\mathbb{F}) = \deg m_\alpha(x),$$

with equality if and only if $\mathbb{D} = \mathbb{F}$.

We show that $\tilde{m}_\alpha(x) = m_\alpha(x)$, thereby proving this equality. Let $m_\alpha(x)$ have distinct roots $\alpha_1, \ldots, \alpha_m$ in $\mathbb{E}$. (As $m_\alpha(x)$ is irreducible in $\mathbb{F}[x]$, it must divide $p(x)$, which is a separable polynomial, so the roots of $m_\alpha(x)$ in $\mathbb{E}$ must be distinct.) Then $m_\alpha(x) = (x - \alpha_1) \ldots (x - \alpha_m)$.

Now $\tilde{m}_\alpha(x)$ divides $m_\alpha(x)$ in $\mathbb{D}[x]$, hence in $\mathbb{E}[x]$, so the roots of $\tilde{m}_\alpha(x)$ must be a subset of $\{\alpha_1, \ldots, \alpha_m\}$. We claim that every α_i is a root of $\tilde{m}_\alpha(x)$.

Let $\mathbb{E}_0$ be a splitting field of $m_\alpha(x)$ in $\mathbb{E}$. Since α_1 and α_i are two roots of the irreducible polynomial $m_\alpha(x)$, by Lemma 4.7.2 there is an isomorphism $\sigma_i \colon \mathbb{F}(\alpha_1) \to \mathbb{F}(\alpha_i)$ with $\sigma_i(\alpha_1) = \alpha_i$, and $\sigma_i|_\mathbb{F} = \mathrm{id}$. Then by Lemma 4.7.3 σ_i extends to an automorphism, which we still denote σ_i, of $\mathbb{E}_0$, $\sigma_i \colon \mathbb{E}_0 \to \mathbb{E}_0$. Now $\mathbb{E}$ is a splitting field of $p(x) \in \mathbb{F}[x]$, so is a splitting field of $p(x) \in \mathbb{E}_0[x]$, so by Lemma 4.7.3 again σ_i extends to an automorphism, which we again still denote by σ_i, of $\mathbb{E}$. Thus $\sigma_i \colon \mathbb{E} \to \mathbb{E}$ with $\sigma_i|_\mathbb{F} = \mathrm{id}$, so $\sigma_i \in \mathrm{Gal}(\mathbb{E}/\mathbb{F})$ and $\sigma_i(\alpha_1) = \alpha_i$. Now by definition $\mathbb{D} = \mathrm{Fix}(\mathrm{Gal}(\mathbb{E}/\mathbb{F}))$, and $\tilde{m}_\alpha(x) \in \mathbb{D}[x]$, so $\sigma_i(\tilde{m}_\alpha(x)) = \tilde{m}_\alpha(x)$. But then (recalling that $\alpha = \alpha_1$)

$$\tilde{m}_\alpha(\alpha_i) = \tilde{m}_\alpha(\sigma_i(\alpha_1)) = \sigma_i(\tilde{m}_\alpha)(\sigma_i(\alpha_1)) = \sigma_i(\tilde{m}_\alpha(\alpha_1)) = \sigma_i(0) = 0$$

as claimed.   □

Corollary 4.9.13. *Let $\mathbb{E}$ be a finite Galois extension of $\mathbb{F}$. If $\mathbb{B}$ is any field intermediate between $\mathbb{E}$ and $\mathbb{F}$, then $\mathbb{E}$ is a Galois extension of $\mathbb{B}$.*

Proof. By Theorem 4.9.12, $\mathbb{E}$ is a splitting field of a separable polynomial $p(x) \in \mathbb{F}[x]$. But $\mathbb{F} \subseteq \mathbb{B}$, so $p(x) \in \mathbb{B}[x]$. Thus $\mathbb{E}$ is a splitting field of the separable polynomial $p(x) \in \mathbb{B}[x]$, so, again by Theorem 4.9.12, $\mathbb{E}$ is a Galois extension of $\mathbb{B}$.   □

Now we return to the question of separability.

Definition 4.9.14. Let $f(x) \in \mathbb{F}[x]$ be a polynomial, $f(x) = a_n x^n + a_{n-1} x^{n-1} + \cdots + a_1 x + a_0$. Its *formal derivative* $f'(x) = Df(x)$ is the polynomial $f'(x) = n a_n x^{n-1} + (n-1) a_{n-1} x^{n-2} + \cdots + a_1$.   ◇

As you can see, the formal derivative is the same as the derivative you learned in calculus. But we are calling it the formal derivative as we are considering this purely algebraically — we have no notion of limit here as we do in calculus.

You should also be familiar with the following result from calculus. We have to prove it separately here, as we can't use limit arguments.

But the purely algebraic proof is in fact much simpler, since we don't have to worry about limits.

Lemma 4.9.15. (a) *The formal differentiation operator* $D\colon \mathbb{F}[x] \to \mathbb{F}[x]$ *is a linear transformation, i.e.,* $D(af(x) + bg(x)) = aDf(x) + bDg(x)$ *for any* $a, b \in \mathbb{F}$ *and any* $f(x), g(x) \in \mathbb{F}[x]$.
(b) *(Leibniz rule) For any* $f(x), g(x) \in \mathbb{F}[x]$,

$$D(f(x)g(x)) = Df(x)g(x) + f(x)Dg(x).$$

Proof.

(a) Is easy to verify.
(b) Given (a), we need only check this when $f(x)$ and $g(x)$ each consist of a single term, say $f(x) = x^i$, $g(x) = x^j$. But then

$$D(f(x)g(x)) = D(x^{i+j}) = (i+j)x^{i+j-1}$$

while

$$
\begin{aligned}
Df(x)g(x) + f(x)Dg(x) &= (Dx^i)x^j + x^i(Dx^j) \\
&= (ix^{i-1})(x^j) + (x^i)(jx^{j-1}) \\
&= (i+j)x^{i+j-1}.
\end{aligned}
$$
$\square$

Lemma 4.9.16. *Let* $f(x) \in \mathbb{F}[x]$ *be an irreducible polynomial. Then* $f(x)$ *is a separable polynomial if and only if* $f(x)$ *and* $f'(x)$ *are relatively prime in* $\mathbb{F}[x]$.

Proof.   Let $g(x)$ and $h(x)$ be any two polynomials in $\mathbb{F}[x]$ and let $\mathbb{E}$ be any extension field of $\mathbb{F}$. Let $d(x)$ be the gcd of $g(x)$ and $h(x)$ as polynomials in $\mathbb{F}[x]$ and $\tilde{d}(x)$ be gcd of $g(x)$ and $h(x)$ as polynomials in $\mathbb{E}[x]$. Recall we showed in Lemma 4.2.17 that $d(x) = \tilde{d}(x)$. In particular, $g(x)$ and $h(x)$ are relatively prime in $\mathbb{F}[x]$ if and only if they are relatively prime in $\mathbb{E}[x]$. We apply this here.

Let $\mathbb{E}$ be a splitting field of $f(x)$.

First suppose that $f(x)$ is separable. Then $f(x) = (x - \alpha_1) \ldots (x - \alpha_n)$ in $\mathbb{E}[x]$, with $\alpha_1, \ldots, \alpha_n$ distinct. Then $f'(x) = (x - \alpha_2) \ldots (x - \alpha_n) + (x - \alpha_1)(x - \alpha_3) \ldots (x - \alpha_n) + \cdots + (x - \alpha_1) \ldots (x - \alpha_{n-1})$. Any nontrivial factor of $f(x)$ must be divisible by $x - \alpha_i$ for some i, but we see that $f'(x)$ is not divisible by any $x - \alpha_i$. Hence $f(x)$

and $f'(x)$ have no nontrivial common factors, i.e., $f(x)$ and $f'(x)$ are relatively prime.

Now suppose that $f(x)$ is not separable. Then $f(x)$ has at least one repeated root α in $\mathbb{E}[x]$, so that $f(x)$ is divisible by $(x - \alpha)^e$ with $e \geq 2$. Write $f(x) = (x-\alpha)^e g(x)$. Then $f'(x) = e(x-\alpha)^{e-1}g(x)+(x-\alpha)^e g'(x)$ and we see that $f(x)$ and $f'(x)$ have the nontrivial common factor $(x - \alpha)^{e-1}$, i.e., $f(x)$ and $f'(x)$ are not relatively prime. $\square$

Corollary 4.9.17. (a) *Let* $\mathbb{F}$ *be a field of characteristic* 0. *Then every polynomial* $f(x) \in \mathbb{F}[x]$ *is separable.*

(b) *Let* $\mathbb{F}$ *be a field of characteristic* p. *Then every irreducible polynomial* $f(x) = \Sigma_{i=0}^n a_i x^i$ *with at least one term* $a_i x^i$ *with* $a_i \neq 0$ *with* i *not divisible by* p *is separable.*

Proof. In case (a), we may assume $f(x)$ is irreducible.

In both of these cases, $f'(x)$ is a nonzero polynomial of lower degree than $f(x)$. Since $f(x)$ is irreducible, $f(x)$ is relatively prime to $f'(x)$. $\square$

Corollary 4.9.18. *Let* $\mathbb{F}$ *be a field of characteristic* p. *If the Frobenius* $\Phi\colon \mathbb{F} \to \mathbb{F}$ *is onto, then every polynomial* $f(x) \in \mathbb{F}[x]$ *is separable. In particular, if* $\mathbb{F}$ *is a finite field, then every polynomial* $f(x) \in \mathbb{F}[x]$ *is separable.*

Proof. It suffices to consider $f(x)$ irreducible. Then by Corollary 4.9.18, the only possible nonseparable polynomials are of the form $f(x) = \Sigma_{j=0}^m a_{pj} x^{pj}$. Now suppose $\Phi\colon \mathbb{F} \to \mathbb{F}$ is onto. Then for every j, we have $a_{pj} = \Phi(b_{pj}) = b_{pj}^p$ for some element b_{pj} of $\mathbb{F}$. Thus,

$$f(x) = \sum_{j=0}^m a_{pj} x_{pj} = \sum_{j=0}^m b_{pj}^p x^{pj} = \left(\sum_{j=0}^m b_{pj} x^j \right)^p$$

is not irreducible. $\square$

We are led to the following definition.

Definition 4.9.19. A field $\mathbb{F}$ is *perfect* if every polynomial $f(x) \in \mathbb{F}[x]$, or, equivalently, every algebraic extension $\mathbb{E}$ of $\mathbb{F}$, is separable.

We can now conclude:

Lemma 4.9.20. *Every field of characteristic* 0, *and every finite field, is perfect.*

Proof. This is immediate from Corollary 4.9.17 and Corollary 4.9.18. □

We close with an example of a nonseparable extension.

Example 4.9.21. Let $\mathbb{F} = \mathbb{F}_p(t)$, the field of rational functions in the variable t over $\mathbb{F}_p$. Let $f(x) = x^p - t \in \mathbb{F}[x]$. By Abel's theorem (Lemma 4.6.9) $f(x)$ is irreducible. Let $\mathbb{E} = \mathbb{F}(s)$ be the field obtained by adjoining a root s of $f(x)$ to $\mathbb{F}$. Then $s^p = t$, and we see $x^p - t = x^p - s^p = (x - s)^p$ is not separable.   ◊

4.10   The fundamental theorem of Galois theory

In this section, we arrive at the fundamental theorem of Galois theory, one of the great theorems of mathematics.

Before we get there we have the following result, which is very important and useful in its own right.

Theorem 4.10.1. *Let $p(x) \in \mathbb{F}[x]$ be a separable polynomial and let $\mathbb{E}$ be a splitting field of $p(x)$. Then*

$$|Gal(\mathbb{E}/\mathbb{F})| = (\mathbb{E}/\mathbb{F}).$$

More generally, if $\mathbb{D}$ is any field intermediate between $\mathbb{E}$ and $\mathbb{F}$, then

$$|Gal(\mathbb{E}/\mathbb{D})| = (\mathbb{E}/\mathbb{D}).$$

Proof. We prove the general case. The particular case is just the special case of the general case when $\mathbb{D} = \mathbb{F}$.

Let $G_\mathbb{D} = \text{Gal}(\mathbb{E}/\mathbb{D})$ and $e = (\mathbb{E}/\mathbb{D})$. We prove the theorem by complete induction on e.

If $e = 1$, then $\mathbb{E} = \mathbb{D}$ and $\text{Gal}(\mathbb{E}/\mathbb{D}) = \{\text{id}\}$ has order 1.

Now suppose the theorem is true for all $e' < e$, and all fields. Let $\mathbb{E}$ be an extension of $\mathbb{F}$ with $e = (\mathbb{E}/\mathbb{D}) > 1$. Since $\mathbb{E}$ is a splitting field of $p(x) \in \mathbb{F}[x]$, we have that $\mathbb{E} = \mathbb{F}(\alpha_1, \ldots, \alpha_n)$ where $\alpha_1, \ldots, \alpha_n$ are the roots of $p(x)$ in $\mathbb{E}$. Then also $\mathbb{E} = \mathbb{D}(\alpha_1, \ldots, \alpha_n)$. We may assume that $\alpha_1, \ldots, \alpha_n \notin \mathbb{D}$ as otherwise $e = 1$.

Consider $\mathbb{B} = \mathbb{D}(\alpha_1)$. There are two cases:

Case 1: $\mathbb{E} = \mathbb{B}$. In this case the element α_1 has degree e over $\mathbb{D}$, so its minimum polynomial $\tilde{m}_{\alpha_1}(x) \in \mathbb{D}[x]$ has degree e. Now $\tilde{m}_{\alpha_1}(x)$

is irreducible in $\mathbb{D}[x]$, and $p(x)$ has α_1 as a root, so $\tilde{m}_{\alpha_1}(x)$ divides $p(x)$ in $\mathbb{D}[x]$. We are assuming that $p(x)$ is separable, so $\tilde{m}_{\alpha_1}(x)$ is separable and hence $\tilde{m}_{\alpha_1}(x)$ has e distinct roots $\alpha_1, \ldots, \alpha_e$ in $\mathbb{E}$. Now by Lemma 4.7.1 any $\sigma \in G_{\mathbb{D}}$ must have $\sigma(\alpha_1) = \alpha_i$ for some i, and by Lemma 4.7.2 there exists $\sigma_i \in G_{\mathbb{D}}$ with $\sigma(\alpha_1) = \alpha_i$ for each i. Furthermore, since $\mathbb{E} = \mathbb{D}(\alpha_1)$, this element σ_i is unique. Thus we see $\text{Gal}(\mathbb{E}/\mathbb{D}) = \{\sigma_1, \ldots, \sigma_e\}$ so $|\text{Gal}(\mathbb{E}/\mathbb{D})| = e = (\mathbb{E}/\mathbb{D})$.

Case 2: $\mathbb{E} \supset \mathbb{B}$. Let $(\mathbb{B}/\mathbb{D}) = b$ and let $G_{\mathbb{B}} = \text{Gal}(\mathbb{E}/\mathbb{B})$. Then $e = (\mathbb{E}/\mathbb{D}) = (\mathbb{E}/\mathbb{B})(\mathbb{B}/\mathbb{D})$ so $(\mathbb{E}/\mathbb{B}) = e/b$. Then $\mathbb{E}$ is a splitting field of $p(x) \in \mathbb{B}[x]$, so is a Galois extension of $\mathbb{B}$ by Theorem 4.9.12. Thus $\text{Fix}(\text{Gal}(\mathbb{E}/\mathbb{B})) = \mathbb{B}$, so by the inductive hypothesis we have that $|G_{\mathbb{B}}| = (\mathbb{E}/\mathbb{B})$.

By the same logic as in case 1, $\tilde{m}_{\alpha_1}(x) \in \mathbb{D}[x]$ is an irreducible polynomial of degree b with distinct roots $\alpha_1, \ldots, \alpha_b$ in $\mathbb{B}$. Let

$$H_i = \{\sigma \in G_{\mathbb{D}} \mid \sigma(\alpha_1) = \alpha_i\}, \quad i = 1, \ldots, b.$$

Then, by Lemma 4.7.1, $G_{\mathbb{D}} = \coprod_i H_i$. Now

$$H_1 = \{\sigma \in G_{\mathbb{D}} \mid \sigma(\alpha_1) = \alpha_1\}$$
$$= \{\sigma \in G_{\mathbb{D}} \mid \sigma|_{\mathbb{B}} = \text{id}\} = G_{\mathbb{B}}.$$

We claim that H_i is a left coset of H_1 for each $i \geq 1$. To see this, let $\sigma_i \in G_{\mathbb{F}}$ be any element with $\sigma(\alpha_1) = \alpha_i$, $i = 1, \ldots, b$. Such elements $\sigma_1, \ldots, \sigma_b$ exist by Lemma 4.7.2. We show that $H_i = \sigma_i H_1$, $i = 1, \ldots, b$.

First, suppose $\sigma \in \sigma_i H_i$. Then $\sigma = \sigma_i \eta$ for some $\eta \in H_1$. But then

$$\sigma(\alpha_1) = \sigma_i \eta(\alpha_1) = \sigma_i(\eta(\alpha_1)) = \sigma_i(\alpha_1) = \alpha_i$$

so $\sigma \in H_i$. Thus $\sigma_i H_1 \subseteq H_i$.

Next, let $\sigma \in H_i$, so that $\sigma(\alpha_1) = \alpha_i$. Now $\sigma_i(\alpha_1) = \alpha_i$ so

$$\sigma_i^{-1}\sigma(\alpha_1) = \sigma_i^{-1}(\sigma(\alpha_1)) = \sigma_i^{-1}(\alpha_i) = \alpha_1$$

so $\sigma_i^{-1}\sigma = \eta \in H_i$. But then $\sigma = \sigma_i \eta \in \sigma_i H_1$, so $H_i \subseteq \sigma_i H_1$.

Thus $H_i = \sigma_i H_1$, and so $H_1 = G_{\mathbb{B}}$ is a subgroup of $G_{\mathbb{D}}$ of index b. But then

$$|G_{\mathbb{D}}| = [G_{\mathbb{D}} : G_{\mathbb{B}}]|G_{\mathbb{B}}| = b|G_{\mathbb{B}}| = b(e/b) = e = (\mathbb{E}/\mathbb{D})$$

as claimed.

Then by induction we are done. $\square$

We observe a very important consequence.

Corollary 4.10.2. *Let* $\mathbb{E}$ *be a finite Galois extension of* $\mathbb{F}$. *Then*

$$|Gal(\mathbb{E}/\mathbb{F})| = (\mathbb{E}/\mathbb{F}).$$

Furthermore, for any field $\mathbb{B}$ *intermediate between* $\mathbb{E}$ *and* $\mathbb{F}$,

$$|(\mathrm{Gal}(\mathbb{E}/\mathbb{B}))| = (\mathbb{E}/\mathbb{B}).$$

Proof.   This follows immediately from Theorem 4.10.1, Theorem 4.9.12, and Corollary 4.9.13.   □

Theorem 4.10.3 (Fundamental theorem of Galois theory (FTGT)).
Let $\mathbb{E}$ *be a finite Galois extension of* $\mathbb{F}$ *and let* $G = Gal(\mathbb{E}/\mathbb{F})$.

(a) *There is a* $1-1$ *correspondence between intermediate fields* $\mathbb{E} \supseteq$ $\mathbb{B} \supseteq \mathbb{F}$ *and subgroups* $\{1\} \subseteq G_\mathbb{B} \subseteq G$ *given by*

$$\mathbb{B} = \mathrm{Fix}(G_\mathbb{B}) \Leftrightarrow G_\mathbb{B} = \mathrm{Gal}(\mathbb{E}/\mathbb{B}).$$

(b) $\mathbb{B}$ *is a normal extension of* $\mathbb{F}$ *if and only if* $G_\mathbb{B}$ *is a normal subgroup of* G. *This is the case if and only if* $\mathbb{B}$ *is a Galois extension of* $\mathbb{F}$. *In this case*

$$\mathrm{Gal}(\mathbb{B}/\mathbb{F}) \cong G/G_\mathbb{B}.$$

(c) *For each* $\mathbb{E} \supseteq \mathbb{B} \supseteq \mathbb{F}$,

$$(\mathbb{B}/\mathbb{F}) = [G\colon G_\mathbb{B}] \quad and \quad (\mathbb{E}/\mathbb{B}) = |G_\mathbb{B}|.$$

Before proving the FTGT, we will make several observations.

Remark 4.10.4.

(a) Note that $G_\mathbb{E} = \{1\}$ and $G_\mathbb{F} = G$. Also, for intermediate fields/subgroups of the Galois group,

$$\mathbb{B}_1 \subseteq \mathbb{B}_2 \Leftrightarrow G_{\mathbb{B}_2} \subseteq G_{\mathbb{B}_1}.$$

(b) Recall from Corollary 4.9.13 that $\mathbb{E}$ is a Galois extension of $\mathbb{B}$ for every intermediate field $\mathbb{B}$.

(c) Since $\mathbb{B}$ is a subfield of $\mathbb{E}$ and $\mathbb{E}$ is a separable extension of $\mathbb{F}$, then $\mathbb{B}$ is certainly a separable extension of $\mathbb{F}$. (If every element of $\mathbb{E}$ is separable, then certainly every element of $\mathbb{B}$ is separable.) This justifies the claim in (b) that $\mathbb{B}$ is a normal extension of $\mathbb{F}$ if and only if $\mathbb{B}$ is a Galois extension of $\mathbb{F}$.

(d) As the proof of the FTGT shows, if $\mathbb{B}$ is a Galois extension of $\mathbb{F}$ the quotient map $\pi \colon G \twoheadrightarrow \mathrm{Gal}(\mathbb{B}/\mathbb{F})$ is given by restriction, i.e., $\pi(\sigma) = \sigma|_{\mathbb{B}}$.

Proof.

(a) For each subgroup H of G, let

$$\mathbb{B}_H = \mathrm{Fix}(H).$$

This gives a mapping

$\Gamma \colon \{\text{subgroups of } G\} \to \{\text{fields intermediate between } \mathbb{E} \text{ and } \mathbb{F}\}.$

We show Γ is a 1–1 correspondence.

Γ is 1–1: Let H_1 and H_2 be subgroups of G. Suppose that $\mathbb{B}_{H_1} = \mathbb{B}_{H_2}$. Let H be the subgroup of G generated by H_1 and H_2. Then $\mathbb{B}_H = \mathbb{B}_{H_1} = \mathbb{B}_{H_2}$. Then, by Theorem 4.10.1,

$$|H| = |\mathrm{Gal}(\mathbb{E}/\mathbb{B}_H)| = |\mathrm{Gal}(\mathbb{E}/\mathbb{B}_{H_1})| = |H_1|$$

$$= |\mathrm{Gal}(\mathbb{E}/\mathbb{B}_{H_2})| = |H_2|$$

But $H_1 \subseteq H$ and $H_2 \subseteq H$, so we must have $H_1 = H = H_2$.

Γ is onto: Let $\mathbb{B}$ be an intermediate field between $\mathbb{E}$ and $\mathbb{F}$. Let

$$H = \{\sigma \in G \mid \sigma|_{\mathbb{B}} = \mathrm{id}\} = \mathrm{Gal}(\mathbb{E}/\mathbb{B}) \subseteq \mathrm{Gal}(\mathbb{E}/\mathbb{F}).$$

As we have observed, $\mathbb{E}$ is a Galois extension of $\mathbb{B}$, so $\mathbb{B} = \mathrm{Fix}(\mathrm{Gal}(\mathbb{E}/\mathbb{B})) = \mathrm{Fix}(H)$.

(c) We have seen in Theorem 4.10.1 that

$$(\mathbb{E}/\mathbb{F}) = |G_{\mathbb{F}}| \quad \text{and} \quad (\mathbb{E}/\mathbb{B}) = |G_{\mathbb{B}}|$$

Now

$$(\mathbb{E}/\mathbb{F}) = (\mathbb{E}/\mathbb{B})(\mathbb{B}/\mathbb{F})$$

and

$$|G_{\mathbb{F}}| = |G_{\mathbb{B}}|[G_{\mathbb{F}} \colon G_{\mathbb{B}}]$$

so

$$(\mathbb{B}/\mathbb{F}) = [G_{\mathbb{F}} \colon G_{\mathbb{B}}].$$

(b) Suppose that $G_{\mathbb{B}}$ is a normal subgroup of G. For any $\sigma \in G$, we have $\sigma(\mathbb{B}) = \text{Fix}(\sigma G_{\mathbb{B}} \sigma^{-1}) = \text{Fix}(G_{\mathbb{B}}) = \mathbb{B}$. Hence, we have a restriction map $R \colon \text{Gal}(\mathbb{E}/\mathbb{F}) \to \text{Gal}(\mathbb{B}/\mathbb{F})$ given by $R(\sigma) = \sigma|_{\mathbb{B}}$, and so we have an isomorphism $\pi \colon G/\text{Ker}(R) \to \text{Im}(R)$.
Now $\text{Ker}(R) = \{\sigma \in \text{Gal}(\mathbb{E}/\mathbb{F}) \mid \sigma|_{\mathbb{B}} = \text{id}\} = \text{Gal}(\mathbb{E}/\mathbb{B})$.
Also, $\text{Im}(R) = \text{Gal}(\mathbb{B}/\mathbb{F})$ as follows: Let $\sigma_0 \in \text{Gal}(\mathbb{B}/\mathbb{F})$, i.e., $\sigma_0 \colon \mathbb{B} \to \mathbb{B}$ is an isomorphism which is the identity on $\mathbb{F}$. Now $\mathbb{E}$ is a Galois extension of $\mathbb{B}$, hence a splitting field of a separable polynomial $p(x) \in \mathbb{F}[x]$, so by Lemma 4.7.3 σ_0 extends to an isomorphism $\sigma \colon \mathbb{E} \to \mathbb{E}$. Thus $\sigma \in \text{Gal}(\mathbb{E}/\mathbb{F})$ with $R(\sigma) = \sigma_0$.
Conversely, suppose that $\mathbb{B}$ is a Galois extension of $\mathbb{F}$. Then $\mathbb{B}$ is a splitting field of a separable polynomial $p(x) \in \mathbb{F}[x]$, and so $\mathbb{B} = \mathbb{F}(\beta_1, \ldots, \beta_r)$ where $\beta_1, \ldots, \beta_r$ are the roots of $p(x)$ in $\mathbb{B}$. Now if σ is any element of $\text{Gal}(\mathbb{E}/\mathbb{F})$, then $\sigma(p(x)) = p(x)$, and so σ permutes $\beta_1, \ldots, \beta_r$, and so σ leaves $\mathbb{B}$ invariant, i.e., $\sigma(\mathbb{B}) = \mathbb{B}$. Then

$$\text{Fix}(G_{\mathbb{B}}) = \mathbb{B} = \sigma(\mathbb{B}) = \text{Fix}(\sigma G_{\mathbb{B}} \sigma^{-1})$$

so by (a), $G_{\mathbb{B}} = \sigma G_{\mathbb{B}} \sigma^{-1}$. Since σ was an arbitrary element of $\text{Gal}(\mathbb{E}/\mathbb{F})$, we conclude that $\text{Gal}(\mathbb{B}/\mathbb{F})$ is a normal subgroup of $\text{Gal}(\mathbb{E}/\mathbb{F})$. $\square$

We can also make the following observation about the relationship between fixed fields and subgroups of the Galois group.

Lemma 4.10.5. *Let $\mathbb{E}$ be a finite Galois extension of $\mathbb{F}$ and let $G = \text{Gal}(\mathbb{E}/\mathbb{F})$. Let H and K be subgroups of G and let $\mathbb{B} = \text{Fix}(H)$ and $\mathbb{D} = \text{Fix}(K)$. Then*

(a) *The composite $\mathbb{B}\mathbb{D} = \text{Fix}(H \cap K)$; and*
(b) *$\mathbb{B} \cap \mathbb{D} = \text{Fix}(L)$, where L is the subgroup of G generated by H and K.*

Proof. If $\sigma \in H \cap K$ then $\sigma \in H$, so σ fixes $\mathbb{B}$, and $\sigma \in K$, so σ fixes $\mathbb{D}$, and hence σ fixes $\mathbb{B}\mathbb{D}$. On the other hand, if σ fixes $\mathbb{B}\mathbb{D}$, then σ fixes $\mathbb{B}$, so $\sigma \in H$, and σ fixes $\mathbb{D}$, so $\sigma \in K$, and hence $\sigma \in H \cap K$.

Also, if $\sigma(\epsilon) = \epsilon$ for every $\sigma \in L$, then $\sigma(\epsilon) = \epsilon$ for every $\sigma \in H$, so $\epsilon \in \mathbb{B}$, and $\sigma(\epsilon) = \epsilon$ for every $\sigma \in K$, so $\epsilon \in \mathbb{D}$, and hence $\epsilon \in \mathbb{B} \cap \mathbb{D}$. On the other hand, if $\epsilon \in \mathbb{B} \cap \mathbb{D}$, then $\epsilon \in \mathbb{B}$, so $\sigma(\epsilon) = \epsilon$ for every

$\sigma \in H$, and $\epsilon \in \mathbb{D}$, so $\sigma(\epsilon) = \epsilon$ for every $\sigma \in K$, so $\sigma(\epsilon) = \epsilon$ for every $\sigma \in L$. □

Our viewpoint here has been to start with a field $\mathbb{F}$ and go to an extension field $\mathbb{E}$. But we could look at things from the reverse viewpoint: Start with a field $\mathbb{E}$ and go to a subfield $\mathbb{F}$. From this reverse viewpoint we see the following result.

Corollary 4.10.6. *Let $\mathbb{E}$ be a field and let G be a finite group of automorphisms of $\mathbb{E}$. Let $\mathbb{F} = Fix(G)$ be the subfield of $\mathbb{E}$ fixed by G. Then,*

(a) $\mathbb{E}$ *is a Galois extension of* $\mathbb{F}$;
(b) $G = Gal(\mathbb{E}/\mathbb{F})$; *and*
(c) *(Artin)* $(\mathbb{E}/\mathbb{F}) = |G|$.

Proof.

(a) Let $\tilde{G}$ be the group of all automorphisms of $\mathbb{E}$ that fix $\mathbb{F}$. By definition, $\tilde{G} = Gal(\mathbb{E}/\mathbb{F})$. Then G is a subgroup of $\tilde{G}$. We claim $G = \tilde{G}$. To see this, note that $\mathrm{Fix}(\tilde{G}) \subseteq \mathrm{Fix}(G)$. But then

$$\mathbb{F} \subseteq \mathrm{Fix}(\tilde{G}) \subseteq \mathrm{Fix}(G) = \mathbb{F}.$$

Hence, $\mathbb{F} = \mathrm{Fix}(\tilde{G})$ and so, by the definition of a Galois extension, $\mathbb{E}$ is a Galois extension of $\mathbb{F}$.
(b) We have that G is a subgroup of $\tilde{G}$ with $\mathbb{F} = \mathrm{Fix}(G) = \mathrm{Fix}(\tilde{G})$, i.e., G and $\tilde{G}$ have the same fixed field. But then, by part (a) of the fundamental theorem of Galois theory, $G = \tilde{G}$.
(c) This is now immediate from (b) and part (c) of the fundamental theorem of Galois theory, or, alternatively, from (b), Theorem 4.9.12 and Theorem 4.10.1. □

Remark 4.10.7. There is a well-known and (justly) well-regarded proof by Artin of the fundamental theorem of Galois theory in which he uses the equality in part (c) of the above corollary as an essential step in the proof. In our proof we have instead used Theorem 4.10.1, so in our approach this equality is a consequence of the fundamental theorem of Galois theory. ◇

Corollary 4.10.8. *Let $\mathbb{E}$ be a finite extension of $\mathbb{F}$. Then $|Gal(\mathbb{E}/\mathbb{F})| \leq (\mathbb{E}/\mathbb{F})$, with equality if and only if $\mathbb{E}$ is a Galois extension of $\mathbb{F}$.*

Proof. Let $\mathbb{B} = \text{Fix}(\text{Gal}(\mathbb{E}/\mathbb{F}))$. Then $\mathbb{F} \subseteq \mathbb{B}$, $\text{Gal}(\mathbb{E}/\mathbb{F}) = \text{Gal}(\mathbb{E}/\mathbb{B})$, and $\mathbb{E}$ is a Galois extension of $\mathbb{B}$. But then by Theorem 4.10.1,

$$|\text{Gal}(\mathbb{E}/\mathbb{F})| = |\text{Gal}(\mathbb{E}/\mathbb{B})| = (\mathbb{E}/\mathbb{B}) \leq (\mathbb{E}/\mathbb{F})$$

with equality if and only if $\mathbb{B} = \mathbb{F}$, i.e., if and only if $\mathbb{E}$ is a Galois extension of $\mathbb{F}$. □

Let us further investigate the structure of Galois groups.

Theorem 4.10.9. *Let $f(x) \in \mathbb{F}[x]$ be a separable polynomial of degree n, and let $\mathbb{E}$ be a splitting field of $f(x)$. Then the Galois group $G = \text{Gal}(\mathbb{E}/\mathbb{F})$ is isomorphic to a subgroup of S_n, the symmetric group on n elements. If $f(x)$ is irreducible, then G is isomorphic to a transitive subgroup of S_n. If $f(x)$ is a product of irreducible polynomials $f(x) = f_1(x) \ldots f_k(x)$ with $\deg f_i(x) = n_i$, then G is isomorphic to a subgroup of $S_{n_1} \times \cdots \times S_{n_k}$.*

Proof. Let $\alpha_1, \ldots, \alpha_n$ be the roots of $f(x)$ in $\mathbb{E}$, so that $\mathbb{E} = \mathbb{F}(\alpha_1, \ldots, \alpha_n)$. We regard S_n as operating on the set $\{\alpha_1, \ldots, \alpha_n\}$. Then we have a homomorphism $\varphi \colon G \to S_n$ defined as follows: If $g \in G$ then $\sigma = \varphi(g)$ is the permutation defined by $\sigma(\alpha_i) = g(\alpha_i)$. Furthermore, φ is 1–1 as if $g(\alpha_i) = \alpha_i$ for every $i = 1, \ldots, n$, then $g \colon \mathbb{E} \to \mathbb{E}$ is the identity automorphism.

Also, we have already seen that if $f(x)$ is irreducible, there is an automorphism g of $\mathbb{E}$ with $g(\alpha_i) = \alpha_j$ for any i, j, and so G acts transitively on $\{\alpha_1, \ldots, \alpha_n\}$ in this case. In any case, G preserves the set of roots of each of the individual irreducible factors of $f(x)$. □

Corollary 4.10.10. *Let $f(x) \in \mathbb{F}[x]$ be a separable polynomial of degree n, and let $\mathbb{E}$ be a splitting field of $f(x)$. If $f(x)$ is irreducible, then $(\mathbb{E}/\mathbb{F})$ is divisible by n and $(\mathbb{E}/\mathbb{F})$ divides $n!$. If $f(x)$ is a product of irreducible polynomials $f(x) = f_1(x) \ldots f_k(x)$ with $\deg f_i(x) = n_i$, then $(\mathbb{E}/\mathbb{F})$ divides $n_1! \ldots n_k!$.*

Proof. This is immediate from Theorem 4.10.1 and Corollary 4.10.6. □

Remark 4.10.11. Note that this is a strengthening of Corollary 4.5.12. ◇

Even before stating the fundamental theorem of Galois theory, we were able to find many Galois groups in Section 4.8. We now revisit some of those examples. But before we look at specific examples, let us look at Remark 4.8.11. We can now see that the reason for the equality there is that every example in that section was an example of a Galois extension, by Theorem 4.9.12, and then the equality in that remark is due to Theorem 4.10.1.

Example 4.10.12. Let $\mathbb{E} = \mathbb{Q}(\sqrt{a_1}, \sqrt{a_2})$ where neither a_1, a_2, nor $a_1 a_2$ are perfect squares. As we saw in Example 4.8.3, in this case $\mathrm{Gal}(\mathbb{E}/\mathbb{Q}) = \{\sigma_{00}, \sigma_{01}, \sigma_{10}, \sigma_{11}\}$ is isomorphic to $\mathbb{Z}_2 \oplus \mathbb{Z}_2$, where $\sigma_{ij}(\sqrt{a_1}) = (-1)^i \sqrt{a_1}$ and $\sigma_{ij}(\sqrt{a_2}) = (-1)^j \sqrt{a_2}$. Then we see that G has three subgroups of order 2, $H_1 = \{\sigma_{00}, \sigma_{01}\}$, $H_2 = \{\sigma_{00}, \sigma_{10}\}$, and $H_3 = \{\sigma_{00}, \sigma_{11}\}$ with fixed fields $\mathrm{Fix}(H_1) = \mathbb{Q}(\sqrt{a_1})$, $\mathrm{Fix}(H_2) = \mathbb{Q}(\sqrt{a_2})$, and $\mathrm{Fix}(H_3) = \mathbb{Q}(\sqrt{a_1 a_2})$. Since G is abelian, each of these subgroups is normal, and so each of these intermediate fields is a Galois extension of $\mathbb{Q}$. Furthermore, we see that these three fields are *all* of the fields strictly intermediate between $\mathbb{E}$ and $\mathbb{Q}$. ◊

Example 4.10.13. Let p be a prime and let $\mathbb{E} = \mathbb{Q}(\zeta_p)$ as in Example 4.8.4. Then, as we have seen, $\mathbb{E}$ is an extension of $\mathbb{Q}$ of degree $p - 1$ and $G = \mathrm{Gal}(\mathbb{E}/\mathbb{Q})$ is isomorphic to $\mathbb{Z}_p^*$, a cyclic group of order $p - 1$. A generator r of $\mathbb{Z}_p^*$ is known as a primitive root (mod p), so that $r^{p-1} \equiv 1 \pmod{p}$ but $r^i \not\equiv 1 \pmod{p}$ for any i with $0 < i < p - 1$. Then G is generated by the automorphism σ_r of $\mathbb{E}$ given by $\sigma_r(\zeta_p) = \zeta_p^r$. Set $\sigma = \sigma_r$.

Now $\mathbb{E}$ is a splitting field of the pth cyclotomic polynomial $\Phi_p(x)$ whose roots are $1, \zeta_p, \ldots, \zeta_p^{p-1}$. Then $\{1, \zeta_p, \ldots, \zeta_p^{p-2}\}$ form a vector space basis for $\mathbb{E}$ over $\mathbb{Q}$, and $1 + \zeta_p + \cdots + \zeta_p^{p-1} = 0$. Thus $\{\zeta_p, \ldots, \zeta_p^{p-1}\}$ is also a vector space basis for $\mathbb{E}$ over $\mathbb{Q}$.

By the fundamental theorem of Galois theory, the fields intermediate between $\mathbb{E}$ and $\mathbb{Q}$ are in 1–1 correspondence with subgroups H of G. Now G is a cyclic group of order $p - 1$, so has a unique subgroup H_d of order d for every d dividing $p - 1$, and these are all the subgroups of G. Let $s = (p - 1)/d$. Then H_d is generated by σ^s.

Then the intermediate fields between $\mathbb{E}$ and $\mathbb{Q}$ are the fields $\mathbb{E}_d = \mathrm{Fix}(H_d)$ for every d dividing $p - 1$.

Noting that G permutes the elements of the basis $\{\zeta_p, \ldots, \zeta_p^{p-1}\}$ of $\mathbb{E}$ over $\mathbb{Q}$, we see that

$$\mathbb{E}_d = \mathbb{Q}(\theta_d) \quad \text{where} \quad \theta_d = \sum_{i=1}^{d} (\sigma^s)^i (\zeta_p) = \sum_{i=1}^{d} \sigma^{is}(\zeta_p)$$

$$= \sum_{i=1}^{d} \zeta_p^{ir(p-1)/d}.$$

We note that $(\mathbb{E}/\mathbb{E}_d) = d$ and $(\mathbb{E}_d/\mathbb{Q}) = s$, and we observe that $K_d = \text{Gal}(\mathbb{E}_d/\mathbb{Q}) \cong \text{Gal}(\mathbb{E}/\mathbb{Q})/\text{Gal}(\mathbb{E}/\mathbb{E}_d)$ is the cyclic group of order s generated by $\bar{\sigma}$, the restriction of σ to $\mathbb{E}_d$.

Note in Example 4.6.8 we performed a special case of this construction. In our notation here, $\zeta_5 = \zeta_5$ and $\theta = \theta_2$, and $(\mathbb{Q}(\zeta_5)/\mathbb{Q}(\theta)) = 2$, $(\mathbb{Q}(\theta)/\mathbb{Q}) = 2$. ◊

Example 4.10.14. We return to Example 4.8.5 and adopt the notation there. We have that $G = \text{Gal}(\mathbb{E}/\mathbb{Q})$ is a group of order 4, isomorphic to $\mathbb{Z}_2 \oplus \mathbb{Z}_2$ with $G = \{\sigma_0, \sigma_1, \sigma_2, \sigma_3\}$, where $\sigma_0(\alpha) = \alpha, \sigma_1(\alpha) = -\alpha, \sigma_2(\alpha) = 1/\alpha$, and $\sigma_3(\alpha) = -1/\alpha$. Thus G has three subgroups $H_1 = \{\sigma_0, \sigma_1\}, H_2 = \{\sigma_0, \sigma_2\}, H_3 = \{\sigma_0, \sigma_3\}$ of order 2, with associated subfields $\text{Fix}(H_1) = \mathbb{Q}(\alpha^2)$, $\text{Fix}(H_2) = \mathbb{Q}(\alpha + \alpha^{-1})$, and $\text{Fix}(H_3) = \mathbb{Q}(\alpha - \alpha^{-1})$. Each of these is a quadratic extension of $\mathbb{Q}$, and these are all the fields strictly intermediate between $\mathbb{E}$ and $\mathbb{Q}$. ◊

Example 4.10.15. We return to Example 4.8.6 and adopt the notation there. We have that $G = \text{Gal}(\mathbb{E}/\mathbb{Q})$ is a group of order 6, isomorphic to the symmetric group S_3, operating as the group of permutations of $\{\alpha, \beta, \gamma\}$, the roots of the irreducible polynomial $p(x)$. From our knowledge of the structure of S_3, we see that G has three subgroups of order 2 (= index 3), and the fixed fields of these subgroups are $\mathbb{F}(\alpha), \mathbb{F}(\beta), \mathbb{F}(\gamma)$. We observe that these subgroups are not normal subgroups of G, so that these fields are not Galois extensions of $\mathbb{Q}$. We also know that G has a normal subgroup of order 3 (= index 2), and so the fixed field of this subgroup is $\mathbb{Q}(\Delta)$ for some $\Delta \in \mathbb{E}$, a quadratic, and Galois, extension of $\mathbb{Q}$. We can choose $\Delta = (\alpha - \beta)(\beta - \gamma)(\alpha - \gamma)$. To see this, note that certainly $\Delta \in \mathbb{E}$, but if $\sigma \in G$ is an odd permutation, then $\sigma(\Delta) = -\Delta$ so

$\Delta \notin \text{Fix}(G) = \mathbb{Q}$. On the other hand Δ^2 is fixed by every element of G so $\Delta^2 \in \mathbb{Q}$. (We will elaborate on this below.) ◊

Example 4.10.16. Let us return to Example 4.8.7. To recapitulate, we considered there $f(x) = x^p - n \in \mathbb{Q}[x]$, for p an odd prime and $n \in \mathbb{Q}$ not a pth power. Then $f(x)$ is irreducible, and we have $\mathbb{E} = \mathbb{Q}(\zeta_p, \sqrt[p]{n})$, a splitting field of $f(x)$, with $(\mathbb{E}/\mathbb{Q}) = p(p-1)$. We found there that $G = \text{Gal}(\mathbb{E}/\mathbb{Q})$ is a nonabelian group; more precisely

$$G = \langle \sigma, \tau \mid \sigma^p = 1, \tau^{p-1} = 1, \tau\sigma\tau^{-1} = \sigma^r \rangle$$

where σ and τ are the automorphisms of $\mathbb{E}$ given by $\sigma(\sqrt[p]{n}) = \zeta_p(\sqrt[p]{n}), \sigma(\zeta_p) = \zeta_p$ and $\tau(\sqrt[p]{n}) = \sqrt[p]{n}, \tau(\zeta_p) = \zeta_p^r$, r a primitive root (mod p). Then $G = N \rtimes H$ is the semidirect product of the normal subgroup N, generated by σ, of order p, and the subgroup H, generated by τ, of order $p - 1$. As we observed in Example 4.10.15, H is cyclic.

Let us find all intermediate fields between $\mathbb{E}$ and $\mathbb{Q}$. Again, we do so by finding all subgroups of G.

Let $g \in G$. Since $G = N \rtimes H$, we see g can be written uniquely as $g = \sigma^i \tau^j$, $0 \leq i \leq p-1$, $0 \leq j \leq p-2$. It is easy to check from the structure of G that if $i \neq 0$, the order of g is divisible by p.

We also know from the Sylow theorems that G has a unique p-Sylow subgroup (and that subgroup is therefore normal). We see that that subgroup is N, and hence every element of order p must belong to N.

With these observations in hand, let us consider a subgroup F of G. There are two possibilities:

Case 1: The order of F is not divisible by p. In this case we must have $F \subseteq H$. But now we proceed as in Example 4.10.14. The subgroup F must be one of the subgroups H_d as in that example. Thus (following the notation there) we obtain the intermediate fields

$$\mathbb{B}_d = \text{Fix}(H_d) = \mathbb{Q}(\theta_d, \sqrt[p]{n}) \quad \text{with} \quad (\mathbb{E}/\mathbb{B}_d) = d$$
$$\text{and} \quad (\mathbb{B}_d/\mathbb{Q}) = p(p-1)/d.$$

Case 2: The order of F is divisible by p. In this case we must have $N \subseteq F$. Now the subgroups of G containing the normal subgroup N are in 1−1 correspondence with the subgroups of $G/N \cong H$.

Hence such a subgroup must be NH_d for some d. Thus we obtain the intermediate fields

$$\mathbb{D}_d = \text{Fix}(NH_d) = \mathbb{Q}(\theta_d) \quad \text{with} \quad (\mathbb{E}/\mathbb{D}_d) = pd$$
$$\text{and} \quad (\mathbb{D}_d/\mathbb{Q}) = (p-1)/d.$$

Since H is abelian, every such subgroup NH_d is normal in G, which agrees with the fact that every such field $\mathbb{D}_d$ is a Galois extension of $\mathbb{Q}$. $\Diamond$

4.11   More on Galois groups

Suppose that $\mathbb{E}$ is a Galois extension of $\mathbb{F}$ and consider the Galois group $G = \text{Gal}(\mathbb{E}/\mathbb{F})$. We have already seen a number of examples in which we were able to completely determine G. In complete generality, this is a subtle and difficult problem. But there are some things we can say with only a moderate amount of effort, and we will say them here.

The first basic theme of this section is that if extensions have a particular structure, we ought to be able to say something particular about their Galois groups.

The second basic theme of this section is in a way the opposite, that if extensions have no particular structure, their Galois groups should be as general as possible. The path to our goal here will take us through symmetric functions, which are themselves very important.

For the next few results we assume that all fields are contained in some large field $\mathbb{A}$, so that it makes sense to talk about composition of field extensions.

Recall Definition 4.3.9.: Two extensions $\mathbb{B}$ and $\mathbb{D}$ of $\mathbb{F}$ are disjoint if $\mathbb{B} \cap \mathbb{D} = \mathbb{F}$.

Theorem 4.11.1. *Let $\mathbb{B}$ and $\mathbb{D}$ be finite extensions of $\mathbb{F}$. Suppose that $\mathbb{B}$ is a Galois extension of $\mathbb{F}$. Then $\mathbb{BD}$ is a Galois extension of $\mathbb{D}$, and $\text{Gal}(\mathbb{BD}/\mathbb{D})$ is isomorphic to $\text{Gal}(\mathbb{B}/\mathbb{B} \cap \mathbb{D})$, a subgroup of $\text{Gal}(\mathbb{B}/\mathbb{F})$, with the isomorphism being given by restriction, $\sigma \mapsto \sigma|\mathbb{B}$.*

In particular, if in this situation $\mathbb{B}$ and $\mathbb{D}$ are disjoint extensions of $\mathbb{F}$, then $\text{Gal}(\mathbb{BD}/\mathbb{D})$ is isomorphic to $\text{Gal}(\mathbb{B}/\mathbb{F})$ with the isomorphism being given by restriction.

Proof. Since $\mathbb{B}$ is a Galois extension of $\mathbb{F}$, it is a splitting field of a separable polynomial $f(x) \in \mathbb{F}[x]$, so $\mathbb{B} = \mathbb{F}(\alpha_1, \ldots, \alpha_n)$ where $\alpha_1, \ldots, \alpha_n$ are the roots of $f(x)$ in $\mathbb{A}$. But then $\mathbb{BD} = \mathbb{D}(\alpha_1, \ldots, \alpha_n)$ so $\mathbb{BD}$ is a splitting field of $f(x) \in \mathbb{D}[x]$, and then $\mathbb{BD}$ is a Galois extension of $\mathbb{D}$.

Let $\sigma \in \mathrm{Gal}(\mathbb{BD}/\mathbb{D})$. Then σ fixes $f(x)$ (as $f(x)$ has coefficients in $\mathbb{D}$), so σ permutes the roots $\{\alpha_1, \ldots, \alpha_n\}$ of $f(x)$, and hence $\sigma(\mathbb{B}) = \mathbb{B}$. Thus the restriction of σ to $\mathbb{B}$ is an element of $\mathrm{Gal}(\mathbb{B}/\mathbb{B} \cap \mathbb{D})$.

Let $R \colon \mathrm{Gal}(\mathbb{BD}/\mathbb{D}) \to \mathrm{Gal}(\mathbb{B}/\mathbb{B} \cap \mathbb{D})$ be the restriction map. We want to show that R is an isomorphism. To do so, we must show that R is 1–1 and onto.

R is 1–1: Suppose that $\bar{\sigma} = R(\sigma)$ is the identity on $\mathbb{B}$. Then in particular $\bar{\sigma}(\alpha_i) = \alpha_i$ for each i, and then $\sigma(\alpha_i) = \alpha_i$ for each i. Also, σ is the identity on $\mathbb{D}$ as it is an element of $\mathrm{Gal}(\mathbb{BD}/\mathbb{D})$, so σ is the identity on $\mathbb{BD}$, i.e., $\sigma = \mathrm{id}$ in $\mathrm{Gal}(\mathbb{BD}/\mathbb{D})$.

R is onto: Let $\bar{\sigma} \in \mathrm{Gal}(\mathbb{B}/\mathbb{B} \cap \mathbb{D})$. Then $\bar{\sigma} \colon \mathbb{B} \to \mathbb{B}$ is an isomorphism, extending the identity on $\mathbb{B} \cap \mathbb{D}$, so in particular $\bar{\sigma}(f(x)) = f(x)$, as $f(x)$ has coefficients in $\mathbb{F} \subseteq \mathbb{B} \cap \mathbb{D}$. As we have observed, $\mathbb{BD}$ is a splitting field of $f(x) \in \mathbb{D}[x]$, so by Lemma 4.7.3 there is an isomorphism $\sigma \colon \mathbb{BD} \to \mathbb{BD}$ extending $\bar{\sigma}$, and so $R(\sigma) = \bar{\sigma}$. $\qquad\square$

Corollary 4.11.2. *Let $\mathbb{B}$ and $\mathbb{D}$ be disjoint finite extensions of $\mathbb{F}$. Suppose that $\mathbb{B}$ is a Galois extension of $\mathbb{F}$. Then*

$$(\mathbb{BD}/\mathbb{D}) = (\mathbb{D}/\mathbb{F}), \quad (\mathbb{BD}/\mathbb{B}) = (\mathbb{D}/\mathbb{F}), \quad \text{and} \quad (\mathbb{BD}/\mathbb{F}) = (\mathbb{B}/\mathbb{F})(\mathbb{D}/\mathbb{F}).$$

Proof. The first equality follows directly from Theorem 4.11.1 and Theorem 4.10.1. Then the others follow from Corollary 4.3.12. $\qquad\square$

Example 4.11.3. Here is a simple example where we do not have equality. Let $\mathbb{B} = \mathbb{Q}(\sqrt[3]{2})$ and $\mathbb{D} = \mathbb{Q}(\zeta_3 \sqrt[3]{2})$, disjoint extensions of $\mathbb{Q}$, both of degree 3, neither of which is Galois. Then $\mathbb{BD} = \mathbb{Q}(\zeta_3, \sqrt[3]{2})$ is an extension of $\mathbb{Q}$ of degree 6. $\qquad\Diamond$

Theorem 4.11.4.

(a) *Let $\mathbb{B}$ and $\mathbb{D}$ be disjoint finite extensions of $\mathbb{F}$, with $\mathbb{B}$ a Galois extension of $\mathbb{F}$, and let $\mathbb{E}$ be the composite $\mathbb{E} = \mathbb{BD}$. Suppose that $\mathbb{E}$ is a Galois extension of $\mathbb{F}$. Let $G = \mathrm{Gal}(\mathbb{E}/\mathbb{F})$, $N = \mathrm{Gal}(\mathbb{E}/\mathbb{B})$, $H = \mathrm{Gal}(\mathbb{E}/\mathbb{D})$. Then N is a normal subgroup of G and G is the semidirect product $G = N \rtimes H$. Also, H is isomorphic to $\mathrm{Gal}(\mathbb{B}/\mathbb{F})$.*

(b) *Let $\mathbb{B}$ and $\mathbb{D}$ be disjoint finite Galois extensions of $\mathbb{F}$, and let $\mathbb{E}$ be the composite $\mathbb{E} = \mathbb{B}\mathbb{D}$. Then $\mathbb{E}$ is a Galois extension of $\mathbb{F}$. Let $G = Gal(\mathbb{E}/\mathbb{F})$, $N_1 = Gal(\mathbb{E}/\mathbb{B})$, $N_2 = Gal(\mathbb{E}/\mathbb{D})$. Then N_1 and N_2 are normal subgroups of G, and G is the direct product $G = N_1 \times N_2$. Also, N_2 is isomorphic to $Gal(\mathbb{B}/\mathbb{F})$ and N_1 is isomorphic to $Gal(\mathbb{D}/\mathbb{F})$.*

(c) *Let $\mathbb{E}$ be a Galois extension of $\mathbb{F}$ and let $G = Gal(\mathbb{E}/\mathbb{F})$. Suppose that $G = N \rtimes H$ is the semidirect product of a normal subgroup N and a subgroup H (resp. that $G = N_1 \times N_2$ is the direct product of normal subgroups N_1 and N_2). If $B = Fix(N)$ and $\mathbb{D} = Fix(H)$ (resp. $\mathbb{B} = Fix(N_1)$ and $\mathbb{D} = Fix(N_2)$) then $\mathbb{B}$ is a Galois extension of $\mathbb{F}$ (resp. $\mathbb{B}$ and $\mathbb{D}$ are Galois extensions of $\mathbb{F}$) and $\mathbb{E} = \mathbb{B}\mathbb{D}$.*

Proof.

(a) N and H are both subgroups of G. Since $\mathbb{B}$ is a Galois extension of $\mathbb{F}$ we have, from the fundamental theorem of Galois theory, that N is a normal subgroup of G and that $Gal(\mathbb{B}/\mathbb{F})$ is the quotient $H_0 = G/N$. But we have an isomorphism from H to H_0 given by restriction, by Theorem 4.11.1, and this is the same as the quotient map, by Remark 4.10.4(d). Hence G is the semidirect product $G = N \rtimes H$.

(b) First we must show that $\mathbb{E}$ is a Galois extension of $\mathbb{F}$. Now N_1 is a subgroup of G, so $\mathrm{Fix}(G) \subseteq \mathrm{Fix}(N_1) = \mathbb{B}$. Also, N_2 is a subgroup of G, so $\mathrm{Fix}(G) \subseteq \mathrm{Fix}(N_2) = \mathbb{D}$. Then

$$\mathbb{F} \subseteq \mathrm{Fix}(G) \subseteq \mathbb{B} \cap \mathbb{D} = \mathbb{F}$$

so $\mathrm{Fix}(G) = \mathbb{F}$ and $\mathbb{E}$ is a Galois extension of $\mathbb{F}$. Now by part (a), G is the semidirect product $G = N_1 \rtimes N_2$ with both N_1 and N_2 normal, so G is the direct product $G = N_1 \times N_2$.

(c) This follows directly from the fundamental theorem of Galois theory and Lemma 4.10.5. □

Example 4.11.5. We have already seen an example of Theorem 4.11.4 (a) in Example 4.8.7. ◇

Example 4.11.6. We have already seen an example of Theorem 4.11.4 (b) in Example 4.8.2. ◇

We now turn our attention to symmetric functions. We fix an arbitrary field $\mathbb{F}_0$ and let

$$\mathbb{E} = \mathbb{F}_0(x_1, \ldots, x_n),$$

the field of rational functions in n commuting variables $x_1, \ldots, x_n$. Then the symmetric group S_n acts on $\mathbb{E}$ by permuting the variables.

Definition 4.11.7. Let $\mathbb{F} = \text{Fix}(S_n)$. Then $\mathbb{F}$ is the field of *symmetric functions* in the n variables $x_1, \ldots, x_n$ over $\mathbb{F}_0$. ◇

Lemma 4.11.8. *Let H be a subgroup of S_n and let $\mathbb{B} = \text{Fix}(H)$. Then $\mathbb{E}$ is a Galois extension of $\mathbb{B}$ with $\text{Gal}(\mathbb{E}/\mathbb{B}) = H$. In particular, $\mathbb{E}$ is a Galois extension of $\mathbb{F}$ with $\text{Gal}(\mathbb{E}/\mathbb{F}) = S_n$.*

Proof. This is just a special case of Corollary 4.9.13. □

Galois theory gives us a more concrete description of $\mathbb{F}$.

Definition 4.11.9. For $I = \{i_1, \ldots, i_k\}$ a subset of $\{1, \ldots, n\}$ let x_I be the product $x_I = x_{i_1} \ldots x_{i_k}$. (If $I = \phi$, $x_I = 1$.)

For $k = 0, \ldots, n$, the kth *elementary symmetric function* s_k of $\{x_1, \ldots, x_n\}$ is defined by

$$s_k = \sum_I x_I \text{ where the sum is over all } k\text{-element}$$
$$\text{subsets } I \text{ of } \{1, \ldots, n\}. \qquad ◇$$

Lemma 4.11.10. *There is a polynomial identity:*

$$f(x) = (x - x_1) \ldots (x - x_n) = \sum_{k=0}^{n} (-1)^k s_k x^{n-k}.$$

Proof. Direct computation. □

Theorem 4.11.11. *The field $\mathbb{F}$ of symmetric functions in n variables over $\mathbb{F}_0$ is*

$$\mathbb{F} = \mathbb{F}_0(s_0, \ldots, s_n)$$

the field of rational functions in the elementary symmetric functions $s_0, \ldots, s_n$.

Proof.   Let $\mathbb{D} = F_0(s_0, \ldots, s_n)$. Certainly $\mathbb{D} \subseteq \mathbb{F}$. By Lemma 4.11.8, $(\mathbb{E}/\mathbb{F}) = |S_n| = n!$. But we see from Lemma 4.11.10 that $\mathbb{E}$ is a splitting field of the polynomial $f(x)$, a polynomial of degree n in $\mathbb{D}[x]$, so $(\mathbb{E}/\mathbb{D}) \leq n!$ Hence $(\mathbb{E}/\mathbb{D}) = (\mathbb{E}/\mathbb{F}) = n!$, so $\mathbb{D} = \mathbb{F}$. $\square$

There is a lot to say about symmetric functions, but this is not the place to say them. Instead, we continue with Galois theory.

Lemma 4.11.12. *Let $p(x)$ be a separable polynomial in $\mathbb{F}[x]$ and let $\mathbb{E}$ be a splitting field of $p(x)$. Let $G = Gal(\mathbb{E}/\mathbb{F})$. Let $\{\alpha_1, \ldots, \alpha_n\}$ be the distinct roots of $p(x)$ in $\mathbb{E}$, and let $\varphi \colon G \to S_n$ be given by $\varphi(\sigma) =$ the permutation $\varphi(\sigma)(\alpha_1, \ldots, \alpha_n) = (\sigma(\alpha_1), \ldots, \sigma(\alpha_n))$.*

Let

$$\delta = \prod_{i<j}(\alpha_j - \alpha_i) \quad \text{and} \quad \Delta = \delta^2.$$

Then $Im(\varphi) \subseteq A_n$, the alternating group, if and only if Δ is a square in $\mathbb{F}$.

Proof.   It is easy to check that for any permutation $\rho \in S_n$,

$$\rho(\delta) = (-1)^{\mathrm{sign}(\rho)}\delta.$$

Thus if $\varphi(G) \subseteq A_n$, $\mathrm{sign}(\varphi(\sigma)) = 1$ for every $\sigma \in G$, so $\delta \in \mathrm{Fix}(G) = \mathbb{F}$ and Δ is a square in $\mathbb{F}$.

On the other hand, if $\varphi(G) \not\subseteq A_n$, let $\sigma_0 \in G$ with $\mathrm{sign}(\varphi(\sigma_0)) = -1$. Then $\varphi(\sigma_0)(\delta) = -\delta$ so $\Delta = \delta^2$ with δ *not* in $\mathbb{F}$ and so Δ is not a square in $\mathbb{F}$. $\square$

Now the point of this lemma is that Δ^2 is a symmetric function of $\{\alpha_1, \ldots, \alpha_n\}$, so by Theorem 4.11.11 can be expressed in terms of the elementary symmetric functions $s_0, \ldots, s_n$, and hence in terms of the coefficients of $p(x)$.

Lemma 4.11.13.

(a) *Let $p(x) = x^2 + ax + b \in \mathbb{F}[x]$ be a separable polynomial. Then $\Delta = a^2 - 4b$.*
(b) *Let $p(x) = x^3 + ax^2 + bx + c \in \mathbb{F}[x]$ be a separable polynomial. Then $\Delta = -4a^3c + a^2b^2 + 18abc - 4b^3 - 27c^2$.*

Proof.   Direct computation. $\square$

Example 4.11.14. Let $\mathbb{E} \subseteq \mathbb{C}$ be a splitting field of the polynomial $p(x) = x^3 + 7x + 1 \in \mathbb{Q}[x]$ of Example. Then $\Delta = -1399$, which is not a square in $\mathbb{Q}$, so $\mathrm{Gal}(\mathbb{E}/\mathbb{Q})$ is isomorphic to S_3, as we saw there. But now we see that the quadratic extension of $\mathbb{Q}$ contained in $\mathbb{E}$, which we simply described there as $\mathbb{Q}(\delta)$, is in fact $\mathbb{Q}(\sqrt{-1399})$. ◊

Example 4.11.15. Let $p(x) = x^3 - 3x + 1$, an irreducible polynomial in $\mathbb{Q}[x]$. Then $\Delta = 81$, which is a square in $\mathbb{Q}$, so $\mathrm{Gal}(\mathbb{E}/\mathbb{Q})$ is isomorphic to A_3, which is itself isomorphic to the cyclic group of order 3. ◊

Now let us return to general theory.

Corollary 4.11.16. *Every finite group is the Galois group $\mathrm{Gal}(\mathbb{E}/\mathbb{F})$ of some finite Galois extension.*

Proof. Every finite group is isomorphic to a subgroup of a symmetric group, by Cayley's theorem (Theorem 2.8.4.), so this follows directly from Lemma 4.11.8. □

Remark 4.11.17. Of course, this shows that every finite group is the Galois group of a Galois extension of *some* field. It is natural to ask whether every finite group is the Galois group of some extension of $\mathbb{Q}$. The answer to this question is unknown. ◊

Suppose now that $f(x) \in \mathbb{F}[x]$ is an irreducible separable polynomial of degree n and let $\mathbb{E}$ be a splitting field of $f(x)$. Let $f(x)$ have roots $\alpha_1, \ldots, \alpha_n$ in $\mathbb{E}$ and consider $G = \mathrm{Gal}(\mathbb{E}/\mathbb{F})$. We may think of constructing G in stages. First, we know that G acts transitively on the roots of $f(x)$, so there is an isomorphism from $\mathbb{F}(\alpha_1)$ to $\mathbb{F}(\alpha_i)$ for any $i = 1, \ldots, n$, and there are n possibilities. Now suppose we have some such isomorphism, and we wish to extend it to an isomorphism from $\mathbb{F}(\alpha_1, \alpha_2)$ to $\mathbb{F}(\alpha_i, \alpha_j)$. We have at most $n - 1$ ways of doing this, and we should expect that "in general" there should be exactly $n - 1$ ways of doing this, that we would have such an isomorphism taking α_2 to α_j for any $\alpha_j \neq \alpha_i$. Of course, this might not be the case. In the most extreme situation, we could have $\mathbb{E} = \mathbb{F}(\alpha_1)$ and then any automorphism of $\mathbb{E}$ is determined by its effect on α_1. But this precisely illustrates our point — in this situation we have some relationship between the roots of $f(x)$ (each α_i is a polynomial in α_1). So if we have no relationship between the roots, we will indeed have

$n-1$ ways of doing this. Then we wish to extend this isomorphism to one from $\mathbb{F}(\alpha_1, \alpha_2, \alpha_3)$ to $\mathbb{F}(\alpha_i, \alpha_j, \alpha_k)$ and "in general" there should be $n-2$ ways of doing this. Thus "in general" we should expect that there will be $n(n-1)(n-2)\ldots 1 = n!$ automorphisms of $\mathbb{E}$ fixing $\mathbb{F}$, or in other words that $G = \mathrm{Gal}(\mathbb{E}/\mathbb{F})$ is isomorphic to the full symmetric group S_n, and so G gives all permutations of the set of roots $\{\alpha_1, \ldots, \alpha_n\}$ of $f(x)$. For example, this was the case in Theorem 4.11.11. Otherwise said, G will be smaller exactly when there are some "hidden relations" between the roots of $f(x)$ which force automorphisms of $\mathbb{E}$ to have certain properties. (For example, if instead of beginning with an irreducible polynomial $f(x)$, we could begin with a product $f_1(x)f_2(x)$ of distinct irreducible separable polynomials $f_1(x)$ and $f_2(x)$, where $f_1(x)$ has roots $\{\alpha_1, \ldots, \alpha_j\}$ and $f_2(x)$ has roots $\{\beta_1, \ldots, \beta_k\}$ in $\mathbb{E}$; then any automorphism of $\mathbb{E}$ would have to leave each of the sets $\{\alpha_1, \ldots, \alpha_j\}$ and $\{\beta_1, \ldots, \beta_k\}$ invariant.)

While we have just expressed the expectation that "in general" if $f(x)$ is an irreducible separable polynomial, $G = \mathrm{Gal}(\mathbb{E}/\mathbb{F})$ should be isomorphic to S_n, we are in the not uncommon situation where we have a general expectation but it requires some work to find an example.

In fact, it is a theorem that for any n, there is a polynomial $f(x) \in \mathbb{Q}[x]$ with $\mathrm{Gal}(\mathbb{E}/\mathbb{Q})$ isomorphic to S_n, where $\mathbb{E} \subseteq \mathbb{C}$ is a splitting field of $f(x)$. We will not prove this in general, but we will prove this when n is a prime.

Lemma 4.11.18. *Let p be a prime. Let $f(x) \in \mathbb{Q}[x]$ be an irreducible polynomial with exactly $p-2$ real roots. Let $\mathbb{E} \subseteq \mathbb{C}$ be a splitting field of $f(x)$. Then $G = \mathrm{Gal}(\mathbb{E}/\mathbb{Q})$ is isomorphic to S_p.*

Proof.   Let $\alpha_1, \ldots, \alpha_{p-2}$ be the real roots of $f(x)$. Then $f(x) = (x - \alpha_1)\ldots(x - \alpha_{p-2})q(x)$ for some quadratic polynomial $q(x)$ with real coefficients, and then from the quadratic formula we see that $q(x)$ has two complex roots β_1 and β_2 with $\beta_2 = \bar{\beta}_1$ (the complex conjugate of β_1).

Since $f(x)$ is irreducible, G acts transitively on $S = \{\alpha_1, \ldots, \alpha_{p-2}, \beta_1, \beta_2\}$. But we also see that complex conjugation leaves $\mathbb{E}$ invariant, and restricted to $\mathbb{E}$ acts as a transposition τ on S. (That is, $\tau(\alpha_i) = \alpha_i$ for each $i = 1, \ldots, p-2$ and $\tau(\beta_1) = \beta_2$, $\tau(\beta_2) = \beta_1$). Thus G is isomorphic to a transitive subgroup of S_p that contains a transposition, and by Lemma 2.10.15 any such subgroup must be the full symmetric group S_p. ☐

It remains to show that there is such a polynomial. We do that now. It is trivial for $p = 2$ so we assume p odd.

Example 4.11.19. Let p be an odd prime. Let $f_0(x)$ be the polynomial

$$f_0(x) = x^2(x - 2)(x - 4)\dots(x - 2(p - 2)).$$

We observe that $f_0(x)$ has roots $x = 0$ (a double root) and $x = 2, 4, \dots, p-2$. Then we see from elementary calculus that $f_0(x)$ has a local maximum at $x = 0$, a local minimum at $x = a_1$ for some value of a_1 with $0 < a_1 < 2$, a local maximum at $x = a_2$ for some value of a_2 with $2 < a_2 < 4$, and alternate local maxima and minima a_i with $\{a_i\}$ "interleaved" between the roots of $f_0(x)$. Then $f_0(x)$ is strictly increasing between $x = a_1$ and $x = a_2$, strictly decreasing between $x = a_2$ and $x = a_3, \dots$.

The local maxima of $f(x)$ are at $x = a_2$ with $2 < a_2 < 4$, $x = a_4$ with $6 < a_4 < 8$, etc. Whatever the exact values of $f(a_2), f(a_4), \dots$, are, we know that $f_0(a_2)$ is the maximum value of $f(x)$ on the interval $[2,4]$, so in particular $f_0(a_2) \geq f_0(3) \geq 3^2$; similarly $f_0(a_4) \geq f_0(7) \geq 7^2$, etc. Thus, if we let

$$f(x) = f_0(x) - 2 = x^2(x - 2)(x - 4)\dots(x - 2(p - 2)) - 2,$$

we see that $f_0(x)$ has exactly $p - 2$ real roots: one between a_1 and a_2, one between a_2 and $a_3, \dots$, one between a_{p-3} and a_{p-2}, and one greater than a_{p-2}.

Finally, we note that $f(x)$ is a monic polynomial of degree p; every coefficient of $f(x)$ other than the coefficient of x^p is divisible by 2, and the constant term -2 is divisible by 2 but not by 4. Hence, by Eisenstein's criterion, $f(x)$ is irreducible. ◊

4.12 Simple extensions

Let $\mathbb{E}$ be a finite extension of $\mathbb{F}$. We know that $\mathbb{E} = \mathbb{F}(\alpha_1, \dots, \alpha_n)$ for some elements $\alpha_1, \dots, \alpha_n$ of $\mathbb{E}$, all of which are algebraic over $\mathbb{F}$. That is, $\mathbb{E}$ is obtained from $\mathbb{F}$ by adjoining finitely many algebraic elements.

We have already asked whether any finite extension of $\mathbb{F}$ can be obtained by adjoining a *single* element α of $\mathbb{E}$ to $\mathbb{F}$, so that $\mathbb{E} = \mathbb{F}(\alpha)$ for some element α of $\mathbb{E}$. In this section we show that is true in very great generality (though not all the time).

Definition 4.12.1. Let $\mathbb{E}$ be a finite extension of $\mathbb{F}$. If there is some element α of $\mathbb{E}$ such that $\mathbb{E} = \mathbb{F}(\alpha)$, then $\mathbb{E}$ is a *simple* extension of $\mathbb{F}$. Such an element α is called a *primitive element* of $\mathbb{E}$ over $\mathbb{F}$. $\Diamond$

We shall abbreviate this to primitive element of $\mathbb{E}$, when $\mathbb{F}$ is understood.

Theorem 4.12.2 (Theorem of the primitive element). *Let $\mathbb{E}$ be a finite separable extension of $\mathbb{F}$. Then $\mathbb{E}$ has a primitive element α, and so $\mathbb{E}$ is a simple extension of $\mathbb{F}$.*

Proof. First we consider the case where $\mathbb{F}$ is a finite field. Then $\mathbb{E}$ is a finite dimensional vector space over $\mathbb{F}$, so has only finitely many elements, i.e., $\mathbb{E}$ is a finite field. But then we know from Corollary 3.3.11 that $\mathbb{E}^*$, the multiplicative group of $\mathbb{E}$, is cyclic. Let α be a generator. Then certainly $\mathbb{E} = \mathbb{F}(\alpha)$.

Now suppose that $\mathbb{F}$ is infinite. Let $\mathbb{E} = \mathbb{F}(\alpha_1, \ldots, \alpha_n)$. First we claim that $\mathbb{E}$ is obtained from $\mathbb{F}$ by adjoining roots of a separable polynomial $f(x) \in \mathbb{F}[x]$. To see this, note that each α_i is a root of some separable irreducible polynomial $f_i(x) \in \mathbb{F}[x]$. Then we may take $f(x)$ to be the product (or, more neatly, the least common multiple) of $f_1(x), \ldots, f_n(x)$.

Let us extend $\mathbb{E}$ to $\mathbb{D}$, a splitting field of $f(x)$. Since $\mathbb{D}$ is the splitting field of a separable polynomial, we know that $\mathbb{D}$ is a Galois extension of $\mathbb{F}$ (Theorem 4.9.12). Let $G = \mathrm{Gal}(\mathbb{D}/\mathbb{F})$. Since $\mathbb{E}$ is an intermediate field between $\mathbb{D}$ and $\mathbb{F}$, by the fundamental theorem of Galois theory we know that $\mathbb{E} = \mathrm{Fix}(H)$ for some subgroup H of G.

Now let $\mathbb{B}$ be any proper subfield of $\mathbb{E}$. Then, again by the fundamental theorem of Galois theory, we know that $\mathbb{B} = \mathrm{Fix}(K)$ for some subgroup K of G that properly contains H. We also observe that $\dim_{\mathbb{F}} \mathbb{B} = (\mathbb{B}/\mathbb{F}) < (\mathbb{E}/\mathbb{F}) = \dim_{\mathbb{F}} \mathbb{E}$.

Now G is a finite group, so has only finitely many subgroups. Thus

$$\bigcup_{\mathbb{B} \subset \mathbb{E}} \mathbb{B} = \bigcup_{H \subset K} \mathrm{Fix}(K)$$

is a finite union of proper subspaces of $\mathbb{E}$ (regarded as an $\mathbb{F}$-vector space). But then, by Theorem B.1, this union cannot be all of $\mathbb{E}$. Let α be any element of $\mathbb{E}$ not in this union. Then $\mathbb{E} = \mathbb{F}(\alpha)$, so α is primitive and $\mathbb{E}$ is simple. □

Remark 4.12.3. A "general" element of $\mathbb{E}$ will not be a member of any finite union of proper subspaces of $\mathbb{E}$ so if we pick α "at random" we should expect that $\mathbb{E} = \mathbb{F}(\alpha)$. ◊

Before proceeding further, we want to make an important observation.

Remark 4.12.4. We have been careful to talk about a Galois group being isomorphic to a subgroup of a symmetric group S_n rather than a Galois group being a subgroup of a symmetric group S_n. Let us see why it has been necessary for us to be so careful. Consider, for example, two distinct irreducible separable polynomials $p_1(x)$ and $p_2(x)$ in $\mathbb{F}[x]$ of degrees n_1 and n_2 respectively. Set $n = n_1 + n_2$. Let $\mathbb{E}$ be a splitting field of the product $p_1(x)p_2(x)$. Then $p_1(x)$ has roots $\{\alpha_1, \ldots, \alpha_{n_1}\}$ in $\mathbb{E}$ and $p_2(x)$ has roots $\{\beta_1, \ldots, \beta_{n_2}\}$ in $\mathbb{E}$, and the action of the Galois group $Gal(\mathbb{E}/\mathbb{F})$ preserves each of these sets, so $Gal(\mathbb{E}/\mathbb{F})$ is isomorphic to a subgroup of $S_{n_1} \times S_{n_2} \subset S_n$; in particular $Gal(\mathbb{E}/\mathbb{F})$ is isomorphic to a *nontransitive* subgroup of S_n. On the other hand, from the theorem of the primitive element we know that $\mathbb{E}$ has a primitive element γ. Let $d = (\mathbb{E}/\mathbb{F})$, and note that d is the degree of γ over $\mathbb{F}$.

Then $m_\gamma(x)$, the minimal polynomial of γ, is an irreducible polynomial of degree d, so has d roots $\{\gamma_1, \ldots, \gamma_d\}$ in $\mathbb{E}$, and the action of the Galois group permutes these roots transitively, so G is isomorphic to a *transitive* subgroup of S_d.

Thus, in general, given an extension $\mathbb{E}$ of $\mathbb{F}$, the Galois group $Gal(\mathbb{E}/\mathbb{F})$ can be realized as a permutation group in very different ways. ◊

Theorem 4.12.2 tells us that, if $\mathbb{E}$ is separable, a primitive element always exists. It is natural to ask how to find one. We begin with a general criterion for an element of $\mathbb{E}$ to be primitive.

Lemma 4.12.5. *Let $\mathbb{E}$ be a finite Galois extension of $\mathbb{F}$. Then $\alpha \in \mathbb{E}$ is primitive if and only if $\sigma(\alpha) \neq \alpha$ for any $\sigma \in Gal(\mathbb{E}/\mathbb{F})$, $\sigma \neq id$, or, equivalently, if $\{\sigma(\alpha) \mid \sigma \in Gal(\mathbb{E}/\mathbb{F})\}$ are all distinct.*

Proof.   We show the contrapositive. By definition, α is not primitive if and only if $\mathbb{F}(\alpha)$ is a proper subfield of $\mathbb{E}$, and by the fundamental theorem of Galois theory, this will be the case if and only if $\mathbb{F}(\alpha)$ is fixed by some nontrivial subgroup H of $\mathrm{Gal}(\mathbb{E}/\mathbb{F})$, which will be the case if and only if α is fixed by H. $\square$

Example 4.12.6. Returning to Example 4.8.3, where $\mathbb{E} = \mathbb{Q}(\sqrt{a_1}, \ldots, \sqrt{a_t})$, from the description of $\mathrm{Gal}(\mathbb{E}/\mathbb{Q})$ we gave there we see that $\alpha = \sqrt{a_1} + \cdots + \sqrt{a_t}$ is a primitive element of $\mathbb{E}$ over $\mathbb{Q}$. $\Diamond$

Here is one situation in which we can readily exhibit a primitive element.

Lemma 4.12.7. *Let $\mathbb{E}$ be a finite Galois extension of $\mathbb{F}$ such that $G = Gal(\mathbb{E}/\mathbb{F})$ is an abelian group. Let $f(x) \in \mathbb{F}[x]$ be any irreducible polynomial having $\mathbb{E}$ as a splitting field, and let α be any root of $f(x)$ in $\mathbb{E}$. Then $\mathbb{E} = \mathbb{F}(\alpha)$, i.e., α is a primitive element of $\mathbb{E}$.*

Proof.   We prove the contrapositive. Let $f(x)$ have roots $\alpha_1, \ldots, \alpha_n$ in $\mathbb{E}$, so that $\mathbb{E} = \mathbb{F}(\alpha_1, \ldots, \alpha_n)$. Let $\alpha = \alpha_1$ and suppose that $\mathbb{E} \neq \mathbb{F}(\alpha)$. For each $i = 1, \ldots, n$, let H_i be the subgroup of the Galois group that fixes α_i, i.e., $H_i = \{\sigma \in G \mid \sigma(\alpha_i) = \alpha_i\}$. Then $H_i = \mathrm{Gal}(\mathbb{E}/\mathbb{F}(\alpha_i))$.

Since $f(x)$ is irreducible, G acts transitively on $\{\alpha_1, \ldots, \alpha_n\}$ and so $\{H_1, \ldots, H_n\}$ are mutually conjugate subgroups of G. Furthermore, these subgroups cannot all be the same, as then we would have $\mathbb{F}(\alpha_i) = \mathrm{Fix}(H_i) = \mathrm{Fix}(H_1) = \mathbb{F}(\alpha_1)$ for each i, in which case $\mathbb{E} = \mathbb{F}(\alpha_1, \ldots, \alpha_n) = \mathbb{F}(\alpha_1) \cdots \mathbb{F}(\alpha_n)$ (the composite) $= \mathbb{F}(\alpha_1)$.

Thus G contains the subgroup H_1 which is not normal, and so G is not abelian. $\square$

Before proceeding further, it is illuminating to introduce a stronger notion.

Definition 4.12.8. Let $\mathbb{E}$ be a finite Galois extension of $\mathbb{F}$ with Galois group $G = \mathrm{Gal}(\mathbb{E}/\mathbb{F})$. If $\theta \in \mathbb{E}$ with $S = \{\sigma(\theta) \mid \sigma \in G\}$ a vector space basis for $\mathbb{E}$ over $\mathbb{F}$, then S is called a *normal basis* for $\mathbb{E}$ over $\mathbb{F}$ and θ is a *normal element* of $\mathbb{E}$ over $\mathbb{F}$. $\Diamond$

Remark 4.12.9.

(a) Normal basis is standard language but normal element is not.

(b) It is a theorem, which we will not prove here, that every finite Galois extension has a normal basis.

(c) Since $|G| = (\mathbb{E}/\mathbb{F})$ (by Theorem 4.10.1), S is a basis for $\mathbb{E}$ if and only if S is linearly independent. ◊

Lemma 4.12.10. *Let $\mathbb{E}$ be a finite Galois extension of $\mathbb{F}$. If $\theta \in \mathbb{E}$ is normal, than θ is primitive.*

Proof. If $\{\sigma(\theta) \mid \sigma \in \mathrm{Gal}(\mathbb{E}/\mathbb{F})\}$ is linearly independent, then certainly all elements of this set are distinct, and so θ is primitive by Lemma 4.12.5. □

Example 4.12.11. Let $a \in \mathbb{Q}$ not be a perfect square, and let $\mathbb{E} = \mathbb{Q}(\sqrt{a})$. Then $\sqrt{a}$ is a primitive but not normal element of $\mathbb{E}$, while $1 + \sqrt{a}$ is a normal element of $\mathbb{E}$. ◊

Our theme for (almost) the remainder of this section will be to see that if we know primitive (or normal) elements of some field extensions, we can use them to obtain primitive (or normal) elements of other field extensions.

Again we shall implicitly assume, when necessary, that all our fields are subfields of some larger field $\mathbb{A}$, so that composition of fields makes sense.

Remark 4.12.12. We see immediately that if α is a primitive (resp. normal) element of $\mathbb{E}$ over $\mathbb{F}$ and $\mathbb{B}$ is any field intermediate between $\mathbb{E}$ and $\mathbb{F}$, then α is a primitive (resp. normal) element of $\mathbb{E}$ over $\mathbb{B}$. ◊

Lemma 4.12.13. *Let $\mathbb{B}$ and $\mathbb{D}$ be disjoint finite extensions of $\mathbb{F}$ with $\mathbb{B}$ a Galois extension of $\mathbb{F}$. Then $\beta \in \mathbb{B}$ is a primitive (resp. normal) element of $\mathbb{B}\mathbb{D}$ over $\mathbb{D}$ if and only if β is a primitive (resp. normal) element of $\mathbb{B}$ over $\mathbb{F}$. Also, $\delta \in \mathbb{D}$ is a primitive element of $\mathbb{B}\mathbb{D}$ over $\mathbb{B}$ if and only if δ is a primitive element of $\mathbb{D}$ over $\mathbb{F}$.*

Proof. The statement about β follows directly from Theorem 4.11.1. As for the statement about δ, let $\mathbb{D}_0 = \mathbb{F}(\delta)$. Since $\mathbb{B}$ and $\mathbb{D}$ are disjoint, certainly $\mathbb{B}$ and $\mathbb{D}_0$ are disjoint. But $\mathbb{D}_0 = \mathbb{D}$ if and only if $(\mathbb{D}_0/\mathbb{F}) = (\mathbb{D}/\mathbb{F})$, and $\mathbb{B}\mathbb{D}_0 = \mathbb{B}\mathbb{D}$ if and only if $(\mathbb{B}\mathbb{D}_0/\mathbb{B}) = (\mathbb{B}\mathbb{D})/\mathbb{B}$, so this follows from Corollary 4.11.2. □

Lemma 4.12.14. *Let $\mathbb{E}$ be a finite Galois extension of $\mathbb{F}$ and let θ be a normal element of $\mathbb{E}$ over $\mathbb{F}$. Let $\mathbb{B}$ be any field intermediate between*

$\mathbb{E}$ and $\mathbb{F}$, so that $\mathbb{B} = Fix(H)$ for some subgroup H of $G = Gal(\mathbb{E}/\mathbb{F})$. Set

$$\alpha_H = \sum_{\sigma \in H} \sigma(\theta).$$

(a) *If $\mathbb{B}$ is a Galois extension of $\mathbb{F}$ (which will be the case if and only if H is a normal subgroup of G) then α_H is a normal element of $\mathbb{B}$ over $\mathbb{F}$.*

(b) *In any case, α_H is a primitive element of $\mathbb{B}$ over $\mathbb{F}$.*

Proof.

(a) Let $H = H_1, \ldots, H_t$ be the left cosets of H in G and let $\alpha_{H_i} = \Sigma_{\sigma \in H_i}\sigma(\theta)$ for $i = 1, \ldots, t$. It is straightforward to check that $\{\alpha_{H_1}, \ldots, \alpha_{H_t}\}$ is a normal basis of $\mathbb{B}$ over $\mathbb{F}$.

(b) Let $\mathbb{B}_0$ be any proper subfield of $\mathbb{B}$. Then $\mathbb{B}_0 = Fix(K)$ for some subgroup K of G properly containing H. Let $\rho \in K, \rho \notin H$. Then $\rho \in H_i$ for some $i \neq 1$, and then $\rho(\alpha_{H_1}) = \alpha_{H_i}$. In particular, $\rho(\alpha_H) \neq \alpha_H$, and so $\alpha_H \notin \mathbb{B}_0$. Thus we must have $\mathbb{F}(\alpha_H) = \mathbb{B}$ and so α_H is a primitive element of $\mathbb{B}$ over $\mathbb{F}$. $\square$

Example 4.12.15. Returning to Example 4.10.13, we began there by considering $\mathbb{E} = \mathbb{Q}(\zeta_p)$. Our argument there shows that ζ_p is a normal element of $\mathbb{E}$. We further showed there that, for each field $\mathbb{E}_d$ intermediate between $\mathbb{E}$ and $\mathbb{Q}$, $\mathbb{E}_d = \mathbb{Q}(\theta_d)$. In fact, the construction of those elements θ_d was just a special case of this lemma, and we see now that each θ_d is not just primitive but also normal. $\lozenge$

Theorem 4.12.16. *Let $\mathbb{B}$ and $\mathbb{D}$ be disjoint finite extensions of $\mathbb{F}$ and let $\mathbb{E}$ be the composite $\mathbb{E} = \mathbb{B}\mathbb{D}$. Suppose that $\mathbb{B}$ and $\mathbb{E}$ are Galois extensions of $\mathbb{F}$.*

(a) *Suppose that $\beta \in \mathbb{B}$ is a primitive element of $\mathbb{E}$ over $\mathbb{D}$, or, equivalently, that β is a primitive element of $\mathbb{B}$ over $\mathbb{F}$, and that $\delta \in \mathbb{D}$ is a primitive element of $\mathbb{E}$ over $\mathbb{B}$, or, equivalently, that δ is a primitive element of $\mathbb{D}$ over $\mathbb{F}$. If $char(\mathbb{F}) \neq 0$ or if $char(\mathbb{F}) = p$ and at least one of $(\mathbb{E}/\mathbb{D})$ and $(\mathbb{E}/\mathbb{B})$ is relatively prime to p, or, equivalently, if at least one of $(\mathbb{B}/\mathbb{F})$ and $(\mathbb{D}/\mathbb{F})$ is relatively prime to p, then $\epsilon = \beta + \delta$ is a primitive element of $\mathbb{E}$ over $\mathbb{F}$.*

(b) *Suppose that $\beta \in \mathbb{B}$ is a normal element of $\mathbb{E}$ over $\mathbb{D}$, or, equivalently, that β is a normal element of $\mathbb{B}$ over $\mathbb{F}$, and that $\delta \in \mathbb{D}$ is a normal element of $\mathbb{E}$ over $\mathbb{B}$, or, equivalently, in the case that $\mathbb{D}$ is a Galois extension of $\mathbb{F}$, that δ is a normal element of $\mathbb{D}$ over $\mathbb{F}$. Then $\epsilon = \beta\delta$ is a normal element of $\mathbb{E}$ over $\mathbb{F}$.*

Proof. First note that the "equivalently" in the statement of the theorem follow from Theorem 4.11.1, Corollary 4.11.2, or Lemma 4.12.13.

(a) We have, by Theorem 4.11.4, that $G = \mathrm{Gal}(\mathbb{E}/\mathbb{F})$ is the semidirect product of its normal subgroup $N = \mathrm{Gal}(\mathbb{E}/\mathbb{B})$ and its subgroup $H = \mathrm{Gal}(\mathbb{E}/\mathbb{D})$. In particular, every element of G can be written uniquely as $\sigma\tau$ with $\sigma \in N$ and $\tau \in H$, and then $\sigma\tau = \tau\sigma'$ with $\sigma' = \tau^{-1}\sigma\tau \in N$. We will apply Lemma 4.12.4 to show that $\epsilon = \beta + \delta$ is a primitive element of $\mathbb{E}$ over $\mathbb{F}$. Suppose $\sigma\tau(\epsilon) = \epsilon$. Then

$$\beta + \delta = \sigma\tau(\beta+\delta) = \sigma\tau(\beta)+\sigma\tau(\delta) = \tau\sigma'(\beta)+\sigma\tau(\delta) = \tau(\beta)+\sigma(\delta)$$

and so

$$\beta - \tau(\beta) = \sigma(\delta) - \delta.$$

But, from Theorem 4.11.1, $\tau(\beta) \in \mathbb{B}$ and $\sigma(\delta) \in \mathbb{D}$. Hence $\beta - \tau(\beta) = \sigma(\delta) - d \in \mathbb{B} \cap \mathbb{D} = \mathbb{F}$.
If $\mathrm{char}(\mathbb{F}) = 0$ choose ρ to be either of σ and τ. If $\mathrm{char}(\mathbb{F}) = p$ then at least one of σ and τ has order k relatively prime to p (as $|N| = (\mathbb{E}/\mathbb{B})$ and $|H| = (\mathbb{E}/\mathbb{D})$); choose ρ to be such an element. If $\rho = \sigma$, choose $\gamma = \delta$, while if $\rho = \tau$ choose $\gamma = \beta$. Then in any case we have $\rho(\gamma) = \gamma + \alpha$ for some $\alpha \in \mathbb{F}$. But then, since $\rho^k = \mathrm{id}$, we have $\gamma = \rho^k(\gamma) = \gamma + k\alpha$ so $\alpha = 0$. But then $\rho = \mathrm{id}$, and then we see that both $\sigma = \mathrm{id}$ and $\tau = \mathrm{id}$, so $\sigma\tau = \mathrm{id} \in G$ and by Lemma 4.12.4 ϵ is a primitive element of $\mathbb{E}$ over $\mathbb{F}$.
(b) This follows directly from our construction of a basis of $\mathbb{E}$ over $\mathbb{F}$ in Lemma 4.3.11. (We observed there in general that we had a spanning set, but given Corollary 4.11.2, here this spanning set is a basis.) $\square$

Example 4.12.17. We return to Example 4.8.3. Beginning with Example 4.12.11, and applying Theorem 4.12.16 inductively, we see

344 An Introduction to Abstract Algebra: Sets, Groups, Rings, and Fields

that $\Sigma_{i=1}^{t}\sqrt{a_i}$ is a primitive element of $\mathbb{E}$ over $\mathbb{Q}$ (as we did in Example 4.12.6) and that $\Pi_{i=1}^{t}(1 + \sqrt{a_i})$ is a normal element of $\mathbb{E}$ over $\mathbb{Q}$. $\diamond$

Example 4.12.18. We consider the field $\mathbb{E}$ of Example 4.8.7, $\mathbb{E} = \mathbb{Q}(\zeta_p, \sqrt[p]{n}) = \mathbb{Q}(\zeta_p)\mathbb{Q}(\sqrt[p]{n})$, with these two fields being disjoint extensions of $\mathbb{Q}$. Then, from Theorem 4.12.16, we see that $\zeta_p + \sqrt[p]{n}$ is a primitive element of $\mathbb{E}$ over $\mathbb{Q}$. $\diamond$

In Theorem 4.12.16 we obtained primitive elements by adding primitive elements, and normal elements by multiplying normal elements. Let us see that the roles of addition and multiplication cannot be interchanged.

Example 4.12.19. Let $\mathbb{E} = \mathbb{Q}(\sqrt{2}, \sqrt{3})$. Then $\sqrt{2}$ is a primitive element of $\mathbb{Q}(\sqrt{2})$, and $\sqrt{3}$ is a primitive element of $\mathbb{Q}(\sqrt{3})$, but $(\sqrt{2})(\sqrt{3}) = \sqrt{6}$ is not a primitive element of $\mathbb{E}$.

Also, $1 + \sqrt{2}$ is a normal element of $\mathbb{Q}(\sqrt{2})$, and $-1 + \sqrt{3}$ is a normal element of $\mathbb{Q}(\sqrt{3})$, but $(1 + \sqrt{2}) + (-1 + \sqrt{3}) = \sqrt{2} + \sqrt{3}$ is not a normal element of $\mathbb{E}$. $\diamond$

Theorem 4.12.16 had a number of hypotheses. First, it required that the extensions $\mathbb{B}$ and $\mathbb{D}$ be disjoint. Here is an example to show we need this hypothesis.

Example 4.12.20. Let $\mathbb{B} = \mathbb{Q}(\sqrt{2}, \sqrt{3})$ and $\mathbb{D} = \mathbb{Q}(\sqrt{3}, \sqrt{5})$, so that $\mathbb{E} = \mathbb{B}\mathbb{D} = \mathbb{Q}(\sqrt{2}, \sqrt{3}, \sqrt{5})$. Then $\sqrt{2} + \sqrt{3}$ is a primitive element of $\mathbb{B}$ over $\mathbb{Q}$, and $-\sqrt{3} + \sqrt{5}$ is a primitive element of $\mathbb{D}$ over $\mathbb{Q}$, but $(\sqrt{2} + \sqrt{3}) + (-\sqrt{3} + \sqrt{5}) = \sqrt{2} + \sqrt{5}$ is not a primitive element of $\mathbb{E}$ over $\mathbb{Q}$.

Also, $(1 + \sqrt{2})(1 + \sqrt{3})$ is a normal element of $\mathbb{B}$ over $\mathbb{Q}$, and $(1 - \sqrt{3})(1 + \sqrt{5})$ is a normal element of $\mathbb{D}$ over $\mathbb{Q}$, but $((1 + \sqrt{2})(1 + \sqrt{3}))((1 - \sqrt{3})(1 + \sqrt{5})) = -2(1 + \sqrt{2})(1 + \sqrt{5})$ is not a normal element of $\mathbb{E}$ over $\mathbb{Q}$. $\diamond$

Next, Theorem 4.12.16 required that at least one of the two intermediate extensions be Galois. Here is an example to show we need this hypothesis.

Example 4.12.21. Let ζ_3 be a primitive cube root of 1 and recall that $1 + \zeta_3 + \zeta_3^2 = 0$. Let $\mathbb{B} = \mathbb{Q}(\sqrt[3]{2})$ and $\mathbb{D} = \mathbb{Q}(\zeta_3\sqrt[3]{2})$. Then $\mathbb{E} = \mathbb{B}\mathbb{D} = \mathbb{Q}(\sqrt[3]{2}, \zeta_3\sqrt[3]{2}) = \mathbb{Q}(\zeta_3, \sqrt[3]{2})$ is a Galois extension of $\mathbb{Q}$ of degree 6. Then $\sqrt[3]{2}$ is a primitive element of $\mathbb{B}$ and $\zeta_3 \sqrt[3]{2}$ is a primitive element of $\mathbb{D}$, but $\sqrt[3]{2} + \zeta_3 \sqrt[3]{2} = -\zeta_3^2 \sqrt[3]{2}$ is not a primitive element of $\mathbb{E}$ over $\mathbb{Q}$. $\Diamond$

Finally, Theorem 4.12.16 had a condition on the degrees of the two intermediate extensions in case $\text{char}(\mathbb{F}) = p$. Here is an example to show we need this hypothesis.

Example 4.12.22. Let $\mathbb{F} = \mathbb{F}_p(t_1, t_2)$, the field of rational functions in two variables t_1 and t_2 over $\mathbb{F}_p$. Let $f_1(x) = x^p - x - t_1$ and $f_2(x) = x^p - x - t_2$, polynomials in $\mathbb{F}[x]$, and let $\mathbb{E}$ be a splitting field of $f_1(x)f_2(x)$. Then $\mathbb{E} = \mathbb{F}(s_1, s_2) = \mathbb{F}(s_1)\mathbb{F}(s_2)$, where $\mathbb{B} = \mathbb{F}(s_1)$ is obtained by adjoining a root of $f_1(x)$ to $\mathbb{F}$ and $\mathbb{D} = \mathbb{F}(s_2)$ is obtained by adjoining a root of $f_2(x)$ to $\mathbb{F}$. Thus $s_1^p - s_1 - t_1 = 0$ and $s_2^p - s_2 - t_2 = 0$. But then

$$(s_1 + 1)^p - (s_1 + 1) - t_1 = s_1^p + 1 - (s_1 + 1) - t_1 = s_1^p - s_1 - t_1 = 0$$

and similarly $(s_2 + 1)^p - (s_2 + 1) = t_2 = 0$. Then the roots of $f_1(x)$ in $\mathbb{B}$ (or $\mathbb{E}$) are $\{s_1, s_1 + 1, \ldots, s_1 + (p - 1)\}$ and the roots of $f_2(x)$ in $\mathbb{D}$ (or $\mathbb{E}$) are $\{s_2, s_2 + 1, \ldots, s_2 + (p - 1)\}$. We then see that $G = \text{Gal}(\mathbb{E}/\mathbb{F})$ is isomorphic to $\mathbb{Z}_p \times \mathbb{Z}_p$, with generators σ and τ acting by $\sigma(s_1) = s_1 + 1, \sigma(s_2) = s_2$ and $\tau(s_1) = s_1, \tau(s_2) = \zeta_2 + 1$. Now s_1 is a primitive element of $\mathbb{B}$ and s_2 is a primitive element of $\mathbb{D}$, but

$$\sigma\tau^{-1}(s_1 + s_2) = \sigma\tau^{-1}(s_1) + \sigma\tau^{-1}(s_2) = \sigma(s_1) + \tau^{-1}(s_2)$$

$$= (s_1 + 1) + (s_2 - 1) = s_1 + s_2$$

so, by Lemma 4.12.5, $s_1 + s_2$ is not a primitive element of $\mathbb{E}$ over $\mathbb{F}$. $\Diamond$

We close this section with an example of an extension that is not simple.

Example 4.12.23. Let $\mathbb{F} = \mathbb{F}_p(t_1, t_2)$ be the field of rational functions in two variables t_1 and t_2 over $\mathbb{F}_p$, as in the last example. Let $f_1(x) = x^p - t_1$ and $f_2(x) = x^p - t_2$, polynomials in $\mathbb{F}[x]$, and let $\mathbb{E}$ be a splitting field of $f_1(x)f_2(x)$. Then $\mathbb{E} = \mathbb{F}(s_1)\mathbb{F}(s_2)$, where $\mathbb{B} = \mathbb{F}(s_1)$ is obtained by adjoining a root of $f_1(x)$ to $\mathbb{F}$ and $\mathbb{D} = \mathbb{F}(s_2)$ is obtained by adjoining a root of $f_2(x)$ to $\mathbb{F}$. Note that neither $\mathbb{B}$ nor $\mathbb{D}$

(nor $\mathbb{E}$) is a separable extension of $\mathbb{F}$: $x^p - t_1 = (x - s_1)^p$ in $\mathbb{B}[x]$ and $x^p - t_2 = (x - s_2)^p$ in $\mathbb{D}[x]$. Now $\mathbb{E}$ is an extension of $\mathbb{F}$ of degree p^2, but it is easy to check that if α is any element of $\mathbb{E}$, then $\alpha^p \in \mathbb{F}$, so $\mathbb{F}(\alpha)$ is an extension of $\mathbb{F}$ of degree 1 or p, and hence $\mathbb{F}(\alpha) \notin \mathbb{E}$. Thus, $\mathbb{E}$ does not contain a primitive element over $\mathbb{F}$, i.e., $\mathbb{E}$ is not a simple extension of $\mathbb{F}$. $\qquad\qquad\qquad\qquad\qquad\qquad\qquad\qquad\qquad\Diamond$

4.13   Finite fields

With the help of Galois theory, we can readily determine the structure of finite fields.

Theorem 4.13.1. *Let p be a prime.*

(a) *For any positive integer n, there is a field $\mathbb{F}_{p^n}$ containing p^n elements. This field is a splitting field of the polynomial $f(x) = x^{p^n} - x \in \mathbb{F}_p[x]$, and is unique up to isomorphism.*

(b) *Let $m \leq n$. If m does not divide n, then $\mathbb{F}_{p^n}$ does not contain a field of p^n elements. If m divides n, $\mathbb{F}_{p^n}$ contains a unique field of p^m elements.*

(c) *The Galois group $\mathrm{Gal}(\mathbb{F}_{p^n}/\mathbb{F})$ is a cyclic group of order n, generated by the Frobenius automorphism $\Phi: \mathbb{F}_{p^n} \to \mathbb{F}_{p^n}$ given by $\Phi(\alpha) = \alpha^p$.*

Proof.

(a) Let $\tilde{\mathbb{E}}$ be any field in which $f(x)$ splits. Then, by Corollary 4.9.17, $f(x)$ has p^n distinct roots in $\tilde{\mathbb{E}}$. Let $\mathbb{E} = \{\alpha \in \tilde{\mathbb{E}} \mid f(\alpha) = 0\}$. We claim that $\mathbb{E}$ is a field. Certainly $0 \in \mathbb{E}$ and $1 \in \mathbb{E}$. If $\alpha_1, \alpha_2 \in \mathbb{E}$ then

$$f(\alpha_1 + \alpha_2) = (\alpha_1 + \alpha_2)^{p^n} - (\alpha_1 + \alpha_2)$$
$$= (\alpha_1^{p^n} - \alpha_1) + (\alpha_2^{p^n} - \alpha_2) = 0 + 0 = 0$$

so $\alpha_1 + \alpha_2 \in \mathbb{E}$. Also,

$$f(\alpha_1 \alpha_2) = (\alpha_1 \alpha_2)^{p^n} - \alpha_1 \alpha_2 = \alpha_1^{p^n} \alpha_2^{p^n} - \alpha_1 \alpha_2 = \alpha_1 \alpha_2 - \alpha_1 \alpha_2 = 0$$

so $\alpha_1\alpha_2 \in \mathbb{E}$. Also, if $\alpha_1 \neq 0$,

$$f(1/\alpha_1) = (1/\alpha_1)^{p^n} - (1/\alpha_1) = 1/\alpha_1^{p^n} - 1/\alpha_1 = 1/\alpha_1 - 1/\alpha_1 = 0.$$

Thus we see that $\mathbb{E}$ is a field. $\mathbb{E}$ consists entirely of the roots of $f(x)$, so $\mathbb{E}$ has exactly p^n elements. Clearly $f(x)$ splits in $\mathbb{E}$, and cannot split in any proper subfield of $\mathbb{E}$, as such a subfield would contain fewer than p^n elements. Thus $\mathbb{E}$ is a field of order p^n that is a splitting field of the polynomial $f(x)$. Then $\mathbb{E}$ is unique up to isomorphism by Lemma 4.7.3.

(b) Let $G = \mathbb{F}_{p^n}^*$ be the multiplicative group of $\mathbb{F}_{p^n}$. Then G is a group of $p^n - 1$ elements, and G is cyclic by Corollary 3.3.11. Let $m \leq n$. It is easy to check that if m does not divide n, $p^m - 1$ does not divide $p^n - 1$, and so G cannot have a subgroup containing $p^m - 1$ elements, and so $\mathbb{F}_{p^n}$ cannot have a subfield containing p^m elements. On the other hand, if m divides n, then $p^m - 1$ divides $p^n - 1$, so G contains a unique subgroup H with $p^m - 1$ elements. Let $\mathbb{B} = H \cup \{0\}$. If $\alpha \in \mathbb{B}$, $\alpha \neq 0$, then $\alpha^{p^m} - \alpha = \alpha(\alpha^{p^m-1} - 1) = \alpha(0) = 0$, and certainly $0^{p^m} - 0 = 0$. Thus, by the same argument as in part (a), $\mathbb{B}$ is a subfield of $\mathbb{F}_{p^n}$ containing p^m elements.

(c) On the one hand, $(\mathbb{F}_{p^n})/\mathbb{F}_p) = \dim_{\mathbb{F}_p} \mathbb{F}_{p^n} = n$. On the other hand, by Theorem 4.10.1, $(\mathbb{F}_{p^n}/\mathbb{F}) = |\mathrm{Gal}(\mathbb{F}_{p^n}/\mathbb{F})|$. Thus this Galois group is a group of order n. But $\Phi \in \mathrm{Gal}(\mathbb{F}_{p^n}/\mathbb{F})$ and Φ is an element of this group of order n by Lemma 4.8.8. Thus we must have that $\mathrm{Gal}(\mathbb{F}_{p^n}/\mathbb{F})$ is the cyclic group of order n generated by Φ. $\qquad\square$

4.14 Cyclotomic fields

We have already investigated the pth cyclotomic field $\mathbb{Q}(\zeta_p)$. We now investigate the nth cyclotomic field $\mathbb{Q}(\zeta_n)$ for an arbitrary positive integer n.

Definition 4.14.1. Let n be a positive integer. Let $\zeta_n = \exp(2\pi i/n)$. Then $\mathbb{Q}(\zeta_n)$ is the nth *cyclotomic field*. $\qquad\diamond$

Definition 4.14.2. Let $\zeta \in \mathbb{C}$ with $\zeta^n = 1$. Then ζ is a *primitive nth root of 1* if $\zeta^m \neq 1$ for any positive integer m less than (or, equivalently, properly dividing) n. $\qquad\diamond$

Definition 4.14.3. The nth *cyclotomic polynomial* $\Phi_n(x)$ is defined by

$$\Phi_n(x) = \prod(x - \zeta) \in \mathbb{C}[x]$$

where the product is taken over the primitive nth roots of 1.   ◊

Lemma 4.14.4. *Let n be an arbitrary positive integer. Then*

$$x^n - 1 = \prod_{d|n} \Phi_d(x).$$

Proof.   The roots of $x^n - 1$ are *all* the complex nth roots of 1. But every complex nth root of 1 is a primitive dth root of 1 for some unique d dividing n, so grouping them together yields the right-hand side.   □

Theorem 4.14.5. *Let n be an arbitrary positive integer. Then $\Phi_n(x)$ is a polynomial with integer coefficients, i.e., $\Phi_n(x) \in \mathbb{Z}[x]$.*

Proof.   We prove this by complete induction on n.

For $n = 1$, $\Phi_1(x) = x - 1 \in \mathbb{Z}[x]$.

Now suppose the theorem is true for every $d < n$. By Lemma 4.14.4, we have that

$$f(x) = x^n - 1 = \left(\Pi' \Phi_d(x)\right) \Phi_n(x) = g(x)\Phi_n(x).$$

where the product inside the parentheses is taken over the proper divisors of n. Clearly $g(x)$ divides $f(x)$ in $\mathbb{C}[x]$. But $f(x) \in \mathbb{Q}[x]$ and by the inductive hypothesis $g(x) \in \mathbb{Q}[x]$. But then, by Lemma 4.2.17, $g(x)$ divides $f(x)$ in $\mathbb{Q}[x]$. Thus, $f(x) = g(x)h(x)$ with $h(x) \in \mathbb{Q}[x]$ so $\Phi_n(x) = h(x) \in \mathbb{Q}[x]$. But then, by Gauss's Lemma $\Phi_n(x) \in \mathbb{Z}[x]$.   □

Remark 4.14.6. Note that Lemma 4.14.4 gives us an inductive procedure for finding $\Phi_n(x)$. Namely,

$$\Phi_n(x) = (x^n - 1)/\left(\Pi' \Phi_d(x)\right)$$

where the product is taken over the proper divisors d of n.   ◊

Theorem 4.14.7. *For every positive integer n, the nth cyclotomic polynomial $\Phi_n(x)$ is irreducible in $\mathbb{Z}[x]$ (or, equivalently, in $\mathbb{Q}[x]$).*

Proof (Dedekind). We prove this theorem by contradiction.

Suppose that $\Phi_n(x)$ is not irreducible. Then $\Phi_n(x)$ has a monic irreducible factor $f(x)$ in $\mathbb{Q}[x]$ with $f(\zeta_n) = 0$ but with $f(\zeta) \neq 0$ where ζ is some primitive nth root of 1. Now every primitive nth root of 1 is of the form $\zeta = \zeta_n^k$ for some k relatively prime to n. Choose the smallest such k with $f(\zeta_n^k) \neq 0$ and let p be any prime factor of k.

Let $\zeta_0 = \zeta_n^{k/p}$ so that $f(\zeta_0) = 0$ but $f(\zeta_0^p) = f(\zeta) \neq 0$.

Then $f(x)$ is an irreducible monic polynomial in $\mathbb{Q}[x]$ with $f(\zeta_0) = 0$, and $k(x) = x^n - 1$ is a monic polynomial in $\mathbb{Q}[x]$ with $k(\zeta_0) = 0$, so $f(x)$ divides $k(x)$ in $\mathbb{Q}[x]$, i.e., $k(x) = f(x)g(x)$ with $f(x), g(x) \in \mathbb{Q}[x]$. But then, by Gauss's Lemma, $f(x), g(x) \in \mathbb{Z}[x]$. Now

$$0 = k(\zeta) = f(\zeta)g(\zeta) \text{ with } f(\zeta) \neq 0, \text{ so } g(\zeta) = 0.$$

Let $g(x) = \Sigma c_i x^i$, so $g(\zeta) = \Sigma c_i \zeta^i = \Sigma c_i (\zeta_0^p)^i$, so $h(\zeta_0) = 0$ where $h(x)$ is the polynomial $h(x) = \Sigma c_i x^{pi}$.

Then $f(x)$ is an irreducible polynomial in $\mathbb{Q}[x]$ with $f(\zeta_0) = 0$, and $h(x)$ is a polynomial in $\mathbb{Q}[x]$ with $h(\zeta_0) = 0$, so $f(x)$ divides $h(x)$ in $\mathbb{Q}[x]$, and by Gauss's Lemma again, $f(x)$ divides $h(x)$ in $\mathbb{Z}[x]$. Observe that $h(x) = g(x^p)$.

Now let $\pi\colon \mathbb{Z}[x] \to \mathbb{Z}_p[x] = \mathbb{F}_p[x]$ be the map given by reducing coefficients (mod p). Then

$$\pi(f(x)) \text{ divides } \pi(h(x)) = \pi(g(x^p)) = \pi(g(x))^p \quad \text{in } \mathbb{Z}_p[x],$$

so $\pi(f(x))$ and $\pi(g(x))$ have an irreducible factor $\bar{m}(x)$ in common in $\mathbb{Z}_p[x] = \mathbb{F}_p[x]$.

Now $k(x) = f(x)g(x)$ so $\pi(k(x)) = \pi(f(x))\pi(g(x))$ so we see that $\pi(k(x))$ is divisible by $\bar{m}(x)^2$ in $\mathbb{F}_p[x]$. But then $\bar{k}(x) = \pi(k(x))$ would have a multiple root in a splitting field $\mathbb{E}$. But $\bar{k}(x) = x^n - 1$ with n relatively prime to p, so $\bar{k}'(x) = nx^{n-1} \neq 0$ in $\mathbb{F}_p[x]$, so by the proof of Lemma 4.9.16 this cannot be the case. $\qquad\square$

Corollary 4.14.8. *Let n be a positive integer. Then $\mathbb{Q}(\zeta_n)$ is an extension of $\mathbb{Q}$ of degree $\varphi(n)$, where $\varphi(n)$ is the Euler totient function.*

Proof. $\mathbb{Q}(\zeta_n)$ is obtained from $\mathbb{Q}$ by adjoining a root ζ_n of the cyclotomic polynomial $\Phi_n(x)$, an irreducible polynomial of degree $\varphi(n)$. $\qquad\square$

We now study the relationship between different cyclotomic fields.

Lemma 4.14.9. *Let m and n be positive integers. Set $g = \gcd(m, n)$ and $\ell = \operatorname{lcm}(m, n)$. Then $\mathbb{Q}(\zeta_m)\mathbb{Q}(\zeta_n) = \mathbb{Q}(\zeta_\ell)$ and $\mathbb{Q}(\zeta_m) \cap \mathbb{Q}(\zeta_n) = \mathbb{Q}(\zeta_g)$. In particular, if m and n are relatively prime then $\mathbb{Q}(\zeta_m)\mathbb{Q}(\zeta_n) = \mathbb{Q}(\zeta_{mn})$ and $\mathbb{Q}(\zeta_m) \cap \mathbb{Q}(\zeta_n) = \mathbb{Q}$.*

Proof.   Since $\zeta_m = (\zeta_\ell)^{\ell/m}$ and $\zeta_n = (\zeta_\ell)^{\ell/n}$ we certainly have that $\mathbb{Q}(\zeta_m)\mathbb{Q}(\zeta_n) \subseteq \mathbb{Q}(\zeta_\ell)$.

On the other hand, we know that we can write $g = mx_0 + ny_0$ for some integers x_0, y_0. We also know that $mn = g\ell$. Thus,

$$1/\ell = g/mn = (mx_0 + ny_0)/mn = x_0/n + y_0/m$$

and so

$$\zeta_\ell = \exp(2\pi i/\ell) = \exp(2\pi i(x_0/n + y_0/m))$$
$$= \exp((2\pi i/n)x_0)\exp((2\pi i/m)y_0)$$
$$= \zeta_n^{x_0}\zeta_m^{y_0}$$

so $\mathbb{Q}(\zeta_\ell) \subseteq \mathbb{Q}(\zeta_m)\mathbb{Q}(\zeta_n)$ and hence they are equal.

Let $\mathbb{E} = \mathbb{Q}(\zeta_m) \cap \mathbb{Q}(\zeta_n)$ and set $e = (\mathbb{E}/\mathbb{Q})$. Since $\zeta_g = (\zeta_m)^{m/g} = (\zeta_n)^{n/g}$ we see that $\mathbb{Q}(\zeta_g) \subseteq \mathbb{E}$. Now $\mathbb{Q}(\zeta_m)$ and $\mathbb{Q}(\zeta_n)$ are disjoint Galois extensions of $\mathbb{E}$, so by Corollary 4.11.2,

$$(\mathbb{Q}(\zeta_\ell)/\mathbb{E}) = (\mathbb{Q}(\zeta_m)\mathbb{Q}(\zeta_n)/\mathbb{E})$$
$$= (\mathbb{Q}(\zeta_m)/\mathbb{E})(\mathbb{Q}(\zeta_n)/\mathbb{E})$$

and so

$$(\mathbb{Q}(\zeta_\ell)/\mathbb{E})(\mathbb{E}/\mathbb{Q})(\mathbb{E}/\mathbb{Q})$$
$$= (\mathbb{Q}(\zeta_m)/\mathbb{E})(\mathbb{E}/\mathbb{Q})(\mathbb{Q}(\zeta_n)/\mathbb{E})(\mathbb{E}/\mathbb{Q}),$$
$$(\mathbb{Q}(\zeta_\ell)/\mathbb{Q})(\mathbb{E}/\mathbb{Q}) = (\mathbb{Q}(\zeta_m)/\mathbb{Q})(\mathbb{Q}(\zeta_n)/\mathbb{Q}),$$

i.e.,

$$\varphi(\ell)e = \varphi(m)\varphi(n).$$

But it follows from our earlier work (see Lemma 3.6.16) that $\varphi(\ell)\varphi(g) = \varphi(m)\varphi(n)$ for any two positive integers m and n, so $e = \varphi(g)$ and $\mathbb{E} = \mathbb{Q}(\zeta_g)$. $\square$

Finally, we look at the Galois group.

Theorem 4.14.10. *Let n be a positive integer. Then the Galois group $Gal(\mathbb{Q}(\zeta_n)/\mathbb{Q})$ is isomorphic to $\mathbb{Z}_n^*$, the multiplicative group of $\mathbb{Z}_n$.*

Proof. The cyclotomic field $\mathbb{Q}(\zeta_n)$ is obtained from $\mathbb{Q}$ by adjoining the element ζ_n, a root of the irreducible polynomial $\Phi_n(x)$, and so there is a unique element σ of $Gal(\mathbb{Q}(\zeta_n)/\mathbb{Q})$ with $\sigma(\zeta_n) = \zeta_n^k$ for every integer $k \pmod{n}$ that is relatively prime to n. $\qquad\square$

4.15 Solvability and unsolvability of equations

Greek mathematicians at the time of Euclid knew how to solve quadratic equations, and the quadratic formula is very old. There matters stood for a long time, until in the 16th century Cardano and Tartaglia derived a formula for solving cubic equations, and Ferrari derived a formula for solving quartic equations.

The next case is quintic equations, which mathematicians worked on for centuries — indeed, this problem spurred on the development of algebra. Then, in 1824, Abel showed that is no quintic formula, i.e., that the general quintic cannot be solved by radicals.

In this section we will develop a criterion for an equation to be solvable by radicals, and apply it to show that the general equation of degree greater than or equal to five cannot be solved by radicals.

We suppose in this section that all fields we are dealing with have characteristic 0. We fix a ground field $\mathbb{F}$.

First we must make the notion of solvability by radicals precise.

Definition 4.15.1. An equation $f(x) = 0$, $f(x) \in \mathbb{F}[x]$, is *solvable by radicals* if there is a sequence of extensions $\mathbb{F} = \mathbb{E}_0 \subset \mathbb{E}_1 \subset \cdots \subset \mathbb{E}_k$ with $\mathbb{E}_k \supset \mathbb{E}$, a splitting field of $f(x)$, where for each $i = 1, \ldots, k$, $\mathbb{E}_i$ is a splitting field of a polynomial $f_i(x) \in \mathbb{E}^{i-1}(x)$ of the form $f_i(x) = x^n - e$.

Remark 4.15.2. Note that, for example, if $n = pq$, then $\sqrt[n]{e} = \sqrt[p]{\sqrt[q]{e}}$. Thus, we may assume in Definition 4.15.1 that each integer n is a prime. $\qquad\lozenge$

Recall that we defined solvable groups in Definition 2.9.16.

Here is our main theorem.

Theorem 4.15.3. *If the equation $f(x) = 0$ is solvable by radicals then the Galois group $G = Gal(\mathbb{E}/\mathbb{F})$ is a solvable group.*

Proof.   Suppose that $f(x) = 0$ is solvable by radicals. In Definition 4.15.1, let the polynomial $f_i(x)$ have degree p_i, where, by Remark 4.15.2, we may assume that each p_i is prime.

Since $\mathbb{E}_i$ is a splitting field of $f_i(x) \in \mathbb{E}_{i-1}[x]$, $\mathbb{E}_i$ is a Galois extension of $\mathbb{E}_{i-1}$, for each $i = 1, \ldots, k$. But $\mathbb{E}_i$ is not necessarily a Galois extension of $\mathbb{F}$. We begin by constructing a new sequence of extensions $\mathbb{F} = \mathbb{E}'_0 \subseteq \mathbb{E}'_1 \subseteq \cdots \subseteq \mathbb{E}'_k$ with each $\mathbb{E}'_i$ a Galois extension of $\mathbb{F}$. We do so inductively as follows:

We begin with $\mathbb{E}'_0 = \mathbb{E}_0 = \mathbb{F}$. Now suppose that $\mathbb{E}'_{i-1}$ is a Galois extension of $\mathbb{F}$. We have, by assumption, that $\mathbb{E}_i$ is a splitting field of a polynomial $f_i(x) = x^{p_i} - e$ with $e \in \mathbb{E}_{i-1}$. Consider instead the polynomial

$$g_i(x) = \prod_{\sigma} (x^{p_i} - \sigma(e))$$

where the product is taken over all $\sigma \in Gal(\mathbb{E}_{i-1}/\mathbb{F})$. Since $g_i(x)$ is invariant under the action of $Gal(\mathbb{E}_{i-1}/\mathbb{F})$, and $\mathbb{E}_{i-1}$ is a Galois extension of $\mathbb{F}$, we have that $g_i(x) \in \mathbb{F}[x]$. Then let $\mathbb{E}'_i = \mathbb{B}\mathbb{E}_{i-1}$ where $\mathbb{B}$ is a splitting field of $g_i(x)$ over $\mathbb{F}$. As both $\mathbb{B}$ and $\mathbb{E}_{i-1}$ are Galois extensions of $\mathbb{F}$, $\mathbb{E}'_i$ is a Galois extension of $\mathbb{F}$; indeed it is a splitting field of the product $g_{i-1}(x)g_i(x) \in \mathbb{F}[x]$. Note $\mathbb{E}_i \subseteq \mathbb{E}'_i$ for each i, and so $\mathbb{E} \subseteq \mathbb{E}'_k$.

Since $\mathbb{E}'_k$ is a Galois extension of $\mathbb{F}$, it is a Galois extension of every intermediate field $\mathbb{E}'_i$. Furthermore each $\mathbb{E}'_i$ is a Galois extension of $\mathbb{E}'_{i-1}$. Thus, if $G = Gal(\mathbb{E}'_i/\mathbb{F})$, $H_i = Gal(\mathbb{E}'_k/\mathbb{E}'_i)$, then by the fundamental theorem of Galois theory $Gal(\mathbb{E}'_i/\mathbb{F})$ is isomorphic to G/H_i, and $Gal(\mathbb{E}'_{i-1}/\mathbb{F})$ is isomorphic to G/H_{i-1}. But then H_{i-1}/H_i is isomorphic to $Gal(\mathbb{E}'_i/\mathbb{E}'_{i-1})$. We now investigate this group.

For simplicity of notation let $\mathbb{B} = \mathbb{E}'_{i-1}$ and $\mathbb{D} = \mathbb{E}'_i$. Let $\mathbb{D}$ be a splitting field of $\Pi^t_{i=1} x^p - \alpha_i$ where p is a prime and $\alpha_i \in \mathbb{B}, i = 1, \ldots, t$. Then $\mathbb{D} = \mathbb{B}(\zeta_p, \beta_1, \ldots, \beta_t)$ where $\beta^p_i = \alpha_i$.

Let $\mathbb{D}_0 = \mathbb{B}(\zeta_p)$ and for $i = 1, \ldots, t$ let $\mathbb{D}_i = \mathbb{D}_{i-1}(\beta_i)$ (so that $\mathbb{D} = \mathbb{D}_t$).

Now $\mathbb{D}_0 = \mathbb{B}\mathbb{Q}(\zeta_p)$ so $\mathrm{Gal}(\mathbb{D}_0/\mathbb{B})$ is isomorphic to a subgroup of $\mathrm{Gal}(\mathbb{Q}(\zeta_p)/\mathbb{Q})$. This group is cyclic so $\mathrm{Gal}(\mathbb{D}_0/\mathbb{B})$ is cyclic as well. For $i = 1, \ldots, t$ consider $\mathrm{Gal}(\mathbb{D}_i/\mathbb{D}_{i-1})$. There are two possibilities:

If $\alpha_i \in \mathbb{D}_{i-1}$ then $\mathbb{D}_i = \mathbb{D}_{i-1}$ and this Galois group is trivial.

If $\alpha_i \notin \mathbb{D}_{i-1}$ then by Abel's theorem the polynomial $x^p - \alpha_i$ is irreducible in $\mathbb{D}_{i-1}$. On the other hand, it splits as $(x - \beta_i) \ldots (x - \zeta_p^{p-1}\beta_i)$ in $\mathbb{D}_i[x]$. Hence, this Galois group is $\{\sigma_0, \ldots, \sigma_{p-1}\}$ where $\sigma_j(\beta_i) = \zeta_p^j \beta_i$, $j = 0, \ldots, p - 1$, and is a cyclic group of order p.

Hence, putting all of these extensions together we see that we have a subnormal series for $G' = \mathrm{Gal}(\mathbb{E}_k'/\mathbb{F})$ of the form

$$G' = G_0' \supset G_1' \supset \ldots \supset G_s' = \{1\}$$

with G_{i-1}'/G_i' cyclic for each $i = 1, \ldots, s$, and so G' is solvable.

Now $\mathbb{E}$, a splitting field of $f(x)$, is a Galois extension of $\mathbb{F}$ and $\mathbb{E} \subseteq \mathbb{F}_k'$. Thus, $\mathrm{Gal}(\mathbb{E}/\mathbb{F}) = G'/\mathrm{Gal}(\mathbb{E}_k'/\mathbb{E})$ is a quotient of a solvable group, and so is solvable as well. $\qquad\square$

Remark 4.15.4. The converse of Theorem 4.15.3 is true as well, but we shall not prove this here. $\qquad\Diamond$

Corollary 4.15.5. *The general equation $f(x) = 0$ of degree greater than or equal to five is not solvable by radicals.*

Proof. For $n \geq 5$, let $\mathbb{F}$ be the field of symmetric functions $\mathbb{F} = \mathbb{F}_0(s_1, \ldots, s_n)$ in the variables $x_1, \ldots, x_n, \mathbb{F}_0$ an arbitrary field of characteristic 0, and let $\mathbb{E}$ be the splitting field of the polynomial $f(x) = (x - x_1) \ldots (x - x_n) \in \mathbb{F}[x]$. Then $\mathrm{Gal}(\mathbb{E}/\mathbb{F})$ is isomorphic to the symmetric group S_n, and S_n is not a solvable group for $n \geq 5$. $\qquad\square$

Remark 4.15.6. Up to sign, $s_1, \ldots, s_n$ are the coefficients of the polynomial $f(x)$ above, so this shows there is no general formula for the roots of a polynomial of degree at least five in terms of radical expressions in its coefficients.

But this leaves open the possibility that for a general polynomial $f(x) \in \mathbb{Q}[x]$, it may be possible to express its roots in terms of radical expressions in its coefficients, even if there is no general formula for doing so. This is also impossible, as we now see. $\qquad\Diamond$

Corollary 4.15.7. *The general equation $f(x) = 0$ with $f(x) \in \mathbb{Q}[x]$ a polynomial of degree greater than or equal to five is not solvable by radicals.*

Proof.   Let $f(x)$ have degree n. If $n = p$ is prime, we exhibited in Example 4.11.19 a polynomial $f(x) \in \mathbb{Q}[x]$ of degree p with $\mathrm{Gal}(\mathbb{E}/\mathbb{Q})$ isomorphic to the symmetric group S_p, where $\mathbb{E}$ is a splitting field of $f(x)$. This is in fact true for n any positive integer, though we will not prove this here. $\square$

4.16   Straightedge and compass constructions

In this section, we consider the question, which goes back to Euclid, of which geometric constructions can be performed by straightedge and compass. To answer this question we first translate it into algebraic terms, and then apply the field theory we have developed to first show that the three classical problems of antiquity — trisecting the angle, doubling the cube, and squaring the circle — cannot be solved by straightedge and compass constructions, and then to determine exactly which regular polygons can be constructed by straightedge and compass (modulo a question in number theory).

We begin by observing that we can certainly draw a line L in the plane just using a straightedge. We do so, and pick two distinct points O and P on L. We declare O to be the origin and the distance from O to P to be 1. Then we may construct the line L' that is perpendicular to L at O by straightedge and compass, use a compass to draw the circle of radius OP centered at O, and let P' be a point at which this circle intersects L'.

We then declare P to be the point with coordinates $(1,0)$, and P' to be the point with coordinates $(0,1)$, so that L is the x-axis and L' is the y-axis. This gives coordinates on the plane.

We next observe that we can perform the four basic arithmetic operations — addition, subtraction, multiplication, and division — by straightedge and compass. We can certainly add and subtract lengths. We may multiply and divide lengths as follows: Given a length x and a length y, draw two intersecting lines L_1 and L_2. Let their intersection point be O_1. Mark off a distance $O_1 P_1$ on L_1 equal to x and a distance $O_1 P_2$ on L_2 equal to y. Mark off a distance $O_1 Q_1$ on L_1 equal to 1, and draw the straight line $Q_1 P_2$. Then draw the line through P_1 parallel to this straight line, and let this line intersect L_2 at the point Q_2. Then, by similar triangles, the line segment O_1Q_2 has length xy. Alternatively, draw the straight line P_1P_2, and then draw the line through Q_1 parallel to this straight line, and let this line intersect L_2 at the point Q_3. Then, by similar triangles, the line segment O_1Q_3 has length y/x.

We may also take the square root of a positive real number x as follows: Draw a line L through a point O_1 and let P_1 be a point on L at a distance x from O_1. Let P_2 be a point on L on the other side of O_1 from P_1 at a distance of 1 from O_1. Let C be the midpoint of the line segment from P_1 to P_2 and draw a semicircle centered at C passing through P_1 and P_2. Construct the line perpendicular to the line L at O_1 and let it intersect this semicircle at the point Q. Then P_1QP_2 is a triangle inscribed in a semicircle, so is a right triangle, and then the triangles Q_1P_1Q and O_1QP_2 are similar. Let w be the length of the line segment O_1Q. Then, by similar triangles, $1/w = w/x$ and so $w = \sqrt{x}$.

Finally, let us identify the point with coordinates (x, y) with the complex number $z = x + iy$. Since addition, subtraction, multiplication, and division of complex numbers may be performed in terms of these operations on their real and imaginary parts, we may perform all of these operations by straightedge and compass constructions. Furthermore, we may take square roots of complex numbers by straightedge and compass: Write the complex number w in polar form as $w = re^{i\theta}$. Then z has a square root $\sqrt{r}e^{i\theta/2}$. We have already seen we can find $\sqrt{r}$, and we can bisect angles as well.

Theorem 4.16.1. *The complex number z can be constructed by straightedge and compass if and only if there is a sequence of fields*

$$\mathbb{Q} = \mathbb{F}_0 \subset \mathbb{F}_1 \subset \cdots \subset \mathbb{F}_k$$

with $z \in \mathbb{F}_k$, or, equivalently, with $\mathbb{Q}(z) \subseteq \mathbb{F}_k$, and with $(\mathbb{F}_i/\mathbb{F}_{i-1}) = 2$ for $i = 1, \ldots, k$.

Proof. First suppose there is such a sequence of fields. We proceed by induction on i. If $i = 0$, every element of $\mathbb{Q}$ can be constructed, as we have shown. Now suppose every element of $\mathbb{F}_{i-1}$ can be constructed. Since $(\mathbb{F}_i/\mathbb{F}_{i-1}) = 2$, we see from the quadratic formula that $\mathbb{F}_i = \mathbb{F}_{i-1}(\sqrt{z_{i-1}})$ for some $z_{i-1} \in \mathbb{F}_{i-1}$. But we have seen that we may construct $\sqrt{w}$ for any complex number w so we may construct $\sqrt{z_{i-1}}$ and then every element of $\mathbb{F}_i$.

Conversely, suppose that z can be constructed by straightedge and compass. Then z is obtained by a sequence of the following operations:

(1) Finding the intersection of two lines.
(2) Finding the intersection of a line and a circle.
(3) Finding the intersection of two circles.

Suppose at any stage of the process the points we are dealing with all lie in a field $\mathbb{F}$. (At the start, $\mathbb{F} = \mathbb{Q}$.) In case of operation (1), both lines are given by linear equations with coefficients in $\mathbb{F}$, so their intersection is a point in $\mathbb{F}$. In case of operation (2), the line is given by a linear equation with coefficients in $\mathbb{F}$, and the circle is given by a quadratic equation with coefficients in $\mathbb{F}$, so their intersection is a point in $\mathbb{F}'$, where either $\mathbb{F}' = \mathbb{F}$ or $(\mathbb{F}'/\mathbb{F}) = 2$. In case of operation (3), each circle is given by a quadratic equation with coefficients in $\mathbb{F}$, but by subtracting a multiple of one of these equations from the other we may obtain a system consisting of one linear equation and one quadratic equation.

Thus, performing these operations repeatedly to obtain the complex number z shows that z is an element of a field $\mathbb{F}_k$ obtained as a sequence of quadratic extensions.   □

Theorem 4.16.2. *It is impossible to perform each of the following by straightedge and compass constructions:*

(1) *Trisecting the angle*
(2) *Doubling the cube*
(3) *Squaring the circle*

Proof.

(1) We may find an angle φ if and only if we can find the point A that is the intersection of the line making an angle of φ with the positive x-axis at the origin with the unit circle. That point A represents the complex number $z = e^{i\varphi} = \cos(\varphi) + i\sin(\varphi)$. Thus, we can construct A by straightedge and compass if and only if we can construct $\cos(\varphi)$ and $\sin(\varphi)$. Now recall the triple-angle formula: $\cos(3\varphi) = 4\cos^3(\varphi) - 3\cos(\varphi)$. Setting $\varphi = \theta/3$, we see

$4\cos^3(\theta/3) - 3\cos(\theta/3) = \cos(\theta)$. Now let $\theta = \pi/3$, an angle we certainly can construct. Then $\cos(\theta) = 1/2$, so that $\cos(\theta/3)$ is a root of the cubic polynomial $4x^3 - 3x = 1/2$, or equivalently $8x^3 - 6x - 1 = 0$. But this polynomial has no linear factor, so is irreducible over $\mathbb{Q}$, and hence $(\mathbb{Q}(\cos(\pi/9))/\mathbb{Q}) = 3$. But 3 does not divide 2^k (for any k), so we cannot have $\mathbb{Q}(\cos(\pi/9)) \subseteq \mathbb{Q}(\mathbb{F}_k)$ for any field $\mathbb{F}_k$ as in Theorem 4.16.1.

(2) Since we can find a cube of volume 1 (the cube whose sides all have length 1), doubling the cube means finding a cube of volume 2, and hence a cube with side length $x = \sqrt[3]{2}$ is a root of the irreducible polynomial $x^3 - 2$ and this is again impossible as in (1).

(3) Squaring the circle means finding a square that has the same area as a given circle. Beginning with a circle of radius 1, that means finding a square of side length $\sqrt{\pi}$. But now we use a famous theorem of Lindemann, that π is transcendental, i.e., that π is not a root of any algebraic equation with coefficients in $\mathbb{Q}$, so this is impossible. □

We now turn to the question of the constructibility of the regular n-gon by straightedge and compass. We see right away that this is equivalent to the constructibility of an angle of $2\pi/n$.

Euclid knew how to construct a regular n-gon for $n = 3, 4, 5$. If we can construct an angle of $2\pi/n_1$ and an angle of $2\pi/n_2$, and n_1 and n_2 are relatively prime, then we can construct an angle of $2\pi/n$ where $n = n_1 n_2$, as follows: By Euclid's algorithm we can write $1 = n_1 a - n_2 b$ for some positive integers a and b, and then $1/(n_1 n_2) = (1/n_2)a - (1/n_1)b$. So we construct a times an angle of $2\pi/n_2$ and subtract from it b times an angle of $2\pi/n_1$ to obtain an angle of $2\pi/n$. Also, since we can bisect an angle, if we can construct an angle of $2\pi/n$, we can construct an angle of $2\pi/(2^k n)$, for any k. Thus, we can construct a regular n-gon for $n = 2^{e_0} 3^{e_1} 5^{e_2}$ with $e_0 \geq 0$, $e_1 = 0$ or 1, $e_2 = 0$ or 1.

There matters stood until Gauss showed how to construct a regular 17-gon. This discovery, the first advance in this problem in over two millennia, played a role in Gauss's decision to become a mathematician.

We now go to work ourselves.

We first reformulate the condition in Theorem 4.16.1.

Lemma 4.16.3. *For a complex number z, the following are equivalent:*

(1) *There is a sequence of fields*

$$\mathbb{Q} = \mathbb{F}_0 \subset \mathbb{F}_1 \subset \cdots \subset \mathbb{F}_k$$

with $z \in \mathbb{F}_k$, or, equivalently, with $\mathbb{Q}(z) \subseteq \mathbb{F}_k$, and with $(\mathbb{F}_i/\mathbb{F}_{i-1}) = 2$ for $i = 1, \ldots, k$.

(2) *If $\mathbb{E}$ is a splitting field of the minimal polynomial $m_z(x)$, then $(\mathbb{E}/\mathbb{Q})$ is a power of 2.*

Proof.   (2) implies (1): Let $G = \text{Gal}(\mathbb{E}/\mathbb{Q})$. Then $|G|$ is a power of 2, say $|G| = 2^k$. But then by Lemma 2.9.6 there is a sequence of subgroups

$$G = G_0 \supset G_1 \supset \cdots \supset G_k = \{1\}$$

with $|G_i| = 2^{k-i}$, or, equivalently, $[G_{i-1}: G_i] = 2$. Let $\mathbb{F}_i = \text{Fix}(G_i)$.

(1) implies (2): We shall show that there is a sequence of fields

$$\mathbb{Q} = \mathbb{E}_0 \subset \mathbb{E}_1 \subset \cdots \subset \mathbb{E}_k$$

with $z \in \mathbb{E}_k$, or equivalently with $\mathbb{Q}(z) \subseteq \mathbb{E}_k$, where each $\mathbb{E}_i$ is a Galois extension of $\mathbb{Q}$, $\mathbb{F}_i \subseteq \mathbb{E}_i$, and with $(\mathbb{E}_i/\mathbb{E}_{i-1})$ a power of 2, for each i. Then $(\mathbb{E}_k/\mathbb{Q}) = (\mathbb{E}_k/\mathbb{E}_{k-1}) \cdots (\mathbb{E}_1/\mathbb{E}_0)$ is a power of 2, and $\mathbb{E} \subseteq \mathbb{E}_k$, so $(\mathbb{E}/\mathbb{Q})$ divides $(\mathbb{E}_k/\mathbb{Q})$ and hence $(\mathbb{E}/\mathbb{Q})$ is a power of 2.

We do so by induction on k. For $i = 0$ this is trivial, and for $i = 1$ we may choose $\mathbb{E}_1 = \mathbb{F}_1$, as $\mathbb{F}_1$ is a quadratic extension of $\mathbb{Q}$ so $\mathbb{F}_1$ is a splitting field of a polynomial $x^2 - a$, $a \in \mathbb{Q}$, and hence is automatically a Galois extension of $\mathbb{Q}$.

Now suppose this is true for $i - 1$. $\mathbb{F}_i$ is a quadratic extension of $\mathbb{F}_{i-1}$, so, as we have just observed, $\mathbb{F}_i$ is a splitting field of a polynomial $f(x) = x^2 - \alpha$ for some $\alpha \in \mathbb{F}_{i-1} \subseteq \mathbb{E}_{i-1}$. Let

$$g_i(x) = \prod_\sigma \sigma(f(x)) = \prod_\sigma (x^2 - \sigma(\alpha))$$

where the product is taken over all $\sigma \in \text{Gal}(\mathbb{E}_{i-1})/\mathbb{Q})$. Let $\mathbb{E}_i$ be a splitting field of $g_i(x) \in \mathbb{E}_{i-1}$. But note that in fact $g_i(x) \in \mathbb{Q}[x]$, as $g_i(x)$ is invariant under $\text{Gal}(\mathbb{E}_{i-1}/\mathbb{Q})$. Then $\mathbb{E}_i$ is a Galois extension of $\mathbb{F}$, as it is a splitting field of the polynomial $g_{i-1}(x)g_i(x)$. It remains to show that $(\mathbb{E}_i/\mathbb{E}_{i-1})$ is a power of 2. To this end, let us denote the elements of $\mathrm{Gal}(\mathbb{E}_{i-1}/\mathbb{Q})$ by $\{\sigma_1, \ldots, \sigma_t\}$, so that $g_i(x) = \prod(x^2 - \sigma_i(\alpha))$. Let $\beta_i \in \mathbb{E}_i$ with $\beta_i^2 = \sigma_i(\alpha)$, so that

$$\mathbb{E}_i = \mathbb{E}_{i-1}(\beta_1, \ldots, \beta_t)$$

and so we have a sequence of fields

$$\mathbb{E}_{i-1} \subseteq \mathbb{E}_{i-1}(\beta_1) \subseteq \mathbb{E}_{i-1}(\beta_1, \beta_2) \subseteq \cdots \subseteq \mathbb{E}_{i-1}(\beta_1, \ldots, \beta_t) = \mathbb{E}_i.$$

Let us examine each of these intermediate extensions

$$\mathbb{E}_{i-1}(\beta_1, \ldots, \beta_{j-1}) \subseteq \mathbb{E}_{i-1}(\beta_1, \ldots, \beta_{j-1})(\beta_j) = \mathbb{E}_{i-1}(\beta_1, \ldots, \beta_j).$$

Either $\beta_j \in \mathbb{E}_{i-1}(\beta_1, \ldots, \beta_{j-1})$, in which case $\mathbb{E}_{i-1}(\beta_1, \ldots, \beta_j) = \mathbb{E}_{i-1}(\beta_1, \ldots, \beta_{j-1})$ and $(\mathbb{E}_{i-1}(\beta_1, \ldots, \beta_j)/\mathbb{E}_{i-1}(\beta_1, \ldots, \beta_{j-1})) = 1$, or $\beta_j \notin \mathbb{E}_{i-1}(\beta_1, \ldots, \beta_{j-1})$. But in this latter case $\mathbb{E}_{i-1}(\beta_1, \ldots, \beta_j)$ is obtained from $\mathbb{E}_{i-1}(\beta_1, \ldots, \beta_{j-1})$ by adjoining β_j, which is a root of the quadratic polynomial $x^2 - \sigma(\alpha_j)$, and so in this case $(\mathbb{E}_{i-1}(\beta_1, \ldots, \beta_j)/\mathbb{E}_{i-1}(\beta_1, \ldots, \beta_{j-1})) = 2$. Thus,

$$(\mathbb{E}_i/\mathbb{E}_{i-1}) = (\mathbb{E}_{i-1}(\beta_1, \ldots, \beta_t)/\mathbb{E}_{i-1}(\beta_1, \ldots, \beta_{i-1})) \cdots (\mathbb{E}_{i-1}(\beta_1)/\mathbb{E}_{i-1})$$

is a power of 2. $\qquad \square$

In preparation for our main result, we do a bit of elementary number theory. We ask when $n = 2^t + 1$ can be prime. We have the algebraic identity, valid for any *odd* a, $x^a + 1 = (x+1)(x^{a-1} - x^{a-2} + \cdots - x + 1)$. Thus, it t has an odd factor greater than 1, say $t = rs$ with r odd greater than 1, then this identity yields

$$2^t + 1 = 2^{rs} + 1 = (2^s)^r + 1 = (2^s + 1)(2^{s(r-1)} - \cdots + 1)$$

and so $2^t + 1$ is composite. Thus, the only possible primes of this form are when t is a power of 2, $t = 2^k$ for some k. We let

$$F_k = 2^{2^k} + 1.$$

If F_k is prime, it is called a *Fermat prime*. This terminology is due to the fact that Fermat believed that F_k is prime for every k. Indeed, $F_0 = 3, F_1 = 5, F_2 = 17, F_3 = 257$, and $F_4 = 65537$ are prime, but

Euler discovered that $F_5 = 4294967297$ is divisible by 641 and so is composite.

Theorem 4.16.4. *A regular n-gon is constructible by straightedge and compass if and only if n is of the form*

$$n = 2^k p_1 \ldots p_j$$

where $p_1, \ldots, p_j$ are distinct Fermat primes.

Proof.   As we have observed, constructing a regular n-gon is the same problem as constructing an angle of $2\pi/n$, and this is the same problem as constructing the complex number $\zeta_n = \exp(2\pi i/n)$. By Lemma 4.16.3 this is possible if and only if the splitting field $\mathbb{E}$ of the minimal polynomial $m_{\zeta_n}(x)$ is an extension of $\mathbb{Q}$ of degree a power of 2. But the polynomial $m_{\zeta_n}(x)$ is just the nth cyclotomic polynomial $\Phi_n(x)$, since ζ_n is a root of this polynomial, and since, as we have shown, this polynomial is irreducible, and $\mathbb{E} = \mathbb{Q}(\zeta_n)$. Thus,

$$(\mathbb{E}/\mathbb{Q}) = \deg \Phi_n(x) = \varphi(n).$$

Now factor n as a product of prime powers, $n = 2^k p_1^{e_1} \ldots p_j^{e_j}$. Then

$$\varphi(n) = \varphi(2^k)\varphi(p_1^{e_1}) \ldots \varphi(p_j^{e_j})$$
$$= (2^{k-1})(p_1^{e_1-1}(p_1 - 1)) \ldots (p_j^{e_j-1}(p_j - 1)).$$

In order to have $\varphi(n)$ be a power of 2, it is necessary and sufficient that $e_1 = \cdots = e_j = 1$, and that $p_i - 1$ is a power of 2 for each i, i.e., that p_i is a Fermat prime for each i. $\square$

Remark 4.16.5. We are almost totally ignorant about Fermat primes. There is no known value of $k > 4$ for which F_k is prime. And in our present state of knowledge any of the following alternatives may be true: There may be no values of $k > 4$ for which F_k is prime; there may be some but only finitely many values of $k > 4$ for which F_k is prime; there may be infinitely many values of $k > 4$ for which F_k is prime and infinitely many values of $k > 4$ for which F_k is composite; there may be only finitely many values of $k > 4$ for which F_k is composite and for all other values of F_k with $k > 4$ F_k is prime. We just don't know. $\diamond$

4.17 The fundamental theorem of algebra

In this section, we prove the fundamental theorem of algebra: Every nonconstant complex polynomial has a complex root.

But first let us note that, despite its name, there can be no purely algebraic proof of this theorem. To see why, let us consider the field of complex numbers $\mathbb{C}$. How is $\mathbb{C}$ defined? It is $\mathbb{C} = \mathbb{R}(i)$ where $i^2 = -1$. Thus $\mathbb{C}$ is an algebraic extension of $\mathbb{R}$ of degree two. But how is the field of real numbers $\mathbb{R}$ defined? The definition of $\mathbb{R}$ involves completeness (i.e., the analytical/topological notion of limits) and that is not an algebraic definition at all. So if we can't even define $\mathbb{R}$ or $\mathbb{C}$ purely algebraically, we certainly can't hope to prove anything about these fields purely algebraically.

But what we can hope to do is to prove the fundamental theorem of algebra with only a minimal use of non-algebraic methods, and that is what we do here. We begin with a result from elementary calculus, and once we have that, we proceed purely algebraically.

(We remark that there are purely analytic proofs of the fundamental theorem of algebra — see any complex analysis textbook.)

Here is the only non-algebraic fact we will use.

Lemma 4.17.1. *Let $f(x) \in \mathbb{R}[x]$ be a polynomial of odd degree. Then $f(x)$ has a real root.*

Proof. This is a familiar result from elementary calculus (to be fair, a result that is always stated in an elementary calculus course, but not proved until a more advanced course). □

Theorem 4.17.2 (Fundamental theorem of algebra). *Let $f(x) \in \mathbb{C}[x]$ be a nonconstant polynomial. Then $f(x)$ splits into a product of linear factors in $\mathbb{C}[x]$, so that $f(x)$ has $n = \deg f(x)$ roots in $\mathbb{C}$.*

Proof. First we observe that it suffices to prove this if $f(x) \in \mathbb{R}[x]$. To see this, consider any polynomial $g(x) \in \mathbb{C}[x]$. Then $f(x) = g(x)\bar{g}(x) \in \mathbb{R}[x]$. If $f(x)$ splits in $\mathbb{C}[x]$, then so does $g(x)$ (and $\bar{g}(x)$).

Thus, let $f(x) \in \mathbb{R}[x]$. Let $\mathbb{E}$ be a splitting field of $f(x)$. Let $(\mathbb{E}/\mathbb{R}) = 2^m d$ with d odd. Let $G = \mathrm{Gal}(\mathbb{E}/\mathbb{R})$. Then $|G| = 2^m d$.

Let H be the 2-Sylow subgroup of G, and let $\mathbb{B}$ be the fixed field of H. Then $(\mathbb{B}/\mathbb{R}) = d$. By the theorem of the primitive element

$\mathbb{B} = \mathbb{R}(\alpha)$ for some α, and then its minimal polynomial $m_\alpha(x) \in \mathbb{R}[x]$ is a polynomial of odd degree d. But then by Lemma 4.17.1 $m_\alpha(x)$ has a root $a \in \mathbb{R}$, i.e., $m_\alpha(x)$ is divisible by $x - a$ in $\mathbb{R}[x]$. Since $m_\alpha(x)$ is irreducible, we must have $m_\alpha(x) = x - a$, so $d = 1$ and $\mathbb{B} = \mathbb{R}$.

Thus, $(\mathbb{E}/\mathbb{R}) = 2^m$, and $H = G$. We prove the theorem by induction on m.

Before doing so, we observe that every complex number has a complex square root. If $z \in \mathbb{C}$, we can write down explicitly a complex number w with $w^2 = z$. If $z = 0$, then $w = 0$. Otherwise, let $c = a/(a^2 + b^2)$. If $a \geq 0$ and $b \geq 0$, then $w = \sqrt{a^2 + b^2}(\sqrt{(1 + c)/2} + i\sqrt{(1 - c)/2})$, with similar formulas for other values of a and b.

What we will actually prove is that if $(\mathbb{E}/\mathbb{R}) = 2^m$, then $\mathbb{E}$ is isomorphic to a subfield of $\mathbb{C}$.

In case $m = 1$, this is trivial: $\mathbb{E} = \mathbb{R}$.

Now suppose this claim is true for all extensions of $\mathbb{R}$ of degree 2^{m-1}, and let $(\mathbb{E}/\mathbb{R}) = 2^m$. Now $G = \mathrm{Gal}(\mathbb{E}/\mathbb{R})$ has order 2^m, i.e., is a 2-group. Choose an element g_0 of G of order 2, and let G_0 be the subgroup of G generated by G_0. Let $\mathbb{E}_0 = \mathrm{Fix}(G_0)$. Then $(\mathbb{E}/\mathbb{E}_0) = 2$.

Now $(\mathbb{E}/\mathbb{R}) = (\mathbb{E}/\mathbb{E}_0)(\mathbb{E}_0/\mathbb{R})$ so $(\mathbb{E}_0/\mathbb{R}) = 2^{m-1}$, i.e., $\mathbb{E}_0$ is an extension of $\mathbb{R}$ of degree 2^{m-1}, so is isomorphic to a subfield of $\mathbb{C}$, by the inductive hypothesis. Choose any isomorphism and let $\tilde{\mathbb{E}}_0$ be the image of $\mathbb{E}$ under this isomorphism.

Then $\mathbb{E}$ is an extension of $\mathbb{E}_0$ of degree 2, so $\mathbb{E} = \mathbb{E}_0(\alpha)$ for some α with $m_\alpha(x)$ a polynomial of degree 2 with coefficients in $\mathbb{E}_0$. But, as we observed, every complex number has a complex square root, so, from the quadratic formula, we see that $\tilde{m}_\alpha(x)$, the image of $m_\alpha(x)$ under the isomorphism from $\mathbb{E}_0$ to $\tilde{\mathbb{E}}_0$, has a root $w \in \mathbb{C}$, so $\mathbb{E}$ is isomorphic to $\tilde{\mathbb{E}}_0(w)$, a subfield of $\mathbb{C}$. Then, by induction, we are done.

Finally, since $f(x)$ splits in $\mathbb{E}$, and $\mathbb{E}$ is isomorphic to $\tilde{\mathbb{E}}_0(w)$, $f(x)$ splits in $\tilde{\mathbb{E}}(w)$, and hence in $\mathbb{C}$. $\square$

4.18   Exercises

1. (a) Let $f(x) = x^3 + 3x^2 + 6x + 3$, an irreducible polynomial in $\mathbb{Q}[x]$. Let $\mathbb{E} = \mathbb{Q}[x]/ < x^3 + 3x^2 + 6x + 3 >$. Let α be a root of $f(x)$ in $\mathbb{E}$. Let $\beta_1 = 2\alpha + 1$, $\beta_2 = \alpha^2 + 4\alpha + 6$. Find β_1^2, $\beta_1\beta_2$, and β_2^2. Also, find the minimal polynomials $m_{\beta_1}(x)$ and $m_{\beta_2}(x)$ and find β_1^{-1} and β_2^{-1}.

(b) Let $g(x) = x^4 + 4x^2 + 2x + 2$, an irreducible polynomial in $\mathbb{Q}[x]$. Let $\mathbb{E} = \mathbb{Q}[x]/ < x^4 + 4x^2 + 2x + 2 >$. Let α be a root of $g(x)$ in $\mathbb{E}$. Let $\beta_1 = \alpha + 2$, $\beta_2 = \alpha^2 + 2\alpha + 5$, $\beta_3 = \alpha^3 + 2\alpha^2 + 1$. Find $\beta_1^2, \beta_2^2, \beta_3^2, \beta_1\beta_2, \beta_1\beta_3$, and $\beta_2\beta_3$. Also, find the minimal polynomials $m_{\beta_1}(x)$, $m_{\beta_2}(x)$, $m_{\beta_3}(x)$ and find β_1^{-1}, β_2^{-1}, and β_3^{-1}.

2. (a) Let $f(x) = x^3 + 3x^2 + 4x + 3$, an irreducible polynomial in $\mathbb{F}_5[x]$. Let $\mathbb{E} = \mathbb{F}_5[x]/ < x^3 + 3x^2 + 4x + 3 >$. Let α be a root of $f(x)$ in $\mathbb{E}$. Let $\beta_1 = 3\alpha + 2$, $\beta_2 = \alpha^2 + 4\alpha + 2$. Find $\beta_1^2, \beta_1\beta_2$, and β_2^2. Also, find the minimal polynomials $m_{\beta_1}(x)$ and $m_{\beta_2}(x)$, and find β_1^{-1} and β_2^{-1}.

   (b) Let $g(x) = x^4 + x^3 + 3x^2 + 2x + 2$, an irreducible polynomial in $\mathbb{F}_5[x]$. Let $\mathbb{E} = \mathbb{F}_5/[x]/ < x^4 + x^3 + 3x^2 + 2x + 2 >$. Let α be a root of $g(x)$ in $\mathbb{E}$. Let $\beta_1 = \alpha + 3$, $\beta_2 = \alpha^2 + 2\alpha + 4$, $\beta_3 = \alpha^3 + 2\alpha + 1$. Find $\beta_1^2, \beta_2^2, \beta_3^2, \beta_1\beta_2, \beta_1\beta_3$, and $\beta_2\beta_3$. Also, find the minimal polynomials $m_{\beta_1}(x)$, $m_{\beta_2}(x)$, $m_{\beta_3}(x)$ and find β_1^{-1}, β_2^{-1}, and β_3^{-1}.

3. (a) Write down the addition and multiplication tables in $\mathbb{F}_7$.
   (b) Write down the addition and multiplication tables in $\mathbb{F}_8$.
   (c) Write down the addition and multiplication tables in $\mathbb{F}_9$.

4. (a) Factor the polynomial $x^4 - x$ into irreducibles in $\mathbb{F}_2[x]$.
   (b) Factor the polynomial $x^8 - x$ into irreducibles in $\mathbb{F}_2[x]$.
   (c) Factor the polynomial $x^{16} - x$ into irreducibles in $\mathbb{F}_2[x]$.
   (d) Factor the polynomial $x^9 - x$ into irreducibles in $\mathbb{F}_3[x]$.
   (e) Factor the polynomial $x^{16} - x$ into irreducibles in $\mathbb{F}_4[x]$.

5. (a) Let $f(x) = x^3 + x + 1$, an irreducible polynomial in $\mathbb{F}_5[x]$. Let $\mathbb{E}_1 = \mathbb{F}_5[x]/ < f(x) >$ and let α be a root of $f(x)$ in $\mathbb{E}_1$. Find all the roots of $f(x)$ in $\mathbb{E}_1$.
   (b) Let $g(x) = x^3 + x^2 + 2$, an irreducible polynomial in $\mathbb{F}_5[x]$. Let $\mathbb{E}_2 = \mathbb{F}_5[x]/ < g(x) >$ and let β be a root of $g(x)$ in $\mathbb{E}_2$. Find all the roots of $g(x)$ in $\mathbb{E}_2$.
   (c) The fields $\mathbb{E}_1$ and $\mathbb{E}_2$ are isomorphic. Find an explicit isomorphism $\varphi\colon \mathbb{E}_1 \to \mathbb{E}_2$.

6. (a) Factor $f(x) = x^3 - 1$ into irreducibles in $\mathbb{F}_3[x]$.
   (b) Factor $f(x) = x^3 - 1$ into irreducibles in $\mathbb{F}_5[x]$.
   (c) Factor $f(x) = x^3 - 1$ into irreducibles in $\mathbb{F}_7[x]$.

7. Let p and q be primes (not necessarily distinct). Show that the number of irreducible polynomials of degree q in $\mathbb{F}_p[x]$ is $(p^q - p)/q$.

8. Suppose that $f(x) = x^{729} - x \in \mathbb{F}_3[x]$ is factored into a product of irreducibles $f(x) = g_1(x) \ldots g_k(x)$ in $\mathbb{F}_3[x]$. For each positive integer d, how many of the polynomials $g_i(x)$ are of degree d?

9. Let $f(x)$ be a polynomial of degree 15 in $\mathbb{F}_p[x]$ and let $\mathbb{E}$ be a splitting field of $f(x)$. Find all possible values of $(\mathbb{E}/\mathbb{F})$.

10. (a) In each case, find the minimal polynomial $m_\alpha(x) \in \mathbb{Q}[x]$:

    (1) $\alpha = \sqrt[3]{27 + 10\sqrt{3}}$

    (2) $\alpha = \sqrt[3]{25 + 22\sqrt{2}}$

    (3) $\alpha = \sqrt{-15 + 10\sqrt[3]{3}}$

    (4) $\alpha = \sqrt{-15 + 12\sqrt[3]{2}}$

    (b) In each case, find the minimal polynomial $m_\alpha(x) \in \mathbb{F}_7[x]$ (where square roots and cube roots are to be interpreted as taken in some extension field of $\mathbb{F}_7$):

    (1) $\alpha = \sqrt[3]{1 + 3\sqrt{5}}$

    (2) $\alpha = \sqrt[3]{3 + 3\sqrt{5}}$

    (3) $\alpha = \sqrt{2 + \sqrt[3]{3}}$

    (4) $\alpha = \sqrt{5 + 2\sqrt[3]{3}}$

11. Let $p(x) \in \mathbb{F}[x]$ be an irreducible monic polynomial of degree n, and let C be the companion matrix of $p(x)$ in $R = M_n(\mathbb{F})$, the ring of n-by-n matrices with entries in $\mathbb{F}$. Let

$$\mathbb{E} = \left\{ \sum_{i=0}^{n-1} a_i C^i \mid a_i \in \mathbb{F} \right\}.$$

    (a) Show that $\mathbb{E}$ is a subring of R.
    (b) Show that $\mathbb{E}$ is field.
    (c) Show that the polynomial $p(x)$ has a root in $\mathbb{E}$.

    (Thus, $\mathbb{E}$ gives a "concrete" realization of the field $\mathbb{F}[x]/ < p(x) >$ of Kronecker's theorem, Theorem 4.2.4.)

12. Let $f(x), g(x) \in \mathbb{F}[x]$ be monic irreducible polynomials. If $f(x)$ and $g(x)$ have a common root in some extension field $\mathbb{E}$ of $\mathbb{F}$, show that $f(x) = g(x)$.

13. (a) Let $\mathbb{E}$ be an extension of $\mathbb{F}$ and let $\alpha, \beta \in \mathbb{E}$. Suppose that $\alpha + \beta$ is algebraic over $\mathbb{F}$ of degree j and that $\alpha\beta$ is algebraic over $\mathbb{F}$ of degree k. Show that each of α and β is algebraic over $\mathbb{F}$ of degree at most $2jk$.

(b) Give an example where each of α and β is algebraic over $\mathbb{F}$ of degree exactly $2jk$.

14. (a) Let $\mathbb{E}$ be an extension of $\mathbb{F}$ of prime degree p and suppose that $\mathbb{E} = \mathbb{F}(\alpha)$. Let $g(x) \in \mathbb{F}[x]$ be any nonconstant polynomial of degree $d < p$ and let $\beta = g(\alpha)$. Show that $\mathbb{E} = \mathbb{E}(\beta)$.

    (b) Let $\mathbb{E}$ be an extension of $\mathbb{F}$ of odd degree and suppose that $\mathbb{E} = \mathbb{F}(\alpha)$. Let $g(x) \in \mathbb{F}[x]$ be any quadratic polynomial and let $\beta = g(\alpha)$. Show that $\mathbb{E} = \mathbb{F}(\beta)$.

15. (a) Let $f(x) \in \mathbb{F}[x]$ be an irreducible polynomial of degree d. Let $\mathbb{E}$ be an extension of $\mathbb{F}$ with $(\mathbb{E}/\mathbb{F})$ relatively prime to d. Show that $f(x)$ is irreducible in $\mathbb{E}[x]$.

    (b) Let α and β be elements of $\mathbb{E}$, an extension field of $\mathbb{F}$, with minimal polynomials $m_\alpha(x)$, $m_\beta(x) \in \mathbb{F}[x]$ of degrees s and t respectively. Let $\tilde{m}_\beta(x) \in \mathbb{B}[x]$ be the minimal polynomial of β over $\mathbb{B} = \mathbb{F}(\alpha)$. If s and t are relatively prime, show that $\tilde{m}_\beta(x) = m_\beta(x)$.

16. (a) Solve the system of equations:

$$x + y = 6$$
$$x^2 + y^2 = -2$$

(1) in the field of complex numbers $\mathbb{C}$.
(2) in the field $\mathbb{F}_{37}$, the finite field with 37 elements.

   (b) Solve the system of equations:

$$x + y + z = 7$$
$$x^2 + y^2 + z^2 = 15$$
$$x^3 + y^3 + z^3 = 31$$

(1) in $\mathbb{C}$.
(2) in $\mathbb{F}_{37}$.

17. Let p be a prime. Show that $s_k(1, 2, , \ldots, p-1) \equiv 0 \pmod{p}$ for $k = 1, \ldots, p-2$ and that $s_{p-1}(1, 2, , \ldots, p-1) \equiv -1 \pmod{p}$. Here s_k is the kth elementary symmetric function.

18. Find the Galois groups of the following polynomials over $\mathbb{Q}$:

   (a) $(x^2 - 3)(x^3 - 1)$
   (b) $(x^2 + 3)(x^3 - 1)$

(c) $(x^3 - 1)(x^3 - 2)$

(d) $(x^3 - 2)(x^3 - 3)$

(e) $x^4 - 4$

(f) $x^4 + 4$

(g) $x^8 - 2$

(h) $x^8 - 3$

19. Let $a \in \mathbb{Q}$ and suppose that $a \neq \pm b^2$ for any $b \in \mathbb{Q}$. Let $\mathbb{E}$ be a splitting field of the polynomial $x^4 - a \in \mathbb{Q}[x]$. Show that $\mathrm{Gal}(\mathbb{E}/\mathbb{Q})$ is isomorphic to D_8. (Compare Example 4.8.7.)

20. Let $\mathbb{E}$ be a splitting field of a polynomial $f(x) \in \mathbb{Q}[x]$ of degree 4. Suppose that $(\mathbb{E}/\mathbb{Q}) = 8$. Show that $f(x)$ is irreducible and that $\mathrm{Gal}(\mathbb{E}/\mathbb{Q})$ is nonabelian.

21. (a) Let n be a positive integer and let $\mathbb{F}$ be a field that contains a primitive nth root of 1, i.e., an element a with $a^n = 1$ but $a^m \neq 1$ for any $1 \leq m < n$. (For example, we could choose $\mathbb{F} = \mathbb{Q}(\zeta_n)$.) Let $\mathbb{E}$ be a Galois extension of degree n with $\mathrm{Gal}(\mathbb{E}/\mathbb{F})$ a cyclic group. If $n = p$ is prime, show that $\mathbb{E} = \mathbb{F}(\beta)$ for some element β of $\mathbb{E}$ with $\beta^n \in \mathbb{F}$. (Thus if $b = \beta^n$, $\mathbb{E}$ is a splitting field of the polynomial $f(x) = x^n - b \in \mathbb{F}[x]$.)

This result is true for any positive integer n, but is more difficult to prove in general.

(b) Let $\gamma_i = \Sigma_{j=0}^{n-1} a^{ij}\beta^j \in \mathbb{E}$. Show that $\{\gamma_0, \ldots, \gamma_{n-1}\}$ is a normal basis for $\mathbb{E}$ over $\mathbb{F}$.

22. Let p be a prime and let $\mathbb{F} = \mathbb{F}_p$. Let a be an integer relatively prime to p. Let $f(x) \in \mathbb{F}[x]$ be the polynomial $f(x) = x^p - x - a$.

(a) If α is a root of $f(x)$ in an extension field $\mathbb{E}$ of $\mathbb{F}$, factor $f(x)$ in $\mathbb{E}$.

(b) Show that $f(x)$ is irreducible in $\mathbb{F}[x]$.

(c) Let $\mathbb{E}$ be a splitting field of $f(x)$. Then $\mathbb{E}$ is an extension of $\mathbb{F}$ of degree p. Describe the action of $\mathrm{Gal}(\mathbb{E}/\mathbb{F})$ on $\mathbb{E}$.

(d) More generally, let $\mathbb{F}$ be a field of characteristic p and let $f(x) = x^p - x - a$ for some $a \in \mathbb{F}$. Suppose that $f(x)$ does not have a root in $\mathbb{F}$. Do parts (a), (b), and (c) in this more general situation.

The field $\mathbb{E}$ is known as an *Artin-Schreier extension* of $\mathbb{F}$.

(e) Let $\mathbb{F} = \mathbb{F}_p$, let a be an integer relatively prime to p, and let $f(x) = x^{p^h} - x - a$ for some positive integer n. Let $\mathbb{B} = \mathbb{F}_{p^n}$

and let $\mathbb{E}$ be a splitting field of $f(x)$. Show that $(\mathbb{E}/\mathbb{B}) = p$. Describe the action of $\mathrm{Gal}(\mathbb{E}/\mathbb{F})$ on $\mathbb{E}$ and identify the subgroup $\mathrm{Gal}(\mathbb{E}/\mathbb{B})$ of $\mathrm{Gal}(\mathbb{E}/\mathbb{F})$.

23. Let $\mathbb{F} = \mathbb{F}_p$. Let k be an integer with $k \not\equiv 0,1 \pmod{p}$. Let d be the order of $k \pmod{p}$, i.e., the smallest positive integer d such that $k^d \equiv 1 \pmod{p}$. Let $f(x)$ be the polynomial $f(x) = x^p - kx - 1 \in \mathbb{F}[x]$. Show that $f(x)$ is the product of a linear factor and $(p-1)/d$ distinct irreducible factors of degree d in $\mathbb{F}[x]$.

24. (a) Let $\mathbb{E}$ be a Galois extension of $\mathbb{F}$ of degree n.
    If n is not a prime, show that there is some field $\mathbb{B}$ strictly intermediate between $\mathbb{F}$ and $\mathbb{E}$ i.e., with $\mathbb{F} \subset \mathbb{B} \subset \mathbb{E}$.
    (b) Give an example of the following situation: $\mathbb{E}$ is a Galois extension of $\mathbb{F}$ of degree n, d is a divisor of n, but there is no field $\mathbb{B}$ intermediate between $\mathbb{F}$ and $\mathbb{E}$ that is an extension of $\mathbb{F}$ of degree d.

25. Let d be an arbitrary positive integer. Show that there is a Galois extension $\mathbb{E}$ of $\mathbb{Q}$ of degree d.

26. (a) Let G be a finite abelian group. Show that for some N, there is a field intermediate between $\mathbb{Q}$ and $\mathbb{Q}(\zeta_N)$ with $\mathrm{Gal}(\mathbb{E}/\mathbb{Q})$ isomorphic to G. In your solution, you may use Dirichlet's famous theorem: Let a and b be relatively prime. Then there are infinitely many primes congruent to $b \pmod{a}$. (Note that this implies the result of the preceding problem.)
    (b) If G is not abelian, show that there is no such N.
    The *Kronecker–Weber theorem* states that any Galois extension of $\mathbb{Q}$ with abelian Galois group is isomorphic to a subfield of some cyclotomic field (i.e., to a subfield of $\mathbb{Q}(\zeta_N)$ for some N).

27. Let $\mathbb{E} = \mathbb{Q}(\zeta_7)$ where ζ_7 is a primitive 7th root of 1. Note that $\mathbb{E}$ is an extension of $\mathbb{Q}$ of degree 6. Note also that $\mathbb{E}$ contains a unique subfield $\mathbb{B}$ that is an extension of $\mathbb{Q}$ of degree 2 and that $\mathbb{E}$ contains a unique subfield $\mathbb{D}$ that is an extension of $\mathbb{Q}$ of degree 3.

    (a) Find an irreducible polynomial $f(x) \in \mathbb{Q}[x]$ such that $\mathbb{B} = \mathbb{Q}(\beta)$ where β is a root of $f(x)$.
    (b) Find an irreducible polynomial $g(x) \in \mathbb{Q}[x]$ such that $\mathbb{D} = \mathbb{Q}(\delta)$ where δ is a root of $g(x)$.

28. Let $D > 1$ be a square-free positive integer, and let $\alpha = \sqrt{D} + i$. Let $m_\alpha(x) \in \mathbb{Q}[x]$ be the minimal polynomial of α, and let $\mathbb{E}$ be a splitting field of $m_\alpha(x)$. For which positive integers n is $\mathbb{Q}(\zeta_n) \subseteq \mathbb{E}$? (Of course, your answer will depend on D.)

29. (a) Let p be a prime and let (k/p) be the Legendre symbol. Let

$$S_p = \sum_{k=1}^{p-1} (k/p)\exp(2\pi i k/p).$$

Show that $(S_p)^2 = (-1/p)p$. Thus $S_p = \pm\sqrt{p}$ if $p \equiv 1 \pmod 4$ and $S_p = \pm i\sqrt{p}$ if $p \equiv 3 \pmod 4$. (The sum S_p is an example of a *Gauss sum*. It is a theorem of Gauss than in each case the sign is $+$. Gauss wrote that after thinking about this problem fruitlessly for a year, the solution came to him "like a bolt of lightning from the sky".)

  (b) Let D be an arbitrary integer. Show that $\mathbb{Q}(\sqrt{D}) \subseteq \mathbb{Q}(\zeta_n)$ for some n.

  (c) Let p be an odd prime and let D be an integer that is not a perfect pth power. Show that $\mathbb{Q}(\sqrt[p]{D}) \nsubseteq \mathbb{Q}(\zeta_n)$ for any n.

30. (a) Let p be a prime and let n be a positive integer. Show that

$$\Phi_{p^n}(x) = \frac{x^{p^n} - 1}{x^{p^{n-1}} - 1}.$$

  (b) Let n be a positive odd integer. Show that

$$\Phi_{2n}(x) = \epsilon\Phi_n(-x)$$

where $\epsilon = -1$ if $n = 1$ and $\epsilon = +1$ if $n > 1$.

  (c) Let p and q be distinct primes. Show that

$$\Phi_{pq}(x) = \frac{(x^{pq} - 1)(x - 1)}{(x^p - 1)(x^q - 1)}.$$

  (d) More generally, let p be a prime and let n be a positive integer not divisible by p. Show that

$$\Phi_{np}(x) = \Phi_n(x^p)/\Phi_n(x).$$

  (e) Let n be a positive integer and let k be the product of the distinct prime factors of n. Show that

$$\Phi_n(x) = \Phi_k(x^{n/k}).$$

31. Let $\mathbb{F} = \mathbb{F}_p$ be the field with p elements and let m be an integer relatively prime to p. Let $\bar{\Phi}_m(x) \in \mathbb{F}[x]$ be the (mod p) reduction of $\Phi_m(x)$, and let $\bar{g}(x) = x^m - 1 \in \mathbb{F}[x]$.

    (a) Show that $\bar{\Phi}_m(x)$ and $\bar{g}(x)$ have the same splitting field $\mathbb{E}$, and furthermore $\mathbb{E} = \mathbb{F}_{p^r}$ where r is the smallest positive integer with $p^r \equiv 1 \pmod{m}$. (In particular, while $\Phi_m(x)$ is always irreducible in $\mathbb{Q}[x]$, $\bar{\Phi}_m(x)$ is irreducible in $\mathbb{F}_p[x]$ if and only if $r = \varphi(m)$.)

    (b) Let $\bar{\Phi}_m(x) = \bar{f}_1(x) \ldots \bar{f}_k(x)$ be a factorization of $\bar{\Phi}_m(x)$ into a product of irreducible polynomials in $\mathbb{F}_p[x]$. Show that each polynomial $\bar{f}_i(x)$ has degree r, and hence that $k = \varphi(m)/r$.

32. Let $f(x) \in \mathbb{F}[x]$ have roots $\{\alpha_1, \ldots, \alpha_n\}$ in some splitting field $\mathbb{E}$. Show that $f(x)$ is irreducible in $\mathbb{F}[x]$ if and only if

$$\prod_{\alpha_i \in T} (x - \alpha_i) \notin \mathbb{F}[x]$$

    for any nonempty proper subset T of $S = \{\alpha_1, \ldots, \alpha_n\}$.

33. Let G be a finite group of automorphisms of a field $\mathbb{E}$ and let $\mathbb{F} \subseteq \mathbb{E}$ be its fixed field. Let $\alpha \in \mathbb{E}$ and let $\{\alpha_1 = \alpha, \ldots, \alpha_r\}$ be its orbit under the action of G. Show that α is algebraic over $\mathbb{F}$ of degree r with minimal polynomial $m_\alpha(x) = (x - \alpha_1) \ldots (x - \alpha_r) \in \mathbb{F}[x]$.

34. Let $n > 2$. Let $\mathbb{E} = \mathbb{Q}(\zeta_n)$ and $\mathbb{B} = \mathbb{E} \cap \mathbb{R}$. Show that $(\mathbb{B}/\mathbb{Q}) = \varphi(n)/2$ and that $(\mathbb{E}/\mathbb{B}) = 2$. Show that $\mathbb{B} = \mathbb{Q}(\cos(2\pi/n))$ and that $\mathbb{E} = \mathbb{B}(i \sin(2\pi/n))$.

35. (a) For $n = 7, 8, 9, 10$ find the minimal polynomial of $\cos(2\pi/n)$ over $\mathbb{Q}$.

    (b) For any n, find the minimal polynomial of $i \sin(2\pi/n)$ over $\mathbb{Q}(\cos(2\pi/n))$.

36. For $n = 7, 8, 9, 10$, let $\mathbb{E} = \mathbb{Q}(\zeta_n)$. Find all fields intermediate between $\mathbb{Q}$ and $\mathbb{E}$. For each intermediate field $\mathbb{B}$, find $\mathrm{Gal}(\mathbb{B}/\mathbb{Q})$ and $\mathrm{Gal}(\mathbb{E}/\mathbb{B})$. Find a primitive element of $\mathbb{B}$. Also, find a polynomial $f(x) \in \mathbb{Q}[x]$ whose splitting field in $\mathbb{B}$.

37. For $n = 7, 8, 9, 10$, let $\mathbb{E}$ be a splitting field of $f(x) = x^p - 7 \in \mathbb{Q}[x]$. Find all fields intermediate between $\mathbb{Q}$ and $\mathbb{E}$. For each intermediate field $\mathbb{B}$, find $\mathrm{Gal}(\mathbb{B}/\mathbb{Q})$ and $\mathrm{Gal}(\mathbb{E}/\mathbb{B})$. Find a primitive element of $\mathbb{B}$. Also, if $\mathbb{B}$ is a Galois extension of $\mathbb{Q}$, find a polynomial $g(x) \in \mathbb{Q}[x]$ whose splitting field is $\mathbb{B}$.

38. In each case, let $m_\alpha(x)$ denote the minimal polynomial of α over $\mathbb{Q}$, and let $\mathbb{E}$ be a splitting field of $m_\alpha(x)$. Let $G = \text{Gal}(\mathbb{E}/\mathbb{Q})$.

    1. (a) Let $\alpha = \sqrt{35 + 10\sqrt{10}}$. Show that $\alpha = c + \sqrt{d}$ for some $c, d \in \mathbb{Q}$. Show that $m_\alpha(x)$ is a quadratic, that $(\mathbb{E}/\mathbb{Q}) = 2$, and that G is isomorphic to $\mathbb{Z}_2$.
       (b) Let $\alpha = \sqrt{20 + 10\sqrt{3}}$. Show that $\alpha = \sqrt{e} + \sqrt{f}$ for some $e, f \in \mathbb{Q}$. Show that $m_\alpha(x)$ is a quartic, that $(\mathbb{E}/\mathbb{Q}) = 4$, and that G is isomorphic to $\mathbb{Z}_2 \oplus \mathbb{Z}_2$.
    2. (a) Let $\alpha = \sqrt{5 + \sqrt{5}}$. Show that $m_\alpha(x)$ is a quartic, that $(\mathbb{E}/\mathbb{Q}) = 4$, and that G is isomorphic to $\mathbb{Z}_4$.
       (b) Let $\alpha = \sqrt{10 + \sqrt{5}}$. Show that $m_\alpha(x)$ is a quartic, that $(\mathbb{E}/\mathbb{Q}) = 8$, and that G is isomorphic to D_8.

    (Note that this problem is a special case of the next problem.)

39. Let $\alpha = \sqrt{a + \sqrt{b}}$, $a, b, \in \mathbb{Q}$, b not a perfect square. Let $m_\alpha(x)$ be the minimal polynomial of α, $m_\alpha(x) \in \mathbb{Q}[x]$. Let $\mathbb{E}$ be a splitting field of $m_\alpha(x)$ over $\mathbb{Q}$, and let $G = \text{Gal}(\mathbb{E}/\mathbb{Q})$. Show the following:

    (1) Suppose that $a^2 - b$ is a perfect square in $\mathbb{Q}$, $a^2 - b = q^2$. (Note q is only defined up to sign)

        (a) If $2a + 2q$ is a perfect square (for some choice of q), them $m_\alpha(x)$ is a quadratic, $(\mathbb{E}/\mathbb{Q}) = 2$, and G is isomorphic to $\mathbb{Z}_2$. (In this case, $\alpha = (c + \sqrt{d})^2$ for some $c, d \in \mathbb{Q}$.)
        (b) If $2a + 2q$ is not a perfect square (for any choice of q), then $m_\alpha(x)$ is a quartic, $(\mathbb{E}/\mathbb{Q}) = 4$, and G is isomorphic to $\mathbb{Z}_2 \oplus \mathbb{Z}_2$. (In this case, $\alpha = \sqrt{e} + \sqrt{f}$ for some $e, f \in \mathbb{Q}$.)

    (2) Suppose that $a^2 - b$ is not a perfect square in $\mathbb{Q}$. Then $m_\alpha(x)$ is a quartic.

        (a) If $(a^2 - b)/b$ is a perfect square, then $(\mathbb{E}/\mathbb{Q}) = 4$ and G is isomorphic to $\mathbb{Z}_4$.
        (b) If $(a^2 - b)/b$ is not a perfect square, then $(\mathbb{E}/\mathbb{Q}) = 8$ and G is isomorphic to D_8.

40. Consider the polynomial $f(x) = x^4 + sx^2 + t \in \mathbb{Q}[x]$. Use the preceding exercise to determine when $f(x)$ is irreducible. In this case, let $\mathbb{E}$ be a splitting field of $\mathbb{Q}[x]$. Determine $(\mathbb{E}/\mathbb{Q})$ and $G = \text{Gal}(\mathbb{E}/\mathbb{Q})$.

41. Let $\mathbb{E}$ be an extension of $\mathbb{Q}$ of degree 4. Show that there is a quadratic extension $\mathbb{B}$ of $\mathbb{Q}$ with $\mathbb{B} \subseteq \mathbb{E}$ if and only if $\mathbb{E} = \mathbb{Q}(\alpha)$ where α is a root of an irreducible polynomial $f(x) = x^4 + sx^2 + t \in \mathbb{Q}[x]$.

42. Let $\mathbb{E}$ be a splitting field of the separable polynomial $f(x) \in \mathbb{F}[x]$, and suppose that the group $G = \mathrm{Gal}(\mathbb{E}/\mathbb{F})$ has no nontrivial proper normal subgroups. Show that $\mathbb{E}$ is a splitting field of some irreducible factor of $f(x)$. Give a counterexample if G does not satisfy this condition.

43. Let $f(x) \in \mathbb{F}[x]$ be a separable irreducible polynomial of prime degree p, and let $\mathbb{E}$ be a splitting field of $f(x)$. Let $G = \mathrm{Gal}(\mathbb{E}/\mathbb{F})$. If G is not isomorphic to $\mathbb{Z}_p$, show that G is not abelian.

44. Let $p(x) \in \mathbb{F}[x]$ be an irreducible polynomial, and let $\mathbb{E}$ be a splitting field of $p(x)$. Let α and β be any two roots of $p(x)$ in $\mathbb{E}$. Let $f(x) \in \mathbb{F}[x]$ be an arbitrary polynomial, and suppose that $f(x)$ factors in $\mathbb{E}[x]$ as $f(x) = g(x, \alpha)h(x, \alpha)$ for some polynomials $g(x, y)$ and $h(x, y)$ in $\mathbb{F}[x, y]$. Show that $f(x)$ also factors in $\mathbb{E}[x]$ as $f(x) = g(x, \beta)h(x, \beta)$.

45. Let $\mathbb{B}$ and $\mathbb{D}$ be any two extensions of $\mathbb{F}$ such that $(\mathbb{B}\mathbb{D})/\mathbb{F} = (\mathbb{B}/\mathbb{F})(\mathbb{D}/\mathbb{F})$. Let $\mathbb{B}_1$ be any field intermediate between $\mathbb{F}$ and $\mathbb{B}$ and let $\mathbb{D}_1$ be any field intermediate between $\mathbb{F}$ and $\mathbb{D}$. Show that $(\mathbb{B}_1\mathbb{D}_1/\mathbb{F}) = (\mathbb{B}_1/\mathbb{F})(\mathbb{D}_1/\mathbb{F})$.

46. Let $\mathbb{B}$ and $\mathbb{D}$ be disjoint Galois extensions of $\mathbb{F}$. Let $\mathbb{B}$ be a splitting field of $f(x) \in \mathbb{F}[x]$ and let $\mathbb{D}$ be a splitting field of $g(x) \in \mathbb{F}[x]$. Let β be a root of $f(x)$ in $\mathbb{B}$ and let δ be a root of $g(x)$ in $\mathbb{D}$. Show that $\beta + \delta$ is a primitive element of $\mathbb{F}(\beta, \delta)$.

47. Let p and q be distinct primes, and let D be an integer that is neither a pth power nor a qth power. Show that the polynomial $f(x) = x^{pq} - D$ is irreducible in $\mathbb{Q}[x]$. (Thus, for example, the polynomial $x^6 - 72$ is irreducible in $\mathbb{Q}[x]$. Note that this polynomial does not satisfy the hypotheses of Eisenstein's criterion.)

48. An extension $\mathbb{B}$ of $\mathbb{F}$ is an abelian extension if it is a Galois extension of $\mathbb{F}$ whose Galois group $\mathrm{Gal}(\mathbb{B}/\mathbb{F})$ is an abelian group.

    (a) Let $\mathbb{E}$ be a finite Galois extension of $\mathbb{F}$. Show that there is a maximal abelian subextension $\mathbb{E}_{ab}$ of $\mathbb{F}$, i.e., that there is an abelian extension $\mathbb{E}_{ab}$ of $\mathbb{F}$ with $\mathbb{E}_{ab} \subseteq \mathbb{E}$, and if $\mathbb{B}$ is any abelian extension of $\mathbb{F}$ with $\mathbb{B} \subseteq \mathbb{E}$, then $\mathbb{B} \subseteq \mathbb{E}_{ab}$.

    (b) If $G = \mathrm{Gal}(\mathbb{E}/\mathbb{F})$ and $G_{ab} = \mathrm{Gal}(\mathbb{E}_{ab}/\mathbb{F})$, identify G_{ab}.

    (c) If G is isomorphic to the alternating group A_n, for $n \geq 5$, show that $\mathbb{E}_{ab} = \mathbb{F}$.

    (d) If G is isomorphic to the symmetric group S_n, for $n \geq 5$, show that $(\mathbb{E}_{ab}/\mathbb{F}) = 2$. In this situation, suppose that $\mathrm{char}(\mathbb{F}) \neq 2$ and $\mathbb{E}$ is a splitting field of an irreducible separable polynomial $f(x) \in \mathbb{F}[x]$. Show that $\mathbb{E}_{ab} = \mathbb{F}(\sqrt{\Delta})$, where Δ is the discriminant of the polynomial $f(x)$.

49. Let $\mathbb{E}$ be a Galois extension of $\mathbb{F}$ and let $\mathbb{B}$ be any field intermediate between $\mathbb{F}$ and $\mathbb{E}$. Show that there is a smallest field $\mathbb{D}$ intermediate between $\mathbb{B}$ and $\mathbb{E}$ that is a Galois extension of $\mathbb{F}$ (i.e., $\mathbb{D}$ is a Galois extension of $\mathbb{F}$ and if $\mathbb{D}'$ is any Galois extension of $\mathbb{F}$ intermediate between $\mathbb{B}$ and $\mathbb{E}$, then $\mathbb{D} \subseteq \mathbb{D}'$.) Identify $\mathrm{Gal}(\mathbb{E}/\mathbb{D})$ and $\mathrm{Gal}(\mathbb{D}/\mathbb{F})$.

50. Let $f(x) \in \mathbb{F}[x]$ be an irreducible polynomial. Let $\mathbb{E}$ be an extension of $\mathbb{F}$ and suppose that $f(x)$ factors as a product of irreducible polynomials $f(x) = f_1(x) \ \ldots \ f_t(x)$ in $\mathbb{E}[x]$.

    (a) If $\mathbb{E}$ is a Galois extension of $\mathbb{F}$, show that for each i, j there is an element $\sigma_{ij} \in \mathrm{Gal}(\mathbb{E}/\mathbb{F})$ such that $\sigma_{ij}(f_i(x)) = f_j(x)$.

    (b) Give a counterexample to this if $\mathbb{E}$ is not a Galois extension of $\mathbb{F}$.

51. Let $\mathbb{E}$ be a splitting field of the irreducible separable polynomial $f(x) \in \mathbb{F}[x]$, and let $\mathbb{B}$ be a Galois extension of $\mathbb{F}$ that is intermediate between $\mathbb{F}$ and $\mathbb{E}$. Let $f(x) = f_1(x) \ \ldots \ f_k(x)$ be a factorization of $f(x)$ into irreducibles in $\mathbb{B}[x]$. Show that each polynomial has the same degree d. Furthermore, if $\alpha \in \mathbb{E}$ is any root of $f(x)$, show that $d = (\mathbb{B}(\alpha)/\mathbb{B})$ and that $k = (\mathbb{B} \cap \mathbb{F}(\alpha)/\mathbb{F})$.

52. Let $f(x) \in \mathbb{F}[x]$ be an irreducible separable polynomial of degree d, and let $\mathbb{E}$ be a splitting field of $f(x)$.

    (a) For a root α of $f(x)$ in $\mathbb{E}$, let $r(\alpha)$ be the number of roots of $f(x)$ in $\mathbb{F}(\alpha)$. Show that $r(\alpha)$ is independent of the choice of α. Call this common value r.

    (b) Let s be the number of distinct fields $\mathbb{F}(\alpha)$, where α is a root of $f(x)$ in $\mathbb{E}$. Show that $rs = d$.

    (c) Give examples where $r = 1$, $1 < r < d$, and $r = d$. (Choose $d > 1$.)

53. Let $\mathbb{E}$ be a splitting field of the irreducible separable polynomial $f(x) \in \mathbb{F}[x]$. Let $d = (\mathbb{E}/\mathbb{F})$. For any root α of $f(x)$ in $\mathbb{E}$, and

any prime p dividing d, show that there is a field $\mathbb{B}$ intermediate between $\mathbb{F}$ and $\mathbb{E}$ with $(\mathbb{E}/\mathbb{B}) = p$ and $\mathbb{E} = \mathbb{B}(\alpha)$.

54. Let $\mathbb{F} \subseteq \mathbb{B} \subseteq \mathbb{E}$ with $\mathbb{E}$ a Galois extension of $\mathbb{F}$. Show that there exist irreducible polynomials $f(x) \in \mathbb{F}[x]$ and $g(x) \in \mathbb{B}[x]$ such that $\mathbb{E}$ is a splitting field of $f(x)$, and that $\mathbb{E}$ is a splitting field of $g(x)$, and $g(x)$ divides $f(x)$ in $\mathbb{B}[x]$.

Chapter 5

Rings of Algebraic Integers and Dedekind Rings

In our investigation of rings, one of our principal interests has been the question of unique factorization. We have seen examples of rings that do have unique factorization, and examples of rings that do not.

Our objective in this chapter is to show that in a very important class of examples, rings of algebraic integers, while we may not have unique factorization of *elements*, we always have unique factorization of *ideals*.

We shall proceed in several stages. We will first define and study rings of algebraic integers. We will next define Dedekind rings, and show that rings of algebraic integers are always Dedekind rings. Then we shall show that we always have unique factorization of ideals in Dedekind rings. Then we will look at a bunch of examples.

As you will see, field theory in general, and Galois theory in particular, plays an essential role in our investigations here.

5.1 Rings of algebraic integers

Definition 5.1.1. An *algebraic number field* $\mathbb{F}$ is a finite extension of $\mathbb{Q}$. ◇

Remark 5.1.2. As we have seen in Theorem 4.5.6, an algebraic number field is obtained from $\mathbb{Q}$ by adjoining a finite number of elements, each of which is algebraic over $\mathbb{Q}$. In this chapter, we shall use algebraic to mean algebraic over $\mathbb{Q}$. ◇

Definition 5.1.3. Let $\mathbb{F}$ be an algebraic number field. An element α of $\mathbb{F}$ is *integral* over $\mathbb{Z}$, or is an *algebraic integer*, if the minimal polynomial $m_\alpha(x) \in \mathbb{Q}[x]$ of α over $\mathbb{Q}$ is a polynomial with coefficients in $\mathbb{Z}$. We set $\mathcal{O}_\mathbb{F} = \{\alpha \in \mathbb{F} \mid \alpha \text{ is an algebraic integer}\}$.

$\mathcal{O}_\mathbb{F}$ is the *ring of algebraic integers* in $\mathbb{F}$. $\Diamond$

Lemma 5.1.4. *An element α of an algebraic number field $\mathbb{F}$ is an algebraic integer if and only if it is a root of some monic polynomial $f(x) \in \mathbb{Z}[x]$.*

Proof.   If $m_\alpha(x) \in \mathbb{Z}[x]$ then we may choose $f(x) = m_\alpha(x)$. On the other hand, if α is a root of $f(x) \in \mathbb{Z}[x]$ then $m_\alpha(x) \in \mathbb{Q}[x]$ divides $f(x)$, so $m_\alpha(x) \in \mathbb{Z}[x]$ by Gauss's lemma. $\square$

Corollary 5.1.5. *Let $\mathbb{F} = \mathbb{Q}$. Then $\alpha \in \mathbb{Q}$ is an algebraic integer if and only if $\alpha \in \mathbb{Z}$, i.e., $\mathcal{O}_\mathbb{Q} = \mathbb{Z}$.*

Proof.   The element $\alpha \in \mathbb{Q}$ has minimal polynomial $m_\alpha(x) = x - \alpha$, and $m_\alpha(x) \in \mathbb{Z}[x]$ if and only if $\alpha \in \mathbb{Z}$. $\square$

Remark 5.1.6. We will need to be careful in distinguishing between the integers (i.e., algebraic integers) in an algebraic number field $\mathbb{F}$ and the integers in $\mathbb{Q}$. Thus, whenever there is the possibility of confusion, we will refer to $\alpha \in \mathbb{Z}$ as a *rational integer* (i.e., an integer in the field of rational numbers). This is standard terminology. $\Diamond$

Definition 5.1.7. In this chapter, we will refer to an abelian group as a $\mathbb{Z}$-*module*. $\Diamond$

Remark 5.1.8. There is a vast, and important, theory of modules over general rings. We do not want to get into that here, as we do not need it for our purposes. But we at least want to introduce the language, so you will become familiar with it. $\Diamond$

By definition, $\mathcal{O}_\mathbb{F}$ is a subset of $\mathbb{F}$. Our first job is to show that $\mathcal{O}_\mathbb{F}$ is a subring of $\mathbb{F}$. In order to do so we develop a criterion for an element of $\mathbb{F}$ to be an algebraic integer.

Definition 5.1.9. Let $\alpha_1, \ldots, \alpha_n$ be arbitrary elements of $\mathbb{F}$. Then $\mathbb{Z}[\alpha_1, \ldots, \alpha_n] = \{p(\alpha_1, \ldots, \alpha_n) \mid p(x_1, \ldots, x_n) \in \mathbb{Z}[x_1, \ldots, x_n]\} \subseteq \mathbb{F}$. $\Diamond$

Lemma 5.1.10. *Let $\alpha \in \mathbb{F}$. Then α is an integer in $\mathbb{F}$ if and only if $\mathbb{Z}[\alpha]$ is a finitely generated $\mathbb{Z}$-module.*

Proof. Suppose α is an integer in $\mathbb{F}$. Let its minimal polynomial $m_\alpha(x) \in \mathbb{Z}[x]$ have degree n. Let $\beta \in \mathbb{Z}[\alpha]$ so that $\beta = p(\alpha)$ for some polynomial $p(x) \in \mathbb{Z}[x]$. Then $m_\alpha(x)$ divides $p(x)$ in $\mathbb{Q}[x]$, and hence in $\mathbb{Z}[x]$, by Gauss's lemma. Thus,

$$p(x) = m_\alpha(x)q(x) + r(x), \; r(x) \in \mathbb{Z}[x] \text{ with } r(x) = 0 \text{ or } \deg r(x) < n,$$

and so $\beta = p(\alpha) = m_\alpha(\alpha)q(\alpha) + r(\alpha) = 0q(\alpha) + r(\alpha) = r(\alpha)$. Thus, we see that $\mathbb{Z}[x]$ is generated by $1, \alpha, \ldots, \alpha^{n-1}$.

On the other hand, suppose that $\mathbb{Z}[x]$ is generated by a finite set of elements $\{\beta_1, \ldots, \beta_k\}$. Let $\beta_i = p_i(\alpha)$, with $p_i(x) \in \mathbb{Z}[x]$ a polynomial of degree d_i, for each $i = 1, \ldots, k$. Choose $n > \max(d_1, \ldots, d_k)$. Then $\alpha^n \in \mathbb{Z}[\alpha]$, so $\alpha^n = \Sigma_{i=1}^k m_i \beta_i = \Sigma_{i=1}^k m_i p_i(\alpha)$ for some integers $m_1, \ldots, m_k$. Thus, if $f(x) = x^n - \Sigma_{i=1}^k m_i p_i(x)$, then $f(x) \in \mathbb{Z}[x]$ is a monic polynomial with integer coefficients of degree n with $f(\alpha) = 0$. Thus, by Lemma 5.1.4, α is an integer. $\quad\square$

Lemma 5.1.11. *Let $\mathbb{F}$ be a field intermediate between $\mathbb{Q}$ and $\mathbb{E}$. Then $\mathcal{O}_{\mathbb{F}} = \mathbb{F} \cap \mathcal{O}_{\mathbb{E}}$.*

Proof. Clear from Definition 5.1.3, as $m_\alpha(x)$ is the same whether α is regarded as an element of $\mathbb{F}$ or an element of $\mathbb{E}$. $\quad\square$

Theorem 5.1.12. *Let $\mathbb{F}$ be an algebraic number field. Then $\mathcal{O}_{\mathbb{F}}$ is a ring with 1.*

Proof. $m_0(x) = x$ so $0 \in \mathcal{O}_F$ and $m_1(x) = x - 1$ so $1 \in \mathcal{O}_{\mathbb{F}}$. If $m_\alpha(x) = \Sigma_{i=0}^n a_i x^i$ then $m_{-\alpha}(x) = \Sigma_{i=0}^n (-1)^{n-i} a_i x^i$ so if $\alpha \in \mathcal{O}_{\mathbb{F}}$ then $-\alpha \in \mathcal{O}_{\mathbb{F}}$.

Now, let $\alpha, \beta \in \mathcal{O}_{\mathbb{F}}$. Then, by Lemma 5.10, $\mathbb{Z}[\alpha]$ is generated by a finite set $\{\alpha_1, \ldots, \alpha_k\}$ and $\mathbb{Z}[\beta]$ is generated by a finite set $\{\beta_1, \ldots, \beta_\ell\}$. Then $\mathbb{Z}[\alpha, \beta]$ is generated by the finite set $\{\alpha_i \beta_j \mid i = 1, \ldots, k, \; j = 1, \ldots, \ell\}$, as any element of $\mathbb{Z}[\alpha, \beta]$ is a sum of product of polynomials in α and β, each of which can be expressed in terms of these two sets.

Thus $\mathbb{Z}[\alpha, \beta]$ is a finitely generated $\mathbb{Z}$-module, and it is certainly torsion-free as $\mathbb{F}$ is a field of characteristic 0. Thus, by Theorem 2.6.32, it is a free $\mathbb{Z}$-module.

378 An Introduction to Abstract Algebra: Sets, Groups, Rings, and Fields

Now $\mathbb{Z}[\alpha + \beta] \subseteq \mathbb{Z}[\alpha, \beta]$ and $\mathbb{Z}[\alpha\beta] \subseteq \mathbb{Z}[\alpha, \beta]$, so each of these is a free finitely-generated $\mathbb{Z}$-module by Corollary 2.6.33, and hence, by Lemma 5.1.10, $\alpha + \beta \in \mathcal{O}_\mathbb{F}$ and $\alpha\beta \in \mathcal{O}_\mathbb{F}$, and so $\mathcal{O}_\mathbb{F}$ is a ring with 1. $\qquad\square$

Example 5.1.13. Let D be a squarefree integer, $D \neq 1$, and let $\mathbb{F} = \mathbb{Q}(\sqrt{D})$, a quadratic extension of $\mathbb{Q}$. Let $\mathcal{O}(\sqrt{D})$ be the ring introduced in Example 3.1.15,

$$\mathcal{O}(\sqrt{D}) = \begin{cases} \{a + b\sqrt{D} \mid a, b \in \mathbb{Z}\} & D \equiv 2, 3 \pmod 4 \\ \{\frac{a+b\sqrt{D}}{2} \mid a, b \in \mathbb{Z}, a \equiv b \pmod 2)\} & D \equiv 1 \pmod 4. \end{cases}$$

Then $\mathcal{O}(\sqrt{D}) = \mathcal{O}_\mathbb{F}$. $\qquad\diamond$

Let $\mathbb{F}$ be an extension of $\mathbb{Q}$ of degree n. Our next goal is to prove that $\mathcal{O}_\mathbb{F}$ is a free $\mathbb{Z}$-module of rank n. This will take some work.

Definition 5.1.14. Let $\mathbb{E}$ be a finite Galois extension of $\mathbb{Q}$, with Galois group $G = \mathrm{Gal}(\mathbb{E}/\mathbb{Q})$.

For $\alpha \in \mathbb{E}$, we let

$$T(\alpha) = \sum_{\sigma \in G} \sigma(\alpha).$$

$T(\alpha)$ is called the *trace* of α. $\qquad\diamond$

Remark 5.1.15. The trace is actually defined more generally. If $\mathbb{E}$ is a Galois extension of $\mathbb{F}$ with Galois group G then $\mathrm{tr}_{\mathbb{E}/\mathbb{F}}(\alpha) = \Sigma_{\sigma \in G}\sigma(\alpha)$. But we will not need this more general definition here. $\qquad\diamond$

Lemma 5.1.16. *Let $\mathbb{E}$ be a finite Galois extension of $\mathbb{Q}$ and let $\alpha \in \mathbb{E}$, $\alpha \neq 0$. Then there is an element $\beta \in \mathbb{E}$ with $T(\alpha\beta) \neq 0$.*

Proof. Let α have degree n and let $\alpha_1 = \alpha, \ldots, \alpha_n$ be the Galois conjugates of α, i.e., the distinct elements of the set $\{\sigma(\alpha) \mid \sigma \in G\}$. Then $m_\alpha(x) = (x - \alpha_1) \ldots (x - \alpha_n)$ and also $m_\alpha(x) = x^n + \Sigma_{i=0}^{n-1} a_i x^i$ with $a_i = (-1)^i s_{n-i}$ where $\{s_1, \ldots, s_n\}$ are the elementary symmetric functions in $\{\alpha_1, \ldots, \alpha_n\}$.

Let $\beta_0 = 1$ and for $i = 1, \ldots, n - 1$, let β_i be the ith elementary symmetric function in $\{\alpha_2, \ldots, \alpha_n\}$. Then for each $i =, \ldots, n$

$$T(\alpha\beta_{i-1}) = k_i s_i \quad \text{for some integer } k_i \neq 0.$$

Thus if $T(\alpha\beta_{i-1}) = 0$ for each $i = 1, \ldots, n$, then $s_i = 0$ for each $i = 1, \ldots, n$, in which case $m_\alpha(x) - x^n$ (and then $n = 1$ and $m_\alpha(x) = x$) and $\alpha = 0$. □

Lemma 5.1.17. *Let $\mathbb{F}$ be an algebraic number field and let α be an arbitrary element of $\mathbb{F}$. Then there is a rational integer N such that $N\alpha$ is an integer in $\mathbb{F}$.*

Proof. Consider $m_\alpha(x) \in \mathbb{Q}[x]$. We may "clear denominators" by multiplying $m_\alpha(x)$ by a nonzero integer N so that $Nm_\alpha(x) \in \mathbb{Z}[x]$. Then

$$Nm_\alpha(x) = Nx^n + \sum_{i=0}^{n-1} b_i x^i \quad \text{with } b_i = Na_i \in \mathbb{Z} \text{ for each } i,$$

and, further multiplying by N^{n-1}, we have

$$N^n m_\alpha(x) = N^n x^n + \sum_{i=0}^{n-1} b_i N^{n-1} x^i$$

$$= N^n x^n + \sum_{i=0}^{n-1} b_i N^{n-1-i} N^i x^i$$

$$= (Nx)^n + \sum_{i=0}^{n-1} c_i (Nx)^i \quad \text{with } c_i = b_i N^{n-1-i} \in \mathbb{Z} \text{ for each } i.$$

Thus if we let $\beta = N\alpha$, $m_\beta(x) = x^n + \Sigma_{i=0}^{n-1} c_i x^i$ and we see that β is an algebraic integer. □

Theorem 5.1.18. *Let $\mathbb{F}$ be an algebraic number field with $(\mathbb{F}/\mathbb{Q}) = m$. Then $\mathcal{O}_\mathbb{F}$ is a free $\mathbb{Z}$-module of rank m.*

Proof. $\mathbb{F}$ is a finite separable extension of $\mathbb{Q}$ so is obtained from $\mathbb{Q}$ by adjoining roots of a separable polynomial $p(x)$. Let $\mathbb{E} \supseteq \mathbb{F}$ be a splitting field of $p(x)$. Then $\mathbb{E}$ is a finite Galois extension of $\mathbb{Q}$. Let $n = (\mathbb{E}/\mathbb{Q})$.

Then $\mathbb{E}$ has a vector space basis $\{\alpha_1, \ldots, \alpha_n\}$ over $\mathbb{Q}$, and then, by Lemma 5.1.17, a vector space basis $\{\beta_1, \ldots, \beta_n\}$ with $\beta_i \in \mathcal{O}_\mathbb{E}$ for each i. Now consider the linear transformation $U \colon \mathbb{E} \to \mathbb{Q}^n$ given by

$$U(\alpha) = \begin{bmatrix} T(\alpha\beta_1) \\ \vdots \\ T(\alpha\beta_n) \end{bmatrix}$$

This is a linear transformation between two $\mathbb{Q}$-vector spaces, both of the same finite dimension n, and by Lemma 5.1.16 $U(\alpha) = 0$ implies $\alpha = 0$, i.e., U is $1 - 1$. Then U must be an isomorphism. Thus for each $i = 1, \ldots, n$, there is a unique element $\gamma_i \in \mathbb{E}$ with $U(\gamma_i) = e_i$, the vector in $\mathbb{Q}^n$ whose ith entry is 1 and whose other entries are all 0, i.e., $T(\gamma_i\beta_j) = 1$ if $j = i$ and 0 if $j \neq i$. Observe also that $\{\gamma_1, \ldots, \gamma_n\}$ is a basis for $\mathbb{E}$ over $\mathbb{Q}$ as $\{U(r_1), \ldots, U(r_n)\}$ is a basis for $\{\mathbb{E}^n\}$.

Now let $\alpha \in \mathcal{O}_\mathbb{E}$ be arbitrary. Then we may write $\alpha = \Sigma_{i=1}^n c_i\gamma_i$ with $c_i \in \mathbb{Q}$ for each i. Since $\beta_j \in \mathcal{O}_\mathbb{E}$ for each j, and $\mathcal{O}_\mathbb{E}$ is a ring, we have that $\alpha\beta_j \in \mathcal{O}_\mathbb{E}$ for each j, and then

$$T(\alpha\beta_j) = \sum_{i=1}^n T(c_i\gamma_i\beta_j) = \sum_{i=1}^n c_i T(\gamma_i\beta_j) = c_j.$$

But if $\alpha\beta_j$ is an element of $\mathbb{E}$ of degree d, with $m_{\alpha\beta_j}(x) = x^d + a_{d-1}x^{d-1} + \cdots$ with $a_{d-1} \in \mathbb{Z}$, then $T(\alpha\beta_j) = (n/d)(-a_{d-1}) \in \mathbb{Z}$. Thus $c_j \in \mathbb{Z}$ for each j. Hence if we let C be the $\mathbb{Z}$-module generated by $\{\gamma_1, \ldots, \gamma_n\}$ and B be the $\mathbb{Z}$-module generated by $\{\beta_1, \ldots, \beta_n\}$ we have that

$$B \subseteq \mathcal{O}_\mathbb{E} \subseteq C.$$

Of course, B and C (and $\mathcal{O}_\mathbb{E}$) are torsion-free since $\mathbb{E}$ is a field of characteristic 0. Thus, B and C are each free of rank n, so by Corollary 2.6.33 $\mathcal{O}_\mathbb{E}$ is free of rank n as well.

Now $\mathcal{O}_\mathbb{F} = \mathbb{F} \cap \mathcal{O}_\mathbb{E}$, so $\mathcal{O}_\mathbb{F}$ is a $\mathbb{Z}$-submodule of a free $\mathbb{Z}$-module of rank n, so is free as well, of some rank m'. But then $m' \geq m$, as otherwise a basis of $\mathcal{O}_\mathbb{F}$ could not span $\mathbb{F}$, and $m \geq m'$, as otherwise a basis of $\mathcal{O}_\mathbb{F}$ could not be linearly independent, so $m' = m$. $\square$

Remark 5.1.19. Of course, Theorem 5.1.18 gives us the additive structure of $\mathcal{O}_\mathbb{F}$, but says nothing about the multiplicative structure of $\mathcal{O}_\mathbb{F}$, which depends on the particular field $\mathbb{F}$ (not just on the degree of $\mathbb{F}$). $\diamond$

5.2 Dedekind rings

In order to define Dedekind rings we will have to first think more about integrality.

Definition 5.2.1. Let R be a subring of the field $\mathbb{F}$. An element α of $\mathbb{F}$ is *R-integral* if there is a monic polynomial $f(x) \in R[x]$ with $f(\alpha) = 0$. A subring A of $\mathbb{F}$ is *R-integral* if every element α of A is R-integral. $\diamond$

Example 5.2.2. If $R = \mathbb{Z}$, and $\mathbb{F}$ is a finite extension of $\mathbb{Z}$, then α is R-integral if and only if α is an algebraic integer.

Also, the ring of algebraic integers $\mathcal{O}_\mathbb{F}$ is R-integral. $\diamond$

Example 5.2.3. Again let $R = \mathbb{Z}$ and consider $\mathbb{F} = \mathbb{Q}(\sqrt{D})$, with D as in Example 5.1.13. Let n be any integer, $n \neq 0, \pm 1$, let $E = n^2 D$, and consider

$$
A = \begin{cases} \left\{ a + b\sqrt{E} \mid a, b \in \mathbb{Z} \right\} & D \equiv 2, 3 \ (\text{mod } 4) \\ \left\{ \frac{a+b\sqrt{E}}{2} \mid a, b \in \mathbb{Z},\ a \equiv b \ (\text{mod } 2) \right\} & D \equiv 1 \ (\text{mod } 4). \end{cases}
$$

Observe that A is a subring of $\mathbb{F}$, that A is $\mathbb{Z}$-integral, and that the quotient field of A is $\mathbb{F}$.

Also, let $D \equiv 1 \ (\text{mod } 4)$ and consider

$$
A = \{ a + b\sqrt{D} \mid a, b \in \mathbb{Z} \}.
$$

Again observe that A is a subring of $\mathbb{F}$, that A is $\mathbb{Z}$-integral, and that the quotient field of A is $\mathbb{F}$. $\diamond$

The point of this example is that we may obtain the *same* field $\mathbb{F}$ as the quotient field of *different* subrings of $\mathbb{F}$. But here there is a "best" choice of subring, the ring of algebraic integers $\mathcal{O}_{\mathbb{F}}$, and we have a criterion to decide what is "best."

Definition 5.2.4. Let $\mathbb{F}$ be a field and let R be a subring of $\mathbb{F}$. A subring A of $\mathbb{F}$ is *R-integrally closed* if every element α of $\mathbb{F}$ that is integral over R is an element of A. $\Diamond$

Example 5.2.5. The ring of algebraic integers $\mathcal{O}_{\mathbb{F}}$ of $\mathbb{F}$ is $\mathbb{Z}$-integrally closed but the subrings A of $\mathbb{Q}(\sqrt{D})$ in Example 5.2.3 are not. $\Diamond$

However, the situation is more complicated. We would like to look at a subring A and have a criterion that only depends on A and not on some other subring R. Here it is:

Definition 5.2.6. Let A be an integral domain and let $\mathbb{F}$ be its quotient field. Then A is *integrally closed* in $\mathbb{F}$ if every $\alpha \in \mathbb{F}$ that is A-integral is an element of A. $\Diamond$

The distinction between A being A-integrally closed and $\mathbb{Z}$-integrally closed is a subtle one. We will need the notion of A-integrally closed to develop the theory of Dedekind rings, which we will define abstractly. But, as we shall see, if $\mathbb{F}$ is an algebraic number field, these notions coincide: $\mathcal{O}_{\mathbb{F}}$ is not only $\mathbb{Z}$-integrally closed (that was its definition) but also $\mathcal{O}_{\mathbb{F}}$-integrally closed.

Now, we come to our main definition.

Definition 5.2.7. An integral domain A is a *Dedekind ring* (or *Dedekind domain*) if:

(a) A is Noetherian;
(b) A is integrally closed in its quotient field; and
(c) Every nonzero prime ideal of A is maximal. $\Diamond$

In fact, we have already seen many Dedekind rings.

Theorem 5.2.8. *Let A be a PID. Then A is a Dedekind ring.*

Proof.   Let A be a PID. From Theorem 3.6.15, we know that A is Neotherian, and from Theorem 3.12.7, we know that every nonzero prime ideal is maximal. So it remains to show that A is integrally closed in its quotient field. Let α be an element of this quotient field that is integral over A. Then α is a root of a polynomial

$$f(x) = x^n + a_{n-1}x^{n-1} + \cdots + a_0 \in A[x].$$

Write $\alpha = \beta/\gamma$ with β, γ relatively prime elements of A. (We can do so because A is a PID.) Then

$$(\beta/\gamma)^n + a_{n-1}(\beta/\gamma)^{n-1} + \cdots + a_0 = 0$$

so, multiplying by γ^n,

$$\beta^n + a_{n-1}\beta^{n-1}\gamma + \cdots + a_0\gamma^n = 0,$$
$$\beta^n = \gamma(-a_{n-1}\beta^{n-1} - \cdots - a_0\gamma^{n-1})$$

and so we see that β^n is divisible by γ. Now, β and γ are assumed to be relatively prime, so β^n and γ are relatively prime as well. Hence, γ must be a unit and $\alpha = \beta \gamma^{-1} \in A$. □

We record the following observation.

Lemma 5.2.9. *Let A be a Dedekind ring. Then A is a PID if and only if A is a UFD.*

Proof. This is a special case of Theorem 3.12.10. □

Here is our next main goal.

Theorem 5.2.10. *Let $\mathbb{F}$ be an algebraic number field and let $\mathcal{O}_\mathbb{F}$ be the ring of algebraic integers of $\mathbb{F}$. Then $\mathcal{O}_\mathbb{F}$ is a Dedekind ring.*

Proof. Let $A = \mathcal{O}_\mathbb{F}$. We must show that A satisfies the three properties of a Dedekind ring.

(a) *A is Noetherian:* Let $I_1 \subseteq I_2 \subseteq I_3 \subseteq \ldots$ be a sequence of ideals of A. We must show this sequence is eventually constant. To this end, let $I = \cup_{i \geq 1} I_i$. Now each I_i is a $\mathbb{Z}$-module, and so is their union I. Now $I \subseteq \mathcal{O}_\mathbb{F}$, which, by Theorem 5.1.18, is a free $\mathbb{Z}$-module of finite rank. Then I itself is a free $\mathbb{Z}$-module of finite rank, by Corollary 2.6.33. (In fact, if $I \neq 0$, the rank of I is equal to the rank of $\mathcal{O}_\mathbb{F}$, but we do not need this fact.) Let $\{\alpha_1, \ldots, \alpha_n\}$ generate I as a $\mathbb{Z}$-module. Then they certainly generate I as an ideal of A. But for each i, $\alpha_i \in I_{k_i}$ for some k_i. If we let $k = \max(k_1, \ldots, k_n)$, then $\alpha_i \in I_k$ for every $i = 1, \ldots, n$, so $I = I_k$ and hence $I_k = I_{k+1} = I_{k+2} = \ldots$.

(b) A is integrally closed in its quotient field: Let $\alpha \in \mathbb{F}$ and suppose that α is integral over A. We must show that $\alpha \in A$. To this end, let α be a root of a monic polynomial $p(x) \in A[x]$. We will show that α is in fact a root of a monic polynomial $q(x) \in \mathbb{Z}[x]$, in which case $\alpha \in A$.

Let $\mathbb{E}$ be a finite Galois extension of $\mathbb{Q}$ containing $\mathbb{F}$. (We found such an extension in the proof of Theorem 5.1.18.) Let $G = \mathrm{Gal}(\mathbb{E}/\mathbb{Q})$ and let

$$q(x) = \prod_{\sigma \in G} \sigma(p(x)).$$

Then α is a root of $q(x)$. Also, $q(x)$ is a product of monic polynomials, so is monic.

Now $q(x)$ is invariant under the action of G, so $q(x) \in \mathbb{Q}[x]$, i.e., every coefficient of $q(x)$ is in $\mathbb{Q}$.

Furthermore, every coefficient of $q(x)$ is a sum of products of coefficients of each polynomial $\sigma(p(x))$, all of which are algebraic integers. (If $\beta \in \mathbb{E}$ is an algebraic integer, then so is $\sigma(\beta)$ for any $\sigma \in G$, as β and $\sigma(\beta)$ have the same minimal polynomial.) But $\mathcal{O}_{\mathbb{F}}$ is a ring, by Theorem 5.1.12, so every coefficient of $q(x)$ is an element of $\mathcal{O}_{\mathbb{F}}$. Thus every coefficient of $q(x)$ is in $\mathbb{Q} \cap \mathcal{O}_{\mathbb{F}}$. But, by Lemma 5.1.11, $\mathbb{Q} \cap \mathcal{O}_{\mathbb{F}} = \mathcal{O}_{\mathbb{Q}}$, and then, by Corollary 5.1.5, $\mathcal{O}_{\mathbb{Q}} = \mathbb{Z}$.

(c) Every nonzero prime ideal of A is maximal: In proving this, we will be repeatedly applying Theorem 3.12.8: Let R be a commutative ring with 1. Then an ideal I of R is prime (resp. maximal) if and only if the quotient R/I is an integral domain (resp. a field).

Let P be a nonzero prime ideal of A. Let $\alpha \in P$, $\alpha \neq 0$. Consider its minimum polynomial $m_\alpha(x) \in \mathbb{Z}[x]$,

$$m_\alpha(x) = x^n + a_{n-1}x^{n-1} + \cdots + a_0.$$

Note that $a_0 \neq 0$ as $\alpha \neq 0$ and $m_\alpha(x)$ is irreducible. Then

$$0 = m_\alpha(\alpha) = \alpha^n + a_{n-1}\alpha^{n-1} + \cdots + a_0$$

so

$$a_0 = -\alpha^n - \cdots - a_1\alpha = \alpha(-\alpha^{n-1} - \cdots - a_1) \in P.$$

Hence, $Q = P \cap \mathbb{Z} \neq \{0\}$. Let $i \colon \mathbb{Z} \to A$ be the inclusion and $\pi \colon A \to A/P$ be the projection. Then $\mathrm{Ker}(\pi i) = \mathbb{Z} \cap P = Q$, so $\mathrm{Im}(\pi i) \cong \mathbb{Z}/Q \subseteq A/P$. But P is a prime ideal of A so A/P is an integral domain and hence $\mathbb{Z}/Q$ is an integral domain. (If A/P has no zero divisors then certainly any subring of A/P has no zero divisors.) Since $\mathbb{Z}/Q$ is an integral domain, Q is a prime ideal in $\mathbb{Z}$. But in $\mathbb{Z}$, prime ideals are maximal, so Q is a maximal ideal and hence $\mathbb{Z}/Q$ is a field.

Now A is integral over $\mathbb{Z}$ so A/P is integral over $\mathbb{Z}/Q$ (by looking at representatives). Also, A is finitely generated over $\mathbb{Z}$ so A/P is finitely generated over $\mathbb{Z}/Q$ (by the images of generators of A over $\mathbb{Z}$). Thus, A/P is an integral domain that is a finite dimensional vector space over the field $\mathbb{Z}/Q$, so by Lemma 4.3.8 A/P is a field. Hence, P is a maximal ideal of A. $\qquad\square$

Corollary 5.2.11. *Let $\mathbb{F}$ be an algebraic number field and let $\mathcal{O}_{\mathbb{F}}$ be the ring of algebraic integers of $\mathbb{F}$. Then $\mathcal{O}_{\mathbb{F}}$ is a PID if and only if $\mathcal{O}_{\mathbb{F}}$ is a UFD.*

Proof. Immediate from Theorem 5.2.10 and Lemma 5.2.9. $\qquad\square$

5.3 Ideals in Dedekind rings

In this section, we reach our goal of establishing unique factorization of ideals in Dedekind rings.

We first recall that if R is any commutative ring with 1, and I and J are ideals of R, their product IJ is the ideal of R generated by $\{\alpha\beta \mid \alpha \in I, \beta \in J\}$; more concretely, $IJ = \{\text{finite sums } \Sigma \alpha_i \beta_i \mid \alpha_i \in I, \beta_i \in J\}$.

We begin with a more general result.

Lemma 5.3.1. *Let R be a Noetherian integral domain. Then every nonzero ideal I of R contains a product of nonzero prime ideals.*

Proof. Suppose that R contains a nonzero ideal I that does not contain a product of nonzero prime ideals. Set $I_1 = I$. Either I_1 is maximal among ideals with this property or it is not. If it is not, let $I_1 \subset I_2$ where I_2 has this property. Either I_2 is maximal among ideals with this property or it is not. If it is not, let $I_2 \subset I_3$ where I_3 has this property. In this way we obtain a sequence of ideals $I_1 \subset I_2 \subset \ldots$. But R is Noetherian, so this sequence stops at some finite stage I_k. Set $J = I_k$. Thus, J does not contain a product of nonzero prime ideals but any ideal J' with $J \subset J'$ does. Now J is not a prime ideal (as then J would contain the prime ideal J). Thus, there are elements α and β of R with $\alpha \notin J$, $\beta \notin J$, but $\alpha\beta \in J$. Then J_1', the ideal generated by J and α, properly contains J, so must contain a product of prime ideals, and J_2', the ideal generated by J and β, properly contains J, so must contain a product of prime ideals. But then $J_1'J_2'$ contains a product of prime ideals. But $J_1'J_2' = (J+\alpha R)(J+\beta R) = J$ as $\alpha\beta \in J$; contradiction. $\qquad\qquad\qquad\qquad\qquad\qquad\qquad\qquad\square$

Definition 5.3.2. Let R be an integral domain and let $\mathbb{F}$ be its quotient field. A subset I of $\mathbb{F}$ is a *fractional ideal* if

(1) $i + j \in I$ whenever $i \in I$, $j \in I$
(2) $ri \in I$ whenever $r \in R$, $i \in I$
(3) There is some nonzero element d of $\mathbb{F}$ such that $dI \subseteq R$.

An element d as in (3) is called a *denominator* of I. $\qquad\qquad\lozenge$

Remark 5.3.3. We see that fractional ideals are generalizations of ideals, as a fractional ideal of R is an ideal of R if (and only if) we can choose its denominator $d = 1$. $\qquad\qquad\qquad\qquad\qquad\lozenge$

Example 5.3.4. Let R be an integral domain with quotient field $\mathbb{F}$. Let $f \in \mathbb{F}$, $f \neq 0$. Write $f = c/d$ with $c, d \in R$, and let $I = \{rf \mid r \in R\}$. Then I is a (principal) fractional ideal of R with denominator d. $\qquad\qquad\qquad\qquad\qquad\qquad\qquad\qquad\lozenge$

Definition 5.3.5. A fractional ideal I of an integral domain R is *invertible* if there is a fractional ideal J of R with $IJ = R$. In this case we write $J = I^{-1}$ (and $I = J^{-1}$). $\qquad\qquad\qquad\qquad\lozenge$

Lemma 5.3.6. *If I is invertible, then I^{-1} is unique.*

Proof. If J_1 and J_2 are both inverses of I, then

$$J_2 = J_1 R = J_1(IJ_2) = (J_1I)J_2 = RJ_2 = J_2. \qquad\qquad\square$$

Example 5.3.7. Let R be an integral domain with quotient field $\mathbb{F}$. Let I be a nonzero principal fractional ideal generated by $f \in \mathbb{F}$. Then I^{-1} is the nonzero principal fractional ideal generated by f^{-1}. $\quad\Diamond$

Lemma 5.3.8. *Let A be a Dedekind ring with quotient field $\mathbb{F}$. Let P be a nonzero prime ideal of A. Set*

$$N = \{\beta \in \mathbb{F} \mid \beta P \subseteq A\}.$$

Then: (a) *N is a fractional ideal of A.*
(b) *$A \subset N$.*
(c) *$PN = A$, and hence $N = P^{-1}$.*
In particular, every nonzero prime ideal of A is invertible.

Proof. (a) It is easy to check that properties (1) and (2) of a fractional ideal hold for N. Let α be any nonzero element of P. By the definition of N, for every $\beta \in N$, $\beta\alpha \in P \subseteq A$. Thus, α is a denominator of N.

(b) Clearly $A \subseteq N$ so we must show $A \neq N$. Choose $\alpha \in P$, $\alpha \neq 0$, and let I be the principal ideal generated by α. By Lemma 5.3.1, I contains a product $P_1 \ldots P_n$ of prime ideals of A. We may suppose that I does not contain a product of fewer than n of these factors. Since $I \subseteq P$, P contains the product $P_1 \ldots P_n$ and hence P contains P_i for some i. (Otherwise, choose $\gamma_i \in P_i$, $\gamma_i \notin P$ for each i. Then $\gamma_1, \ldots, \gamma_n \in P$, which is impossible, as P is a prime ideal.) We may suppose that P contains P_1. But then $P = P_1$ as prime ideals of A are maximal.

If $n = 1$, set $J = A$. If $n > 1$, set $J = P_2 \ldots P_n$. Then I does not contain J. Let γ be any element of J with γ not an element of I. Then, recalling that $P = P_1$,

$$P\gamma \subseteq PJ = P_1 P_2 \ldots P_n \subseteq I = A\alpha,$$

so $P(\gamma\alpha^{-1}) \subseteq A$. Thus if we set $\beta = \gamma\alpha^{-1}$, then from the definition of N we see that $\beta \in N$. (Here we take $\alpha^{-1} \in \mathbb{F}$.) But $\beta \notin A$, as if $\beta \in A$ then $\gamma = \beta\alpha \in A\alpha = I$; contradiction.

(c) By the definition of N we have that $PN \subseteq A$. Since $P \subset A$, we have that $1 \in N$ so $P \subseteq PN$. Now P is a prime ideal of A, hence a maximal ideal of A, so we must have $PN = P$ or $PN = A$. We will show that $PN = P$ is impossible. Then $PN = A$ as claimed.

Suppose that $PN = P$. Let d be a denominator of the fractional ideal N. For any β in N, we have $P\beta \subseteq PN = P$. But then

$$P\beta^2 = (P\beta)\beta \subseteq (PN)\beta = P\beta \subseteq PN = P$$

and by induction we have that $P\beta^k \subseteq P$ for every positive integer k, and hence $\beta^k \in N$ for every positive integer k, so $d\beta^k \in A$ for every positive integer k. Thus if we let M be the subset of $\mathbb{F}$ generated by A and $\{\beta^k \mid k = 1, 2, \ldots\}$, then M has denominator d and so is a fractional ideal of A.

Now dM is an ideal of A, and A is Noetherian, so dM is finitely generated as an ideal of A, and hence M is finitely generated as a fractional ideal of A. (If $\{\gamma_1, \ldots, \gamma_k\}$ generates dM, then $\{d^{-1}\gamma_1, \ldots, d^{-1}\gamma_k\}$ generates M.)

In the special case that $A = \mathcal{O}_{\mathbb{F}}$, we have seen that A is finitely generated as a $\mathbb{Z}$-module, and hence M is finitely generated as a $\mathbb{Z}$-module, and so β is an element of A.

For A a general Dedekind ring, let $\{\delta_1, \ldots, \delta_k\}$ generate M as a fractional ideal of A. Then, by the definition of M, $\delta_i = p_i(\beta)$ for some polynomial $p_i(x) \in A[x]$, for each $i = 1, \ldots, k$. Choose $n > \max(\text{degree}(p_1(x)), \ldots, \text{degree}(p_k(\gamma)))$. Then $\beta^n \in M$, so $\beta^n = \Sigma_{i=1}^{k} \alpha_i p_i(\beta)$ for some elements $\alpha_1, \ldots, \alpha_k$ of A. But then, if

$$f(x) = x^n - \sum_{i=1}^{k} \alpha_i p_i(x),$$

$f(x)$ is a monic polynomial in $A[x]$ with $f(\beta) = 0$. Hence, β is integral over A. But A is integrally closed in $\mathbb{F}$, so β is an element of A.

Now β was an arbitrary element of N, so we have $N = A$. But this contradicts (b). $\qquad \Box$

Here is our main result.

Theorem 5.3.9. *Let A be a Dedekind ring with quotient field $\mathbb{F}$. Then any nonzero fractional ideal I of A can be expressed as a product*

$$I = P_1^{e_1} \cdots P_k^{e_k}$$

for some mutually distinct prime ideals $P_1, \ldots, P_k$ of A and nonzero integers $e_1, \ldots, e_k$, and this expression is unique up to the order of the factors. (Here we regard A as the empty product.)

Proof. First we prove the existence of the factorization of I as a product of prime ideals and then we prove uniqueness.

We begin with the case that I is an ideal of A, and we will show in this case that each of the exponents is positive.

Suppose there is some ideal I_1 that cannot be expressed in this way. If I_1 is maximal among ideals of A with this property, stop. Otherwise $I_1 \subset I_2$ where I_2 has this property. If I_2 is maximal among ideals of A with this property, stop. Otherwise $I_2 \subset I_3$ where I_3 has this property. Keep going.

In this way we obtain a sequence of ideals $I_1 \subset I_2 \subset I_3 \subset \dots$. But A is a Noetherian ring, so this sequence must stop at some I_k. Set $I = I_k$. Then I cannot be expressed as a product of prime ideals, but every ideal J properly containing I can be.

Now $I \neq A$, as A is the empty product. Since A is Noetherian, by a similar sort of argument I is contained in a maximal ideal P_1. Since A is a Dedekind ring, P_1 is a prime ideal, and then by Lemma 5.3.8 P_1 is invertible.

Since $I \subseteq P_1$, $IP_1^{-1} \subseteq P_1 P_1^{-1} = A$. Also, $A \subseteq P_1^{-1}$ so $I = IA \subseteq IP_1^{-1}$. We claim that in fact $I \subset IP_1^{-1}$. This follows by the same argument as in the proof of part (c) of Lemma 5.3.8: If $I = IP_1^{-1}$ then for every $\beta \in P_1^{-1}$, $\beta^k I \subseteq I$ for every positive integer k, and looking at the fractional ideal generated by I and $\{\beta^k \mid k = 1, 2, \dots\}$ we see that β is integral over A so $\beta \in A$; contradiction.

Thus, $J = IP_1^{-1}$ is an ideal of A properly containing I, so J can be expressed as as product of prime ideals $J = P_2 \dots P_n$. Then $I = P_1 P_2 \dots P_n$. This product may have some repeated factors; if so we can group them together and renumber to obtain an expression for I as in the statement of the theorem. Note that in this case each of the exponents is positive.

Now let I be a fractional ideal of A. Let d be a denominator for I. Then $I' = I(Ad)$ is an ideal of A, so $I' = P_1^{e_1} \dots P_j^{e_j}$. Also, Ad is an ideal of A, so $Ad = P_{j+1}^{e_{j+1}} \dots P_k^{e_k}$. But then

$$I = I'(Ad)^{-1} = P_1^{e_1} \dots P_j^{e_j} P_{j+1}^{-e_{j+1}} \dots P_k^{-e_k}$$

(where if there is any duplication in the prime ideal factors of I' and Ad, we combine terms.)

Now for uniqueness. Suppose that we have a fractional ideal I of A with I expressed as a product of prime ideals in two ways:

$$I = P_1^{e_1} \cdots P_k^{e_k} = Q_1^{f_1} \cdots Q_\ell^{f_\ell}$$

Multiplying both sides by suitable powers of these prime ideals and renumbering as necessary, we may assume that we have an ideal I' of A with $e_i > 0$ for each i and $f_j > 0$ for each j, and that the sets of prime ideals $\{P_1, \ldots, P_k\}$ and $\{Q_1, \ldots, Q_\ell\}$ are disjoint. Then

$$P_1 \supseteq P_1^{e_1} \cdots P_k^{e_k} = Q_1^{f_1} \cdots Q_\ell^{f_\ell}$$

and each of these ideals is prime, so in particular $P_1 \supseteq Q_j$ for some j. (Otherwise, let $q_j \in Q_j$, $q_j \notin P_1$ for each j. Then $q_1^{f_1} \cdots q_\ell^{f_\ell} \in P_1$, which is impossible as P_1 is a prime ideal.) But Q_j, being prime, is also maximal, and so $P_1 = Q_j$; contradiction.   □

Corollary 5.3.10. *In the situation of Theorem 5.3.9, then I is an ideal of a Dedekind ring A if and only if $e_i > 0$ for each i.*

Proof.   Certainly if each $e_i > 0$ then I is an ideal of A, and we observed the converse in the course of proving Theorem 5.3.9.   □

Corollary 5.3.11. *The set of nonzero fractional ideals of a Dedekind ring A forms an abelian group under multiplication of fractional ideals.*

Proof.   Multiplication of fractional ideals is commutative and associative, and A is the identity. Also, every fractional ideal I has an inverse: If $I = P_1^{e_1}, \ldots, P_k^{e_k}$ as in Theorem 5.3.9, then $I^{-1} = P_1^{-e_1}, \ldots, P_k^{-e_k}$.   □

Theorem 5.3.12. *The group of nonzero fractional ideals of a Dedekind ring A is a free abelian group with basis $\{$nonzero prime ideals of $A\}$.*

Proof.   Referring back to Definition 2.6.22 and Definition 2.6.24 of what it means for an abelian group G to be free with a basis B, we see that this immediately follows from Theorem 5.3.9. (Note here that we are writing this group multiplicatively rather than additively.)   □

Theorem 5.3.13. *Let $\mathbb{F}$ be an algebraic number field and let $A = \mathcal{O}_{\mathbb{F}}$ be the ring of algebraic integers of $\mathbb{F}$. Then Theorem 5.3.9, Corollary 5.3.10, Corollary 5.3.11, and Theorem 5.3.12 hold for A.*

Proof. By Theorem 5.2.10, $A = \mathcal{O}_{\mathbb{F}}$ is a Dedekind ring. $\qquad\square$

5.4 Examples

In this section, we do a bunch of examples. We do both families of examples and individual examples, and at the end we do a "nonexample".

Here will be our general set-up throughout. We will let $\mathbb{F} = \mathbb{Q}(\sqrt{D})$ be a quadratic extension of $\mathbb{Q}$, and let $A = \mathcal{O}(\sqrt{D})$, the ring of algebraic integers of $\mathbb{F}$. Then $G = \mathrm{Gal}(\mathbb{F}/\mathbb{Q})$ is a group of order 2, with the nontrivial element of G being $\sigma\colon \mathbb{F} \to \mathbb{F}$ defined by $\sigma(\sqrt{D}) = -\sqrt{D}$. If I is an ideal of A, we let $\bar{I} = \sigma(I)$.

We will denote the ideal I generated by elements $\alpha_1, \ldots, \alpha_k$ of A by $< \alpha_1, \ldots, \alpha_k >$. Often parentheses are used for this, but we will be using parentheses to group elements for multiplication, as usual, so we choose this notation to avoid ambiguity.

We will begin by considering the case $D < 0$, and afterwards consider the case $D > 0$.

Lemma 5.4.1. *Let $D < 0$ and suppose $|D|$ is a squarefree composi-tive positive integer $|D| = p_1 \ldots p_k$ with $k \geq 2$. Then*

$$|D| = (p_1)(p_2) \ldots (p_k) = -(\sqrt{D})^2$$

are two factorizations of $|D|$ into irreducibles. None of these irreducibles are prime.

If $I_i = < p_i, \sqrt{D} >$, $i = 1, \ldots, k$, then each I_i is a prime ideal, with $I_i^2 = < p_i >$, a principal ideal. Furthermore $\bar{I}_i = I_i$.

Then, we have factorizations

$$< D >= I_1^2 \ldots I_k^2 \quad and \quad < \sqrt{D} >= I_1 \ldots I_k.$$

Proof. Recall that we have a multiplicative norm on A given by $N(a + b\sqrt{D}) = |a^2 - b^2 D|$ and it is easy to check from Lemma 3.9.1 that each of $p_1, \ldots, p_k$ and $\sqrt{D}$ is irreducible. The factorization of $|D|$ given in the statement of the lemma shows that none of these are prime.

Let $I = <p, \sqrt{D}>$ for any p dividing D. Clearly $\bar{I} = I$. We observe that I is a proper ideal of A, as $N(\alpha)$ is divisible by p for every $\alpha \in I$. We now show that I is a maximal ideal of A, and hence prime.

To this end, let $I' \supset I$ and let $\alpha \in I'$, $\alpha \notin I$. Write $\alpha = a + b\sqrt{D}$. If $D \equiv 1 \pmod 4$ replace α by 2α if necessary, so that $a, b \in \mathbb{Z}$. Since I' is an ideal, $\beta = \alpha + (-b\sqrt{D}) = a \in I'$, and since I is an ideal, $\beta \notin I$. Thus $a \in \mathbb{Z}$ is not divisible by p, and so, since p is a prime in $\mathbb{Z}$, there are integers x and y with $ax + py = 1$. Thus $1 \in I'$ and so $I' = A$.

Now

$$I^2 = <p, \sqrt{D}> <p, \sqrt{D}> = <p^2, p\sqrt{D}, p\sqrt{D}, D>$$
$$= <p^2, p\sqrt{D}, p(D/p)> \subseteq <p>.$$

But p and D/p are relatively prime so there are integers x and y with $px + (D/p)y = 1$ and so $p \in I^2$. Thus, $I^2 = <p>$. Then, we immediately see that $<D> = <p_1> \ldots <p_k> = I_1^2 \ldots I_k^2$.

More subtly, let us consider

$$J = I_1 \ldots I_k = <p_1, \sqrt{D}> <p_2, \sqrt{D}> \ldots <p_k, \sqrt{D}>.$$

Set $q_i = D/p_i$, for convenience.

Generators of the product ideal J are given by multiplying generators of each of the factors I_i. If we choose any generator to be $\sqrt{D}$, then this product is divisible by $\sqrt{D}$, so is an element of $<\sqrt{D}>$. The only other choice is $p_1 \ldots p_k = -D$, which is also divisible by $\sqrt{D}$, so is also an element of $<\sqrt{D}>$. Thus $J \subseteq <\sqrt{D}>$.

On the other hand, we can choose $\sqrt{D}$ from the first factor and $p_2, \ldots, p_k$ from the remaining factors, so $q_1\sqrt{D} \in J$. Similarly, $q_i\sqrt{D} \in J$ for each i. But $\{q_1, \ldots, q_k\}$ is relatively prime, so, as above, $\sqrt{D} \in J$ and so $<\sqrt{D}> \subseteq J$. Thus, $J = <\sqrt{D}>$. $\square$

Lemma 5.4.2. *Let $D < 0$ and suppose $|D| = p$ a prime with $p \equiv 1$ (mod 4). Then*

$$p + 1 = 2((p+1)/2) = (1 + \sqrt{D})(1 - \sqrt{D})$$

are two factorizations of $p + 1$ into irreducibles. None of these irreducibles are prime.

If $I_1 = <2, 1 + \sqrt{D}>$ and $I_2 = <(p+1)/2, 1 + \sqrt{D}>$, then I_1 and I_2 are nonprincipal prime ideals, with $I_1^2 = <2>$ and $I_2^2 = <(p+1)/2>$, both principal ideals. Furthermore, $\bar{I}_1 = I_1$ but $\bar{I}_2 \neq I_2$. Then we have factorizations

$$<p+1> = I_1^2 I_2 \bar{I}_2 \quad \text{and} \quad <1 + \sqrt{D}> = I_1 I_2,$$
$$<1 - \sqrt{D}> = I_1 \bar{I}_2.$$

Proof. For simplicity, set $r = (p+1)/2$ and note that r is an odd integer with $r > 1$.

Again each of 2, r, and $1 \pm \sqrt{D}$ is irreducible by Lemma 3.9.1, but not prime.

Note that $D = -p \equiv 3 \pmod 4$. Thus $A = \{a + b\sqrt{D} \mid a, b \in \mathbb{Z}\}$. Clearly $\bar{I}_1 = I_1$, but $\bar{I}_2 \neq I_2$, as if $\bar{I}_2 = I_2$, then $r \in I_2$, $2 = (1 + \sqrt{D}) + (1 - \sqrt{D}) \in I_2$, and then $I_2 = A$. But I_1 (resp I_2) is a proper ideal of A, as $N(\alpha)$ is divisible by 2 (resp. r) for every $\alpha \in I_1$ (resp. $\alpha \in I_2$).

I_1 and I_2 are both maximal, and hence prime, ideals by an argument similar to the proof of Lemma 5.4.1. It is easy to check that they are not principal.

Now

$$I_1^2 = I_1 \bar{I}_1 = <2, 1 + \sqrt{D}> <2, 1 - \sqrt{D}>$$
$$= <4, 2(1 - \sqrt{D}), 2(1 + \sqrt{D}), (1 + \sqrt{D})(1 - \sqrt{D})>$$
$$= <4, 2(1 + \sqrt{D}), 1 - D> = <4, 2(1 + \sqrt{D}), 2r>$$
$$= <2>$$

as r is odd.

Also,

$$I_2 \bar{I}_2 = <r, 1 + \sqrt{D}> <r, 1 - \sqrt{D}>$$
$$= <r^2, r(1 - \sqrt{D}), r(1 + \sqrt{D}), 2r> = <r>$$

again as r is odd.

Thus,

$$<p+1> = <2> <r> = I_1^2 I_2 \bar{I}_2.$$

More subtly, let us consider

$$I_1 I_2 = \; <2, 1+\sqrt{D}> \, <r, 1+\sqrt{D}>$$
$$= \; <2r, 2(1+\sqrt{D}), r(1+\sqrt{D}), (1+\sqrt{D})(1+\sqrt{D})>$$
$$= \; <(1+\sqrt{D})(1-\sqrt{D}), 2(1+\sqrt{D}),$$
$$r(1+\sqrt{D}), (1+\sqrt{D})(1+\sqrt{D})>$$
$$= \; <1+\sqrt{D}>$$

once again as r is odd, and similarly $I_1 \bar{I}_2 = \; <1-\sqrt{D}>$.   □

Remark 5.4.3. We saw in Chapter 3 that for $D = -1, -2, -3$, $A = \mathcal{O}(\sqrt{D})$ is a Euclidean domain and hence a PID. So this leaves the cases $D = -p$ for $p \equiv 3 \pmod 4$, $p > 3$, open. If $p \equiv 3 \pmod 4$ then $p \equiv 3$ or $7 \pmod 8$. It can be shown by arguments similar to that in Chapter 3 that for $D = -7$, $A = \mathcal{O}(\sqrt{D})$ is a Euclidean domain, and hence a PID. It can also be shown by arguments similar to that of Lemma 5.4.2 that if $D = -p$, for $p \equiv 7 \pmod 8$, $p > 7$, then $A = \mathcal{O}(\sqrt{D})$ is not a PID.

This takes care of all negative values of D except $D = -p$ for p a prime, $p \equiv 3 \pmod 8$. It can be shown that for $D = -11, -19, -43, -67, -163$, $A = \mathcal{O}(\sqrt{D})$ is a PID.

Assembling these results, we see that $A = \mathcal{O}(\sqrt{D})$ is a PID for the following nine values of D: $D = -1, -2, -3, -7, -11, -19, -43, -67, -163$. This was known to Gauss. Gauss also conjectured that these are the *only* negative values of D for which $A = \mathcal{O}(\sqrt{D})$ is a PID. This turns out to be true. That is a very deep fact, and one of the great theorems of 20th century mathematics.   ◊

Example 5.4.4. (a) Let $D = -5$ and let us reconsider the factorizations in Example 3.9.5(a). We had

$$6 = 2 \cdot 3 = (1+\sqrt{-5})(1-\sqrt{-5}),$$
$$9 = 3 \cdot 3 = (2+\sqrt{-5})(2-\sqrt{-5}).$$

The first of these is a special case of Lemma 5.4.2, so we concentrate on the second.

Let $I = \; <3, 1+\sqrt{-5}>$ so that $\bar{I} = \; <3, 1-\sqrt{-5}>$.

You may check that these are both maximal, and hence prime, ideals. Then

$$I\bar{I} = <3, 1+\sqrt{-5}><3, 1-\sqrt{-5}>$$
$$= <9, 3(1+\sqrt{-5}), 3(1-\sqrt{-5}), 6> = <3>$$

so we see

$$<9> = <3><3> = (I\bar{I})(I\bar{I}).$$

We also compute

$$I^2 = <3, 1+\sqrt{-5}><3, 1+\sqrt{-5}>$$
$$= <9, 3(1+\sqrt{-5}), 3(1+\sqrt{-5}), -4+2\sqrt{-5}>$$
$$= <9, 3(1+\sqrt{-5}), -4+2\sqrt{-5}>$$
$$= <(2-\sqrt{-5})(2+\sqrt{-5}), (2-\sqrt{-5})$$
$$(-1+\sqrt{-5}), (2-\sqrt{-5})(-2)>$$
$$= <2-\sqrt{-5}>$$

and similarly

$$\bar{I}^2 = <2+\sqrt{-5}>$$

so we see

$$<9> = <2+\sqrt{-5}><2-\sqrt{-5}> = (\bar{I})^2(I)^2$$

and so both expressions give the same factorization of the ideal $<9>$ into a product of prime ideals.

Note that I and $\bar{I}$ are not principal ideals, but $I\bar{I}$, I^2, and $\bar{I}^2$ are principal ideals.

(b) Let $D = -6$ and let us reconsider the factorizations in Example 3.9.5 (b). We had

$$6 = 2 \cdot 3 = -1(\sqrt{-6})^2,$$
$$10 = 2 \cdot 5 = (2+\sqrt{-6})(2-\sqrt{-6}).$$

The first of these is a special case of Lemma 5.4.1, so we concentrate on the second.

Let $I_1 = <2, \sqrt{-6}>$ so that $\bar{I}_1 = I_1$, and let $I_2 = <5, 2+\sqrt{-6}>$ so that $\bar{I}_2 = <5, 2 - \sqrt{-6}>$.

You may check that these are both maximal, and hence prime, ideals. Then

$$I_1^2 = <2, \sqrt{-6}><2, \sqrt{-6}> = <4, 2\sqrt{-6}, 2\sqrt{-6}, -6> = <2>$$

and

$$I_2\bar{I}_2 = <5, 2+\sqrt{-6}><5, 2-\sqrt{-6}>$$
$$= <25, 5(2+\sqrt{-6}), 5(2-\sqrt{-6}), 10> = <5>$$

so we see

$$<10> = <2><5> = I_1^2 I_2\bar{I}_2.$$

We also compute

$$I_1 I_2 = <2, \sqrt{-6}><5, 2+\sqrt{-6}>$$
$$= <10, 4+2\sqrt{-6}, 5\sqrt{-6}, -6+2\sqrt{-6}>$$
$$= <10, 4+2\sqrt{-6}, 5\sqrt{-6}, -10>$$
$$= <10, 4+2\sqrt{-6}, 5\sqrt{-6}>$$
$$= <10, 4+2\sqrt{-6}, 5\sqrt{-6}, 10+5\sqrt{-6}>$$
$$= <(2+\sqrt{-6})(2-\sqrt{-6}), (2+\sqrt{-6})(2),$$
$$(2+\sqrt{-6})(5)) >$$
$$= <2+\sqrt{-6}>$$

and similarly

$$I_1\bar{I}_2 = <2-\sqrt{-6}>$$

so we see

$$<10> = <2+\sqrt{-6}><2-\sqrt{-6}> = (I_1 I_2)(I_1\bar{I}_2)$$

and so both expressions give the same factorization of the ideal $<10>$ into a product of prime ideals.

We further compute

$$I_2^2 = \ < 5, 2 + \sqrt{-6} > < 5, 2 + \sqrt{-6} >$$
$$= \ < 25, 5(2 + \sqrt{-6}), 5(2 + \sqrt{-6}), -2 + 4\sqrt{-6} >$$
$$= \ < 25, 5(2 + \sqrt{-6}), -2 + 4\sqrt{-6} >$$
$$= \ < (-1 + 2\sqrt{-6})(-1 - 2\sqrt{-6}), (-1 + 2\sqrt{-6})$$
$$(2 - \sqrt{-6}), (-1 + 2\sqrt{-6})(2) >$$
$$= \ < -1 + 2\sqrt{-6} >$$

and similarly

$$\bar{I}_2^2 = \ < -1 - 2\sqrt{-6} >.$$

We note that I_1, I_2, and $\bar{I}_2$ are not principal ideals but $I_1^2, I_1 I_2, I_1 \bar{I}_2, I_2^2$, and $\bar{I}_2^2$ are principal ideals.

(c) Let $D = -26$ and observe that

$$27 = (3)^3 = (1 + \sqrt{-26})(1 - \sqrt{-26})$$

You may check that these are two distinct factorizations of 27 into a product of irreducibles.

Let $I = \ < 3, 1 + \sqrt{-26} >$ so that $\bar{I} = \ < 3, 1 - \sqrt{-26} >$. You may check that these are both prime ideals. Then

$$I\bar{I} = \ < 3, 1 + \sqrt{-26} > < 3, 1 - \sqrt{-26} >$$
$$= \ < 9, 3(1 - \sqrt{-26}), 3(1 + \sqrt{-26}), 27 >$$
$$= \ < 9, 6, 3(1 + \sqrt{-26}), 27 > = \ < 3 >$$

so we see

$$< 27 > = (I\bar{I})^3.$$

We also compute

$$I^2 = \ < 3, 1 + \sqrt{-26} > < 3, 1 + \sqrt{-26} >$$
$$= \ < 9, 3(1 + \sqrt{-26}), 3(1 + \sqrt{-26}), -25 + 2\sqrt{-26} >$$
$$= \ < 9, 3(1 + \sqrt{-26}), -25 + 2\sqrt{-26} >$$
$$= \ < 9, 3(1 + \sqrt{-26}), 2(1 + \sqrt{-26}) >$$
$$= \ < 9, 1 + \sqrt{-26} >$$

and further that

$$I^3 = I^2 I = \; < 9, 1 + \sqrt{-26} > < 3, 1 + \sqrt{-26} >$$
$$= \; < 27, 9(1 + \sqrt{-26}), 3(1 + \sqrt{-26}), -25 + 2\sqrt{-26} >$$
$$= \; < 27, 3(1 + \sqrt{-26}), -25 + 2\sqrt{-26} >$$
$$= \; < 27, 3(1 + \sqrt{-26}), 2(1 + \sqrt{-26}) >$$
$$= \; < 27, 1 + \sqrt{-26} >$$
$$= \; < (1 + \sqrt{-26})(1 - \sqrt{-26}), 1 + \sqrt{-26} > = \; < 1 + \sqrt{-26} >$$

and similarly $\bar{I}^2 = \; < 9, 1 - \sqrt{-26} >$ and $\bar{I}^3 = \; < 1 - \sqrt{-26} >$ so we see

$$< 27 > = (I^3)(\bar{I}^3)$$

and so both expressions give the same factorization of the ideal $< 27 >$ into a product of prime ideals.

Note that I, I^2, $\bar{I}$, $\bar{I}^2$ are not principal ideals but $I\bar{I}, I^3$, and $\bar{I}^3$ are principal ideals.   ◇

Example 5.4.5. Let $D = 6$ and consider the factorizations

$$6 = (2)(3) = (\sqrt{6})^2.$$

At first glance this appears to be an example of two distinct factorizations of 6 into a product of irreducibles, just as in Example 5.4.4 (b), but in this case appearances are deceiving. It is *not*.

We have that

$$2 = -(2 + \sqrt{6})(2 - \sqrt{6}),$$
$$3 = (3 + \sqrt{6})(3 - \sqrt{6})$$

so that

$$6 = (2)(3)$$
$$= -(2 + \sqrt{6})(2 - \sqrt{6})(3 + \sqrt{6})(3 - \sqrt{6})$$
$$= -(2 - \sqrt{6})(3 + \sqrt{6})(2 + \sqrt{6})(3 - \sqrt{6})$$
$$= -(\sqrt{6})(\sqrt{6})$$

so these are the same factorization.

In fact, $\mathcal{O}(\sqrt{6})$ is known to be a PID.   ◇

On the other hand, we have the following class of examples, which generalizes Example 3.9.5(c).

Lemma 5.4.6. *Let $D = 2p$ where p is a prime with $p \equiv 5$ (mod 8). Then*

$$D = (2)(p) = (\sqrt{D})^2$$

are two factorizations of D into irreducibles. None of these irreducibles are prime.

If $I_1 = <2, \sqrt{D}>$ and $I_2 = <p, \sqrt{D}>$ then I_1 and I_2 are both prime ideals, with $I_1^2 = <2>$ and $I_2^2 = <p>$, both principal ideals. Furthermore, $\bar{I}_1 = I_1$, $\bar{I}_2 = I_2$. Then we have factorizations:

$$<D> = I_1^2 I_2^2 \quad and \quad <\sqrt{D}> = I_1 I_2.$$

Proof. We have a multiplicative norm on A given by $N(a+b\sqrt{D}) = |a^2 - b^2 D|$. Then $N(2) = 4$, $N(p) = p^2$, and $N(D) = 2p$. We will show that A does not have an element of norm 2, and also does not have an element of norm p, and then it follows from Lemma 3.9.1 that 2, p, and $\sqrt{D}$ are all irreducible.

Suppose $\alpha \in A$ with $N(\alpha) = 2$. Writing $\alpha = a + b\sqrt{D}$ with $a, b \in \mathbb{Z}$,

$$|a^2 - b^2(2p)| = 2$$
$$a^2 - 2pb^2 = \pm 2$$

and reducing mod p,

$$a^2 \equiv \pm 2 \ (\text{mod } p)$$

which has no solution by Corollary 2.7.15 and Corollary 2.7.16.

Suppose $\alpha \in A$ with $N(\alpha) = p$. Writing $\alpha = a + b\sqrt{D}$ with $a, b \in \mathbb{Z}$,

$$|a^2 - b^2(2p)| = p$$
$$a^2 = \pm p + b^2(2p) = p(2b^2 \pm 1)$$

Hence, a must be divisible by p, in which case $2b^2 \pm 1$ is divisible by p, i.e.,

$$2b^2 \pm 1 \equiv 0 \ (\text{mod } p)$$

which again has no solution by Corollary 2.7.15 and Corollary 2.7.16.

The rest of the proof of this lemma is identical to the proof of Lemma 5.4.1. □

Remark 5.4.7. If $A = \mathcal{O}(\sqrt{D})$, then A has norm $N(a + b\sqrt{D}) = |a^2 - b^2 D|$. Note that if $D < 0$, $-D > 0$, so it is easy to determine whether A has an element of any given norm n: There are only finitely many possibilities for a and b, as we must have $|a| \leq \sqrt{n}$ and $|b| \leq \sqrt{n/D}$.

On the other hand, if $D > 0$, $-D < 0$ and it is a much more subtle question to determine whether A has an element of norm n: There are infinitely many possibilities for a and b. (Compare the case $D = 6$, where in Example 5.4.6 we saw that A has an element of norm 2, with the case $D = 10$, where as a special case of Lemma 5.4.6 A does not have an element of norm 2.) Compare also the generality of Lemma 5.4.1 and Lemma 5.4.2, both for $D < 0$, with the much more specific case of Lemma 5.4.6, for $D > 0$, and note we had to use some specific number-theoretical information to prove Lemma 5.4.6.

In fact, much more is known about $\mathcal{O}(\sqrt{D})$ for $D < 0$ than is known for $D > 0$. Gauss conjectured that there are infinitely many values of $D > 0$ for which $\mathcal{O}(\sqrt{D})$ is a PID (though he did not state the conjecture in this language) and this conjecture is still completely open. (Contrast this with Remark 5.4.3.) ◊

We conclude with an example that illustrates what can go wrong when we are dealing with a ring that is not a Dedekind ring.

Example 5.4.8. Let $R = \mathbb{Z}[\sqrt{-3}] = \{a + b\sqrt{-3} \mid a, b \in \mathbb{Z}\}$. Then R has quotient field $\mathbb{F} = \mathbb{Q}(\sqrt{-3})$, but $R \neq \mathcal{O}_\mathbb{F}$ and R is not a Dedekind ring. We have the factorizations

$$4 = (2)(2) = (1 + \sqrt{-3})(1 - \sqrt{-3})$$

and you may check that these are two distinct factorizations of 4 into a product of irreducibles in R.

You may also check that $I = <2, 1 + \sqrt{-3}> = <2, 1 - \sqrt{-3}>$ is a maximal, and hence prime, ideal, and furthermore that I is the unique prime ideal with $I \supset <2>$, I is the unique prime ideal with $I \supset <1 + \sqrt{-3}>$, and I is the unique prime ideal with $I \supset <1 - \sqrt{-3}>$.

You may further compute that $I^2 = <4, 2(1 + \sqrt{-3})> = <4, 2(1 - \sqrt{-3})>$ and so

$$I \supset <2> \supset I^2, \quad I \supset <1 + \sqrt{-3}> \supset I^2, \quad I \supset <1 - \sqrt{-3}> \supset I^2.$$

Thus, the ideals $< 2 >$, $< 1 + \sqrt{-3} >$ and $< 1 - \sqrt{-3} >$ do not have factorizations into a product of prime ideals.

Also, note that $I^2 = < 2 > I$ so we do not have "cancellation" in multiplying products of nonzero prime ideals, as we do in Dedekind rings.

Finally, observe that for any positive integer k, $I^k = < 2^{k-1} > I$, so that no power of I is a principal ideal. $\Diamond$

5.5 Further developments

In this section, we want to mention some further results in algebraic number theory that would take us beyond the bounds of this book to prove, but which we feel the reader should be aware of.

Definition 5.5.1. Let A be a Dedekind ring and let G be the group of nonzero fractional ideals of A under multiplication. Let H be the subgroup of G that is the group of nonzero principal fractional ideals of A under multiplication. Then the *ideal class group* $Cl(A)$ is the quotient $Cl(A) = G/H$. $\Diamond$

Theorem 5.5.2. *Let $\mathbb{F}$ be an algebraic number field, and let $A = \mathcal{O}_{\mathbb{F}}$. Then $Cl(A)$ is a finite group.*

Definition 5.5.3. Let $\mathbb{F}$ be an algebraic number field. Then the *class number* $h(\mathbb{F})$ of $\mathbb{F}$ is $h(\mathbb{F}) = |Cl(\mathcal{O}_{\mathbb{F}})|$, the order of the ideal class group of $\mathcal{O}_{\mathbb{F}}$. $\Diamond$

Remark 5.5.4. We see that $h(\mathbb{F}) = 1$ if and only if $\mathcal{O}_{\mathbb{F}}$ is a PID. $\Diamond$

Corollary 5.5.5. *Let I be any ideal in $\mathcal{O}_{\mathbb{F}}$. Then there is a positive integer k for which I^k is a principal ideal.*

Proof. Any element of a finite group has finite order. $\square$

Theorem 5.5.6. *For any algebraic number field $\mathbb{F}$, there is an effective procedure for finding the group $Cl(\mathcal{O}_{\mathbb{F}})$ (and hence for finding $h(\mathbb{F})$).*

Remark 5.5.7. In the special case when $\mathbb{F}$ is a quadratic extension of $\mathbb{Q}$, i.e., $\mathbb{F} = \mathbb{Q}(\sqrt{D})$, there is a formula, due to Dirichlet, for $h(\mathbb{F})$. $\Diamond$

5.6 Exercises

1. Let A be a Dedekind ring and let I and J be nonzero fractional ideals of A. Let I and J have prime factorizations $I = P_1^{e_1}, \ldots, P_k^{e_k}$ and $J = P_1^{f_1}, \ldots, P_k^{f_k}$. (Here we allow some e_i or f_i to be 0 so that the same prime ideals appear in both factorizations.) Show that $I \subseteq J$ if and only if $f_i \geq e_i$ for each $i = 1, \ldots, k$.

2. Let A be a Dedekind ring and let I and J be nonzero fractional ideals of A with $I^n \subseteq J^n$ for some positive integer n (resp. $I^n = J^n$ for some nonzero integer n). Show that $I \subseteq J$ (resp. $I = J$).

3. Let I and J be nonzero ideals of a Dedekind ring A. We say that I *divides* J if there is an ideal K of A with $J = IK$. Show that I divides J if and only if $J \subseteq I$.

4. We say that two nonzero ideals I and J of a Dedekind ring A are *relatively prime* if there is no prime ideal P that appears in both of the factorizations of I and J into products of prime ideals.

   (a) Suppose that I and J are relatively prime. Show that I and J are coprime, i.e., that $I + J = A$.

   (b) Suppose that I and J are relatively prime. Show that $I \cap J = IJ$.

5. Let I and J be nonzero ideals of a Dedekind ring A. We say that a nonzero ideal G of A is a *greatest common divisor* (gcd) of I and J if

   (i) G divides both I and J; and

   (ii) If K is any nonzero ideal that divides both I and J, then K divides G.

   We say that a nonzero ideal L of A is a *least common multiple* (lcm) of I and J if

   (i) Both I and J divide L; and

   (ii) If K is any nonzero ideal that is divisible by both I and J, then L divides K.

   (a) Show that any two nonzero ideals I and J of A have a gcd G and an lcm L, and express the prime factorizations of G and L in terms of the prime factorizations of I and J.

   (b) If I and J are relatively prime, show that $A = \gcd(I, J)$ and that $IJ = \text{lcm}(I, J)$.

(c) Let H be any nonzero ideal of A. Show that

$$\gcd(HI, HJ) = H \gcd(I, J) \quad \text{and}$$
$$\operatorname{lcm}(HI, HJ) = H \operatorname{lcm}(I, J).$$

6. Let I and J be nonzero ideals of a Dedekind ring A. Show that A/I and J/IJ are isomorphic.

7. (a) Let A be a Dedekind ring and let I be a nonzero ideal of A. Show that there is a nonzero ideal J of A, with I and J relatively prime, such that the ideal IJ of A is principal.
   (b) Let K be any nonzero ideal of A. Show that we may choose the ideal J in (a) to be relatively prime to K.

8. Let A be a Dedekind ring with quotient field $\mathbb{F}$, and let I and J be fractional ideals of A. Show that there are elements α and β of $\mathbb{F}$ such that αI and βJ are relatively prime ideals of A.

9. Let A be a Dedekind ring with only finitely many prime ideals. Show that A is a PID.

10. Let A be a Dedekind ring and let P be a nonzero prime ideal of A. Let R be the localization of A at P. Show that R is a discrete valuation ring.

11. Let A be a Dedekind ring and let I be a nonzero ideal of A. Let $\alpha \neq 0$ be an arbitrary nonzero element of I. Show that there is an element β of I such that $I = <\alpha, \beta>$. Thus, in particular, in a Dedekind ring every ideal can be generated by at most two elements.

12. Let A be a Dedekind ring and let I be a nonzero ideal of A.

   (a) Show that A/I has only finitely many ideals.
   (b) Show that every ideal of A/I is principal.

   (Of course, if I is prime then A/I is a field. But if I is not prime, A/I is not an integral domain.)

13. Let A be an integral domain in which every nonzero ideal factors uniquely as a product of prime ideals. Show that A is a Dedekind ring.

   (As a matter of historical fact, Dedekind proved that the ring of algebraic integers of any algebraic number field has unique factorization as in Theorem 5.3.9. Later, E. Noether abstracted the properties of these rings that make this work and defined a Dedekind ring as in Definition 5.2.7.)

14. Let $S = \{p_1, \ldots, p_k\}$ be a set of district primes, and let $D = -p_1, \cdots, p_k$. Then the subsets of S form a group G of order 2^k under the operation of symmetric difference, which we denote by $*$. Let G_0 be the subgroup $G_0 = \{\phi, S\}$ of order 2. For any subset T of S, let d_T be the product of the elements of T (with $d_T = 1$ if $T = \phi$) and let I_T be the ideal $I_T = <d_T, \sqrt{D}>$ of $\mathcal{O}(\sqrt{D})$.

   (a) Let $\pi \colon G \to Cl(\mathcal{O}(\sqrt{D}))$ be the map given by $\pi(T) = [I_T]$, where $[I_T]$ is the equivalence class of I_T in $Cl(\mathcal{O}(\sqrt{D}))$. Show that $[I_{T_1 * T_2}] = [I_{T_1}][I_{T_2}]$, so that π is a group homomorphism.

   (b) Show that I_T is a principal ideal if and only if $T \in G_0$. Hence, $\mathrm{Ker}(\pi) = G_0$.

   Thus, we see that $\mathrm{Im}(\pi)$ is a subgroup of $Cl(\mathcal{O}(\sqrt{D}))$ isomorphic to G/G_0, a group isomorphic to $(\mathbb{Z}/2)^{k-1}$.

   (c) Suppose that $D \equiv 3 \pmod 4$. Let $J = <2, 1 + \sqrt{D}>$. Show that J is not a principal ideal of $\mathcal{O}(\sqrt{D})$, and also that JI_T is not a principal ideal of $\mathcal{O}(\sqrt{D})$ for any $T \subseteq S$, while J^2 is a principal ideal of $\mathcal{O}(\sqrt{D})$.

   Thus, we see that in this case $\mathrm{Im}(\pi)$ and $[J]$ generate a subgroup of $Cl(\mathcal{O}(\sqrt{D}))$ isomorphic to $(\mathbb{Z}/2)^k$.

   Thus, we conclude that if $\{p_1, \ldots, p_k\}$ is a set of distinct primes, and $D = -p_1, \ldots, p_k$, then $h(\mathcal{O}(\sqrt{D}))$, the class number of $\mathcal{O}(\sqrt{D})$, is divisible by 2^{k-1}, and if $D \equiv 3 \pmod 4$, $h(\mathcal{O}(\sqrt{D}))$ is divisible by 2^k.

   Note that this exercise generalizes the work we did in Lemma 5.4.1.

15. Let n be a positive integer and let $q > 1$ be an odd integer. Let a be a positive integer that is relatively prime to q. Let D be the unique squarefree integer that is defined by $b^2 D = a^2 - q^n$, where b is a positive integer. Let I be the ideal

$$I = <q, a + b\sqrt{D}>$$

of $\mathcal{O}(\sqrt{D})$.

   (a) If $\bar{I} = <q, a - b\sqrt{D}>$, show that $I\bar{I} = <q>$, a principal ideal.

   (b) Show that $I^k = <q^k, a + b\sqrt{D}>$ for every positive integer k.

   (c) Show that $I^n = <a + b\sqrt{D}>$, a principal ideal.

Conclude that there is a homomorphism $\pi_I\colon \mathbb{Z}_n \to Cl(\mathcal{O}(\sqrt{D}))$ given by $\pi_I(k) = [I^k]$. Observe that $\mathrm{Im}(\pi_I) = \mathrm{Im}(\pi_{\bar{I}})$. Hence, $Cl(\mathcal{O}(\sqrt{D}))$ has a subgroup $\mathrm{Im}(\pi_I)$ isomorphic to $\mathbb{Z}_j$ for some integer j dividing n, and hence that $h(\mathcal{O}(\sqrt{D}))$ is divisible by this integer j.

(d) Show that j is the smallest positive integer such that q^j and $a + b\sqrt{D}$ have a common nonunit factor in $\mathcal{O}(\sqrt{D})$. In particular, if q^k and $a+b\sqrt{D}$ do not have a common nonunit factor in $\mathcal{O}(\sqrt{D})$ for any proper divisor k of n, then $j = n$, and so in this case $h(\mathcal{O}(\sqrt{D}))$ is divisible by n.

(e) Suppose that $D < 0$. Let p be the smallest prime factor of n, and set $m = n/p$. (If n is prime, then $m = 1$.) Suppose that $|D| \geq q^m$ if $D \equiv 2$ or $3 \pmod 4$, and $|D| \geq 4q^m$ if $D \equiv 1 \pmod 4$. Show that $j = n$ in (d), so that in this case $\mathrm{Im}(\pi_I)$ is isomorphic to $\mathbb{Z}_n$ and hence, $h(\mathcal{O}(\sqrt{D}))$ is divisible by n.

(f) Give examples of D with $\mathcal{O}(\sqrt{D})$ divisible by n for each $n = 2, \ldots, 10$.

(g) Let $q = 3$ and let $a = 2 \cdot 5 \cdot 7 \cdot 11 = 770$. Let $p \geq 11$ be a prime and let b^2 be the largest perfect square dividing $q^p - a^2$. (Note that $3^{11} > 770^2$.) Let $D = (a^2 - q^p)/b^2$. Show that $Cl(\mathcal{O}(\sqrt{D}))$ has an element of order p, and hence that $h(\mathcal{O})\sqrt{D}))$ is divisible by p.

16. Let $\mathbb{F}$ be an algebraic number field and let $A = \mathcal{O}_{\mathbb{F}}$ be the ring of algebraic integers of $\mathbb{F}$.

(a) Let I be a nonzero ideal of A. Show that $I \cap \mathbb{Z}$ is a nonzero ideal of $\mathbb{Z}$.

(b) Let P be a nonzero prime ideal of A. Show that $P \cap \mathbb{Z}$ is a nonzero prime ideal of $\mathbb{Z}$.

(c) Let P be a nonzero prime ideal of A. Show that $P = <p, \beta>$ for a unique prime $p \in \mathbb{Z}$ and some element $\beta \in P$.

17. Let $\mathbb{F}$ be an algebraic number field that is an extension of $\mathbb{Q}$ of degree n, and let $A = \mathcal{O}_{\mathbb{F}}$ be the ring of algebraic integers of $\mathbb{F}$. Let I be a nonzero ideal of A. Define the *norm* $N(I)$ by

$$N(I) = \#(A/I),$$

the cardinality of the quotient A/I.

(a) Let $a \in \mathbb{Z}$ be a nonzero element and let $I = <a>$ be the ideal of A generated by a. Show that $N(I) = |a|^n$.

(b) Show that $N(I)$ is finite for every nonzero ideal I of A.

(c) Let I and J be nonzero ideals of A. Show that $N(IJ) = N(I)N(J)$.

18. Let $\mathbb{F} = \mathbb{Q}(\sqrt{D})$ be a quadratic extension of $\mathbb{Q}$, and let $A = \mathcal{O}_{\mathbb{F}}$ be the ring of algebraic integers in $\mathbb{F}$. Let p be a prime in $\mathbb{Z}$ and let $I = <p>$ be the ideal of A generated by I. Show there are exactly three possibilities:

   (i) I is a prime ideal of A and $N(I) = p^2$.

   (ii) $I = J^2$ for some prime ideal J of A and $N(J) = p$.

   (iii) $I = J_1 J_2$ for some distinct prime ideals J_1 and J_2 of A and $N(J_1) = N(J_2) = p$.

   (Here $N(I)$ denotes the norm of the ideal I.)

   The prime p is said to be *inert*, to *ramify*, or to *split* in cases (i), (ii), and (iii) respectively.

19. (a) Let $\mathbb{F} = \mathbb{Q}(\sqrt{-1})$. Determine the behavior of p, as in the previous problem, for every prime p in $\mathbb{Z}$.

   (b) Let $\mathbb{F} = \mathbb{Q}(\sqrt{-2})$. Determine the behavior of p, as in the previous problem, for $p = 2, 3, 5, 7$.

   (c) Let $\mathbb{F} = \mathbb{Q}(\sqrt{-5})$. Determine the behavior of p, as in the previous problem, for $p = 2, 3, 5, 7$.

   (d) Let $\mathbb{F} = \mathbb{Q}(\sqrt{-6})$. Determine the behavior of p, as in the previous problem, for $p = 2, 3, 5, 7$.

20. Let $\mathbb{F}$ be an algebraic number field and let $A = \mathcal{O}_{\mathbb{F}}$ be the ring of algebraic integers of $\mathbb{F}$.

   (a) Let I be a nonzero ideal of A, and let $m = N(I)$. Show that $m \in I$ and hence $<m> \subseteq I$.

   (b) Show that there are only finitely many nonzero ideals of A with any given norm.

   (c) We have the following theorem, which we shall not prove here.

   Theorem. *Let $\mathbb{F}$ be an algebraic number field and let $\mathcal{O}_{\mathbb{F}}$ be the ring of algebraic integers of $\mathbb{F}$. Then there is a constant N with the property that if I is any nonzero ideal of $\mathcal{O}_{\mathbb{F}}$, there is an ideal J of $\mathcal{O}_{\mathbb{F}}$, with $[I] = [J]$ in the ideal class group $Cl(\mathcal{O}_{\mathbb{F}})$, and with $N(J) \leq N$.*

   Show that this theorem implies that $Cl(\mathcal{O}_{\mathbb{F}})$ is a finite group.

21. In this exercise, we will use the fact that for any algebraic number field $\mathbb{F}$, the ideal class group $Cl(\mathcal{O}_\mathbb{F})$ is finite. In particular, if $Cl(\mathcal{O}_\mathbb{F})$ has order n, then for any nonzero ideal I of $\mathcal{O}_\mathbb{F}$, I^d is a principal ideal for some d dividing n, and for every nonzero ideal I of $\mathcal{O}_\mathbb{F}$, I^n is a principal ideal.

    Let I be an ideal of $\mathcal{O}_\mathbb{F}$ and let $\mathbb{E}$ be a finite extension of $\mathbb{F}$. We let $\hat{I}$ be the ideal of $\mathcal{O}_\mathbb{E}$ generated by I.

    (a) If I_1 and I_2 are two ideals of $\mathcal{O}_\mathbb{F}$, show that $\widehat{I_1 I_2} = \hat{I}_1 \hat{I}_2$.

    (b) Let I be a nonzero ideal of $\mathcal{O}_\mathbb{F}$ with $I^d = <a>$, a principal ideal of $\mathcal{O}_\mathbb{F}$. Let $\mathbb{E} = \mathbb{F}(\alpha)$ where α is a root of the polynomial $x^d - a \in \mathbb{F}[x]$. Let $\hat{J} = <\alpha>$ be the principal ideal of $\mathcal{O}_\mathbb{E}$ generated by α. Show that $\hat{I} = \hat{J}$. Such an element α of $\mathcal{O}_\mathbb{E}$ is called an *ideal element* of I. (Note that for any $b \in I$, the quotient b/α is an algebraic integer.)

    (c) Show that there is an extension $\mathbb{E}$ of $\mathbb{F}$ of degree at most n such that for every ideal I of $\mathcal{O}_\mathbb{F}$, the ideal $\hat{I}$ of $\mathcal{O}_\mathbb{E}$ is principal. (Note this does not imply, and it is not in general true, that every ideal of $\mathcal{O}_\mathbb{E}$ is principal.)

22. Let $\mathbb{F} = \mathbb{Q}(\sqrt{-5})$. As we have seen, we have the following non-principal ideals of $\mathcal{O}_\mathbb{F}$:

$$I = (2, 1 + \sqrt{-5}) = (1 + \sqrt{-5}, 1 - \sqrt{-5}) = (2, 1 - \sqrt{-5}) = \bar{I}$$
$$I_1 = (3, 1 + \sqrt{-5}), I_2 = (3, 1 - \sqrt{-5}) \text{ with } I_2 = \bar{I}_1, I_1 = \bar{I}_2.$$

    (a) Show that $\alpha = \sqrt{2}$ is an ideal element of I, that $\alpha_1 = \sqrt{2 - \sqrt{-5}}$ is an ideal element of I_1, and that $\alpha_2 = \sqrt{2 + \sqrt{-5}}$ is an ideal element of I_2.

    (b) Verify that α, α_1, and α_2 are indeed algebraic integers.

    (c) Verify that $2/\alpha$, $(1 + \sqrt{-5})/\alpha$, and $(1 - \sqrt{-5})/\alpha$ are algebraic integers, that $3/\alpha_1$ and $(1 + \sqrt{-5})/\alpha_1$ are algebraic integers, and that $3/\alpha_2$ and $(1 - \sqrt{-5})/\alpha_2$ are algebraic integers.

Appendix A: Some Properties of the Integers

In this appendix we simply list some properties of the integers, most, if not all, of which you are probably familiar with. We list these in order to be able to get started, as we will need them in our study of group theory, with which we begin. But we do not prove these here, as we will prove all of them in a more general context in our study of ring theory.

Theorem A.1 (The division algorithm). *Let a and b be integers with $b \neq 0$. Then there are unique integers q and r such that*

$$a = bq + r \quad \text{with} \quad 0 \leq r < |b|.$$

Theorem A.2. *Let a and b be integers, not both 0. Then there is a unique positive integer g such that:*

(1) *g divides both a and b; and*
(2) *if d is any integer that divides both a and b, then d divides g.*

Furthermore, there are integers x_0 and y_0 (not unique) such that

$$g = ax_0 + by_0.$$

Definition A.3. The integer g in Theorem A.2 is the *greatest common divisor* of a and b, $g = \gcd(a, b)$. ◇

Definition A.4. Two integers a and b, not both 0, are *relatively prime* if $\gcd(a, b) = 1$. ◇

Lemma A.5 (Euclid's lemma). *Let a, b, and c be nonzero integers. Suppose that a divides bc. If a and b are relatively prime, then a divides c.*

Corollary A.6. *Let a, b, and c be nonzero integers. Suppose that a divides c and b divides c. If a and b are relatively prime, then ab divides c.*

Corollary A.7. *Let a, b, and c be nonzero integers. Suppose that a and b are relatively prime, and that a and c are relatively prime. Then a and bc are relatively prime.*

Lemma A.8. *Let a and b be integers, not both 0. Let d be a common divisor of a and b. Then $\gcd(a/d, b/d) = \gcd(a, b)/d$. In particular, if d is a positive integer, then a/d and b/d are relatively prime if and only if $d = \gcd(a, b)$.*

Definition A.9. Let n be a positive integer.

If $n = 1$, then n is a *unit*.

If $n > 1$, and the only positive integers dividing n are 1 and n, then n is a *prime*.

Otherwise, n is *composite*. $\Diamond$

Theorem A.10 (Fundamental theorem of arithmetic). *Let n be a positive integer. Then n can be written as a product of primes in a unique way up to order, i.e.,*

$$n = p_1 p_2 \ldots p_k \quad \text{for some primes } p_1, \ldots, p_k$$

and if also

$$n = q_1 q_2 \ldots q_l \quad \text{for some primes } q_1, \ldots, q_l$$

then $k = l$, and after possible reordering, $p_i = q_i$, $i = 1, \ldots, k$.

Equivalently,

$$n = p_1^{e_1} \ldots p_k^{e_k} \quad \text{for some distinct primes } p_1, \ldots, p_k$$
$$\text{and positive integers } e_1, \ldots, e_k$$

and if also

$$n = q_1^{f_1} \ldots q_l^{f_l} \quad \text{for some distinct primes } q_1, \ldots, q_l$$
$$\text{and positive integers } f_1, \ldots, f_l$$

then $k = l$, and after possible reordering, $p_i = q_i$ and $e_i = f_i$, $i = 1, \ldots, k$.

Remark A.7. In the fundamental theorem of arithmetic, we regard the integer 1 as having the empty factorization. $\Diamond$

Appendix B: A Theorem from Linear Algebra

In this appendix we prove a theorem that we have used in our development of field theory.

Theorem B.1. *A vector space V over an infinite field $\mathbb{F}$ is not the union of finitely many proper subspaces.*

Proof. We first consider the case where V is finite dimensional. Let $n = \dim(V)$. We proceed by induction on n.

If $n = 0$ there is nothing to prove (as a 0-dimensional vector space has no proper subspaces).

If $n = 1$ the theorem is trivial (as the only proper subspace of a 1-dimensional vector space is the subspace $\{0\}$).

Let $n = 2$. Choose a basis $B = \{b_1, b_2\}$ of V. For $f \in \mathbb{F} \cup \{\infty\}$, let

$$a_f = b_1 + f b_2 \quad \text{if } f \in \mathbb{F}$$
$$a_f = b_2 \quad\quad\;\; \text{if } f = \infty$$

and let W_f be the 1-dimensional subspace of V spanned by a_f. $\quad\square$

It is easy to check that $\{W_f\}$ are all of the 1-dimensional subspaces of V, and they are all distinct (with $W_{f_1} \cap W_{f_2} = \{0\}$ if $f_1 \neq f_2$). Note there are infinitely many of these.

Now let $\{U_1, \ldots, U_k\}$ be any set of finitely many proper subspaces of V. We may assume that none of them is the zero subspace. Then $U_1 = W_{f_1}, \ldots, U_k = W_{f_k}$ in above notation, for some subset $\{f_1, \ldots, f_k\}$ of $\mathbb{F} \cup \{\infty\}$. But now choose any f_0 in $\mathbb{F} \cup \{\infty\}$ that is not in this subset. Then $W_{f_0} \cap W_{f_i} = \{0\}, i = 1, \ldots, k$, so in particular $a_{f_0} \in V$ with $a_{f_0} \notin W_1 \cup \ldots \cup W_k$, and hence $W_1 \cup \ldots \cup W_k \neq V$.

Now let $n \geq 3$. Suppose the theorem is true for any $(n-1)$-dimensional vector space and let V be n-dimensional. Choose a basis $B = \{b_1, b_2, \ldots, b_n\}$ of V. With a_f as above, let W_f be the $(n-1)$-dimensional vector space with basis $\{a_f, b_3, \ldots, b_n\}$. There are infinitely many of these subspaces (though they are certainly not all of the $(n-1)$-dimensional subspaces of V).

Now let $\{U_1, \ldots, U_k\}$ be any set of finitely many proper subspaces of V. Since there are infinitely many W_f, there is some $f_0 \in \mathbb{F} \cup \{\infty\}$ with $U_i \neq W_{f_0}$ for any $i = 1, \ldots, k$. (We are not assuming that each U_i is $(n-1)$-dimensional, as if U_i has dimension $< n - 1$, certainly $U_i \neq W_{f_0}$.)

Let $U_i' = U_i \cap W_{f_0}$. Since $U_i \neq W_{f_0}$, U_i' is a proper subspace of W_{f_0}, so $\{U_1', \ldots, U_k'\}$ is a finite set of proper subspaces of W_{f_0}, a vector space of dimension $n - 1$, so by the inductive hypothesis there is some element $a \in W_{f_0}$ with $a \notin U_1' \cup \cdots \cup U_k'$. But then $a \notin U_1 \cup \cdots \cup U_k$, so $V \neq U_1 \cup \cdots \cup U_k$. (If $a \in U_i$, then, since $a \in W_{f_0}$, $a \in U_i \cap W_{f_0} = U_i'$; impossible.)

Then by induction, in the finite dimensional case, we are done.

Now let V be arbitrary, and let $\{U_1, \ldots, U_k\}$ be any set of finitely many proper subspaces of V. Since each U_i is a proper subspace of V, there is an element b_i of V with $b_i \notin U_i$, for each $i = 1, \ldots, k$.

Let V_0 be the subspace of V spanned by $\{b_1, \ldots, b_k\}$, and note that V_0 is finite dimensional (indeed, $\dim V_0 \leq k$). Let $U_i' = U_i \cap V_0$ for $i = 1, \ldots, k$ and note that each U_i' is a proper subspace of V_0 (as $b_i \in V_0$ but $b_i \notin U_i'$). Then, by the finite dimensional case, there is some $a \in V_0$ with $a \notin U_1' \cup \cdots \cup U_k'$. But then (as above) $a \notin U_1 \cup \cdots \cup U_k$ so $V \neq U_1 \cup \cdots \cup U_k$.

Index

Printed in the United States
by Baker & Taylor Publisher Services